Plasmonic Metal Nanostructures

Plasmonic Metal Nanostructures

Preparation, Characterization, and Applications

Caixia Kan

WILEY-VCH

Author

Prof. Caixia Kan
Nanjing University of Aeronautics and Astronautics
29 Yudao Street
Qinhuai District
Nanjing
China
210016

Cover Image: Courtesy of Caixia Kan

Library of Congress Card No.: applied for

British Library Cataloguing-in-Publication Data:
A catalogue record for this book is available from the British Library.

Bibliographic information published by the Deutsche Nationalbibliothek
The Deutsche Nationalbibliothek lists this publication in the Deutsche Nationalbibliografie; detailed bibliographic data are available on the Internet at http://dnb.d-nb.de.

Print ISBN: 978-3-527-35174-9
ePDF ISBN: 978-3-527-84083-0
ePub ISBN: 978-3-527-84084-7
oBook ISBN: 978-3-527-84085-4

Typesetting: Straive, Chennai, India
Printing and Binding: CPI Group (UK) Ltd, Croydon, CR0 4YY

C9783527351749_120324

Contents

Preface

In nanophotonics, plasmonics are primarily concerned with the manipulation of light at the nanoscale based on the properties of propagating and localized surface plasmons. Plasmons can turn impossibility into reality through modulating the energy flow, trapping light, and injecting hot electrons for semiconductors. Rich phenomena related to plasmonics have inspired practical uses in many technologies. You may find out why colloidal gold is more precious than bulk pretty gold. You will understand that nanogold does not glitter but that its future looks bright.

This book is mainly focused on noble metal nanostructures. Nanometer-sized particles have existed more than 15 centuries, decorated in historical art of the Lycurgus glass cup, although people could not recognize them at that time. Mie's electromagnetic theory, derived from Maxwell's equations 100 years ago, can be applied to explain the color of colloidal metals. The term plasmon was initially proposed by Pines and Bohm in 1952. After a period of tepid development, plasmonics got a mighty rebirth after the demonstration of surface-enhanced Raman scatting (SERS) in the 1990s. As a "hot" academic direction, plasmonics has expanded and penetrated into various interdisciplinary research fields in the past two decades, involving materials science, optoelectronics, analytical chemistry, energy, and biomedicine. For the methodology and material fabrication, the advanced characterization, precise growth, and tunable properties of plasmonic metal nanostructures have also attracted tremendous attention.

There are many outdated and latest reviews about the precise preparation, characterization, and application of plasmonic nanostructures. To contribute to plasmonics, we need many references to go beyond our research work. I'm extremely grateful to all the members of our group for their contribution to the chapters of this book One striking feature of the book is that more than half of the chapters come from our recent research work.

Plasmonic nanostructures provide the fundamental basis and practical device for rapid sensing and detection of chemical and biological objects, such as colloidal gold, which is applied in pregnancy tests in family planning and the 2019 novel coronavirus infection. Many of the techniques and concepts described in this book will come true in the near future. Ideally, readers will be inspired to improve the methods described here or even to develop fundamentally new methods.

This book was compiled with a lot of newly published references and books. There are inevitably some improper descriptions and errors; any criticism and corrections are welcome.

Basely, Jun. 2023 *Caixia Kan*
Nanjing

I

Introduction of This Book

I.1 Human Beings Never Stop Exploring Light

As an important branch of physics, optics is an important foundation for the development of science and technology. According to the historical records in the ancient Chinese book, Mozi did the first small hole imaging experiment about 2500 years ago and explained the inverted standing image using the linear propagation of light, as indicated by Figure I.1 for the concept map.

Throughout the history of the Nobel Prize over one hundred years, there have been more than 40 achievements, about 40% of physics Awards, related directly or indirectly to light. Since the 1960s, the development of laser technology has brought great changes to industries such as industry and medical treatment. Optics brings new development opportunities to other fields. Particularly, two Nobel Prizes in chemistry (1999 and 2014) are also related to optics. They are breakthrough contributions of femtosecond spectroscopy used in chemical reactions and super-resolution optical microscopes.

According to the diffraction of light in Optics, we know that the resolution of optical instruments is dependent on the size of the circular screen and circular hole. When the focal length of the lens and the size of the lens are in the same order, the energy of Airy spot produced by parallel light diffracted by the convex lens is focused on the size equivalent to wavelength scale, let alone the fact that the focal length of the lens is usually larger than the size of the lens. Therefore, in the field of optical imaging, the diffraction limit of light limits the size of the focusing region when we focus the light field with a convex lens, which in turn limits the imaging accuracy and resolution of optical microscopy.

How can we improve the resolution of imaging? There are two extreme directions, radio telescope and electron microscope, in improving the imaging resolution from the Optics principles. They are now being used respectively to observe distant stars and study the structure of tiny substances such as viruses and atoms. Microscope and telescope are two expensive imaging technologies, enabling us to know the world to the boundless large and infinitesimal region.

In the Nano Era, nanotechnology has penetrated many fields of national economic and social development. With the development of electron microscopy and

Figure I.1 Concept map for the linear propagation of light through the small hole.

micromachining, understanding of the material properties and device construction are also in the advance of micro direction. Internationally, a series of basic research projects and major engineering plans have been carried out in the design and development of new photoelectric devices with high performance and high integration. With the development of nano-optics and optoelectronics, significant progress has been made on the nanometer-sized devices with the assistance of electron microscopy and micromachining. At the same time, the development of high-performance micro devices also depends on the preparation of new functional structures and modification of existing materials. Nowadays, materials with desired properties are in great demand, and they are more or less related to the optical properties of materials in small sizes. For example, information and data storage require the medium to meet larger storage density and higher transmission speed. In the biomedicine field, it is required that bacteria and viruses can be quickly detected with high efficiency. And drug molecules can load and release intelligently under irradiation. In terms of energy, we should make full use of solar radiation as much as possible and effectively produce new clean energy sources.

What new effects and phenomena will be discovered during the interaction between light and matter or materials on the microscopic scale? Here, we will start the story of the interaction between light and matter from the color of metals in different sizes.

I.2 Charming Metal Color

Do you know the color of gold (Au)? Yes, of course, we do know the glitter of BRICs wealth, as shown in Figure I.2. We are very familiar with some metals with specific colors, such as golden gold and silvery silver (Ag). Can you imagine the Au BRICs are the same materials as the red colloidal matter dispersed in the inserted two flasks? Do you believe the Ag coin is the same as the black powder on the plate? Moreover, it is known that the row of colored dispersions are the Au/Ag mixing particles of various ratios. Why do the metals present so many puzzling colors?

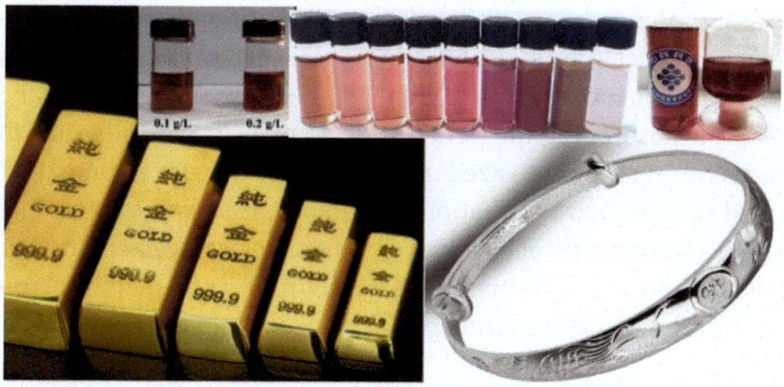

Figure I.2 Bulk Au, Ag and their nanometer-sized materials in colloidal states. Source: The colloid product is from the website: Zhongke Keyou/https://www.zhongkekeyou.cn//last accessed July 14, 2023.

The optical properties of materials are related to the basic principles of solid-state physics, including the energy band structure inter- and intra-band transition. Here, we briefly describe the color of the bulk metals. There are a large number of electrons that can move freely in metals, forming an electron sea. The collective electrons, titled plasma, respond to the electromagnetic field, resulting in a plasma frequency far greater than that of visible light. Therefore, bulk metals can reflect visible light and form metallic luster. From the view of energy band theory, metals are zero band gaps due to the overlap of valence and conduction bands. The electrons excited by light absorption jump to higher unoccupied energy levels above the Fermi surface, generating photocurrent, which rapidly discharges to emit photons of the same wavelength. As a result, the absorbed light is immediately emitted to the metal surface. That is why a thin metal coating has high reflection. Due to the special electron arrangement of d-orbit relative to the outer s-orbit for Au and Cu metals, they can selectively absorb green light, exhibiting golden and red colors.

As for the metals in small nanometer sizes, their physical and chemical properties are different from those of the bulk metals and atomic states. The photoelectric characteristics can be described by the property of the dispersive medium, absorbing and scattering electromagnetic waves of specific bands. Therefore, different colors can be achieved if they strongly absorb specific visible light.

There is a legendary story about one magic Lycurgus cup collected in the British Museum from ancient Rome related to the Au and Ag nanoparticles. Actually, the history of Au nanoparticles can be dated back to the sixteenth century. It is said that Paracelsus (1493–1541) a Swiss chemist, prepared “drinking gold” to treat mental illness. Then, in 1857, Faraday (1791–1867), a British scientist, studied the wonderful color of ultrathin gold leaf under light and also discovered that colorless liquid turned to red when extremely thin Au tablets were dispersed in water. As an experimental physicist, Faraday did many experiments to reduce gold chloride precursor solution with phosphorus and obtain ruby-colored liquid. In the case of the matter state in a colored liquid, he found a clear beam through the liquid by irradiating the liquid with a beam of light. This effect is known as the Dindal phenomenon for colloids with light irradiation, which is an experimental method for directly distinguishing colloids from solution. Then, 50 years later, the scattering theory of light by small particles was established by Mie by solving Maxwell's equations. The color of Au nanoparticles colloid is theoretically explained by low reflection and strong absorption for light in different visible bands due to the dielectric properties of nanometer-sized metals. With the development of colloidal chemistry and optical characterization technology, it is well known that the response wavelength and intensity of particles to light are closely related to their morphology, size, composition, aggregation state, and environmental medium. You can achieve various colorful metal nanoparticle colloids.

I.3 Natural Development of Plasmonics

Unlike conventional optics, plasmonics enables unrivaled concentration of optical energy well beyond the diffraction limit of light. The physical mechanism behind this striking functionality is the excitation of surface plasmon polaritons. Along with the ongoing efforts and development of scientific technologies, it is important to find and synthesize better plasmonic materials. Another strategy relies on hybrid photonic-plasmonic devices by coupling plasmonic nanostructures to resonant optical elements. It is confirmed that nanostructures and newly developed plasmonics are two effective methods for light diffraction limitation. Plasmonics is a scientific branch of photonics involving the study of the localization, propagation and guidance of surface plasmons to manipulate light in desired structures of different dimensions. For the nanometer-sized metals, such as Au, Ag, Cu, and Al, they are also titled "plasmonic metals" in many reports. Surface plasmons can realize a series of unique optical properties. Based on the resonance effect of surface plasmon, the electromagnetic field can be localized in the subwavelength scale, forming a booming new field in nanophotonics. It is widely studied and applied in the fields of information transmission, optical sensing, photoelectric detection, display and so on. It is highly evaluated that "Nanogold does not glitter, but its future looks bright."

Surface plasmon involves conducting surface plasmon and local surface plasmon. Their properties are both closely related to materials, in which the concentration of free electrons (carrier of semiconductor), mobility and the damping by interband transition are three main factors. Although conducting surface plasmon can be used to fabricate novel subwavelength waveguides, the transverse propagation distance is not only determined by the electron's Ohmic loss, but also limited by the dispersive property. It cannot be directly excited by light incident on the metal surface because of wave vector mismatch. In order to excite the surface plasmon, additional structures or devices must be applied to achieve the wave vector matching conditions.

Compared with conducting surface plasmon, it is relatively easy to excite the localized surface plasmon of metal nanostructures. In fact, we know that a convex lens can focus sunlight for local heating. Similarly, if the light energy is focused on a scale much smaller than the wavelength, it can not only produce a strong field but also greatly improve the resolution of an optical microscope. On the other hand, this localization can significantly enhance the interaction between light and matter,

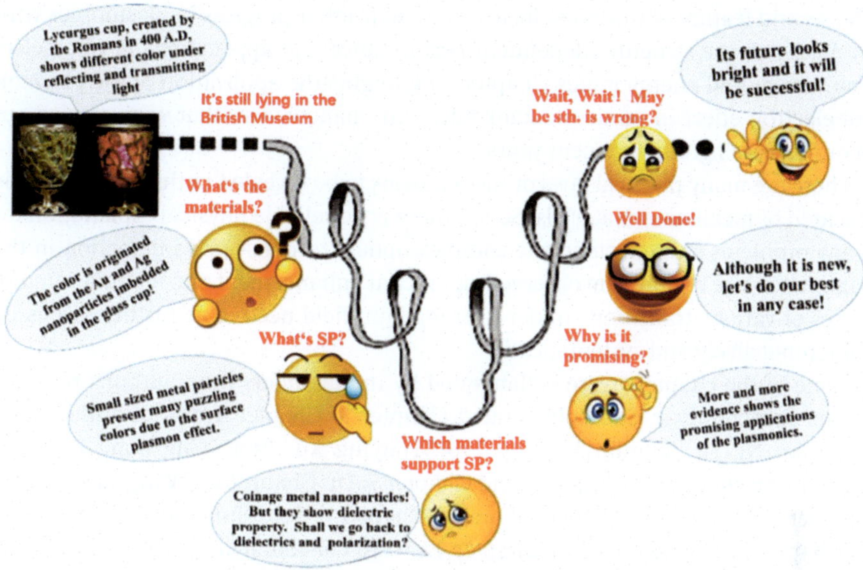

Figure I.3 Cognition attitude towards the plasmonics field.

which is similar to the nonlinear effect of macro media under laser irradiation. The nonlinear optical effect of strong field will induce many unique physicochemical properties. How do we further focus the light field in the microscopic size? Surface plasmon provides an effective solution.

Localized surface plasmonic nanostructures can efficiently collect light propagating in free space and converge it to nanoscale hotspots in the near field region, enabling efficient excitation of molecules. Conversely, spectra information about molecules in the hotspots can be broadcast to far fields. This process is accompanied by the enhancement of light absorption, radiation, scattering, light force resonance migration and photo-thermal effects. The rich phenomena associated with localized surface plasmonic nanostructures have resulted in a series of applications in many fields. At the same time, the complexity of its behavior also brings challenges to physics. Thus, confronting this reality spurred many new directions in the field of plasmonics; research on plasmonics involves several development stages, like the Kubler-Ross grief cycle.

I.4 About This Book

The book is compiled in a combination of teaching work with the research of the authors' group in the past two decades. Based on the plasmonic metal nanostructures, the optical effects of plasmon, precise preparation, properties, and applications of different plasmonic nanostructures will be presented. We first arrange the progress in surface plasmon physics to understand the common problems of plasmon nanostructures and updated characterization technologies.

The second feature of this book focused on the frontier progress in plasmonics with emphasis on the structure-dependent performance and applications in photoelectron, sensor and energy issues Chapter 9 highlights the application of the plasmon hot electron effect in semiconductor micro and nanodevices that were carried out in our research group in recent years.

There are many physical laws or mechanisms to be revealed in the optoelectronic research of metal nanomaterials. So far, there is no reliable physical explanation for many problems and effects in the complex optical response and interaction in the nanometer-size region. The observed significant enhancement and modulation processes of various transitions under a strong light field need to be further discussed comprehensively and systematically.

Some of the chapters were collaborated by the members from author's research group. Prof. Shuiyan Cao collaborated Chapter 3. Chapter 4 was co-authored by Dr. Yuan Ni, Dr. Shanlin Ke, and Prof. Haiying Xu. Prof. Xingzhong Zhu and Dr. Juan Xu completed Chapter 5 and Chapter 7. Dr. Changshun Wang collaborated Chapter 6. Dr. Tingcha Wei collaborated Chapter 8. Prof. Junfeng Lu contributed to Chapter 9. Chapter 10 and Chapter 11 were collaborated by Dr. Yuan Ni and Dr. Juan Xu.

In the process of editing this book, a large number of references and literature have been cited. We only list the main references and books with great thanks. With the never-ending research proceeding, some viewpoint descriptions and conclusions also need to be revised. Thanks to the readers. We apologize for errors and would be thankful for suggestions and corrections if any.

Part I

Fundamental and Latest Development in the Plasmonics

1

Theoretical Backgrounds and Advances of Plasmonics

The field of nanoplasmonics is young but rich in phenomena that have inspired practical uses in physics, biomedicine, environmental monitoring, and national security

1.1 Introduction

When light irradiates on a solid materials, two kinds of electrons excitation will be formed in the materials. One is electron-hole pair excitation, which is called exciton. The other is the collective oscillation of electrons, which is known as plasmon. In physics, plasmon is a quantum of plasma oscillation that can be considered as a quasiparticle. Thus, plasmons are collective (a discrete number) oscillations of the free electron gas density [1, 2]. As a new type of exciton formed by the interaction between light and condensed matter, plasmon-related topic has become focus research in the frontier fields of micro-nano photonics. Ideally, plasmon is easy to excite without decay. However, all the materials exist in a certain medium environment, particularly for the low dimensional structure with rich surface and interfaces. When the dimensions of materials is comparable to the mean path of free electrons, Landau damping becomes the dominant loss source in plasmonics.

For the well-known metals, such as gold (Au) and silver (Ag), plasmons of these metals have significant spatial confinement and propagation loss due to the strong Landau damping effect and the scattering between plasmon and phonon [3]. In this chapter, some basic theory and experimental advances about plasmon will be presented, involving an overview of plasmon, dielectric function, propagation behavior, excitation and applications of surface plasmon.

1.2 Drude Model for Free Electron Gas

As one of the most basic models of solid physics, Drude (Drude-Lorentz) model described that the electrons in the metal are free around the fixed ion solid and

Plasmonic Metal Nanostructures: Preparation, Characterization, and Applications,
First Edition. Caixia Kan.

dispersed in the whole space of the metal, forming the so-called free electron gas [4]. The physical significance of the excitation at ω_p can be understood by considering the collective longitudinal oscillation of the conduction electron gas versus the fixed positive background of the ion cores in a plasma slab. To visualize a plasmon oscillation, it can be imagined that a metal was placed in an external electric field. A collective displacement of the electron cloud by a distance x leads to a surface charge density $\sigma = \pm n_e ex$ at the slab boundaries (n_e is number density of electrons). This establishes a homogeneous electric field $\boldsymbol{E}$. Thus, the displaced electrons experience a restoring force, and their movement can be described by the equation of Newtonian motion. If the electric field is removed, electrons oscillate back and forth at the plasma frequency until the energy is lost in some kind of resistance or damping. Inserting the expression for the electric field, we get the motion function (ε_0 is electric permittivity or dielectric constant of vacuum).

$$m\ddot{\boldsymbol{x}} = -e\boldsymbol{E} = -{}^{n_e e^2 x}/_{\varepsilon_0} \tag{1.1}$$

Equation (1.1) can also be expressed by $\ddot{\boldsymbol{x}} + \frac{n_e e^2}{m\varepsilon_0}\boldsymbol{x} = 0$, in which plasmon vibrates at the same plasma frequency (ω_p) with $\omega_p^2 = \frac{n_e e^2}{m\varepsilon_0}$. ω_p is the plasma frequency of bulk materials. It exists in metals, semiconductors and insulators. But it cannot be excited nor can be observed by transverse light wave.

The plasma frequency ω_p can thus be recognized as the natural frequency of a free oscillation of the electron sea. The quanta of these charge oscillations are called plasmons (or volume plasmons) to distinguish them from surface and localized plasmons, which will be discussed in the next part. For most metals, the ω_p is in the ultraviolet regime ($\sim 10^{16}$ Hz), depending on details of the electrons density (n_e) and band structure. Light with frequencies less than ω_p will be reflected by the metals because the electrons in the metals screen the electric field of the light. If the light with frequencies above ω_p, the light will be transmitted through the metals because the electrons in the metals cannot respond fast enough to screen it. For most metals, the ω_p is in the ultraviolet region, making them shiny (reflective) in the visible range. Some metals, such as Cu and Au, have electronic interband transitions in the visible range, whereby specific light energies (colors) are absorbed, yielding their distinct color.

Due to the longitudinal nature of the excitation, volume plasmons do not couple to transverse electromagnetic waves. Plasmons can only be observed when high-speed electrons pass through a metal film or photons are reflected on the surface of the film. Another consequence is that their decay occurs only via energy transfer to single electrons, a process known as Landau damping.

Although Drude model is one classical free electron theory with Boltzmann distribution law under four approximations of independent electron, free electron, collision hypothesis and relaxation time, it can be well applied to explain the physical mechanism of surface plasmon and many experimental phenomena. Then the revised Drude-Sommerfeld model was proposed by taking Fermi-Dirac distribution law of quantum mechanics into Drude model.

1.3 Dielectric Function of the Free Electron Gas

For the materials in bulk, film or low dimensions, dielectric function is one important parameter in describing the photoelectronic properties of these structures. For some of the noble metals, interband effects already start to occur for energies of 1 eV. As examples, the real part (ε_r) and the imaginary part (ε_i) of the dielectric function for Au and Ag and Drude model fits to the data [5]. Clearly, this model is not adequate for describing either ε_r or ε_i at high frequencies. For the case of Au, its validity breaks down at the boundary between the near-infrared and the visible region.

With introduction of damping coefficient (γ), electron movement under the irradiation of electric field (ω) can be described by the equation of motion:

$$m\ddot{\boldsymbol{x}} + m\gamma\dot{\boldsymbol{x}} = -e\boldsymbol{E} \tag{1.2}$$

$$\boldsymbol{x} = \frac{e}{m(\omega^2 + i\gamma\omega)}\boldsymbol{E}(t) \tag{1.3}$$

According to the polarization of dielectric $\boldsymbol{D} = \varepsilon_0(1+\chi)\boldsymbol{E} = \varepsilon_0\varepsilon\boldsymbol{E}$, the polarization can be achieved.

$$\boldsymbol{P} = -n_e e\boldsymbol{x} = \chi\mathbf{E} = -\frac{n_e e^2}{m(\omega^2 + i\gamma\omega)}\boldsymbol{E}(t) \tag{1.4}$$

$\boldsymbol{P}$ describes the electric dipole moment per unit volume inside the material, caused by the alignment of microscopic dipoles with the electric field. The linear relationship between $\boldsymbol{D}$ and $\boldsymbol{E}$ is often implicitly defined using the dielectric susceptibility χ, particularly in quantum mechanical treatments of the optical response. The linear relationship between $\boldsymbol{D}$ and $\boldsymbol{E}$ via:

$$\boldsymbol{D} = \varepsilon_0(1+\chi)\boldsymbol{E} = \varepsilon_0\left(1 - \frac{\omega_p^2}{\omega^2 + i\gamma\omega}\right)\boldsymbol{E}(t) \tag{1.5}$$

Here the dielectric property as a function of frequency can be described as:

$$\varepsilon(\omega) = 1 - \frac{\omega_p^2}{\omega^2 + i\gamma\omega} \tag{1.6}$$

Then we can get real and imaginary parts of dielectric functions $\varepsilon_r(\omega) = 1 - \frac{\omega_p^2}{\omega^2+\gamma^2}$, $\varepsilon_i(\omega) = \frac{\omega_p^2\gamma}{\omega(\omega^2+\gamma^2)}$. Therefore, the rcfractive index of electrolyte is expressed in the form of $\tilde{N} = n + ik$. Here, n is refractive index coefficient, indicating the permittivity $\varepsilon_r = n^2 - k^2$. k is extinction coefficient related to the conductivity of the medium ($2nk = \sigma/\omega\varepsilon_0$). k presents the material loss during the relaxation (or Landau damping), in which the polarization cannot keep up with the change of external high-frequency electric field. Under the conditions that $\omega \gg \gamma$ or weak damping $\gamma = 0$, Eq. (1.6) can be written:

$$\varepsilon(\omega) = 1 - \frac{\omega_p^2}{\omega^2} \tag{1.7}$$

It can be taken as the dielectric function of the undamped free electron plasma. $\gamma > 0$ leads to damping of transmitted wave.

In general, $\varepsilon(\omega) = \varepsilon_r(\omega) + \varepsilon_i(\omega)$ complex-valued functions of frequency ω. ε can be experimentally determined for example via reflectivity studies and the determination of the complex refractive index $\tilde{N}$. As defined $\tilde{N} = \sqrt{\varepsilon}(\omega)$, $\tilde{N}$ is imaginary for $\omega < \omega_p$. A material with negative dielectric permittivity does not support propagation of electromagnetic waves; instead the electromagnetic field decays inside the materials with a certain depth and most of the incident energy is reflected. This behavior is the characteristic of metals, where the depth is typically around tens of nanometers. For this case, there is a strong interaction between metal and incident electromagnetic wave, known as absorption and scattering in the following parts of this chapter and Chapter 2.

Reflectivity of metal can be expressed by $R = \left|\frac{\tilde{N}-1}{\tilde{N}+1}\right|^2 = 1$, showing that metals have high reflectivity. Then $\tilde{N}$ is positive for $\omega > \omega_p$, and metal is only a conventional dielectric material for incident light.

For transverse waves, $\boldsymbol{K} \cdot \boldsymbol{E} = 0$, yielding the generic dispersion relation:

$$K^2 = \varepsilon(\boldsymbol{K}, \omega)\frac{\omega^2}{c^2} \tag{1.8}$$

From Eqs. (1.7) and (1.8), we can get the dispersion relation of volume plasmon $\omega^2 = \omega_p^2 + K^2c^2$. It is clear that this value is always larger than that in free space, as shown in Figure 1.1.

As it is discussed above, $\tilde{N}$ is real in value for $\omega > \omega_p$, while $\tilde{N}$ is imaginary for $\omega < \omega_p$, and $K = \frac{i}{C}\sqrt{\omega_p^2 - \omega^2}$, leading to rapid decay of electromagnetic waves in the material. The skin depth is defined as $\delta = c/\sqrt{\omega_p^2 - \omega^2}$, which determines whether the electromagnetic wave can propagate through the medium. For example, in

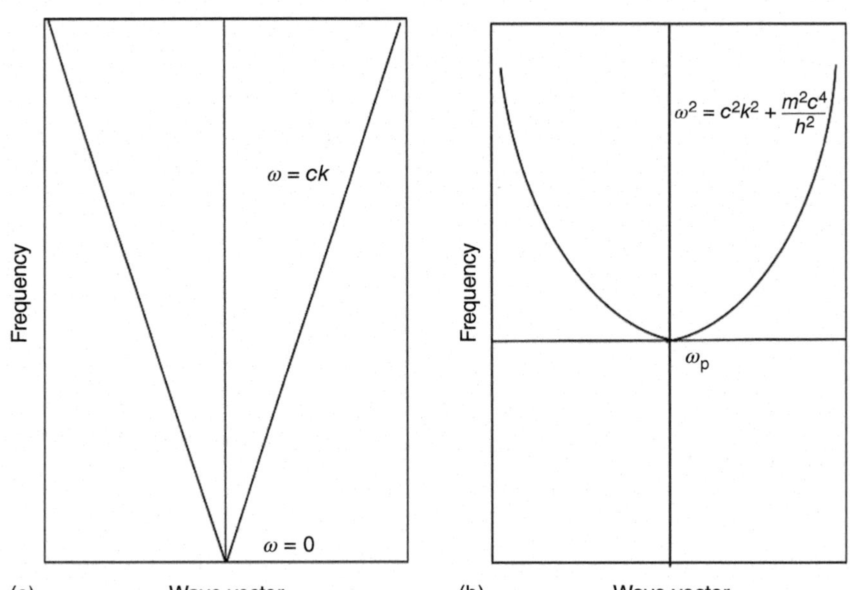

Figure 1.1 Dispersion of light in free space (a) and metals (b).

the ionosphere (50–500 km) above our ground, the plasmon density in the order of $10^{12-15}/\text{m}^3$, and the cut-off frequency (~10 MHz) is less than the TV band. Therefore, the TV electromagnetic signal can penetrate into this ionosphere and spread signal from the satellites. The free electrons in metals can also be regarded as plasmon. However, the frequency of visible light (~10 THz) is less than ω_p of metals (>100 THz), and visible light cannot penetrate most metals. As a result, most metals are silvery white for high reflectivity of visible light. In the metal-medium (or vacuum) system, the plasmons are confined strongly with light to surfaces, resulting in a surface polariton.

Actually, the electrons of metals are not free as described in the Drude model. The electrons have inherent frequency ω_0 and form restoring force that depends on the existing space or local structure. Therefore, the electron movement was revised into Eq. (1.9)

$$m\ddot{\boldsymbol{x}} + m\gamma\dot{\boldsymbol{x}} + m\omega_0^2\boldsymbol{x} = -e\boldsymbol{E} \tag{1.9}$$

We can get Eq. (1.10)

$$\varepsilon\left(\frac{1}{c}\frac{\partial}{\partial t}\right) = 1 - \frac{\omega_\text{p}^2}{\omega^2 + i\gamma\omega - \omega_0^2} \tag{1.10}$$

For practical purpose, the advantage of the Drude model is that it can easily be incorporated into time-domain based numerical solvers for Maxwell's equations. Its inadequacy in describing the optical properties of Au and Ag at visible frequencies can be overcome by replacing Eq. (1.6) with Eq. (1.10). Here, the interband transitions are thus described using the classical picture of a bound electron with resonance frequency ω_0.

In the Drude model for free electrons, the collective electrons vibration is a longitudinal wave mode due to the electrons vibration caused by electrostatic field force. The electric field caused by surface plasmons decays quickly after penetrating the interface, localizing in the range of a few nanometers on the surface.

1.4 Surface Plasmon Polaritons

Similar to the complex wave of water surface caused by liquid surface tension, there will be rich physical phenomena on the interface between metal and medium due to the breaking of symmetry. Surface plasmon is a special mode of plasmon due to the special boundary conditions on the surface.

Since plasmon was proposed as the quantization of classical plasma oscillation from a Hamiltonian for the long-range electron–electron correlations, most of their properties can be derived directly from Maxwell's equations [6]. When electromagnetic wave incidents on the interface between metal and medium, the coupling between collective electron vibration and incident light leads to the formation of near-field electromagnetic wave propagating along the metal surface. And resonance will occur if these two frequencies are the same. Under the resonance state, the energy of electromagnetic field is effectively transformed into the collective vibration energy of free electrons on the metal surface. That is the well-known surface plasmon resonances (SPR).

There are also two modes of surface plasmon. One is longitudinal surface plasmon wave mode caused by electron oscillation, and the other one is confined to the surfaces that can interact with light to form propagating surface plasmon polaritons (SPPs). Polariton is the elementary excitation of the coupling vibration of photons and other quasiparticles. SPP has two excitation motions, charge motion (plasmon) and electromagnetic waves decayed at surface.

The electromagnetic field of SPP at a dielectric-metal interface is obtained from the solution of Maxwell's equations and the associated boundary conditions. To introduce the main characteristics of SPP, we consider a system consisting of a dielectric material, characterized by an isotropic dielectric constants ε_2 (real and positive), in the half-space $z > 0$, and a metal in the half-space $z < 0$, characterized by complex dielectric function $\varepsilon_1(\omega) = \varepsilon_r(\omega) + \varepsilon_i(\omega)$.

Using the continuity of electric field at the interface, we can get the electromagnetic field of SPP at the dielectric-metal interface with the polarization direction of the incident light parallel to the incident plane. From the Maxwell's electromagnetic field theory, we can obtain the following equations:

$$\begin{cases} \nabla \times H_i = \varepsilon_i \dfrac{\partial}{\partial t} E_i \\ \nabla \times E_i = -\mu_0 \dfrac{\partial}{\partial t} H_i \\ \nabla \cdot \varepsilon_i E_i = 0 \\ \nabla \cdot H_i = 0 \end{cases} \tag{1.11}$$

$i = 1$ at $z < 0$, and $i = 2$ at $z > 0$

Then we can get the s-polarized wave (TE mode) with $\frac{\partial}{\partial t} = i\omega$

$$\nabla^2 \mathbf{E} + K^2 \varepsilon \mathbf{E} = 0 \tag{1.12}$$

We then consider a p-polarized (transverse magnetic or TM) wave in this structure that propagates in the x-direction.

$$\begin{cases} \dfrac{\partial^2 \mathbf{E}_y}{\partial z^2} + \left(K^2 \varepsilon - \mathbf{K}_x{}^2\right) \mathbf{E}_y = 0 \\ \mathbf{H}_x = i \dfrac{1}{\omega \mu_0} \dfrac{\partial \mathbf{E}_y}{\partial z} \\ \mathbf{H}_z = \dfrac{K'}{\omega \mu_0} \mathbf{E}_y \end{cases} \tag{1.13}$$

$$\begin{cases} \dfrac{\partial^2 \mathbf{H}_y}{\partial z^2} + \left(K^2 \varepsilon - \mathbf{K}_x{}^2\right) \mathbf{H}_y = 0 \\ \mathbf{E}_x = -i \dfrac{1}{\omega \varepsilon \mu_0} \dfrac{\partial \mathbf{H}_y}{\partial z} \\ \mathbf{E}_z = -\dfrac{K'}{\omega \varepsilon \varepsilon_0} \mathbf{H}_y \end{cases} \tag{1.14}$$

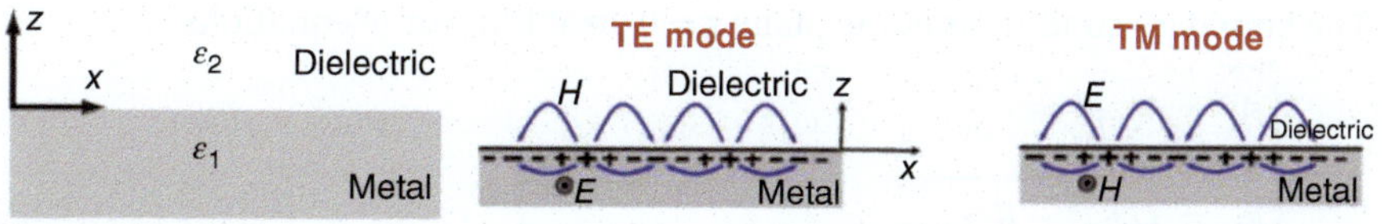

Figure 1.2 Dispersion of transverse electric TE mode and transverse magnetic TM mode on the dielectric-metal interface.

For the TE case, k_1 and k_2 are wave vectors in metal and medium along z direction. We now consider an s-polarized (transverse electric or TE) wave in the structure depicted in Figure 1.2. In a wave of this polarization, it is the electric vector that is perpendicular to the plane of incidence. The solutions of Maxwell's equations that are wavelike in the x-direction localized to the interface can be written as:

$$\begin{cases} \boldsymbol{E}_y(z) = A_1 \mathrm{e}^{i\boldsymbol{K}_x x} \mathrm{e}^{k_{1z} z} \\ \boldsymbol{H}_x(z) = iA_1 \frac{1}{\omega\mu_0} k_{1z} \mathrm{e}^{i\boldsymbol{K}_x x} \mathrm{e}^{k_{1z} z} \quad Z < 0 \\ \boldsymbol{H}_z(z) = A_1 \frac{K'}{\omega\mu_0} \mathrm{e}^{i\boldsymbol{K}_x x} \mathrm{e}^{k_{1z} z} \end{cases} \tag{1.15}$$

$$\begin{cases} \boldsymbol{E}_y(z) = A_2 \mathrm{e}^{i\boldsymbol{K}_x x} \mathrm{e}^{-k_{2z} z} \\ \boldsymbol{H}_x(z) = -iA_2 \frac{1}{\omega\mu_0} k_{2z} \mathrm{e}^{i\boldsymbol{K}_x x} \mathrm{e}^{-k_{2z} z} \quad Z > 0 \\ \boldsymbol{H}_z(z) = A_2 \frac{K'}{\omega\mu_0} \mathrm{e}^{i\boldsymbol{K}_x x} \mathrm{e}^{-k_{2z} z} \end{cases} \tag{1.16}$$

The continuity of the tangential components of the electric ($\boldsymbol{E}_y$) and magnetic fields ($\boldsymbol{H}_x$) across the interface $z = 0$ yields the pair of equations.The solution of this equation is $A_1 = A_2 = 0$. Thus an s-polarized SPP (TE mode) cannot exist in the structure depicted.

Therefore, we only consider a p-polarized (transverse magnetic or TM) wave in this structure that propagates in the x direction. In a wave of this polarization the magnetic vector is perpendicular to the plane of incidence – the plane defined by the direction of propagation and the normal to the surface. The solutions of Maxwell's equations that are wavelike in the x direction and whose amplitudes decay exponentially with increasing distance into each medium from the interface $z = 0$ can be written as:

$$\begin{cases} \boldsymbol{H}_y(z) = A_1 \mathrm{e}^{i\boldsymbol{K}_x x} \mathrm{e}^{k_{1z} z} \\ \boldsymbol{E}_x(z) = -iA_1 \frac{1}{\omega\varepsilon_1\varepsilon_0} k_{1z} \mathrm{e}^{i\boldsymbol{K}_x x} \mathrm{e}^{k_{1z} z} \quad Z < 0 \\ \boldsymbol{E}_z(z) = -A_1 \frac{K'}{\omega\varepsilon_1\varepsilon_0} \mathrm{e}^{i\boldsymbol{K}_x x} \mathrm{e}^{k_{1z} z} \end{cases} \tag{1.17}$$

$$\begin{cases} \boldsymbol{H}_y(z) = A_2 \mathrm{e}^{i\boldsymbol{K}_x x} \mathrm{e}^{-k_{2z} z} \\ \boldsymbol{E}_x(z) = iA_2 \frac{1}{\omega\varepsilon_2\varepsilon_0} k_{2z} \mathrm{e}^{i\boldsymbol{K}_x x} \mathrm{e}^{-k_{2z} z} \quad Z > 0 \\ \boldsymbol{E}_z(z) = -A_1 \frac{K'}{\omega\varepsilon_2\varepsilon_0} \mathrm{e}^{i\boldsymbol{K}_x x} \mathrm{e}^{-k_{2z} z} \end{cases} \tag{1.18}$$

The boundary conditions at the plane $z = 0$ yield the pair of equations:

$$\begin{cases} A_1 = A_2, \frac{k_{2z}}{k_{1z}} = -\frac{\varepsilon_2}{\varepsilon_1} \\ k_{1z} = \sqrt{K_x{}^2 - K^2\varepsilon_1} \\ k_{2z} = \sqrt{K_x{}^2 - K^2\varepsilon_2} \end{cases} \tag{1.19}$$

Here we can write wave vector in the x direction.

$$K_x = K\sqrt{\frac{\varepsilon_1\varepsilon_2}{\varepsilon_1 + \varepsilon_2}} = \frac{\omega}{c}\sqrt{\frac{\varepsilon_1\varepsilon_2}{\varepsilon_1 + \varepsilon_2}} \tag{1.20}$$

In the region of $z < 0$, k_z determines the decay of the electromagnetic field with increasing distance from the surface.

Surface plasmon occurs at the interface of a material exhibiting positive part in a dielectric constant (such as vacuum, air, glass, and other dielectrics) and a metal (or heavily doped semiconductor) whose real part of permittivity is negative at the given frequency of light. The dispersions of bulk plasmon, SPP at a dielectric-metal interface, light in air and in the dielectric medium is shown in Figure 1.3. In addition to opposite sign of the real part of the permittivity, the magnitude of the real part of the permittivity in the negative permittivity region should typically be larger than the magnitude of the permittivity in the positive permittivity region, otherwise the light is not bound to the surface (i.e. the surface plasmons do not exist).

For the existence of surface plasmon wave, the wave vector must be a positive value. Due to the facts of $\varepsilon_1 < 0$, $\varepsilon_2 > 0$, and $\lfloor\varepsilon_1\rfloor > \varepsilon_2$, it can be seen that the surface wave vector $\boldsymbol{K}_x$ is always larger than $\boldsymbol{K}$ for the incident light of any frequency. $\boldsymbol{K}_z$ must be an imaginary number,and electromagnetic field of the SPP decays

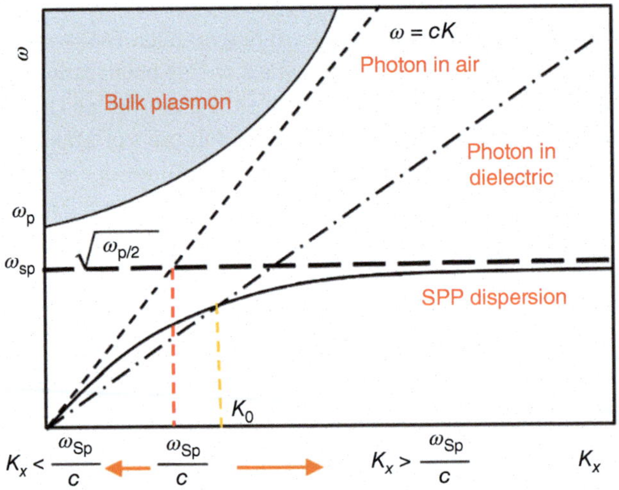

Figure 1.3 Dispersion of SPP at a dielectric-metal interface. Also plotted is the dispersion of light in air and in the dielectric medium, and the corresponding surface plasmon frequency.

exponentially in the z direction into the dielectric medium in contact with the metal. Consequently, the SPP cannot radiate light into the dielectric medium, and cannot be excited with conventional illumination from the adjacent dielectric. Moreover, the frequency of the surface plasmon is also the limiting frequency of the SPP with $\boldsymbol{K}_x \rightarrow \infty$.

Thus, the existence of SPs depends entirely on the fact that dielectric function $\varepsilon(\omega)$ has a negative real part ($\varepsilon_1 < 0$). The SPs are well pronounced as resonances when the losses are small enough, i.e. $\lfloor \varepsilon_1 \rfloor \gg \varepsilon_2$. This is the well-known property of a good plasmonic metal. We will define the substance as a good plasmonic metal if these two properties are satisfied simultaneously.

1.5 Plasmon at Metal-Vacuum Interface

For the metal-vacuum interface with $\varepsilon_2 = 1, K = \frac{\omega}{c}$, when $\boldsymbol{K}_x \longrightarrow \infty$, $\varepsilon_1 + \varepsilon_2 = 0$, we can see the relation between the bulk plasmon and SPP in the Drude model Eq. (1.7)

$$\omega_{\mathrm{sp}} = \sqrt{\frac{\omega_{\mathrm{p}}}{2}} \tag{1.21}$$

For the metal-dielectric interface, $\varepsilon_2 = \varepsilon_{\mathrm{d}}, K = \frac{\omega}{c}\sqrt{\varepsilon_{\mathrm{d}}}$ in dielectric, and $\omega = cK/\sqrt{\varepsilon_{\mathrm{d}}}$. This wave vector k_0 equals to one value of SPP.

As discussed above, the metal film is not treated as a conductor as bulk metal in electromagnetic field. The metal films and nanostructures in the following chapters will be treated as special dispersive medium with polarization property in electromagnetic field. Under this case, the SPP mode for metals is obtained.

1.5.1 Propagation Length

As the electromagnetic field of the SPP decays exponentially along the z direction into the dielectric medium in contact with the metal. It is necessary to know the propagation length and depth.

From Eq. (1.10), it is clear that $\varepsilon(\omega) = \varepsilon_{\mathrm{r}}(\omega) + \varepsilon_{\mathrm{i}}(\omega)$ and wave vector has real and imaginary parts for the metals. For the metal-vacuum interface, we can see that the real and imaginary parts of wave vectors are as following:

$$K_{x\mathrm{r}} \approx \frac{\omega}{c}\left(\frac{\varepsilon_{1\mathrm{r}}}{\varepsilon_{1\mathrm{r}}+1}\right)^{1/2} \tag{1.22}$$

$$K_{x\mathrm{i}} \approx \frac{\omega}{c}\frac{\varepsilon_{1\mathrm{i}}}{2\varepsilon_{1\mathrm{r}}^2}\left(\frac{\varepsilon_{1\mathrm{r}}}{\varepsilon_{1\mathrm{r}}+1}\right)^{3/2} \tag{1.23}$$

Due to Ohmic losses in the metal, characterized by the imaginary part of the dielectric function of the metal, the energy carried by a SPP decays exponentially as the SPP propagates along the planar dielectric-metal interface. The decay length, called the energy propagation length L_{sp} is determined by the imaginary part of the SPP wavenumber. The propagation length along the metal–vacuum interface is defined as $L_{sp} = (2K_{xi})^{-1}$. For example, the propagation lengths for Au and Ag are $L_{Au} = 14\ \mu m$ and $L_{Ag} = 44\ \mu m$ with the irradiation 633 nm light (the ε of metals of this frequency can be referred to Ref. [5]).

Plasmons or bulk plasmons cannot be coupled with light waves and can only be excited by particle collisions. In metals (as mentioned Au and Ag) with large number of free electrons, plasmons have very high propagation loss due to the strong Landau damping effect and the scattering between plasmons and phonons. While for two-dimensional materials, the plasmons attenuation is greatly suppressed because they are confined to the plane. As typical representative two-dimensional materials, graphene and black phosphorus become extremely popular plasmon materials [7].

1.5.2 Propagation Depth (or Skin Depth)

From Eq. (1.19) the wave vector in the z direction can be written as:

$$\begin{cases} k_{1z} = \dfrac{\omega}{c}\sqrt{\dfrac{-\varepsilon_1^2}{\varepsilon_1 + \varepsilon_2}} \\ k_{2z} = \frac{\omega}{c}\sqrt{\dfrac{-\varepsilon_2^2}{\varepsilon_1 + \varepsilon_2}} \end{cases} \tag{1.24}$$

The skin depth is defined as $\delta_i = {}^1\!/_{k_i}$. It is clear that the penetration depth in vacuum is greater than that in metals. SPP is an evanescent wave along the normal direction of the interface. Figure 1.4 shows the schematic diagram and FDTD simulation diagram for SPP propagation in medium, metals and their interfaces [8]. The finite skin depth of real metals leads to an effect related to nanoplasmonic confinement. For example, the skin depths for most metals (δ_{metal}) are about several to tens of nanometers [2], δ_{diel} is in the range of 250–1000 nm, and δ_{SPP} is about 2–20 μm with the irradiation 500 nm light. The characteristic lengths of Ag plasmons are slightly larger than those of other metals. For electromagnetic wave with frequency of 50 MHz ($\lambda = 6$ m), the skin depth is more than one hundred meters for dry soil of poor conductivity, while the skin depth is greatly reduced to millimeters for sea water. Thus, submarines under the sea cannot receive radio waves, and we can receive the signal under the subway.

The common presumption that electromagnetic radiation cannot be concentrated or confined to a region whose size is less than half the incident wavelength. It is obvious that the minimum volume of an optical field localized in a cavity – bounded by three pairs of ideal mirrors with size of $\lambda/2$. How can an optical field be localized 6–7 orders of magnitude tighter on the scale of 1 nm? From the above skin depth, the particles should be smaller than the metal's skin depth – how deeply the optical radiation penetrates the metal. That depth is roughly 25 nm for Au, Ag, and Cu over the entire optical and most of the IR region [2]. On the minimum end, the particles should not be smaller than the so-called nonlocality length about 2 nm. On the scale of 2–20 nm, an incident light wave is always a plane wave, and no light focusing in the optical far-field is possible.

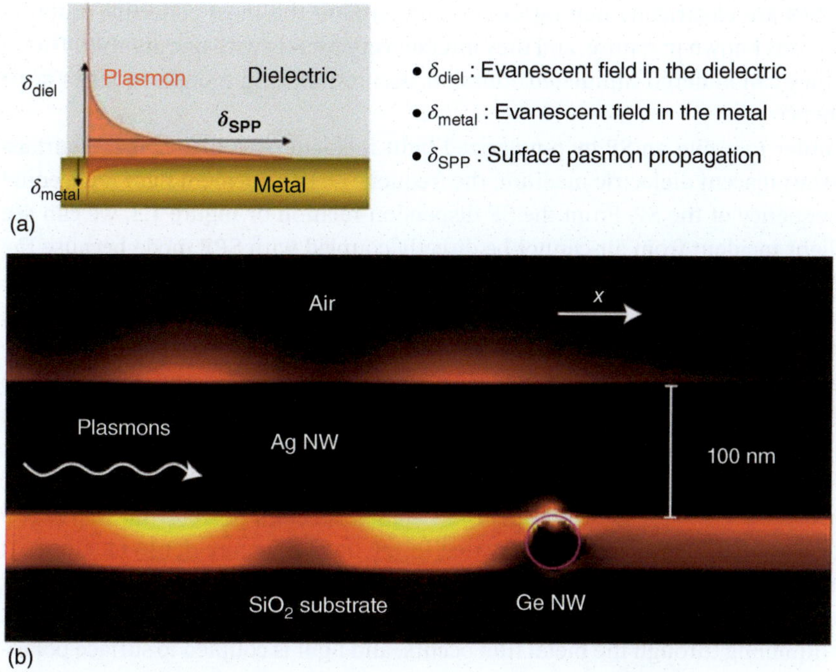

Figure 1.4 Schematic diagram and simulation diagram for SPP propagation in medium-metals (a) and their interfaces. δ_{diel} and δ_{metal} evanescent wave (b) FDTD simulation of electric-field intensity. Ag nanowire diameter = 100 nm, Ge nanowire diameter = 40 nm. Source: Falf et al. [8]/Reproduced with permission from © 2009 Springer Nature.

1.6 Excitation and Detection of SP

Generally, surface plasmon has momentum as a quasiparticle. SP can be excited by electrons or photons. The response function of material to external field is dielectric function. The very small momentum of light can be ignored. The momentum conservation also limits the coupling between plasmon and light. Usually, the bulk plasmon is the quantum of longitudinal electron-density wave within the bulk metal, and it cannot be excited by transverse electromagnetic plane waves. It can only be excited by electron beam. While the SP, also quantum of longitudinal electron-density waves, confined to the metal-dielectric interface, and it is often excited using dielectric prism coupled with a thin metal film, near-field optics and other methods. Here optical and electrical excitation of SP will be presented. As an important content of this book, localized surface plasmon (LSP) of metal nanoparticles will also be delivered.

1.6.1 Optical Excitation of SP

The reason that the SPs can be optically excited is because their relevant physical property are the same. SPs are Bosons with spin of 1 as photons do. At the same time, SPs are electrically neutral excitations. SPs are the most collective material oscillations known in nature, and they interact very weakly with one another. Therefore, they can undergo stimulated emission, accumulating in a single mode in large numbers.

In order to excite an SP by p-polarized light incident on a planar metal surface from the adjacent dielectric medium, the frequency of the incident light must equal the frequency of the SP. From the SP dispersion relation or Figure 1.3, we can see that light incident from air cannot be directly coupled with SPP mode because the wave vector of SP is larger than that in air. The SP can be excited with the assistance of special experimental arrangements or structures for matching the wave vector. Therefore, several devices (Figure 1.5) have been designed to provide the necessary wave vector conservation, including reflection geometry (Kretschmann and Otto configurations), scanning near-field optical microscopy (SNOM), diffraction effects (grating couples) surface roughness and scattering particles [9].

In the Kretschmann configuration, a metal film is illuminated through a dielectric prism at an angle of incidence greater than the critical angle for total internal reflection. The wavevector of light is increased in the optically more dense medium. At the angle of incidence at which the in-plane component of the photon wave-vector in the prism coincides with the SPP wave vector at the air-metal interface, resonant light tunneling through the metal film occurs, and light is coupled to surface polaritons. Figure 1.6a presents the detailed wave vector in the Kretschmann geometry. Under these resonant conditions, a sharp minimum is observed in the reflectivity from the prism-metal interface as light can be coupled to SPP. Since the SPP field is concentrated close to a metal surface, it is significantly enhanced at the surface. With the increase of the metal film thickness the efficiency of the SPP excitation (and the

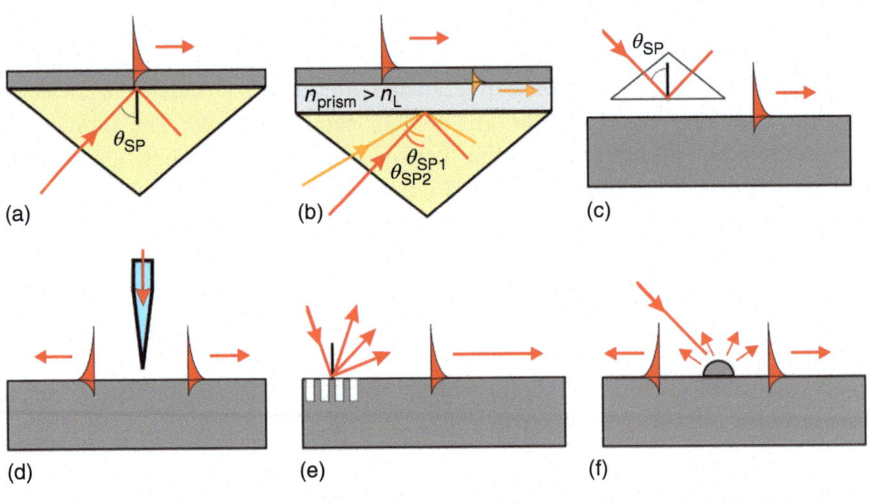

Figure 1.5 SP excitation configurations: (a) Kretschmann geometry, (b) two-layer Kretschmann geometry, (c) Otto geometry, (d) excitation with a SNOM probe, (e) diffraction on a grating, and (f) diffraction on surface features. Source: Reproduced from Zayatsa et al. [9]/with permission of Elsevier.

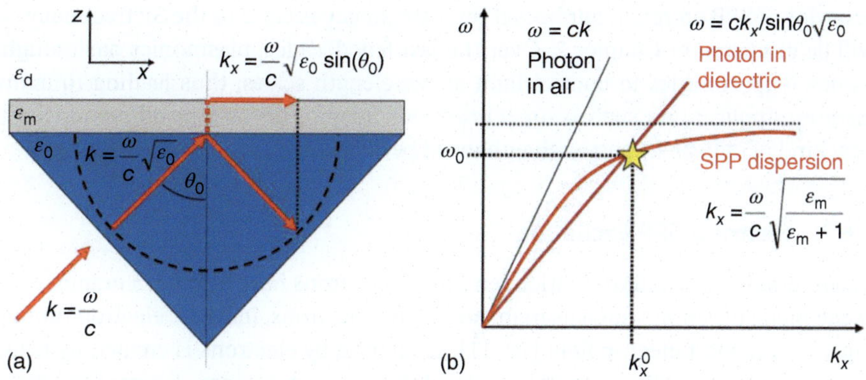

Figure 1.6 (a) Wave vector dispersion in the Kretschmann geometry. (b) Typical SPP dispersion at the metal-dielectric interface.

field enhancement) decreases with increasing the distance. Usually, an additional dielectric layer with a refractive index smaller than the one of the prism should be deposited between the prism and the metal film. In this two-layer geometry, the photon tunneling through this additional dielectric layer can also provide excitation of SPP on the inner interface, as shown in Figure 1.6b.

With the development of scanning probe techniques, it become possible to study SPs locally on a surface using SNOM. In particular, SNOM provides an opportunity to probe the SP field directly over the SP existing surface. The application of SNOM to SP studies introduced a new technique that provides the possibility to excite SP locally at a given surface. Using illumination through a SNOM fiber tip, circular SP waves can be locally launched at the surface. This configuration can be treated as either a diffraction or a tunneling mechanism of SP excitation. This technique allows local and precise excitation of SPs at the SNOM tip.

Plasmon can be excited by particle collisions without any special arrangements, although it cannot be coupled with light waves. Light scattering and absorption by a sphere surely constitute the most evolved model for light scattering by obstacles. On a randomly rough surface, the SP excitation conditions can be achieved. The rough surface can be consisted of small sized particles, and the SP is highly localized near the particles, as known LSPs. Here we briefly present the concept of LSP. LSP can be supported by subwavelength features on a small sized metal particle surface where the LSP can decay into SP. As a result, LSP can be excited by SPP. These results in a significant enhancement of SPP scattering by surface defects if the frequency of SP and the resonant frequency of LSP are close to each other.

This is possible since in the near-field region diffracted components of light are all wave vectors (Figure 1.5f). The problem with random roughness is ill-defined nonresonant SP excitation, resulting in a low efficiency of light-to-SP coupling. Such nonresonant SP excitation processes result in a complex field distribution over the surface due to the interference of the SPPs excited on different interfaces of the film and the non-coupled illuminating light. LSP resonances play a significant role in the behavior of SPP on rough surfaces if their frequency is close to the SP frequency, as will be discussed in Chapter 2. Over the last few decades, plasmonics has brought us new opportunities to control light at wavelength scales, thus leading to many inspiring applications, such as super resolution imaging beyond the diffraction limit, new kind of optical sensors with enhanced performance.

1.6.2 Electrical SPP Excitation

In addition to optical illumination, energized electrons have also been employed to excite metal plasmons, such as high energy free electrons, inelastic electron tunneling, and a modest electric field [10, 11]. Excitation by electrons is created by firing electrons into the bulk of a metal. As the electrons scatter, energy is transferred into the bulk plasmon. The component of the scattering vector parallel to the surface results in the formation of SPP. With the development of nano-fabrication techniques, on-chip electrically-driven plasmonic circuitries have gained great attention in the post-Moore era for they hold the promise to combine both the small device footprint of electronic circuitry and the large information capacity. Recently, direct electrical excitation of SPP via inelastic electron tunneling in metal–insulator–metal junctions has reemerged as a promising ultrafast source to drive the integrated plasmonic circuitries.

Since 2003, plasmonics has facilitated the realization of coherent nanolasers [12], plasmon enhance diodes [13]. Naturally, the next step is to realize the electrically driven plasmon sources that can be integrated with optoelectronic circuitry. In 2012, Fan et al. present an experimental demonstration of electrically driven GaAs nanowire light sources integrated with plasmonic Ag nanostrip waveguides [14]. SPPs are experimentally demonstrated and analyzed with the help of full-field electromagnetic simulations, as indicated by Figure 1.7. They also found that light leaking out of plasmonic waveguide bend and emerges from the Y-splitter junction, and suggested that making splitting junction as "sharp" as possible can minimize splitting losses.

In 2019, Yang et al., designed and experimentally demonstrated an electrically driven UV plasmonic laser at room temperature in a p-GaN/MgO/ZnO/MgO/Ag structure, as shown in Figure 1.8. [15] This structure includes an Ag/MgO/ZnO MIS hybrid plasmonic waveguide, in which strong optical confinement and relatively low metal loss can be satisfied simultaneously. In the proposed structure, holes are injected into ZnO layer from p-GaN, and electrons are injected from Ag film through MgO gap layer. An additional electron blocking MgO layer has been introduced at the interface between the gain region and the GaN layer to enhance the emission from ZnO by blocking electrons in the ZnO from entering into the p-GaN. This plasmonic laser device represented the state-of-the-art performances inelectrically driven plasmonic lasers.

Neutens et al. demonstrated an integrated electrical source of confined SPPs in LED-MIM device [16]. The electrical injection of SPPs inside an MIM waveguides can combine high spatial field confinement and propagation lengths of several micrometers. Since the metallic cladding layers ensure that the fields are confined

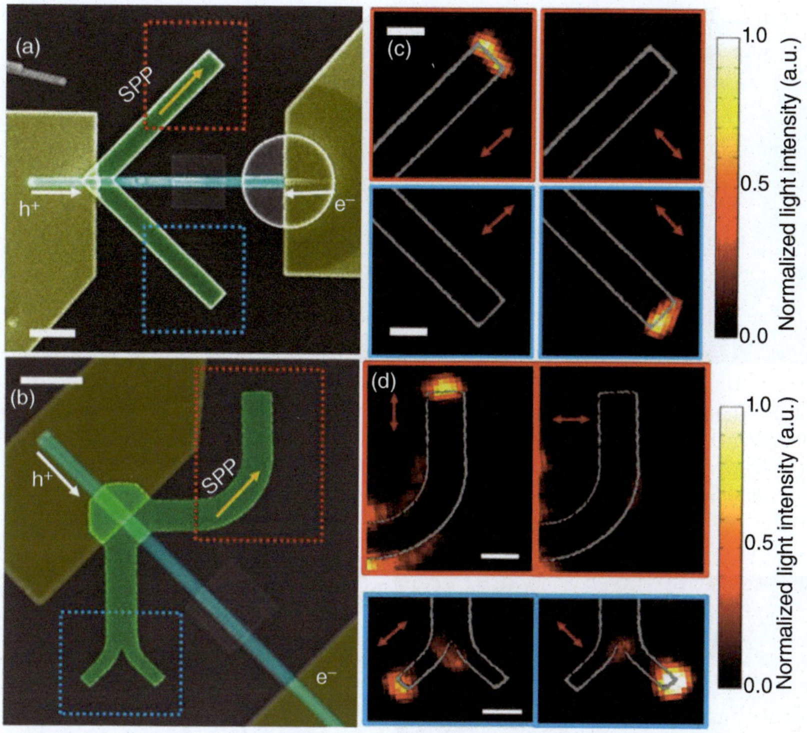

Figure 1.7 (a, b) Scanning electron microscopy (SEM) image of the structures (straight strip, bend, and Y-shape) collected by CCD camera under lamp illumination with false-color highlighting the different components: yellow is for electrical contacts, cyan is for GaAs nanowire, and green indicates the Ag strip waveguides. Scale bar is 2 μm. (c, d) Detailed spatial distribution of the scattered light intensity near the terminations of waveguides with different shapes when only collecting light with the electric field component polarized along and perpendicular to the waveguides, respectively. Scale bars are 1 μm. Source: Fan et al. [14]/Reproduced with permission from © 2012, American Chemical Society.

within the waveguide, one can easily separate plasmonic effects from any other nonplasmonic background radiation. The working principle of the LED-MIM device was clearly demonstrated. When the polarization of the dipole is set in plane and perpendicular to the slit axis, light generated in the quantum well excites a local plasmonic mode inside the injection slit. This local mode couples efficiently to a propagating mode inside the MIM waveguide.

The injection and detection slits in the waveguide partially reflect surface plasmons, resulting in a Fabry–Perot cavity inside the MIM waveguide. Only light with transverse magnetic polarization can be emitted at the detection slit. They also observed theoretically expected an exponential decay of the SPPs along the waveguide, caused by the Ohmic losses inside the metal layers of the metallic slot waveguide. It is well-known that the nonradiative plasmonic decay can also generate hot carriers from metal nanostructures. Sun et al. reported on the incandescent-type light source composed of Au NRs decorated with single Ga-doped ZnO microwire, as illustrated by Figure 1.9 [17].

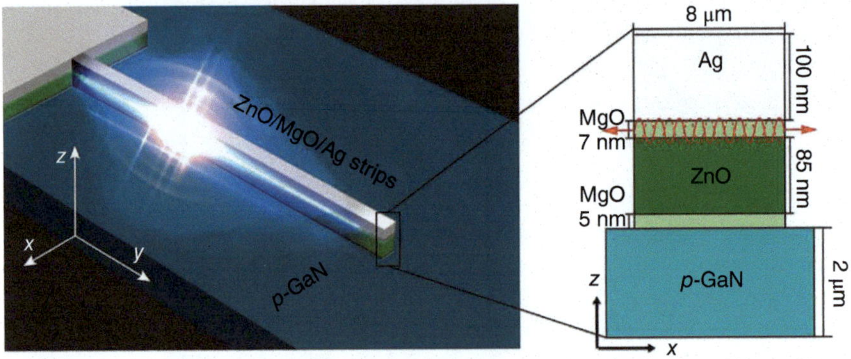

Figure 1.8 The schematic diagram of electrically driven plasmonic laser based on p-GaN/MgO/ZnO/MgO/Ag structure. Inset illustrates the SPPs feedback diagram in the ZnO/MgO/Ag cavity. Source: Reproduced from Yang et al. [15] © 2019/with permission of John Wiley and Sons Inc.

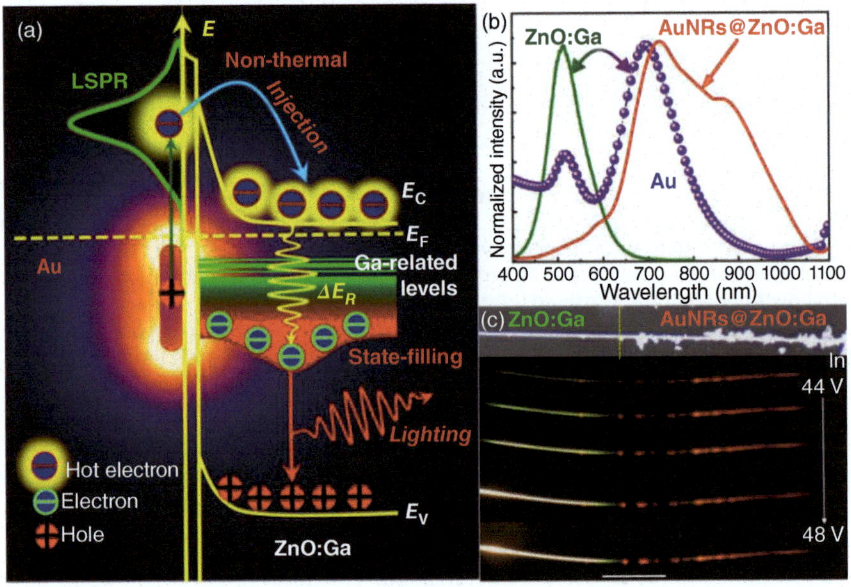

Figure 1.9 (a) Schematic diagram of the working principle of bright visible light emitting from electrically biased single AuNRs@ZnO:Ga MW-based incandescent-type light source. (b) Normalized intensities of the EL spectrum from single bare ZnO:Ga MW-based fluorescent light source, the EL spectrum from single AuNRs@ZnO:Ga MW-based fluorescent light source, and the extinction spectrum of the deposited Au NRs. (c) Optical microscopic images of bright visible light emitting from electrically driven single ZnO:Ga MW prepared with partial Au NRs decoration (scale bar, 300 μm). Source: Sun et al. [17]/Reproduced with permission from © 2020, Optica Publishing Group.

Under a sufficiently strong electric field (~10^6V m^{-1}), hot spots could be produced owing to the extremely poor heat dissipation of semiconductor. Then collective oscillation of the free electrons in Au NRs can be electrically excited. Through nonradiative plasmonic decay, Au NR plasmons can transfer their energies into oscillating higher-energy electrons, leading to formation of hot electrons. Especially once the hot electron possessed the energy up to plasmon frequency, Au-nanorod plasmons inducing the generation of nonequilibrium distribution of hot electrons can lead to the state-filling effect in the energy-level configuration of ZnO:Ga, which can be employed to dominate the near-infrared emission characteristics from single AuNRs@ZnO:Ga microwire based incandescent light source, as shown in Figure 1.9b,c. Nonradiative decay of surface plasmons generates hot carriers in metal nanostructures that can be injected into neighboring semiconductor micro/nanostructures. Thus wavelength-tunable incandescent-type emissions can be achieved from visible to near-infrared region adjusting the aspect ratios of the Au NRs. Such working principle on the generation of hot carriers has been utilized to construct and further enhance the electronic and optoelectronic devices. The plasmon-enhanced optoelectronic device will be discussed in Chapter 9.

Optically excited SPs are featured by dipolar charge oscillations and they can be understood as bright SPs due to their dipole-active character. Bright SPs can be further tailored in Vis-NIR regime by controlling the size and morphology of noble metal nanoparticles. But dark surface-plasmon modes cannot be excited by optical means. Although SNOM can excite SPP locally at a given surface, this technique can hardly reach a spatial resolution. In contrast, electron-energy-loss spectroscopy (EELS) performed in a scanning transmission electron microscope (STEM) routinely provides nanometer spatial resolution by focusing a subnanometer electron beam on nanoparticles and analyzing the energy losses of the electrons. Chu et al. reported that STEM-EELS can excite bright and dark SP modes in metal monomer and coupled nanochain structures [18]. Upon external illumination with the electric field polarized along the dimer axis, the induced charge oscillation thus involves a longitudinal coupled SP along the long axis and a transversal coupled SP perpendicular to that axis. Compared to the SP excitation energy of a single spherical nanoparticle, the SP excitation of coupled dimer behaviors red- and blue-shifting, which will be discussed in the following part of "detection of surface plasmon" using EELS.

These types of SP-distribution features should be observable by tilting the sample and measuring the spectra at various angles of electron incidences. In such cases, STEM-EELS should then be able to provide information on the local polarization of the SP near fields, as described in Chapter 3.

1.6.3 Localized Surface Plasmon of Nanoparticles

In addition to SPP and surface plasmons at a planar dielectric-metal interface, as shown in Figure 1.5f for the photon excitation of SPP, it can be seen that the isolated surface defects, such as a groove, a particle, a slit or a corrugation on a planar surface provides a mechanism by which free-space radiation and SPs can exchange energy and hence couple. In the localized space, electromagnetic excitations can exist in nanometer sized metal objects of various shapes.

As discussed above, for materials with negative dielectric permittivity (characteristic behavior of metals), the most of the incident irradiation is reflected and incident electromagnetic field decays inside the materials with a depth of a few tens nanometers. But for a nanoparticle with size smaller than the depth size, optical fields are able to penetrate its entire volume and drive SP oscillations. It is the skin depth that determines the characteristic length scale in nano-optics and makes the nanoscale important. Plasmonic nanoparticles can couple with electromagnetic radiation of wavelengths far larger than the particle due to the nature of the dielectric-metal interface.

For the case of SP oscillations, conduction electrons in metal or metal-doped semiconductor nanomaterials can be coherently excited by incident light to collectively oscillate at metal/dielectric interfaces. For metal nanoparticles, surface waves can be excited directly by optical excitation, leading to the conception of metallic or "plasmonic" nanostructures as nanoscale optical components. Such surface excitations in bounded geometries are the well-known LSPs. Plasmons are the oscillations of free electrons that are the consequence of the formation of a dipole or multipole in the material under electromagnetic waves. The electrons migrate in the material to restore its initial state. However, the light waves oscillate, leading to a constant shift in the dipole or multipole that forces the electrons to oscillate at the same frequency as the light. This coupling only occurs when the frequency of the light is equal to or less than the plasma frequency and is greatest at the plasma frequency that is therefore called the resonant frequency. Under the resonant frequency, plasmonic nanoparticles exhibit interesting scattering, absorbance, and coupling properties.

For metal objects of various shapes with sizes smaller than the light wavelength, absorption and scattering response of light by particles can be characterized after solving Maxwell's equations. In 1908, Gustav Mie obtained an analytical solution for single spherical particles by solving the scattering of a monochromatic plane wave by a homogeneous dielectric sphere of arbitrary size in a homogeneous medium [19]. The theory of plane-wave scattering by a homogeneous isotropic sphere is also called Lorenz–Mie theory or Lorenz–Mie–Debye theory for their contributions to solving this problem [20, 21]. Briefly, the polarization of dipole when one uniform spherical particle with a volume of V was irradiated under the plane light.

$$\chi = 3V\frac{\varepsilon - \varepsilon_{\mathrm{m}}}{\varepsilon + 2\varepsilon_{\mathrm{m}}} \tag{1.25}$$

Then the expressions for the optical cross, such as extinction section σ_{ext}, scattering section σ_{sca}, and absorption section σ_{abs}, can be obtained ($\sigma_{\mathrm{ext}} = \sigma_{\mathrm{sca}} + \sigma_{\mathrm{abs}}$).

$$\sigma_{\mathrm{sca}} = \frac{k^4}{6\pi}|\chi|^2 = C_1\frac{a^6}{\lambda^4}\left|\frac{\varepsilon - \varepsilon_{\mathrm{m}}}{\varepsilon + 2\varepsilon_{\mathrm{m}}}\right|^2 = C_1\frac{a^6}{\lambda^4}\frac{(\varepsilon_1 - 2\varepsilon_{\mathrm{m}})^2 + \varepsilon_2^2}{(\varepsilon_1 + 2\varepsilon_{\mathrm{m}})^2 + \varepsilon_2^2} \tag{1.26}$$

$$\sigma_{\mathrm{abs}} = k\mathrm{Im}(\chi) = C_2\frac{a^3}{\lambda}\left|\frac{\varepsilon - \varepsilon_{\mathrm{m}}}{\varepsilon + 2\varepsilon_{\mathrm{m}}}\right|^2 = C_2\frac{a^3}{\lambda}\frac{\varepsilon_{\mathrm{m}}\varepsilon_2}{(\varepsilon_1 + 2\varepsilon_{\mathrm{m}})^2 + \varepsilon_2^2} \tag{1.27}$$

where $C_1 = 128\pi^3/3$, $C_2 = 32\pi^2$, a presents the radius of particle. When the particle size is smaller than the light wavelength, we can see $\frac{a^6}{\lambda^4} = a^2\frac{a^4}{\lambda^4} \ll a^2\frac{a}{\lambda}$. Therefore, σ_{abs} is dominated for small sized particles, and σ_{sca} can not be ignored for large sized nanoparticles (note that $a \ll \lambda$). Additionally, the absorption and scattering peaks are highly sensitive to the medium. The resonance absorption (or scattering) will occur when $\varepsilon + 2\varepsilon_m \to 0$. From Drude mode with $\gamma \ll \omega$, the resonant frequency of spherical nanoparticles is derived.

$$\omega_{max} = \frac{\omega_p}{\sqrt{1 + 2\varepsilon_m}} \tag{1.28}$$

The resonant frequency is approximately linear with the refractive index (n), red-shifting and blue-shifting with increasing and decreasing the refractive index. For metal nanoparticles, the SPR are characterized by discrete, complex frequencies, depending on the size and shape of the object. LSPs are confined to curved metal particles. In contrast to the SPP mode that can be excited only if both the frequency and wave vector of the exciting light match the frequency and wave vector of the SPP, LSPs of nanoparticles with $d \ll \lambda$ can be resonantly excited with light of appropriate frequency irrespective of the wave vector of the exciting light, which is the well-known localized SPR. The resonant energy depends on the particle size and shape, and the permittivity (dielectric constant) of the metal nanoparticle itself and surrounding materials, and can be tuned by these factors. Different modes of localized SPR exhibit enhanced near-field amplitude at the resonance wavelength. This field is highly localized at the nanoparticle and decays rapidly away from the nanoparticle/dielectric interface into the dielectric background, though far-field scattering by the particle.

Therefore, with the adsorption or modification of molecular on the particle surface, the local medium environment changes or the particles will be assembled into a certain configuration, which will lead to the obvious shifting of resonant peak position [22]. Using this feature of SPR, scientists can realize highly sensitive biological and chemical sensors.

The localized SPR is the fundamental principle behind many technology applications. Another important characteristic of SP in application is quality factor (Q), which is the number of electron oscillations that occur coherently, and during which the mode is able to sustain its phase and accumulate energy from the external excitation field. The Q ($Q = \frac{\omega}{2\gamma}$) of plasmonic metals ranges from 10 to 100. The SP spectral width (or relaxation rate) γ, quality factor Q and lifetime of SP depends explicitly only on the optical frequency ω and the image part of permittivities of the metal. This lifetime is in the range 10–60 fs for silver and 1–10 fs for gold in the plasmonic region. These data show that nanoplasmonic phenomena are ultrafast (femtosecond) [2].

In response to an external excitation, a plasmonics nanoparticle can generate local fields that are enhanced by Q with respect to the external field. The region occupied by these enhanced fields is determined solely by the size of the nanoparticle. Another route to enhancing the local field is through adjusting the morphology. For instance, the sharp tip can create intense local fields, which will be delivered in Chapters 4 and 7 for nanostructures based on the Au nanobipyramids.

The above LSP is a quasi electrostatic model, in which metal nanoparticles are treated as dipoles. The delay effect and radiation attenuation effect of electric field in particles are not considered in this model. And it can only be used to calculate particles with diameter much smaller than the wavelength. When the particle size large and comparable to the wavelength, electric field in the particle is no longer uniform because of the small penetration depth of electric field in particles, as discussed in Section 1.5.2. For this case, quadrupole and octupole must be considered, and the polarizability of large sized particle must modified according to the long wavelength approximation, as widely reported in many literatures. On the other extreme direction, influence of strong coupling effect and quantum tunneling effect on plasmons should be considered when the particle size is smaller than the critical size.

Since an LSP is confined to a particle (or a curved surface), it results in a significant electromagnetic field enhancement at small metallic particles due to the small volume of the LSP mode. This effect contributes to numerous phenomena such as light emission from STM tunnel junctions, enhanced scattering, surface enhanced Raman scattering, and second-harmonic generation, and finds applications in active photonic elements and apertureless scanning near-field microscopy.

1.6.4 Other Surface Plasmon Supports

Because of their large negative values of the real part of permittivity, metals are excellent materials for plasmonic applications, particularly noble metals of Au and Ag. These two materials have been used frequently in plasmonics due to their chemical stability and relatively low Ohmic loss in the Vis-NIR wavelength region. Other materials such as Na, Al, Ni, Pt as well as oxides and metallic alloys have demonstrated plasmonic properties.

Different from plasmon mode supported by the isotropic metal or heavily doped semiconductor surface [23], the plasmon in two-dimensional materials is confined to a plane, and its attenuation is greatly suppressed. As a typical two-dimensional material, graphene has become a very popular plasmon support. However, as a semi metal, plasmon frequency of graphene located in the infrared region due to its low carrier concentration. Other two-dimensional materials are also mostly semi-metallic (such as $TiSe_2$) or semi-conductive (such as BN and MoS_2) with low carrier concentration. Thus, it is urgent to find suitable two-dimensional metal materials for integrated photonics device. Bornene is a new two-dimensional material with intrinsic metallicity, and it can be a promising plasmon support with low attenuation.

As two dimensional materials with in-plane anisotropy, it is predicted to be the second kind of wail semi metal with series of novel electrical properties, such as ultra-high electron mobility and adjustable carrier concentration. Recently, two dimensional hyperbolic plasmon resonance mode in far infrared band was detected in the WTe_2 thin film through optimizing Fourier transform infrared spectroscopy [24].

1.6.5 Detection of Surface Plasmon

Localized SPR of metal nanoparticles were historically probed on UV-Vis-NIR spectrometer by illuminating the sample with far-field plane-wave light and studying the optical response by measuring absorption, scattering spectra. The spectra are an average effect of many particles dispersed in one sample. In 2001, Nagao et al. reported, for the first time, energy and the linewidth dispersion of a plasmon in a dense two-dimensional electron system on a silicon surface. The plasmon energy dispersion exhibited an excellent agreement with the nearly free-electron theory [25].

After SP excitation, it will decay according to two categories: radiative SP decay (studied seldom) and nonradiative SP decay in most cases. Nonradiative decay of SP is usually considered an unwanted process. Much effort was focused on avoiding or mitigating nonradiative SP decay. However, recent studies have proven that hot carriers generated from nonradiative SP decay can be used in many practical applications are plagued by the plasmonic loss – SP diphase, such as photothermal effect, surface imaging, and hot carrier photodetection [26].

Photodetection for the SP includes hot electron and hot hole photodetectors. Compared with hot hole (lower than Fermi level in d band) detection, hot electron (higher than Fermi level) detectors have been widely studied. For example, optical antenna can realize effectively photoelectric conversion and detection, in which the hot electrons in metal nanoantennas can be injected into adjacent semiconductors. However, nanoantennas are usually in subwavelength size, which often requires complex and expensive processing technology, such as electron beam exposure and focused ion beam etching. Another coupling structure of surface plasmon and low dimensional nanomaterials can also realize local electric field enhancement. The low dimensional nanomaterials, such as nanowires, quantum dots, and two-dimensional materials, act as excellent thermionic acceptor and can be applied to construct plasmon hot electron detectors.

Since the plasmons can be excited by both photons and electrons, they can also be detected by means of electrons. In 1957, the existence of surface plasmons in a thin metal film was theoretically predicted by Ritchie [27] for electron energy losses in thin metal film below the bulk plasmons and soon experimentally demonstrated in 1959 by Powell [28]. Stern and Ferrell, in 1960, used transmission experiments to probe the dispersion of the electron energy loss as a function of scattering angle [29]. The experimental observation of LSPRs in metal spheres excited by 50 keV electrons was reported in later of 1960s [30].

With the emergence of energy-filtered TEM and spectrum-image mode in the STEM in the 1980s of metallic nanoparticles, the SP modes and charge distribution were well characterized. Oftentimes, one particular energy is selected for examination and the spectral intensity is plotted as a function of position, resulting in an energy-filtered spectrum-image, which is commonly known as an electron energy loss spectroscopy (EELS) map. The early EELS experiments was the first observation of the bulk plasmon in thin Al film [31]. However, the early EELS experiments were not able to reveal the spatial profiles of LSPR, and it was not until the emergence of the energy-filtered TEM and STEM in the 1990s that mapping studies became possible typically for high-energy atomic transitions beyond the visible region (400–700 nm). The first EELS mapping experiments represented a significant step toward SPR mapping were reported by Nelayah et al., in 2007 [32]. In their work, mappings the localized SPR modes in Au and Ag nanostructures down to region of 1.75–3 eV.

EELS performed in an STEM has demonstrated unprecedented power in the characterization of surface plasmons. For example, Song et al. investigated the plasmon excitation in individual Ag nanoparticles and plasmon maps of coupled Ag nanoparticles [33]. They found that the intensity maxima are distributed along the circular edge of the particle, where the electron beam excites the plasmon effectively. The line profile of the plasmon intensities is shown in Figure 1.10a.

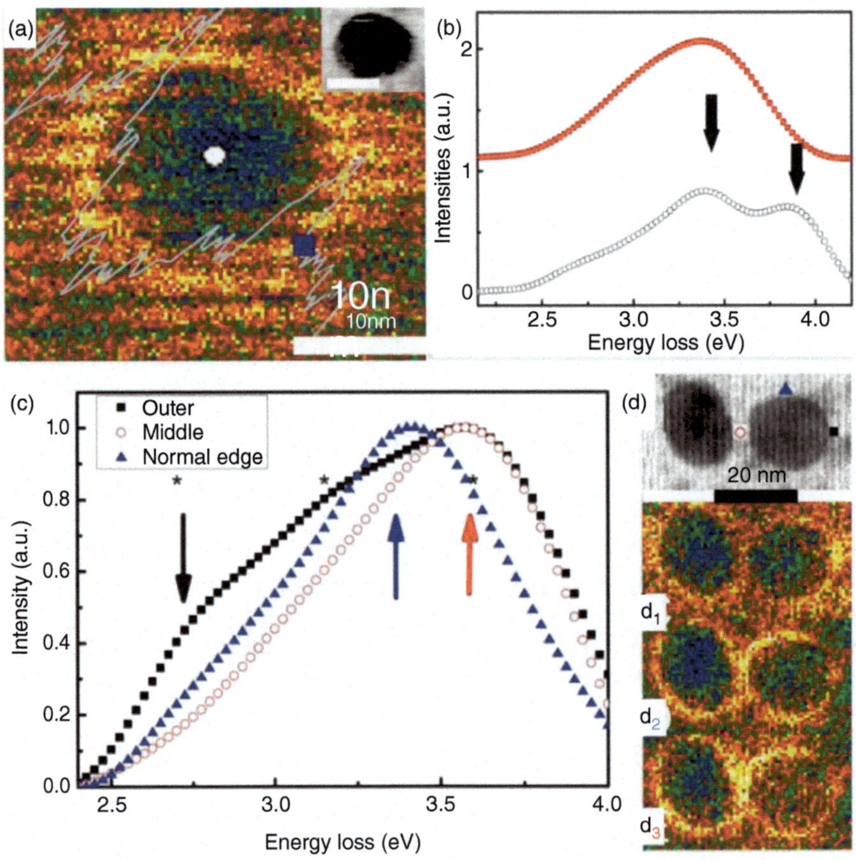

Figure 1.10 (a) EEL intensity mapping of an Ag particle (energy window: 3.3–3.6 eV). The gray curve is the line profile and the top right is the STEM image. The two spots mark the extraction positions of the EEL spectra in (b). EEL spectra from the spots in (a). The top curve is from the white-circle spot and the bottom curve is from the blue-box spot. (c) EELS from the spots marked in the STEM image of (d), where the red-circle, black-box, and blue-triangle curves refer to the middle, outer edge, and perpendicular edge of the cluster dimer, respectively. The black asterisks mark three features of the spectrum from the black spot. Source: Song et al. [33]/Reproduced with permission from ©2010, JOHN WILEY & SONS, INC.

The maximum in plasmon intensity in the spectrum is located at 3.4 eV, which is assigned to the spherical mode of Ag nanoparticles. When the electron beam is in the middle of the nanoparticle, another feature emerges at 3.8 eV due to the interband transition of the bulk Ag, as shown in Figure 1.10b. For a given dimer with two identical spheres, the EELS extracted from the spots turned out as expected and are shown in Figure 1.10c,d. The distinct spectra from three regions, as shown by the spots in Figure 1.10d, that is, the edge parallel to the center axis (blue triangles), the edge perpendicular to the axis (black boxes), and the center (red circles). The maximum intensities of EELS in Figure 1.10d1–d_3 fall around the black, blue, and red spots, respectively, corresponding to the EELS mappings with energy windows of (d_1) 2.6–2.9, (d_2) 3.2–3.5, and (d_3) 3.5–3.8 eV. And they assigned the in-phase mode at 2.7–2.9 eV and the anti-phase mode at 3.6 eV. These positions agree with those predicted theoretically.

Nowadays, the subangstrom spatial resolution achieved in EELS and its capability of exciting the full set of LSPR modes supported by a metallic nanostructure makes STEM/EELS an ideal tool in the study of LSPR. The plasmonic properties characterized using EELS can be associated with geometric or structural features collected simultaneously in a STEM to achieve a deeper understanding of the plasmonic response.

1.7 Surface Plasmon Effects

Under excitation, SP can either propagate on a planar metallic surface as SPP or be confined to the particles known as LSP, providing a novel means of enhancing light-matter interaction at the nanoscale. Especially for the LSP of metal nanostructures, it has the characteristics or advantages of selective light scattering and absorption, electric field localization and strong enhancement. More and more achievements show that surface plasmons are promising in the fields of biology, chemistry, materials, energy, catalysis, and so on. Here, we present some typical surface plasmon effects, such as field enhancement and hot electron effect that are of great important for catalytic reaction, surface enhanced spectroscopy, photothermal conversion and photoelectric device.

1.7.1 Electric Field Enhancement

Surface plasmons are oscillations confined to the surfaces of conducting materials and interact strongly with light. It is well-known that a metal particle of Au or Ag having an intense localized surface plasmon resonance (LSPR) peak can act as a nanoantenna for trapping light to form a locally enhanced electric field in the proximity of metal nanoparticles. After surface plasmon excitation, resonant energy transfer is a nonradiative energy transfer process. Based on the energy transfer mechanism, light-harvesting structures have been designed to implement the performance of detectors or photovoltaic structures [34].

Irradiating metal nanoparticles with light at their plasmon frequency generates intense local electric fields at the surface of the nanoparticles. Electromagnetic simulations using the FDTD method have shown that the electric field intensity of local plasmonic "hot spots" can reach as much as 1000 times that of the incident electric field [35]. In these "hot spot" regions, the electron-hole pair generation rate is 1000 times that of the incident electromagnetic field. When the metal nanoparticles were coated on the surface of semiconductor, such as the commonly applied TiO_2 and CdS, photo-induced charge is generated locally in the semiconductors due to the local near-field enhancement of nanoparticles LSPR effect. Torimoto et al. found that the LSPR-induced electric field and photocatalytic activity of Au/SiO_2/CdS for photocatalytic H_2O splitting greatly depend on the distance between CdS and Au nanoparticles, which further supports the near-field electromagnetic enhancement mechanism [36]. It is found that direct deposition of Au on CdS photocatalysts is not suitable for the photocatalytic H_2O splitting due to the direct photoinduced electron transferring from CdS to Au particles. At the same time, simple particle mixture of CdS@SiO_2 and Au@SiO_2 did not cause any enhancement of H_2 evolution rate because the intensity of the LSPR-induced electric field decays with increase in distance from the metal surface. Therefore, photocatalysts of semiconductor nanoparticles are expected to be effectively photoexcited only when the distance between metal and semiconductor nanoparticles is appropriately adjusted.

Ho et al. investigated the plasmonic response of Au/TiO_2-dumbbell-on-Au-film nanocavities and explore the physical mechanism of dumbbell-film coupling mediated hot-carrier generation and transfer at the two TiO_2-Au interfaces, as shown in Figure 1.11 [37]. From Figure 1.11a, hot carriers can be generated only through localized SPR excitation in the Au NR in the Au/TiO_2/ITO system. However, both localized SPR excitation in the Au NR and direct optical excitation of d-band transitions in the Au film can induce hot carriers in the Au/TiO_2/Au system. The dual-channel carrier generation mechanism associated with the Au/TiO_2/Au system leads to an intriguing carrier transport process, as illustrated in Figure 1.11b. FDTD Solutions was applied to investigate their extinction response and electric near-field distribution profile of the longitudinal and transverse resonance modes for the two systems, as displayed in Figure 1.11c,d. Compared to the Au/TiO_2 dumbbell on the ITO substrate, the longitudinal localized SPR of the Au-film-coupled nanocavity exhibits a linewidth shrinking. Moreover, the extinction intensity of the two resonance modes in the Au-film-coupled nanocavities is obviously enhanced compared to that for the Au/TiO_2 dumbbells on ITO, which is originated from the plasmonic coupling between the Au/TiO_2 dumbbell and the Au film. They found that Au/TiO_2 dumbbell nanostructures assembled on a thin Au film serve as an efficient optical absorber and a hot-carrier generator in the visible region. Upon excitation of LSPs in such coupled particle-on-film nanocavities, the energetic conduction electrons in Au can be injected over the Au/TiO_2 Schottky barrier and migrated to TiO_2, participating in the chemical reaction occurring at the TiO_2 surface.

Plasmonic enhancement has also been demonstrated in photocatalytic methane formation by the reduction of CO_2 with H_2O. Usually, photocatalytic CO_2 reduction is a more challenging problem than solar water splitting due to its requirement of conduction band edge, additional photovoltage, and multielectron process. Similar work was carried out by Hou et al. They found the quantum efficiency to be greatly

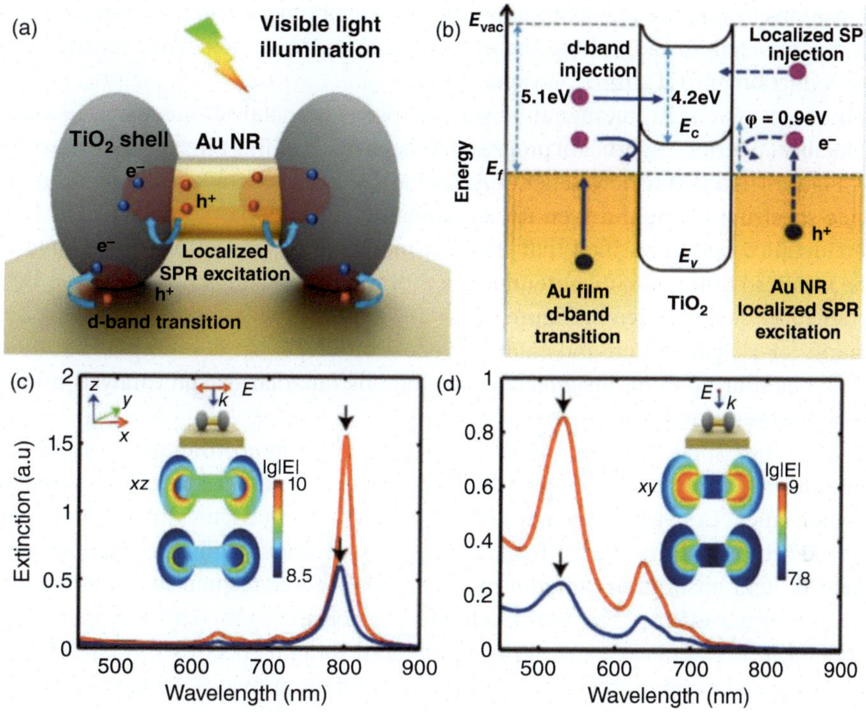

Figure 1.11 (a) Pictorial sketch of the hot-carrier generation in an Au/TiO_2-dumbbell-on-Au-film nanocavity through both localized SPR excitation and d-band transitions. (b) Schematic band diagram for an Au/TiO_2-dumbbell-on-Au-film nanocavity, illustrating the pathways of hot-carrier excitation and injection. The work function of Au, and affinity of TiO_2 are indicated in the band diagram. Simulated extinction spectra of an Au/TiO_2 dumbbell nanoparticle with experimentally determined geometrical dimensions on a 100 nm thick Au film (red lines) and on a semi-infinite ITO substrate (blue lines): (c) the incident E-field is parallel to the x-axis and (d) parallel to the y-axis both under normal incidence. Each inset shows corresponding excitation configuration and simulated electric near-field distribution profiles at the resonance wavelengths as labeled by the arrows. Source: Reproduced from Ho et al. [37]/with permission of John Wiley & Sons.

improved by integrating Au nanoparticles with TiO_2 when the photon energy (532 nm) matches the plasmon resonance of the Au nanoparticles [38]. To obtain a more rigorous understanding of the mechanism of catalytic enhancement, they performed numerical simulations of the charge and electric potential distributions at the Au nanoparticle-TiO_2 interface under plasmon resonance excitation. The electromagnetic response of the film was dominated by local hot spots where the electric field intensity reaches 1000 times that of the incident electric field at the TiO_2 surface. They also found that the plasmon-excited electrons in the Au nanoparticles cannot transfer directly from the Au to the TiO_2. Only when the photon energy is high enough to excite the d band electrons of Au to a conduction band that lies above the conduction band of TiO_2, does direct charge transfer occur between these two materials. When the incident photon energy is high enough (254 nm in UV region), an additional mechanism involving the interband electric transitions in Au produces a number of additional photocatalytic reaction products. In this wavelength range, both the excited electrons in Au and TiO_2 contribute to the reduction of CO_2 with H_2O vapor.

In recent years, CO_2 methanation via heterogeneous catalysts has attracted a considerable attention. Significant progress has been made in this direction, as expressed by Figure 1.12a in a review paper of Aziz et al. [39]. For example, Fu et al. designed wide-spectrum sunlight-driven Rh/Al nanoantenna as photothermal catalyst [40]. The detailed results verified that the CO_2 methanation process was facilitated by the localized SPR and nanoantenna effects of the Rh/Al nanostructure under light irradiation, as illustrated in Figure 1.12b–d. Combined with the photocatalytic oxidation for H_2 generation and catalytic reduction of CO_2, it is expected that these work could bring more insights and inspirations for solar-driven catalysis toward clean and sustainable fuel conversion.

Now, "hot spots" created around the plasmonic nanostructures can greatly promote the interaction between light and matter, enabling efficient excitation of molecules. Conversely spectral information about molecules in the hot spots can be broadcasted to the far field. This process is accompanied by the enhancement of light absorption radiation scattering light force resonance migration and photo-thermal effects. The rich phenomena associated with LSP nanostructures resulted in the application of surface enhanced Raman spectroscopy (SERS) and tip enhanced Raman spectroscopy (TERS), which provides "fingerprints" for

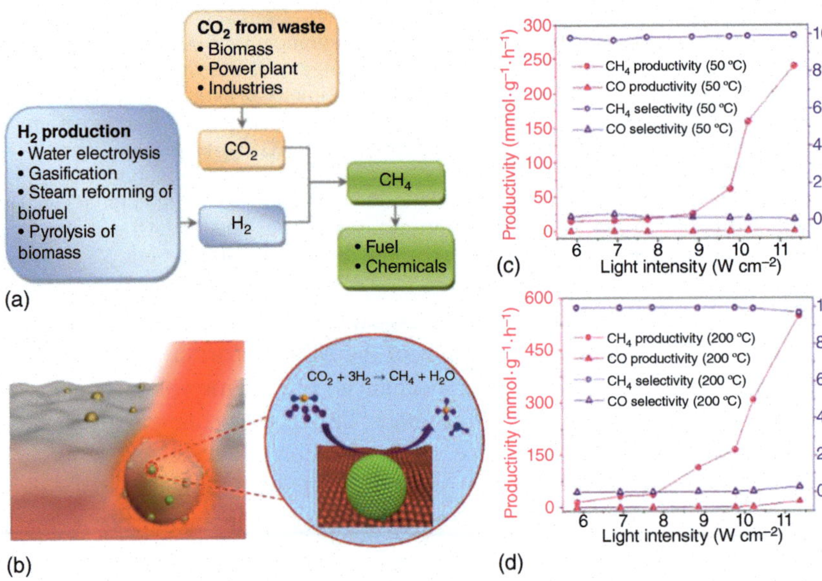

Figure 1.12 (a) Overview of CO_2 methanation. Source: Adapted from Aziz et al. [39]. (b) Schematic diagrams of photothermal CO_2 methanation on the Rh/Al nanoantenna catalyst. CH_4 and CO productivity and selectivity under different simulated sunlight intensities at the chamber temperatures of (c) 50 and (d) 200 °C. Source: Reproduced from Fu et al. [40]/with permission of American Chemical Society.

recognizing chemical species in neighboring surfaces, making it a promising tool for in-situ observation of catalytic process [41–47].

In addition to the tunable absorption/scattering, large extinction cross section and local field enhancement effect metal nanostructures, the LSP resonant position is highly dependent on the medium, size and shape of the nanostructure. With the adsorption or modification of molecular on the particle surface, the local medium environment changes or the particles will be assembled into a certain configuration. These changes will lead to the obvious shifting of resonant wavelength of plasmonic nanostructure. This feature enables metal nanostructures to be promising for the rapid sensing and detection of chemical and biological objects [48, 49]. With the epidemic outbreak of global Novel Coronavirus 2019 (abbreviated as 2019-ncov), the State Food and Drug Administration approved the novel coronavirus antibody detection reagent.

1.7.2 Hot Electrons Effect

In metals with high concentration of free electrons, the proportion of radiation recombination in metal microstructure is very weak, which is quite different from the excitation and radiation recombination of semiconductor. When the incident light on the nanostructure is resonant with LSP, the photon energy drives the surface of plasmon to produce collective oscillation within 10 fs, during which the electrons energy is higher than the Fermi level and becomes high-energy electrons. In the following 100 fs, the internal damping effects, such as phonon scattering, lattice defects and Landau damping, lead to excitation dephasing to produce electron hole pairs. Through the scattering between electrons and optical phonons, the energy of electrons is finally transferred to phonon vibration energy after excitation (~1 ps). Then the phonons energy is transformed into the heat energy of lattice vibration (~10 ps), generating localized heat on the nanometer sized region. The localized energy dissipated into the environmental medium through the process of heat radiation and heat conduction after ~1000 ps. The mechanism and energy conversion process is presented by Figure 1.13 [50]. The photothermal conversion can be regulated and localized in a nanogap or nanometer sized cavity. Each stage of these energy transfer processes can be used for a specific application. Many research papers and some review articles have dealt with the plasmon related applications [51–53].

1.7.2.1 For New Energy

Under optical excitation, plasmon quantum can decay either radiatively into a photon or nonradiatively into an electron-hole pair. The most likely electron-hole pair formation occurs when the plasmon quantum excites an electron from the Fermi level to an occupied state below the vacuum level. These hot electrons can transfer into specific electronic states of molecules adsorbed on the metal surface, inducing processes such as molecular dissociation and chemical reactions.

Mukherjee et al. [54] reported experimental evidence of room temperature photocatalytic dissociation of H_2. The reaction can only be enabled by hot electrons

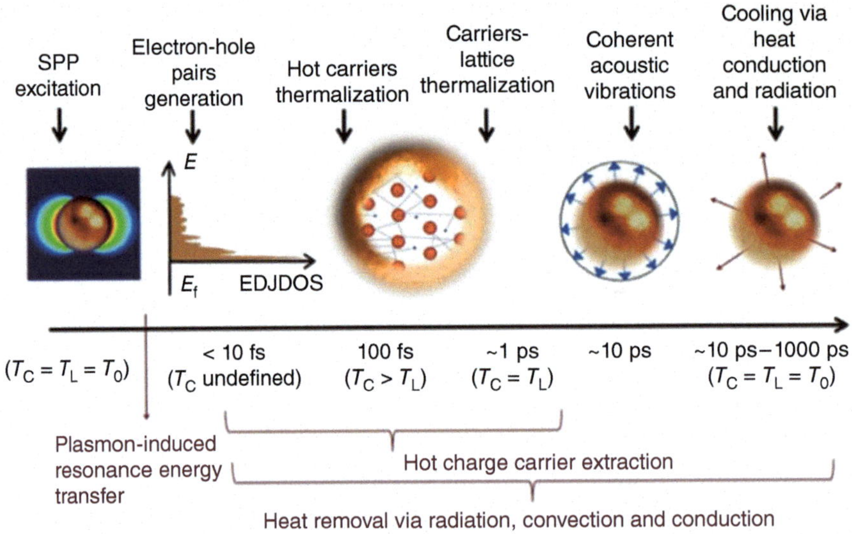

Figure 1.13 The timescales and mechanism and energy conversion process of LSP after irradiation. Source: Reproduced from Boriskina et al. [50]/with permission of Chinese Laser Press.

generated by surface plasmon decay from small Au nanoparticles supported on TiO_2 matrix. The dissociation of H_2 process was probed by detecting the formation of HD molecules from the dissociations of H_2 and D_2. The photocatalysis was confirmed with and without laser illumination. They also observed a steady increase of photocatalytic rate with increasing Au loading, showing a strong dependence on Au nanoparticle concentration for HD generation. As shown in Figure 1.14, the hot electrons are initially optically excited at the Au nanoparticle surface. Then hot electrons soon lose coherence and form a nonequilibrium Fermi-Dirac type distribution which thermalizes by Auger scattering. The narrow bonding and broad antibonding states of adsorbed H_2 are denoted as B and AB, respectively. The hot electrons in the high energy tail of this distribution have sufficient energy to transfer into the antibonding state of the H_2 molecule, creating a transient negative ion. Then the transient negative ion transfers the electron back to the AuNP surface, where it returns to the ground-state potential energy continuum and dissociates.

Although silicon or metal oxides are still under development for solar energy devices. These materials cannot meet the requirements for practical applications due to one or more shortcomings, such as large bandgap, low absorption of solar radiation, low charge mobility, high-density surface trap states, and indirect bandgap feature. To extend the light absorption spectral range and to improve the energy conversion efficiency of semiconductors, plasmonic metal nanostructures with broad absorption are incorporated to form metal-semiconductor Schottky junction. Plasmonic hot electrons can enter semiconductor across the Schottky junction or through the Schottky emission. Through introduction of an ultrathin passivate insulator or organic interface, the power conversion efficiency were greatly improved.

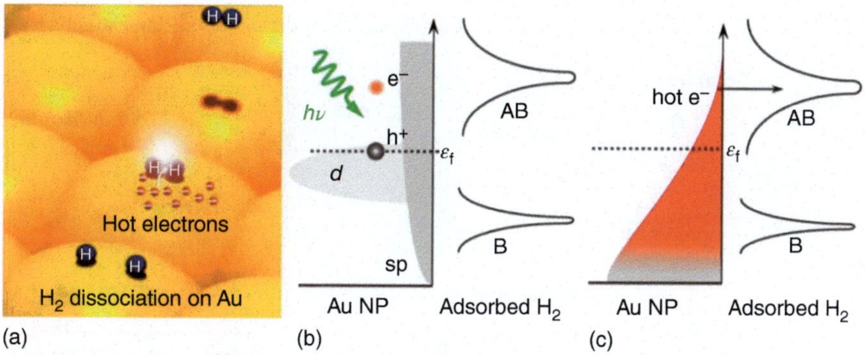

Figure 1.14 (a) Mechanistic representation of H_2 dissociation on the Au nanoparticle surface. (b) Schematic of hot electron excitation in Au nanoparticle showing: d-band electron-hole pair excited above the Fermi level upon plasmon decay. The narrow bonding and broad antibonding states of adsorbed H_2 are denoted as B and AB, respectively. (c) Schematic of Fermi-Dirac type distribution of hot electrons permitting hot electron transfer into the antibonding state of H_2. Source: Mukherjee et al. [54]/Reproduced with permission from © 2013, American Chemical Society.

One potential of the hot electrons effect of LSP is the solar photothermal conversion. Due to the unique optical absorption and thermal localization characteristics of plasmon absorber, researchers have greatly improved the efficiency of solar energy generation through material selection, structural design and system optimization. It is expected that solar photothermal effect can be widely used in the fields of seawater desalination, sewage treatment, high-temperature steam sterilization, and power generation. At the same time, some scientific problems, such as the photothermal conversion mechanism of thermal and dynamic processes and the precise control of thermal effects in time and space, can be solved with the development of ultrafast spectroscopy technology.

Most recently, due to unique capability of broadband light harvesting and pronounced electromagnetic field enhancement, plasmonic nanostructures have been demonstrated for promising applications in a new sustainable field of plasmon enhanced water purification. Chen et al. [55] demonstrated Ag-based asymmetric plasmonic structure for dual functions of solar water purification and pollution detection, as shown in Figure 1.15. Such impressive performance in solar thermal conversion and chemical sensing are stemmed from the broadband and efficient light absorption (solar absorptivity > 90%) as well as pronounced local field enhancement, respectively. As an example of functional integration, this dual functional Ag-based plasmonic structure can enable on-site pollution detection and purification of water.

The concept of complex plasmonic structures based on Au and Ag nanostructures with integrated functionalities opened up exciting opportunities for advanced plasmonic science [56]. In 2016, through physical vapor deposition and assembly of Au nanoparticles in the Al_2O_3 template, Zhou et al. achieved plasmon absorber with high efficiency in the ultra-wide band range of 400 nm to 10 μm [57]. Through depositing Pb, Au, Ag particles mixture evenly into the pores of natural wood, Zhu

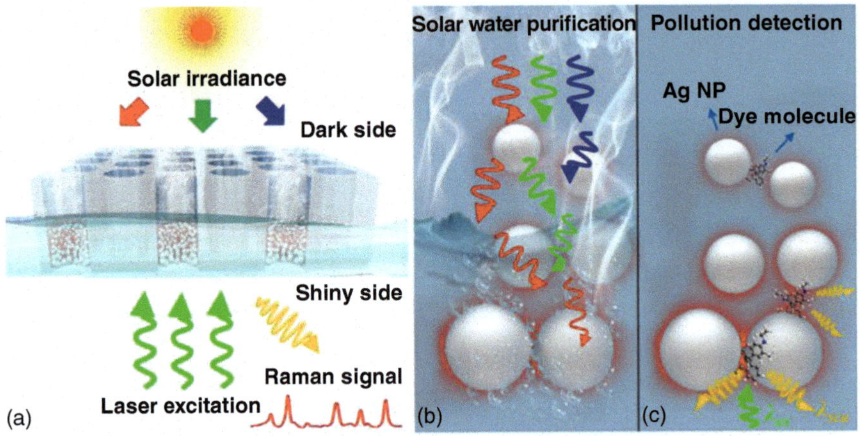

Figure 1.15 Schematic of the integrated dual-functions of an Ag-based plasmonic structure: Plasmonic heating and spectroscopic amplification. (a) The schematic of structure, the dark side (facing up toward the light) for the solar water purification and the shiny side (facing down toward the laser illumination) for the pollution detection. (b) Mechanism for the dark side of the structure enabled solar water purification. (c) Mechanisms for the shiny side of the structure enabled organic pollution detection. Source: Chen et al. [55]/Reproduced with permission from ©2018, ELSEVIER.

obtained plasmonic wood absorbers with excellent performance of solar steam generation [58].

1.7.2.2 For Photoelectric Device

Another important application of hot electron effect of LSP is enhancement or tunability in photoelectric device. Metal nanostructures generate "hot electrons" from light by creating surface plasmons [59]. There are several proposed metal-to-semiconductor charge-separation pathways. Based on the conventional plasmon-induced hot-electron transfer (PHET) mechanism (Figure 1.16a), the reported efficiencies of devices based on this PHET concept are often low (1–2%) because the plasmon decays into a hot electron-hole pair within the metal via Landau damping on time scales of a few to tens of femtoseconds. Hot-electron transfer competes with electron relaxation through rapid electron-electron scattering on time scales of hundreds of femtoseconds [60, 61]. Metal-to-semiconductor hot-electron transfer efficiencies can be enhanced if the competition with ultrafast electron-electron scattering in the metal can be avoided. One approach is to create a direct metal-to-semiconductor interfacial charge-transfer transition (DICTT) that can be directly excited to promote an electron from the metal into the semiconductor CB. However, these interfacial transitions are often too weak as compared with bulk metal transitions [60]. A desirable photoinduced hot-electron transfer pathway would combine the strong light-absorbing power of plasmonic transitions with the superior charge-separation.

In 2015, Wu et al. first proposed and experimentally demonstrated a plasmon-induced metal-to-semiconductor interfacial charge-transfer transition (PICTT)

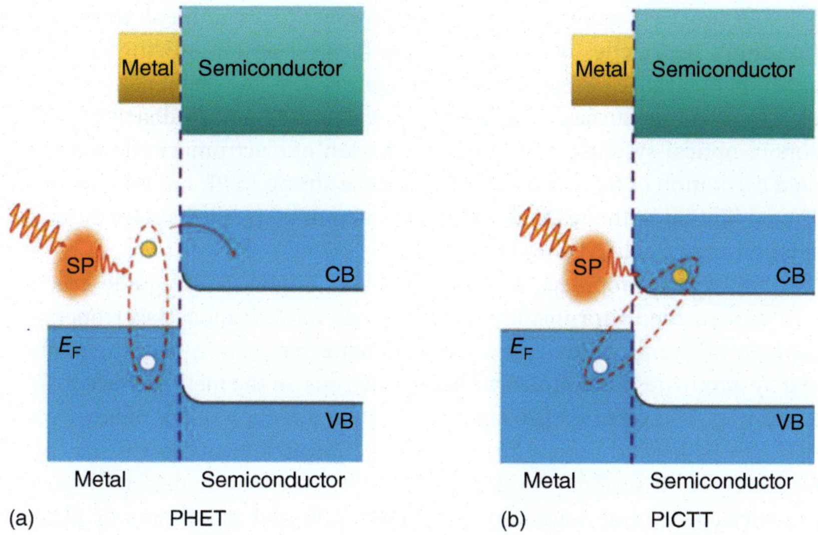

Figure 1.16 Metal-to-semiconductor charge-separation pathways. (a) Conventional PHET mechanism, in which a photoexcited plasmon in the metal decays into a hot electron-hole pair (solid and open red circles in the dotted ellipsoids) through Landau damping, followed by injection of the hot electron into the CB of the semiconductor. (b) The PICTT pathway, in which the plasmon decays by directly generating an electron in the conduction band of the semiconductor and a hole in the metal. VB is the semiconductor valence band and hv indicates the excitation photons. Source: Zhu et al. [26]/with permission of AIP Publishing LLC.

pathway in CdSe-Au NR heterostructure, as presented in Figure 1.16b [62]. In this PICCT pathway, strong Au–CdSe interactions led to strong plasmon damping in the Au tip. When the plasmon was excited along the nanorods axis, transient absorption anisotropy showed more efficient (>24%) charge transferred from Au to CdSe, and the charge-separation efficiencies were independent of excitation photon energy. The damped plasmon decays via direct excitation of an interfacial electron-hole pair (with an electron in CdSe and a hole in Au). The electronic structure of CdSe–Au NRs can be viewed as a combination of the strongly interacting tip region, which resembles the CdSe-Au dimers with strongly damped plasmon and exciton bands. The center part is similar to the unperturbed CdSe nanorods. Direct evidence to support this proposed mechanism was obtained through ultrafast transient absorption study. A pump laser with a photon energy below the CdSe band gap was used to excite the Au tip, and the electron transferred to the CdSe domain was probed through the bleaching of the exciton.

1.7.2.3 For Plasmonic Nanolaser

The plasmonic nanolaser (also called SPASER abbreviated for "surface plasmon amplification by stimulated emission of radiation") is an active nanosystem including a gain medium and a plasmonic metal nanoparticle core. It theoretically predicted in 2003 by Bergman and Stockman from Georgia State University [63]. Then kinds of surface plasmon nanolasers with new structures have been constructed. Nanolasers applied the idea of SPP effect that can break through the diffraction limit. Researches of nanolaser were focused on minimizing metal loss and realization of electric pump at room temperature and low threshold.

Nanolasers based on surface plasmon are not pumping and feedback of optical resonators in optical systems. Under the irradiation of electromagnetic wave, the stimulated oscillation of free electrons can produce strong LSPR on the nanometer scale. The oscillation in the resonant cavity was amplified by the gain medium and laser emission can be achieved [64].

In 2008, Oulton et al. proposed a novel scheme and designed a waveguide structure composed of semiconductor-insulator-metal (SIM) for F-P mode laser operation [65]. The gain medium of the laser is semiconductor nanowires. And the photons generated by nanowires are coupled with free electrons on the metal surface to form surface plasmon. The plasmon propagates along the direction of the nanowire and oscillates in the Fabry-Perot cavity formed by reflection at both ends of the nanowire. In 2009, Zhang et al. verified the design feasibility in experiments and prepared nanowire surface plasmon nanolaser [66]. Then different nanolasers of plate or multilayer structures were prepared, such as metal–semiconductor–metal (MSM), MIM, metal–oxide–semiconductor (MOS), as summarized by Ma et al. [67]. In their review paper, the plasmonic confinement effect on the plasmon laser through the enhancement of spontaneous emission in sub-wavelength-scale plasmonic cavities, ultrafast behavior of plasmon laser, as well as future prospects and challenges were discussed.

In 2014, Lu et al. reported broadband tunable, single-mode plasmonic nanolasers, based on a single MOS nanostructure platform comprising of InGaN/GaN semiconductor nanorods supported on an Al_2O_3-capped epitaxial Ag film [68]. They successfully reduced the continuous-wave lasing thresholds to ultrasmall values for all three primary RGB colors and possibility of "thresholdless" lasing operated at 7 K for the blue plasmonic nanolaser. Although the lasing line width is not in the lasing region as predicted by the classical Schawlow-Townes theory, temperature evolution of the spectral for the device with plasmonic cavity indicated the line width is in the lasing regime. Photon bunching is characteristic of the spontaneous emission of the gain medium. And the line width alone is not be a conclusive indication of lasing.

This new type lasers can not only greatly reduce the energy loss in metal, but also realize long-distance surface plasmon wave transmission. At the same time, the threshold of laser is reduced, which makes it possible for the nano laser work at room temperature. Once the nanolaser with low threshold and high integration is successfully developed, the size of photonics devices will be greatly reduced. Thus, the mismatch between photonic devices and electronic devices will be improved, and the product size in the optoelectronic industry will be reduced. Nanolaser with high performance will be great useful in biosensors, precision measurement and so on. Among the potentials of nanolaser, the "killer" application will be ultrafast, high-density on-chip communications for future information processing. Some of our recent work about plasmon enhanced photoelectric devices, including plasmonic nanolaser with single mode, can be seen in Chapter 9.

1.8 Summary of this Chapter

In this chapter, we present the theoretical backgrounds and research advances of plasmonics, including plasmon originated from the collective oscillation of free electron based on Drude mode, SPPs at metal-vacuum interface, propagation characteristic of SPP. Then advances in the electric and optical excitation and detection of SPP, as well as the LSP of nanostructures were delivered. Rapid development in synthesis and fabrication techniques resulted in the availability of systematically designed plasmonic materials and devices, which not only widened the scope of applications of plasmon and hot carriers but also advanced experimental discoveries.

In recent years, surface plasmon effects, such as energy transfer, hot electron transfer, and field enhancement effect, can be used to enhance photocatalysis and realize plasmon capture light energy. Although there are many hot topics in the plasmon related fields, the internal physical mechanism remains to be revealed. More researches were limited on the plasmonic metals of Au, Ag, Cu, and Al, as well as some two dimensional materials, working on certain bands with own advantages and limitations. It is found that heavily doped semiconductors can also generate hot carriers by themselves with more high energy transfer efficiency due to less energy loss via the indirect transfer pathway across the interface or via the direct transfer pathway with easier orbital hybridization. Plasmonic semiconductor nanostructures are promising candidates that can manifest SPR under visible-light or infrared-light excitation. Moreover, they have been used for SERS, photocatalysis, and water splitting.

However, fully understanding the enhancement effect of plasmon and its internal physical process can effectively promote the application, especially for the coupling structures with controllable energy exchange and transfer path. Among these issues, some related works should be developed simultaneously, such as theoretical models and calculation methods, design and preparation, characterization and testing, ultrafast time resolution and high resolution space detection technology.

References

1. Pines, D. and Bohm, D. (1952). A collective description of electron interactions: II. Collective vs individual particle aspects of the interactions. *Phys. Rev.* 85: 338.
2. Stockman, M.I. (2011). Nanoplasmonics: past, present, and glimpse into future. *Opt. Express* 19 (22): 22029–22106.
3. Bohren, C.F. and Huffman, D.R. (1983). *Absorption and Scattering of Light by Small Particles*. Wiley.
4. Drude, P. (1900). Zur Elektronentheorie der Metalle. *Ann. Phys.* 1: 566–613.
5. Johnson, P.B. and Christy, R.W. (1972). Optical constants of noble metals. *Phys. Rev. B* 6: 4370–4379.
6. Blackstock, A.W., Ritchie, R.H., and Birkhoff, R.D. (1955). Mean free path for discrete electron energy losses in metallic foils. *Phys. Rev.* 100: 1078.

7 Xu, M.S., Liang, T., Shi, M.M., and Chen, H.Z. (2013). Graphene-like two dimensional materials. *Chem. Rev.* 113: 3766–3798.

8 Falk, A.L., Koppens, F.H.L., Yu, C.L. et al. (2009). Near-field electrical detection of optical plasmons and single-plasmon sources. *Nat. Phys.* 5: 475–479.

9 Zayatsa, A.V., Smolyaninov, I.I., and Maradudin, A. (2005). Nano-optics of surface plasmon polaritons. *Phys. Rep.* 408: 131–314.

10 Zhou, Y.J. and Eda, G. (2017). Efficient carrier-to-exciton conversion in field emission tunnel diodes based on MIS-type van der Waals heterostack. *Nano Lett.* 17: 5156–5162.

11 Palash, B., Alexandre, B., and Lukas, N. (2011). Electrical excitation of surface plasmons. *Phys. Rev. Lett.* 106: 226802.

12 Ma, R.M., Oulton, R.F., Sorger, V.J. et al. (2011). Room-temperature sub-diffraction-limited plasmon laser by total internal reflection. *Nat. Mater.* 10: 110–113.

13 Kojori, H.S., Yun, J.H., Paik, Y. et al. (2016). Plasmon field effect transistor for plasmon to electric conversion and amplification. *Nano Lett.* 16: 250–254.

14 Fan, P., Colombo, C., Huang, K.C.Y. et al. (2012). An electrically-driven GaAs nanowire surface plasmon source. *Nano Lett.* 12: 4943–4947.

15 Yang, X., Ni, P.N., Jing, P.T. et al. (2019). Room temperature electrically driven ultraviolet plasmonic lasers. *Adv. Opt. Mater.* 7: 1801681.

16 Neutens, P., Lagae, L., Borghs, G., and Van Dorpe, P. (2010). Electrical excitation of confined surface plasmon polaritons in metallic slot waveguides. *Nano Lett.* 10: 1429–1432.

17 Sun, Z.P., Jiang, M.M., Mao, W.Q. et al. (2020). Nonequilibrium hot-electron-induced wavelength tunable incandescent-type light sources. *Photonics Res.* 8: 91–102.

18 Chu, M.W., Myroshnychenko, V., Chen, C.H. et al. (2009). Probing bright and dark surface-plasmon modes in individual and coupled noble metal nanoparticles using an electron beam. *Nano Lett.* 9: 399–404.

19 Mie, G. (1908). Beiträge zur optik trüber medien, Speziell kolloidaler metallösungen. *Ann. Phys.* 330 (3): 377–445.

20 Lorenz, L. (1898). *Sur La lumière réfléchie Et réfractée par Une sphère (Surface) Transparente.* Tome Premier, 403–529. Copenhague: libraire lehmanń & stage.

21 Debye, P. (1909). Der Lichtdruck Auf Kugeln Von Beliebigem Material. *Ann. Phys.* 335: 57–136.

22 Halas, N.J., Lal, S., Chang, W.S. et al. (2011). Plasmons in strongly coupled metallic nanostructures. *Chem. Rev.* 111: 3913–3961.

23 Agrawal, A., Cho, S.H., Zandi, O. et al. (2018). Localized surface plasmon resonance in semiconductor nanocrystals. *Chem. Rev.* 118: 3121–3207.

24 Wang, C., Huang, S., Xing, Q. et al. (2020). Van der Waals thin films of WTe_2 for natural hyperbolic plasmonic surfaces. *Nat. Commun.* 11: 1158.

25 Nagao, T., Hildebrandt, T., Henzler, M., and Hasegawa, S. (2001). Dispersion and damping of a two-dimensional plasmon in a metallic surface-state band. *Phys. Rev. Lett.* 86 (25): 5747.

26 Zhu, Y., Xu, H.X., Yu, P., and Wang, Z.M. (2021). Engineering plasmonic hot carrier dynamics toward efficient photodetection. *Appl. Phys. Rev.* 8: 021305.

27 Ritchie, R.H. (1957). Plasma losses by fast electrons in thin films. *Phys. Rev.* 106: 874–881.

28 Powell, C.J. and Swan, J.B. (1959). Origin of the characteristic electron energy losses in aluminum. *Phys. Rev.* 115: 869–875.

29 Stern, E.A. and Ferrell, R.A. (1960). Surface plasma oscillations of a degenerate electron gas. *Phys. Rev.* 120: 130–136.

30 Fujimoto, F. and Komaki, K.I. (1968). Plasma oscillations excited by a fast electron in metallic particle. *J. Phys. Soc. Jpn.* 25: 1679–1687.

31 Pettit, R.B., Silcox, J., and Vincent, R. (1975). Measurement of surface plasmon dispersion in oxidized aluminum films. *Phys. Rev. B* 11: 3116–3123.

32 Nelayah, J., Kociak, M., Stephan, O. et al. (2007). Mapping surface plasmons on a single metallic nanoparticle. *Nat. Phys.* 3: 348–353.

33 Song, F., Wang, T., Wang, X. et al. (2010). Visualizing plasmon coupling in closely spaced chains of Ag nanoparticles by electron energy-loss spectroscopy. *Small* 6 (3): 446–451.

34 Lunz, M., Gerard, V.A., Gunko, Y.K. et al. (2011). Surface plasmon enhanced energy transfer between donor and acceptor CdTe nanocrystal quantum dot monolayers. *Nano Lett.* 11 (8): 3341–3345.

35 Le, F., Brandl, D.W., Urzhumov, Y.A. et al. (2008). Metallic nanoparticle arrays: a common substrate for both surface-enhanced Raman scattering and surface-enhanced infrared absorption. *ACS Nano* 2: 707–718.

36 Torimoto, T., Horibe, H., Kameyama, T. et al. (2011). Plasmon-enhanced photocatalytic activity of cadmium sulfide nanoparticle immobilized on silica-coated gold particles. *J. Phys. Chem. Lett.* 2: 2057–2062.

37 Ho, K.H.W., Shang, A., Shi, F. et al. (2018). Hot carriers: plasmonic Au/TiO_2-dumbbell-on-film nanocavities for high-efficiency hot-carrier generation and extraction. *Adv. Funct. Mater.* 28: 1800383.

38 Hou, W., Hung, W.H., Pavaskar, P. et al. (2011). Photocatalytic conversion of CO_2 to hydrocarbon fuels via plasmon-enhanced absorption and metallic interband transitions. *ACS Catal.* 1: 929–936.

39 Aziz, M.A.A., Jalil, A.A., Triwahyono, S., and Ahmad, A. (2015). CO_2 methanation over heterogeneous catalysts: recent progress and future prospects. *Green Chem.* 17: 2647–2663.

40 Fu, G., Jiang, M., Liu, J. et al. (2021). Rh/Al nanoantenna photothermal catalyst for wide-spectrum solar-driven CO_2 methanation with nearly 100% selectivity. *Nano Lett.* 21: 8824–8830.

41 Fleischmann, M., Hendra, P.J., and McQuillan, A.J. (1974). Raman spectra of pyridine adsorbed at a silver electrode. *Chem. Phys. Lett.* 26: 163–166.

42 Jeanmaire, D.L. and Van Duyne, R.P. (1977). Surface Raman spectroelectro-chemistry. Part I. Heterocyclic, aromatic, and aliphatic amines adsorbed on the anodized silver electrode. *J. Electroanal. Chem.* 84: 1–20.

43 Nie, S.M. and Emory, S.R. (1997). Probing single molecules and single nanoparticles by surface-enhanced Raman scattering. *Science* 275: 1102–1106.

44 Kneipp, K., Wang, Y., Kneipp, H. et al. (1997). Single molecule detection using surface-enhanced Raman scattering (SERS). *Phys. Rev. Lett.* 78: 1667–1670.

45 Panneerselvam, R., Liu, G.K., and Wang, Y.H. (2018). Surface enhanced Raman spectroscopy bottle necks and future directions. *Chem. Commun.* 54 (10): 10–25.

46 Ke, S., Kan, C., Zhu, X. et al. (2021). Effective fabrication of porous Au-Ag alloy nanorods for in-situ Raman monitoring catalytic oxidation and reduction reactions. *J. Mater. Sci. Technol.* 91: 262–269.

47 Zhang, W.H., Cui, X.D., Yeo, B.S. et al. (2007). Nanoscale roughness on metal surfaces can increase tip enhanced Raman scattering by an order of magnitude. *Nano Lett.* 7 (5): 1401–1405.

48 Dantham, V.R., Holler, S., Barbre, C. et al. (2013). Label-free detection of single protein using a nanoplasmonic-photonic hybrid microcavity. *Nano Lett.* 13: 3347–3351.

49 Stockman, M.I. (2015). Nanoplasmonic sensing and detection. *Science* 348: 287–288.

50 Boriskina, S.V., Cooper, T.A., Zeng, L. et al. (2017). Losses in plasmonics from mitigating energy dissipation to embracing loss-enabled functionalities. *Adv. Opt. Photonics* 9: 775–827.

51 Clavero, C. (2014). Plasmon-induced hot-electron generation at nanoparticle/metal-oxide interfaces for photovoltaic and photocatalytic devices. *Nat. Photonics* 8 (2): 95–103.

52 Zhang, Y., He, S., Guo, W. et al. (2018). Surface-plasmon-driven hot electron photochemistry. *Chem. Rev.* 118: 2927–2954.

53 Tang, H., Chen, C.J., Huang, Z. et al. (2020). Plasmonic hot electrons for sensing, photodetection, and solar energy applications: a perspective. *J. Chem. Phys.* 152: 220901.

54 Mukherjee, S., Libisch, F., Large, N. et al. (2013). Hot electrons do the impossible: plasmon-induced dissociation of H_2 on Au. *Nano Lett.* 13: 240–247.

55 Chen, C.L., Zhou, L., Yu, J.Y. et al. (2018). Dual functional asymmetric plasmonic structures for solar water purification and pollution detection. *Nano Energy* 51: 451–456.

56 Liang, J., Liu, H.Z., Yu, J.Y. et al. (2019). Plasmon enhanced solar vapor generation. *Nanophotonics* 8 (5): 771–786.

57 Zhou, L., Tan, Y.L., Ji, D.X. et al. (2016). Self-assembly of highly efficient broad band plasmonic absorbers for solar steam generation. *Sci. Adv.* 2 (4): e1501227.

58 Zhu, M.W., Li, Y.J., Chen, F.J. et al. (2018). Plasmonic wood for high efficiency solar steam generation. *Adv. Energy Mater.* 8 (4): 170102.

59 Brongersma, M.L., Halas, N.J., and Nordlander, P. (2015). Plasmon-induced hot carrier science and technology. *Nat. Nanotechnol.* 10: 25–34.

60 Hartland, G.V. (2011). Optical studies of dynamics in noble metal nanostructures. *Chem. Rev.* 111: 3858–3887.

61 Kale, M.J., Avanesian, T., Xin, H. et al. (2014). Controlling catalytic selectivity on metal nanoparticles by direct photoexcitation of adsorbate-metal bonds. *Nano Lett.* 14: 5405–5412.

62 Wu, K., Chen, J., McBride, J.R., and Lian, T. (2015). Efficient hot-electron transfer by a plasmon-induced interfacial charge-transfer transition. *Science* 349: 632–635.

63 Stockman, M.I. (2020). Brief history of spaser from conception to the future. *Adv. Photonics* 2: 054002.

64 Okamoto, K., Niki, I., Shvartser, A. et al. (2004). Surface-plasmon-enhanced light emitters based on InGaN quantum wells. *Nat. Mater.* 3 (9): 601–605.

65 Oulton, R.F., Sorger, V.J., Genov, D.A. et al. (2008). A hybrid plasmonic wave guide for subwavelength confinement and long-range propagation. *Nat. Photonics* 28: 496–500.

66 Oulton, R.F., Sorger, V.J., Zentgraf, T. et al. (2009). Plasmon lasers at deep subwavelength scale. *Nature* 461: 629–632.

67 Ma, R., Oulton, R.F., Sorger, V.J., and Zhang, X. (2013). Plasmon lasers: coherent light source at molecular scales. *Laser Photonics Rev.* 7 (1): 1–21.

68 Lu, Y.J., Wang, C.Y., Kim, J. et al. (2014). All-color plasmonic nanolasers with ultralow thresholds: autotuning mechanism for single-mode lasing. *Nano Lett.* 14: 4381–4388.

2

Dielectric Modification and Fundamental of Plasmonic Nanostructure

Nanogold does not glitter,
but its future looks bright.

2.1 Introduction

The art of using colloidal Au and Ag for coloring glass has a long history dating back to the Ancient World. The famous Lycurgus glass cup, created by the Romans 1600 years ago, is a prototypical early example of a dichroic material. The Lycurgus cup is extraordinary since its color depends on whether the light is reflected or transmitted. At places where light is transmitted through the glass, it appears ruby red, whereas at places where light is scattered near the surface, the scattered light appears opaque green. It is also one of the earliest known objects in which the nanometer-sized metal particles embedded in the glass that exhibit what we now know as a localized surface plasmon resonance (SPR) of Ag and Au contained in the glass in very small amounts.

At that time, people knew how to use colloids for coloring glasses, but it was unknown that nanoparticles (NPs) were the reason for these coloring effects. In 1663, Andreas Cassius described the preparation of reddish-purple-colored glass with Au by the reaction of Au precursor solutions and Sn precursor solutions. Up to date, colloidal Au is also called Cassius Au purple.

Centuries later, it was Michael Faraday who not only described the first systematic preparation of Au NPs by chemical reduction of Au salts with phosphorus but also described the interaction of light with the Au particles using the new theory of waves [1]. Then half a century later, a series of papers were published to explain the scattering of light by small spherical particles in the early 1900s. The first rigorous theoretical description was by G. Mie [2], as described in Chapter 1. The first numerical evaluations of this theory applied to Au and Ag NPs in colloidal suspensions were carried out by M. Quinten, as summarized in his book [3]. U. Kreibig and M. Vollmer [4] have given a comprehensive overview of the optical properties of metallic NPs. From the theoretical description, it can be seen that the optical property of the particles is the result of the interaction between light and particles that largely depends on the dielectric data of NPs.

Plasmonic Metal Nanostructures: Preparation, Characterization, and Applications,
First Edition. Caixia Kan.

Now, it is known that the resonant frequency of localized surface plasmon depends on the size and morphology of metal nanostructure, as well as the external dielectric environment. The metal nanostructure can selectively scatter and absorb light of different frequencies. That is why we can prepare colorful colloidal of noble metal NPs when they respond to different visible light. Both calculations and measurements have shown that the frequency and amplitude of the SPR modes are sensitive to particle shape, which determines how the free electrons are polarized and distributed on the surface. As a result, controlling the shape of a plasmonic NP represents the most powerful means of tailoring and fine-tuning its optical resonance properties.

With the development of synthesis methods for plasmon nanostructures and the requirements of nanomaterials in application fields, various nanostructures have been fabricated in the past two decades. Beyond Mie's theory for nanospheres, models for both light scattering by nonspherical particles and electromagnetic interactions among neighboring nanoparticles are available. This chapter will focus on the local surface plasmon properties of metal nanostructures, involving dielectric properties of nanometer-sized metal, numerical calculation, and simulation of the optical properties of coupling structures.

2.2 Drude–Lorentz Model of Metal Nanoparticles

The dielectric function of metal-free electrons has been proposed in the 1900s under the model that the electrons in the metal are in the state of free electron gas, as described in Section 1.3. In fact, the dielectric function of noble metal nanomaterials is much more complex than the Drude model.

Generally, the dielectric function of noble metal nanomaterials can be expressed by the combination of the Lorentz model and the Drude model. They correspond to the electronic forms of the interband transition (bonding electron) and the intraband transition (free electron) in the material, respectively [5]. There are many expressions for the interband transition contribution, and the most commonly used is the Lorentz model [6], which is known as the Drude–Lorentz model. The dielectric function is expressed as:

$$\varepsilon(\omega) = \varepsilon_{\mathrm{f}}(\omega) + \varepsilon_{\mathrm{b}}(\omega) = \varepsilon_{\infty} - \frac{\omega_{\mathrm{p}}^2}{\omega^2 + i\omega\gamma} + \sum_{j=1}^{n} \frac{A_j}{\left(\omega_j^2 - \omega^2\right) - i\omega\Gamma_j} \tag{2.1}$$

where $\varepsilon_{\mathrm{f}}(\omega)$ is the Drude contribution describing the intraband processes with consideration of damping constant γ, ε_{∞} is the high-frequency limit dielectric constant, and ω_{p} is the plasma frequency. The second term $\varepsilon_{\mathrm{b}}(\omega)$ is the interband contribution, which describes the excitations from the occupied to unoccupied bands in the metal. Γ_j is the damping term of bulk plasma, which is related to the average free path (λ_{e}) of electrons in bulk materials. $\Gamma_j = \frac{V_{\mathrm{F}}}{\lambda_{\mathrm{e}}}$, V_{F} is Fermi velocity of electron. n is the number of vibrators with intensity A_j, lifetime $1/\Gamma_j$, and electron interband transition frequency ω_j.

For the dielectric function of metal nanomaterials, it is usually modeled by Drude–Lorentz method. Drude term is the contribution of intraband transition to dielectric function, while the interband transition contribution is usually modeled by one or more Lorentz oscillators. However, it is necessary to apply a large number of arbitrary Lorentz terms to fit a real dielectric function (any shaped curves), which leads to the model losing physical meaning. Experimentally, the strength of interband transitions in metals can be measured on the nanofilms. Then the dielectric function can be calculated after measuring these parameters, together with some physical constants [7]. The interband transitions allow for frequencies above the band edge of the metals Al, Cu, Ag, and Au, which is important for the application solar spectrum.

A majority of the energy in the solar spectrum resides in the spectral range of 300–1750 nm. Capturing the largest possible fraction of this spectrum is essential for any solar application. Figure 2.1a shows the regions where interband transitions occur in these metals. Figure 2.1b shows the real part of dielectric function for these metals. The dashed line indicates the resonance condition for nanospheres in air, which is accordant with experimental results. Compared with Cu, both Ag and Au are excellent plasmonic materials, with Ag having slightly lower loss, and Au having lower reactivity. Figure 2.1c clearly demonstrates over what range of the solar spectrum they can be expected to be strong absorbers. The band structure of Al

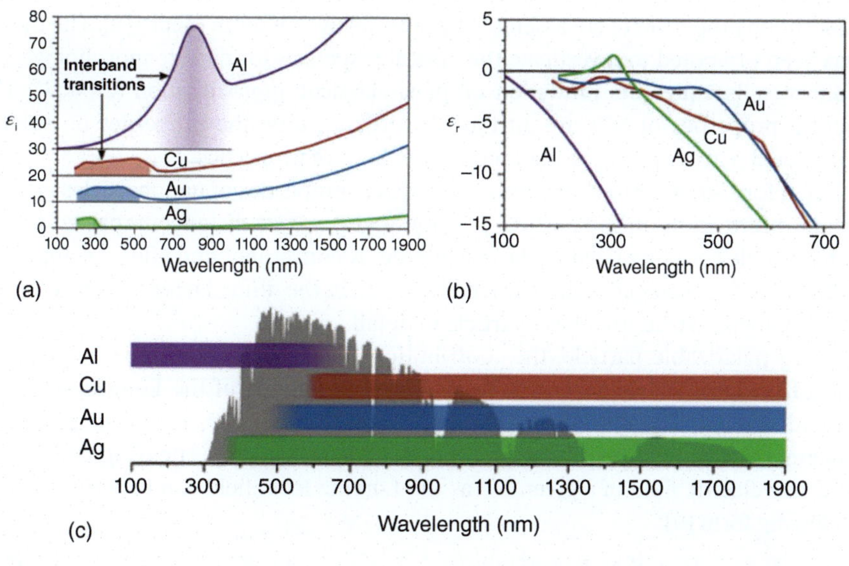

Figure 2.1 Dielectric functions of plasmonic metals. (a) The imaginary part of the dielectric functions of Al, Cu, Ag, and Au with shaded sections indicates the presence of interband transitions. (b) The real part of the dielectric function. The dashed line indicates the resonance condition for a sphere at −2. (c) The energy flux of solar spectrum (gray) overlayed with the regions where the metals have good plasmonic properties. Source: Reproduced with permission from Shahbazyan and Stockman [7]; © 2013, Springer Dordrecht Heidelberg New York London.

shows the frequency of the band edge is ~0.8 eV, and interband transitions occur in the vis–NIR region. However, there are no interband transitions in the 100–500 nm regime, indicating that Al NPs may be of some utility in the 300–400 nm wavelength range. Because of interband transitions, Au and Cu are poor plasmonic materials in the wavelength range of 300–550 nm; Ag is poor below about 350 nm. This suggests that a combination of these NPs of various sizes could allow for SPRs from the UV to the NIR while minimizing the interference of interband transitions.

2.3 Dielectric Properties of Complex Nanostructures

2.3.1 Maxwell Garnett Effective Medium Theory

Mie proposed the optical responses of single metallic NPs by classical dynamic model measurement. This model explained qualitatively the effect of structural size and surrounding medium on the optical properties and assumed that there is no interaction between particles.

The experimental and theoretical research studies illustrate that electromagnetic coupling of interparticles plays a critical role in the extinction spectra, when the separation distance (d) of two Au NPs is five times less than the radius (R) of NP ($d \leq 5R$). Especially, unique resonance modes can be induced. Therefore, the Mie theory is unsuitable for the discussion of extinction property in the nanosystem of aggregations. Date back to 1904, Maxwell Garnett achieved to declare the light absorption properties of NP assembly by the effective medium theory [8]. The theory has been promoted to investigate the optical responses for diverse morphologies of nanocrystals under the limitation of quasi-classical approximation ($2R \ll \lambda$). The optical properties of NPs are attributed to two factors: the properties of isolated individual particles and the interaction of collective properties.

Therefore, in order to characterize the optical properties of nanoparticles system, detail information on the particle position and pairwise distance impacted on the interparticle coupling needs to be considered. A convenient and effective approach is to obtain the averaged volume fraction (f), that is, the filling factor, which describes the topological structure of the particle in detail.

For a polarizable particle, the contributions of the neighboring particles to the local field have to be concerned. In this case, the model of the Lorentz sphere is extended to explore the interaction field from all other NPs in the metallic NPs system. In consideration of Lorentz's theory of dielectric properties (Figure 2.2), the effective electric field of a given NP as the Lorentz local field can be written in the following form [9]:

$$E_{\text{local}} = E_0 + E_{\text{d}} + E_{\text{S}} + E_{\text{near}} \tag{2.2}$$

where E_0 is the external field, E_{d} is the field derived from the charges at the surface, E_{S} is the field attributed to the polarization charges around the sphere, and E_{near} is the field from dipoles inside the sphere. Putting the field of E_{Maxwell} in the above equation, we get the expression.

$$E_{\text{local}} = E_{\text{Maxwell}} + E_{\text{S}} + E_{\text{near}} \tag{2.3}$$

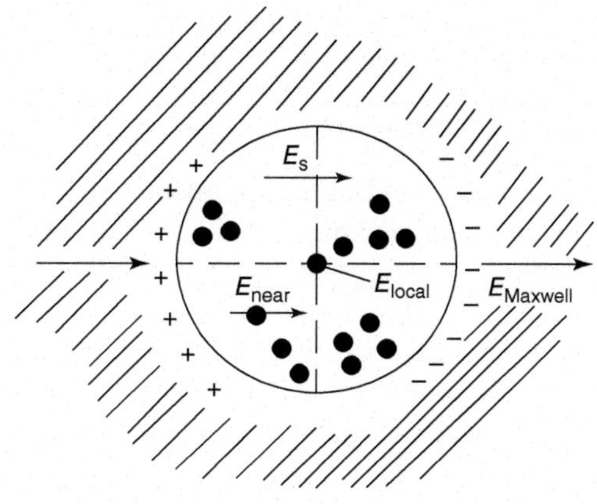

Figure 2.2 The Lorentz sphere scheme applied to cluster matter.

where $E_{Maxwell}$ is the incident field in the external matrix (defined as the average field $E_{Maxwell} = E_0 + E_d$).

To explore the effect of the field on the plasmons in a system of contacting colloidal spheres, a new embedding approach is applied to solve Maxwell's equations, as shown in Figure 2.2. At low filling fractions, the interaction of the Lorentz sphere can be illustrated by Maxwell Garnett approximation. In this situation, E_{near} is zero because of the cancellation by the interference of the scattering fields for all the particles inside the sphere at the sphere center. However, the field from polarization charges around the sphere (E_S) is obtained by:

$$E_S = \int_0^{\pi} \left(\frac{P\cos\theta}{R^2}\right)\cos\theta(2R^2\sin\theta)\mathrm{d}\theta = \frac{4\pi}{3}P \tag{2.4}$$

Therefore, the local field (E_{local}) is connected with the average field and polarization:

$$E_{local} = E + 4\pi P/3 \tag{2.5}$$

Subsequently, the local field can be used to calculate the dipole moment (μ) and the polarization (P) of an atom or a particle, which can be given by

$$\mu = \alpha E_{local} \tag{2.6}$$

$$P = \sum_i N_i\mu_i = \sum_i N_i\alpha_i E_{local}(i) \tag{2.7}$$

where N_i and α_i are corresponding to the number density and the polarizability of the ith particle, respectively. Furthermore, we can get polarization.

$$P = \sum_i N_i\mu_i = \sum_i N_i\alpha_i\left(E + \frac{4\pi}{3}P\right) \tag{2.8}$$

The susceptibility (χ) is defined as:

$$\chi = \frac{P}{E} = \frac{\sum_i N_i a_i}{1 - \frac{4\pi}{3}\sum_i N_i a_i} \tag{2.9}$$

The polarizability of the spherical NP is related to the permittivity (ε) and radius (R), which is given as

$$\alpha = \frac{\varepsilon - \varepsilon_m}{\varepsilon + 2\varepsilon_m} R^3 \tag{2.10}$$

In the quasi-classical approximation, the system can be replaced by an effective medium, approximated to a homogeneous material. The effective dielectric function (ε_{eff}, $\varepsilon_{eff} = \varepsilon_{1,\,eff} + i\varepsilon_{2,\,eff}$) of the whole composite material represents the linear response of the sample to the external field, which can be defined as the invariable physical properties through the replacement of heterogeneous samples into the effective medium. The relationship of Clausius–Mossotti can describe the correlation between ε_{eff} in macroscopy and the parameter of α in microscopy in the form:

$$\frac{\varepsilon_{eff} - \varepsilon_m}{\varepsilon_{eff} + 2\varepsilon_m} = \frac{4\pi}{3} \sum_i N_i \frac{\varepsilon - \varepsilon_m}{\varepsilon + 2\varepsilon_m} R^3 \tag{2.11}$$

Then, the introduction of the volume fraction (f) can be simplified as:

$$\frac{\varepsilon_{eff} - \varepsilon_m}{\varepsilon_{eff} + 2\varepsilon_m} = f \frac{\varepsilon - \varepsilon_m}{\varepsilon + 2\varepsilon_m} \tag{2.12}$$

The above equation is a popular expression of the Lorentz–Lorentz formula, which can be translated into the following form:

$$\varepsilon_{eff} = \varepsilon_m \frac{1 + 2f\Lambda}{1 - f\Lambda} \tag{2.13}$$

wherein

$$\Lambda = \frac{1}{\varepsilon_m} \frac{\varepsilon - \varepsilon_m}{\varepsilon + 2\varepsilon_m} = \frac{\alpha}{3\varepsilon_0 \varepsilon_m V} \tag{2.14}$$

When the system is composed of two equivalent inclusions, the Maxwell Garnett formula is transformed into the form:

$$\frac{\varepsilon_{eff} - \varepsilon_m}{\varepsilon_{eff} + 2\varepsilon_m} = f_1 \frac{\varepsilon_1 - \varepsilon_m}{\varepsilon_1 + 2\varepsilon_m} + f_2 \frac{\varepsilon_2 - \varepsilon_m}{\varepsilon_2 + 2\varepsilon_m} \tag{2.15}$$

The collective effects induced by the homogeneous polarization of the Lorentz sphere do not express the position- and size-depended properties of individual particle. Equation (2.13) is in the condition of quasi-classical approximation. If $\alpha_i(\omega)$ presents the size dependence due to quantum size effects, the particle size has been implicitly considered. If the particles are NPs, Eq. (2.13) is named as the Maxwell Garnett formula. The Maxwell Garnett theory is limited to NP ensembles with low filling factors and interaction through far field. That is, the theory is unsuitable to the condition that the near-field interaction is possessed in the closed NPs.

2.3.2 Drude-Critical Points Model for Dielectric Function of Alloy

Multimetallic nanostructures composed of different metal elements that exhibit better physicochemical properties than single metallic nanostructures in specific applications and will also exhibit novel properties and enhanced performance. In general, the addition of a second metal to form bimetallic nanocrystals can significantly alter their electronic structure and spatial arrangement pattern, which further enhances the structural properties. A typical example is the Au/Ag alloy nanostructure. Au/Ag alloys can keep Au inert and retain optical properties of Ag. Furthermore, the Raman scattering signal and sensing properties of Au nanostructures can be enhanced by the incorporation of Ag.

In some cases, the multimetallic nanostructures may exhibit properties completely different from that of the component, even though they have the same composition and stoichiometric ratio. Because the optoelectronic properties are highly sensitive to the spatial atom arrangement, it is not sufficient to describe bimetallic NPs with the term component. The extinction spectrum for a mixture of different components will likely show a broad absorption band or two bands, whereas the extinction spectrum of the actual alloy NPs shows a relatively narrow plasmonic peak. Therefore, different from the mixture of two metal NPs, the spectral calculation can be calculated by a weighted method. However, for the dielectric value of the alloy nanostructure, the simple weighted average method is lack of scientific meaning.

Etchegoin et al. proposed a different approach by replacing the Lorentz term with an analytical model called critical points (CPs), which has long been used to analyze interband transitions in semiconductors [10]. Unlike simple Lorentz term, the advantage of using CPs is that the model allows simple adjustments to asymmetric shapes. Lu et al. have successfully used this model with two terms to represent the optical properties of various noble and transition metals [11]. However, there are some issues with this model because the phase parameter makes the imaginary part of one (or both) in the units modifying the Lorentz function to be a negative value, which is a nonphysical effect. Then Rioux and coworkers improved this model based on the same approach, but they solved the problems in Etchegoin's model by considering the energy band structure of metals [12].

In the improved model, the interband transitions can be represented by means of two CPs of M1 and M2 type:

$$\begin{aligned}\varepsilon_{\mathrm{CP1}} = A_1 \Bigg[&-\frac{\sqrt{\omega_{\mathrm{g1}}-\omega_{01}}}{2(\omega+i\Gamma_1)^2}\,\mathrm{In}\left(1-\left(\frac{\omega+i\Gamma_1}{\omega_{01}}\right)^2\right)+\frac{2\sqrt{\omega_{\mathrm{g1}}}}{(\omega+i\Gamma_1)^2}\tanh^{-1} \\ &\times\left(\sqrt{\frac{\omega_{\mathrm{g1}}-\omega_{01}}{\omega_{\mathrm{g1}}}}\right)-\frac{\sqrt{\omega-i\Gamma_1-\omega_{\mathrm{g1}}}}{(\omega+i\Gamma_1)^2}\tan^{-1}\left(\sqrt{\frac{\omega_{\mathrm{g1}}-\omega_{01}}{\omega+i\Gamma_1+\omega_{\mathrm{g1}}}}\right) \\ &-\frac{\sqrt{\omega-i\Gamma_1-\omega_{\mathrm{g1}}}}{(\omega+i\Gamma_1)^2}\tanh^{-1}\left(\sqrt{\frac{\omega_{\mathrm{g1}}-\omega_{01}}{\omega+i\Gamma_1+\omega_{\mathrm{g1}}}}\right)\Bigg]\end{aligned} \tag{2.16}$$

$$\varepsilon_{\mathrm{CP2}}(\omega) = \frac{A_2}{2(\omega+i\Gamma_1)^2}\ln\left(1-\left(\frac{\omega+i\Gamma_2}{\omega_{02}}\right)^2\right) \tag{2.17}$$

where for each transition $j = 1, 2$, A_j is the amplitude, ω_{0j} the threshold, ω_{gj} the gap, and Γ_j the broadening factor. Finally, the frequency-dependent dielectric function in this model is:

$$\varepsilon(\omega) = \varepsilon_\infty - \frac{\omega_{\mathrm{p}}^2}{\omega^2+i\omega\Gamma_{\mathrm{p}}} + \varepsilon_{\mathrm{CP1}}(\omega) + \varepsilon_{\mathrm{CP2}}(\omega) \tag{2.18}$$

The main contribution to the shape of the dielectric function is therefore the shape of the joint density of states, which itself is mainly affected by the CPs. The band structures of both Au and Ag can be calculated by density functional theory. Consequently, the contribution of the CPs to the dielectric function can be obtained. Although the CPs are important for the calculation of the dielectric function, it is found that two CPs have no obvious contribution to the real part of the electric function from the Drude-critical points (DCP) prediction and the data measured from the alloyed thin films.

In general, the dielectric functions of the nanoalloys are calculated by the composition-weighted average method. Unfortunately, this method is not able to reproduce the variations of the onset of the interband transitions in the alloys that shifts roughly linearly as a function of the composition. The DCP model represents the dielectric functions ($\varepsilon_{\text{AuAg}}(\omega)$) of the $\text{Au}_x\text{Ag}_{1-x}$ alloy nanostructures, in which the Drude term represents the intraband transitions and two CP terms account for the interband transitions, providing a better understanding of the optical properties of plasmonic alloyed nanostructures. Figure 2.3 shows the real and imaginary dielectric data of Au–Ag alloy nanostructures as functions of wavelength through the DCP method [13]. From Figure 2.3c,d, it can be seen that the modeled dielectric constants of pure Au and Ag metals are well agreed with the widely used data reported by Johnson and Christy [14], especially in the range of 300–800 nm for LSPR peaks location. Therefore, the high agreement ensures

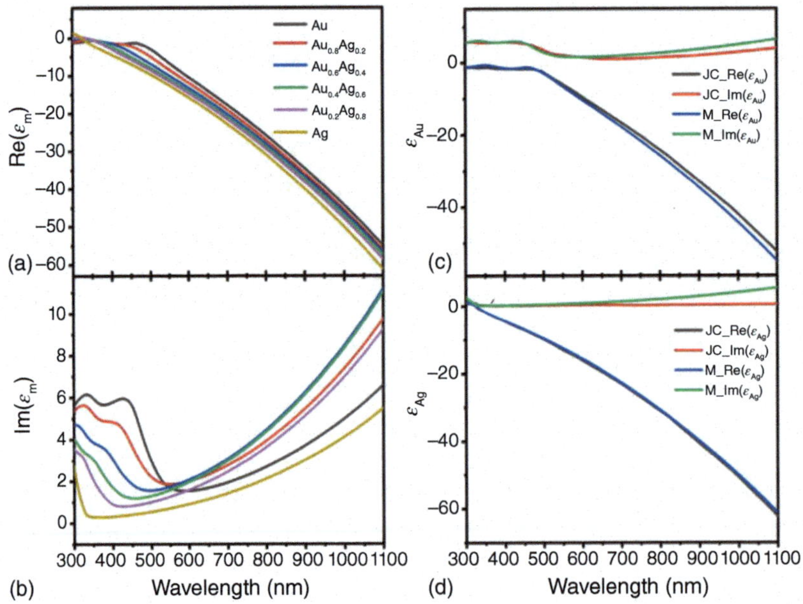

Figure 2.3 The calculated real (a) and imaginary (b) parts of the dielectric function (c) and (d) comparison between the modeled complex refractive index of pure Au and Ag metals (M) with the data reported by Johnson and Christy. Source: Reproduced with permission from Wang et al. [13]; © 2020, Royal Society of Chemistry.

the reliability of the application of the modeled dielectric function data in the subsequent finite-difference time-domain (FDTD) simulation and the accuracy of the simulation results.

2.3.3 Effect of Dielectric Function on Properties

Here, we perform theoretical calculations on the SPR refractive index (RI) sensitivity of Au–Ag alloy nanorods with different Ag contents to investigate the effect of the dielectric function on the results. For randomly oriented ellipsoidal particles, the absorption cross-section can be obtained from the quasi-static approximation. In this approximation, the electron magnetic phase is assumed to be constant as the particle is much smaller than the wavelength of light. In this case, light scattering is very weak, and the extinction can be expressed by the absorption, as given by [15–17]:

$$\sigma_{\text{abs}}(\omega) = \frac{V}{3c}\varepsilon_{\text{m}}^{3/2}\sum_{i=1}^{3}(Y_i+1)^2\frac{\omega\varepsilon_2(\omega)}{(\varepsilon_1(\omega)+Y_i\varepsilon_{\text{m}})^2+\varepsilon_2^2(\omega)} \tag{2.19}$$

where V is the particle volume, c is the speed of light in a vacuum, $\varepsilon(\omega) = \varepsilon_1(\omega) + i\varepsilon_2(\omega)$ is complex dielectric functions of metal NPs with real and imaginary parts, and surrounding medium ε_{s} is real. Here, the defined parameter Y_i related to the shape or depolarization factor (L_i along the axis i of nanostructures) is expressed as $Y_i = 1/L_i - 1$. The shape factor $L = 0$ (Y goes to infinity) represents an infinite columnar structure along the axis of ellipsoid symmetry. $L = 1/3$ ($Y = 2$) represents a sphere. And $L = 1$ ($Y = 0$) represents a flat disk. Therefore, the Y values of ellipsoid nanostructures with elongated geometries are greater than 2, increasing with the aspect ratio.

The resonance frequency (ω_{res}) is determined by the condition of the minimum of the denominator (2.19), and this condition can be further simplified for the noble metals, especially in a vis–NIR frequency region where $\varepsilon_2(\omega)$ is small.

$$\varepsilon_1(\omega) + Y_i\varepsilon_{\text{s}} = 0 \tag{2.20}$$

With the simplified real part of the dielectric function $\varepsilon_1(\omega) = \varepsilon_\infty - \omega_{\text{p}}^2/\omega^2$ derived from Eq. (2.18), ω_{res} can be expressed as:

$$\omega_{\text{res}} = \frac{\omega_{\text{p}}}{\sqrt{\varepsilon_\infty + Y_i\varepsilon_{\text{s}}}} \tag{2.21}$$

Introducing the RI $n_{\text{s}} = \sqrt{\varepsilon_{\text{s}}}$ of the surrounding medium and the bulk plasmon wavelength $\lambda_{\text{p}} = 2\pi c/\omega_{\text{p}}$, we can express the resonance wavelength $\lambda_{\text{res}} = 2\pi c/\omega_{\text{res}}$ from Eq. (2.21):

$$\lambda_{\text{res}} = \lambda_{\text{p}}\sqrt{\varepsilon_\infty + Y_i n_{\text{s}}^2} \tag{2.22}$$

SPR peak wavelengths are linearly related to RIs for a short range of RIs n_{s}. A plot of Eq. (2.22) for λ_{p} across a suitable range of n_{s} can be obtained.

In general, changes in the RI of the surrounding medium have an essentially linear effect on the SPR peak wavelength. So, the RI sensitivity for a given NP type is commonly expressed as nanometers of peak shift per RI unit (nm RIU^{-1}). There are many different types of nanostructures, each with its own unique shape and metal type, that may affect the SPR sensitivity to changes in RI. Because even a tiny change in the RI of the surrounding medium may induce a noticeable spectral shift and a visible change in intensity at the same wavelength, a high figure of merit (FOM) resulting from a narrow line width is very desirable. A greater magnitude of the real dielectric function and a smaller imaginary dielectric function of Ag makes the plasmon resonance of Ag nanostructure more sensitive to changes in the index than that of Au nanostructure.

The mathematical RI sensitivity (S) was defined as the relative shift in resonance wavelength (λ_{res}) with respect to the RI (n_s) change of the surrounding materials, as follows:

$$S = \frac{d\lambda_{res}}{dn_s} = \frac{\lambda_p n_s Y_i}{\sqrt{n_\infty^2 + n_s^2 Y_i}} = \frac{2\pi c n_s Y_i}{\omega_p \sqrt{\varepsilon_\infty + n_s^2 Y_i}} \tag{2.23}$$

Equation (2.23) clearly shows that the RI sensitivity of the SPR depends not only on the nanostructure geometry Y_i but also on the metal composition, that is, the bulk plasma frequency (ω_p) and ε_∞. Note that plasma frequency (ω_p) of Au_xAg_{1-x} alloy nanostructures are almost constants, while ε_∞ ($\varepsilon_\infty = 2.614 - 0.747x_{Ag}$) decrease obviously with increasing Ag concentrations. When the nanostructures are spherical or elongated rods (infinitely long nanorods are not considered, i.e. $L = 0, Y_i \to +\infty$), the effect of the bulk plasma frequency ω_p on the RI sensitivity is negligible, while that of ε_∞ cannot be ignored. Therefore, for nanostructures with the same shape, the metal composition plays a key role in the RI sensitivity. As in the case of the Au–Ag alloy nanostructures discussed herein, the higher the Ag content in the nanostructures, the more sensitive owing to its decreasing ε_∞ value.

In addition to the above physical properties, chemical interface properties such as charge transfer between nanostructures and the environment also affect the RI sensitivity of SPR. The full-width half maximum (FWHM) of the resonance absorption band can be derived by exploiting the Taylor expansion of $\sigma_{abs}(\omega)$ near the SPR [18]. It is abundantly obvious from Eq. (2.23) that the RI sensitivity of the SPR is dependent not only on the nanostructure geometry (Y_i) but also on the metal composition, that is, the bulk plasma frequency (ω_p) and ε_∞. The SPR spectroscopy relies on the response of localized plasmon bands to changes in the surrounding dielectric medium. The FOM is defined by the sensitivity of the RI and the FWHM [19]. The accuracy of a change of nm is dependent on both S and the maximum line width ($\Delta\lambda$) [20].

$$\text{FOM} = \frac{\text{RI (nm RIU}^{-1})}{\text{FWHM (nm)}} = \frac{S}{\Delta\lambda} \tag{2.24}$$

This equation indicates that the smaller imaginary dielectric function of metal and the real part with a steeper gradient would be a favor for the narrower FWHM.

The normalized extinction spectra of AuBP@Au_xAg_{1-x} nanorods dispersed in water (n = 1.3334) synthesized by the above method are shown in Figure 2.4a. Figure 2.3b illustrates the calculated extinction spectra of AuBP@Au_xAg_{1-x} nanorods (x: 0.2, 0.4, 0.6, and 0.8) dispersed in a surrounding medium with a RI of 1.33 (the simulation method is FDTD, which will be introduced in the following part). It is worth noting that the characteristic SPR absorption peaks of Au and

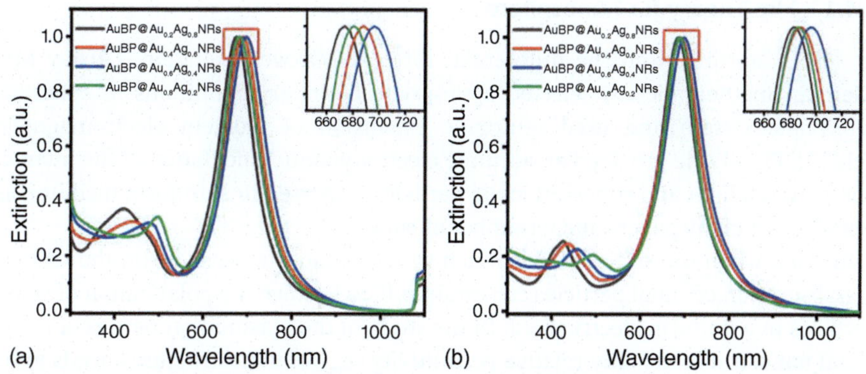

Figure 2.4 (a) The normalized extinction spectra of AuBP@Au_xAg_{1-x} nanorods (x: 0.2, 0.4, 0.6, and 0.8) dispersed in water synthesized by a wet-chemical method. (b) The FDTD calculated normalized extinction spectra of AuBP@Au_xAg_{1-x} nanorods ($R = 2.55$) in a surrounding medium of 1.3334. The insets are the enlarged extinction spectra from the marked area in pane. Source: Reproduced with permission from Wang et al. [13]; © 2020, Royal Society of Chemistry.

Ag nanospheres are 520 and 380 nm, respectively. According to Figure 2.4a,b, the experimental extinction spectra of AuBP@Au_xAg_{1-x} nanorods agree well with that calculated by FDTD simulation. Specifically, the SPR_T of AuBP@Au_xAg_{1-x} nanorods red shift from 380 to 520 nm as the decrease of Ag concentrations in the alloy shells both in experiments and simulation. Furthermore, the experimental and FDTD calculated SPR_L of AuBP@Au_xAg_{1-x} nanorods are both locate close to ~685 nm, red-shifting slightly with decreasing Ag, except for the AuBP@$Au_{0.8}Ag_{0.2}$ nanorods. Therefore, using the modeled alloy dielectric data, the FDTD simulated result agrees well with that from experimental measurements even for the CP of the change process (as marked with the frame), indicating the accuracy of the modeled dielectric function.

2.4 Optical Property Analysis of Isolated Nanoparticles

In an experimental study, a metal nanostructure colloid is a highly dispersed object system, and its color or spectrum is the average effect of light response of a large number of NPs. Therefore, the homogeneity of sample in morphology and size is very important. Otherwise, in order to understand the optical properties of samples, we should analyze the optical properties of different nanostructures. With the achievement of dielectric function of metal NPs, one can now calculate or simulate the optical property of isolated NPs of arbitrary shapes using numerical analytic formula or numerical simulation. The localized surface plasmon describes the absorption and scattering responses of metal nanostructures to electromagnetic fields by solving Maxwell equations with special boundary conditions. Here, we mainly introduce the analytical expression of optical properties of isolated nanosphere and ellipsoidal NP.

2.4.1 Mie Theory for Nanosphere

In 1908, starting from the differential form of Maxwell equation, Gustav Mie (German mathematician) selected appropriate boundary conditions in spherical coordinate system and used multipole expansion of incident electromagnetic field [2]. Finally, he provided an accurate electrodynamic calculation of the interaction between light and spherical metal particles dispersed in a uniform medium, as described in many papers and a number of books.

In brief, when the size of NPs is much smaller than the wavelength, the scattering/absorption of small particles can be simplified to solve the polarization of small particles in a uniform electric field. In the study of the plasmon properties of noble metal nanostructures, only relative permittivity (ε_m) of nanostructures needs to be considered due to nonmagnetic property of noble metals. As expressed by Eqs. (1.26) and (1.27) described in Section 1.6 for expressions for the optical cross of NPs, it can also be obtained by Eq. (2.19). For small spheres with a size less than the light wavelength, only the dipole-active excitation ($n = 1$) is important. In the chapter of "Mie's Theory for Single Spherical Particles," σ_{sca} and σ_{abs} can be obtained in Ref. [3]. Here, the coefficient in the formula is further simplified by constants of C_i:

$$\sigma_{sca} = C_1 \frac{a^6}{\lambda^4} \frac{(\varepsilon_1 - 2\varepsilon_m)^2 + \varepsilon_2^2}{(\varepsilon_1 + 2\varepsilon_m)^2 + \varepsilon_2^2} \tag{2.25}$$

$$\sigma_{abs} = C_2 \frac{a^3}{\lambda} \frac{\varepsilon_2}{(\varepsilon_1 + 2\varepsilon_m)^2 + \varepsilon_2^2} \tag{2.26}$$

where $C_1 = 128\pi^3/3$, $C_2 = 32\varepsilon_m\pi^2$, a presents the radius of particle. Therefore, σ_{abs} is dominated for small-sized particles, and σ_{sca} cannot be ignored for big particles (note that $a \ll \lambda$).

Once the dielectric data of metal NPs were obtained, one can calculate the optical property of isolated NPs and plot the scattering and absorption spectra. In 1972, by measuring the reflection and transmission of thin films with nanometer thickness formed by evaporation, Johnson and Christy achieved the dielectric data for several metals (Cu, Ag, and Au) of a certain frequency regime [14]. More and more dielectric data of many different kinds of materials were obtained, and the database was widely applied in simulations and calculations.

Now, the scattering, absorption, and extinction spectra can be obtained by using FORTRAN, MATLAB, and other scientific calculation software. For a small-sized nanosphere, it is clear that condition for the occurrence of plasmon resonance in NPs is that $\varepsilon_1 = -2\varepsilon_{2m}$ (note that ε_2 is usually very small). For Au nanospheres in water, the resonance absorption peak is theoretically at about 520 nm consistent with the experimental observations.

Mie theory is the strict solution of Maxwell equations, and the calculation accuracy depends only on the order of polarization (n). Although it is aimed to solve the optical response of small particles, many results show that the numerical results are in good agreement with the experiments' spectra when solving the plasmon absorption of slightly larger particles. As the size of the sphere increases, the contributions of higher multipoles become more and more significant. In the limit of a very large sphere $n \to \infty$, the LSP frequency approaches that of a surface plasmon at a planar dielectric–metal interface.

2.4.2 Gan Equation for Optical Properties of Ellipsoidal Particles

Gustav Mie applied Maxwell's equations and appropriate boundary conditions to calculate the optical response of metal micro-nanospheres (Mie theory). In 1912, based on Mie theory, Gans broke the limitations of Mie theory and extended it to Gans–Mie theory (Gans equation) [21]. This equation can be used to describe and solve the absorption and scattering problems of anisotropic ellipsoidal NPs. Equation (2.27) is the absorption cross-section [22].

$$\sigma_{\text{abs}} = C_3 \frac{a^3}{\lambda} \sum_j \frac{\varepsilon_2 / P_j^2}{(\varepsilon_1 + (1 - P_j)\varepsilon_{\text{m}}/P_j)^2 + \varepsilon_2^2} \tag{2.27}$$

Here, C_3 is a constant related to the ε_{m}, P_j is the depolarization factor in three directions, and j represents each axis of an ellipsoidal particle. There are three axes for an ellipsoid ($A > B = C$). P_A, P_B, and P_c can be expressed:

$$P_A = \left(\frac{1 - e^2}{e^2}\right)\left[\frac{1}{2e}\ln\left(\frac{1 + e}{1 - e}\right) - 1\right] \tag{2.28}$$

$$P_B = P_C = \frac{1 - P_A}{2} \tag{2.29}$$

$$e = \sqrt{1 - \left(\frac{B}{A}\right)^2} \tag{2.30}$$

A/B is the aspect ratio of ellipsoid, which is very important for the optical property of ellipsoid and other heterogeneous nanostructures. Therefore, for a long ellipsoidal NP, there are two resonant peaks in the extinction spectrum, namely, the transverse plasmon resonance (SPR_T) and the longitudinal plasmon resonance (SPR_L). They correspond to free electron oscillations perpendicular and parallel to the long axis of the ellipsoid (or rod). And the position of SPR_L is highly related to the aspect ratio of particles. With increasing (or decreasing) the aspect ratio of the NPs or the dielectric constant of the medium, it will cause the absorption peak to move toward the long (or short) wave direction, that is, redshift or blueshift.

Figure 2.5 shows the calculated absorption spectra of Au nanorods (treated as ellipsoids) with different aspect ratios and dispersed in different mediums. Two absorption peaks can be found, one weak SPR_T appeared in the visible (~530 nm) and a strong SPR_L linear redshift with increasing aspect ratio in the longer wavelength regions [23]. On the other hand, when the Au nanorods (Au NRs) were dispersed in solutions of different refractive indices, the SPR_L peak position is significantly red shifted. The SPR_L wavelength shift is proportional to the change in the RI of the medium [24]. Many experimental measurements have confirmed the linear relationship between the two variables. Using this feature of Au NRs to medium or absorbed molecule, highly sensitive chemical/biological sensors were constructed, as presented in Chapter 1.

The numerical methods based on Mie theory and Gans equation can be used to calculate the absorption and scattering of single NPs. In actual colloidal NPs that

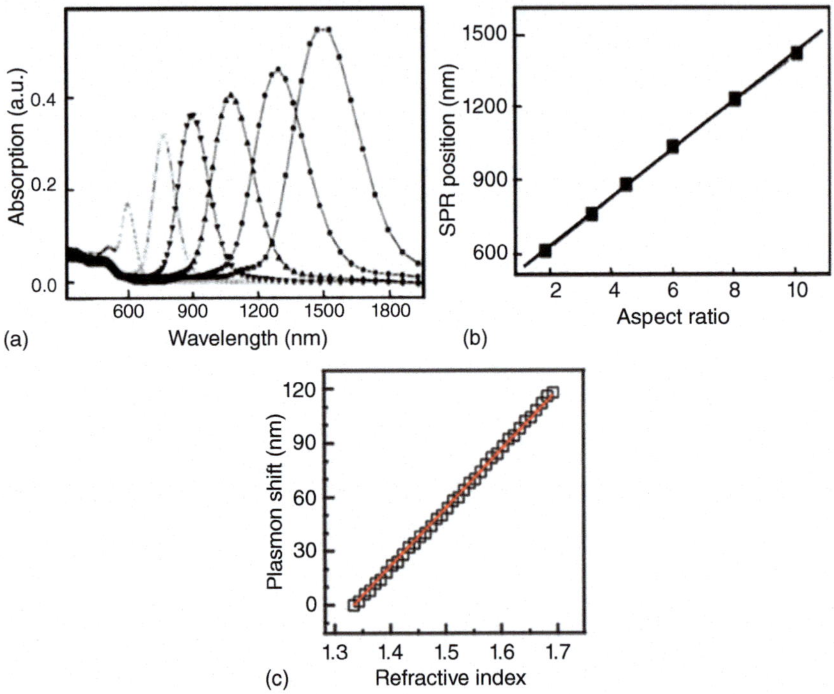

Figure 2.5 (a) Theoretical optical absorption spectra of Au NRs using Gans equation. (b) The linear relationship of peak position of SPR_L with aspect ratio [23]. Source: Adapted from Li et al. [23]. (c) The shifting of SPR_L with refractive indices [24]. Source: Adapted from Draine and Flatau [24].

are highly dispersed in solution, the particle spacing is large, and the interaction or coupling effect among particles can be not considered. Usually, the numerical results are also in good agreement with most experiments. However, Mie theory and Gans equation cannot be used to calculate the assembly system of NPs, or even to explain the tunable SPR property of spherical NPs assembly, such as dimers, trimers, and chains. In the realization of nanomaterials in extreme sizes, the quantum confinement effect should be taken into the simulation work when the particle size is smaller than the critical size. When the spacing of NPs is smaller than the subnanometer, the strong coupling effect and quantum tunneling effect cannot be ignored.

2.5 Numerical Simulation of Optical Properties

With the development of nanomaterial science, anisotropic NPs with controllable shapes, such as rods, plates, and multipods, have been successfully prepared. Morphology and optical properties can be characterized and determined by relevant instruments. Before characterizations, it was difficult to calculate the response of the complex surface plasmons to light by solving Maxwell equations. And in most cases, complex nanostructures may be composed of monomers assembled in different functional units. Without the expensive near-field optical technology, the broad spectra from far-field spectroscopy are the average effect of the optical response of nanostructures with different sizes or shapes. Moreover, other effects on the spectra should also be considered, involving substrate, surfactant molecules absorbed on the NP's surface, and heterogeneity of size and morphology. When the metal nanostructures are in complex shapes or in the case of quasi-static approximate failure, the simulation requires the developed methods, such as discrete dipole approximation (DDA), T-matrix method, finite element method, and FDTD. These methods are essential for numerical simulation calculation, including the establishment of the structural model and transferring dielectric value of the material in the database and physical parameter fitting.

DDA has proven to be a powerful and flexible tool for computing the scattering of electromagnetic waves by arbitrary-shaped particles [25, 26]. However, it is restricted to particles consisting of only one component and can only be applied to particles in a vacuum or air as a surrounding medium. The most serious limitation of the DDA is its restriction to only dipole–dipole coupling, meaning that the spherical subregions are always small compared with the incident wavelength as only a dipole moment (the TM dipole) is induced. Furthermore, the outgoing dipole re-radiation that is incident at the site of a neighboring dipole is expressed only in terms of dipole radiation. Extensions of the DDA including the TE dipole and appropriate discretization scheme were improved by Mulholland and Piller [27, 28].

The T-matrix solution begins with an expansion of the incident, scattering, and internal electromagnetic fields into corresponding vector spherical harmonics, as the Mie theory for spheres. Most implementations of this method are restricted to axisymmetric particles, such as spheroids and finite circular cylinders. Then extensions to nonaxisymmetric particles were made by Scheider and Peden [29].

As one of the simplest exact methods, there is a clever finite differencing algorithm in FDTD developed by Yee, using grids for the electric field **E** and magnetic field **H** shifted by half a grid spacing relative to each other. At present, as a commercial software, FDTD (product from Canadian Lumerial Solutions) can be used to simulate the interaction between electromagnetic waves from ultraviolet to microwave and complex structures with typical subwavelength dimensions. Now it is widely used to simulate the optical properties of artificial micro-nanostructures related to photonics and device design. FDTD method simulates the electromagnetic field in the time domain. It substitutes directly the finite difference expression for the differential expression in Maxwell's time-domain equations and obtains the finite difference expression about the field component. The electromagnetic field in the space domain is calculated alternately, updating in the time domain. In recent years, FDTD has been widely used in many fields, such as radiation antenna analysis, research on microwave devices and guided wave structures, propagation and scattering of electromagnetic pulse, periodic structure analysis, light propagation and diffraction characteristics, and so on.

FDTD discretizes the space and time of the calculation area (including the target) into small cubes as defined "Yee" cell [30], as shown in Figure 2.6. Then the field

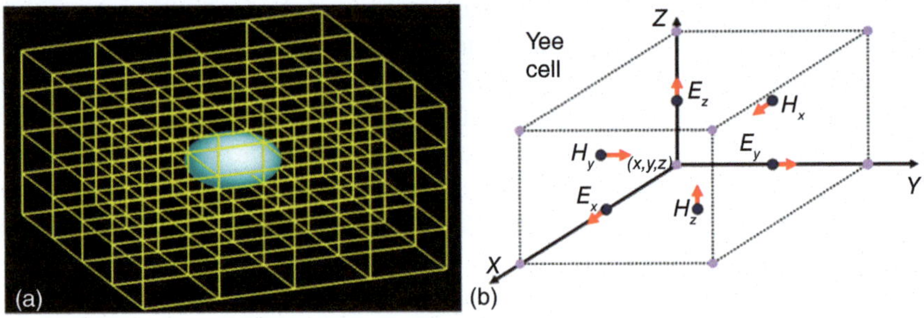

Figure 2.6 (a) Difference grids in FDTD discretization and (b) Yee cell. Source: Adapted from Yee [30].

distribution in the whole space is obtained by calculating the field components on each Yee cell. The electric field and magnetic field are selected alternately in time sequence, and Maxwell's rotation equation forms a difference scheme after difference dispersion. The time step is related to the time taken for light to pass through the mesh size because the Maxwell equation is solved in a discrete time domain. By decreasing the mesh size, this method can accurately describe Maxwell's equation. Therefore, the distribution of the electromagnetic field in the space at each time can be calculated step by step.

Figure 2.7a shows the FDTD-simulated optical absorption spectra of one-dimensional (1D) Au nanostructures with the same aspect ratio and different morphologies, indicating the SPR of 1D Au nanostructures is highly related to the ending structure. Spectra in Figure 2.7b are the simulated optical spectra of Au NRs with aspect ratio using DDA theory [31]. Figure 2.7c is the FDTD-simulated longitude SPR of Au NRs with different aspect ratios [32]. With increasing the aspect ratios, the calculated SPR results are in good agreement with the experimental results (Figure 2.7d) [33]. These results not only show the high response sensitivity of NPs to light irradiation but also prove the accuracy of the theory's development and application based on Maxwell equations.

Ag nanostructures have also been the source of extensive research for their wide-spread use as substrates for SERS, optical labeling, and near-field optical probing. The number of resonances increases as the symmetry of a structure decreases. Compared with the optical properties of Au nanostructures, the SPR modes for anisotropic Ag nanostructures are more complex. For example, the spectrum of the sphere only shows one SPR peak [34], whereas the spectrum of the Ag nanocubes (NCs) contains a couple of peaks or shoulders beyond the strong dipole peak near 450 nm. Figure 2.8 is the FDTD-calculated absorption spectra of Ag NCs with sizes of 30, 50, 70, and 90 nm. From the theoretical spectra, more than three SPR peaks can be observed with increasing the size of NCs, in which the absorption peaks located at 350, 400–425 nm, and one strong SPR peak shifting in a large region (450–550 nm) are assigned to the out-of-plane quadrupole resonance, in-plane quadrupole resonance, and dipole resonances of Ag NCs, respectively [35].

With the development of colloidal chemistry, more and more complex plasmonic nanostructures were obtained, such as core–shell, Yolk–shell, or (rattle like) nanostructures. Simultaneously, study on the local fields coherently coupled to

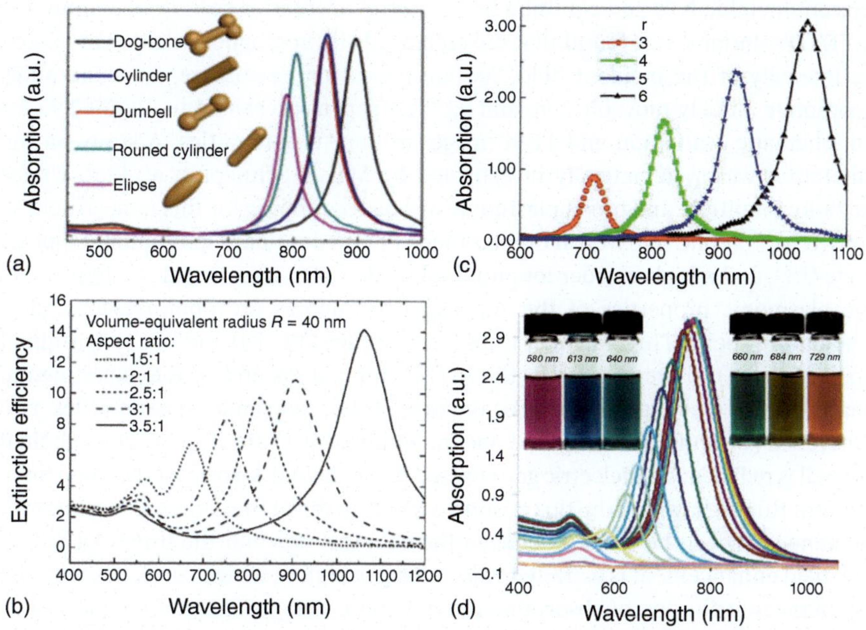

Figure 2.7 (a–c) Simulated optical absorption spectra of Au NRs with different aspect ratios, using DDA and FDTD solutions [31, 32]. Source: Reproduced with permission from Yin et al. [31]; © 2006, The Korean Physical Society. Reproduced with permission from Li et al. [32]; © 2014, Beijing University Press. (d) Colloids photograph and UV-vis absorbance spectra of Au NRs with SPR peaks labeled on the bottle [33]. Source: Adapted from [33].

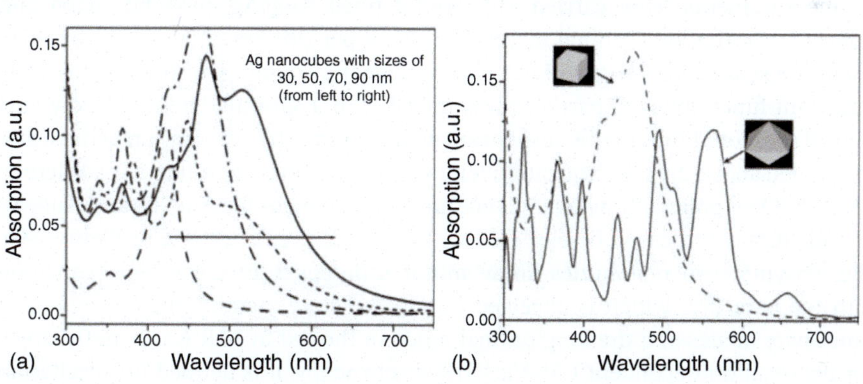

Figure 2.8 (a) FDTD calculations of the absorption spectra for Ag nanocubes with sizes of 100 nm. (b) FDTD calculations of the absorption spectra for Ag nanocube and nanoctahedron with a size of 50 nm [35]. Source: Reproduced with permission from Ding et al. [35]; © 2012, Springer Netherlands.

adjacent metallic NPs and confined in a nanogap or interior cavity were simulated by FDTD solutions. It is found that the enhanced electromagnetic fields can exceed the intensity of the incident field. As one typical nanostructure, Au nanorattles incorporate a freely movable core and variable gap, as presented in Figure 2.9a for the schematic evaluation and TEM images of the Au nanorattles. A study on the nanorattles was systematically investigated by Max J. Schnepf et al. [36]. Under excitation, multiple and broad plasmonic resonances appear for this nanostructure. Au NRs exhibit one transversal dipolar mode (T1), a dominant longitudinal dipolar mode (L1), and a higher-order longitudinal mode (L2), as indicated in Figure 2.9b. The plasmonic properties of the Au@Ag nanocuboids are dominated by their cuboidal Ag shells. Three distinct transversal modes (T1–T3) and one longitudinal dipolar mode (L1) are present in the FDTD simulations and in the experimental measurements (see Figure 2.9c). For the case of the axisymmetric nanorattles, two transversal (T1 and T2) modes and one longitudinal (L1) are excited. The cuboidal Ag shell is reduced to a dielectric gap covered by a cuboidal Au box. At the same time, the ultra-thin Ag layer on the tips ensures a fixed Au rod at the center. Consequently, the nanocavity can be excited between the cuboidal box and the inner Au NR. A high field enhancement is expected within the gap region, which correlates with the dominant contribution of absorption cross-section. The positions of the plasmonic modes, as well as the intensity ratios, are also in good agreement with the results from FDTD electromagnetic simulations (see Figure 2.9b–d). The broadening resonances were attributed to size and wall thickness variations, as well as the random nature of these wall defects that cannot be included in the FDTD simulations.

Corresponding to the schematic representation of preparation process and numerical simulation, experiments for the nanostructures preparation and optical measurements were carried out. Well-defined Au@Ag cuboidal shape and elongation in the longitudinal direction (~2 nm per tip) are obtained. In order to orient the synthesized NPs without electromagnetic coupling, a suspension of the surfactant-functionalized particles was mixed with a saturated aqueous polyvinyl alcohol solution. Films of this suspension were prepared by evaporation of the water and heated above the glass transition temperature, followed by uniaxially stretching by 200%. Then polarization dependence and color changes for Au NRs and Au@Ag cuboids film could easily be observed with the bare eye, as shown in Figure 2.9e. The axisymmetric nanorattles show minor differences in color by observation with the bare eye, which is expected due to the close vicinity of the transversal nanocavity modes and the longitudinal mode in the red to NIR wavelength range. For polarization-dependent UV–Vis–NIR spectroscopy, it is defined two excitation possibilities similar to those of the FDTD simulations: parallel and perpendicular to the stretching axis. Perpendicular excitation of the Au NRs leads to a transversal mode at 510 nm (T1) only. For parallel excitation, the higher-order dipolar mode at 605 nm (L2) and the dipolar longitudinal mode at 1095 nm (L1) are observable. For Au@Ag cuboids, three resonances at 448 nm, 401 nm, and 348 nm (T1–T3) were found during the polarization-dependent measurement, corresponding to the FDTD modeling results. For parallel excitation, only the longitudinal dipolar resonance (L1) at 670 nm was observed. For this excitation state, weak plasmonic modes were

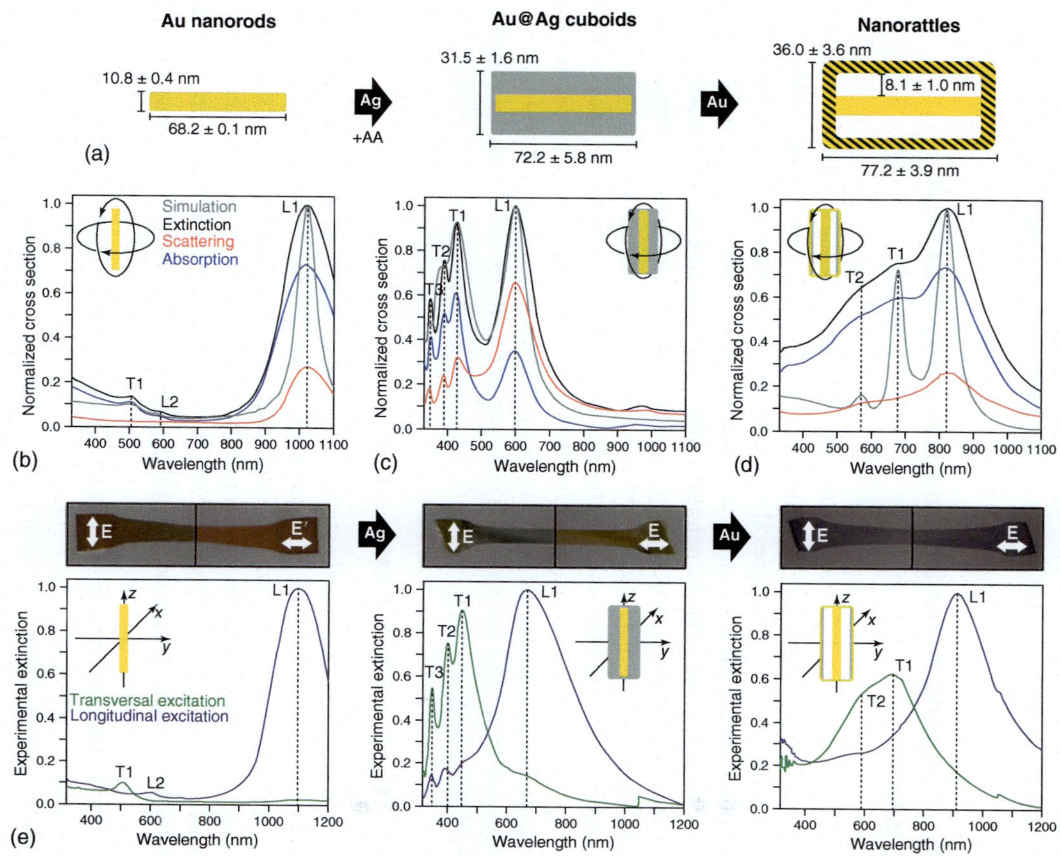

Figure 2.9 (a) Schematic representation of the nanoparticle dimensions and morphology evaluated by three-dimensional modeling. Numerical simulation (gray line) of the ensemble-averaged geometric model and comparison with absorption, scattering, and extinction UV–Vis–NIR spectroscopy for Au NRs (b), Au@Ag cuboids (c), and axisymmetric nanorattles (d). (e) Polarization imaging and polarized UV–Vis–NIR spectroscopy reveal the separate excitation of the transversal (green) and the longitudinal (purple) plasmonic modes for Au NRs, Au@Ag cuboids and axisymmetric nanorattles [36]. Source: Reproduced with permission from Schnepf et al. [36]; © 2017, Royal Society of Chemistry.

also present in the UV region, which match the cubic modes from perpendicular excitations. For the axisymmetric nanorattles, extinction along the stretching axis shows a pronounced plasmon resonance at 911 nm. For perpendicular excitation, two cavity modes (T1 at 690 nm, and T2 at 600 nm) could be clearly identified.

Compared with the monometallic nanostructure, bimetallic core–shell or rattle-like nanostructures exhibit more unique plasmonic properties, as described above for the yolk–shell nanostructures or nanorattles. They are one kind of typical coupling nanostructures that will be discussed in the following part.

2.6 Coupling Nanostructure Assembly with High Sensitivity

It is known that metal nanostructures can significantly enhance the local electromagnetic field owing to the excitation of SPR. This plasmonic effect yields giant signal enhancement for surfaces absorbed by materials, especially in the nanoscale "hotspots" of nanogaps, nanotips, nanoholes, and nanostructures with grooves or ridges. Among these hotspots, nanogaps generated from the coupled nanostructures with controllable inter-particle size have attracted both fundamental research and application in trace-molecule detection. When the plasmonic NPs are aggregated into a dimer, chains, core–shell, cavity, and other complex structures, the coupling effect produced by electromagnetic interaction will bring two novel properties: the electromagnetic field energy is greatly concentrated in the nanogap, and new SPR modes will be formed due to strong coupling effect. Here, some typical assemblies will be described.

2.6.1 Coupling Effect of Nanospheres

The spectrum of localized surface plasmons associated with an ensemble of metallic particles is determined by the interaction between the individual LSP resonances. The resulting spectrum and the nonlinear electromagnetic field enhancement depend significantly on the shape and size of the individual particles and on the particles' distance. In such ensembles of interacting metallic NPs, a very strong concentrated electromagnetic field enhancement can be observed in the nanogaps, which is the coupling effect produced by electromagnetic interaction. Localized surface plasmons associated with assembly configuration are sometimes called gap plasmons.

With the development of micro spectral characterization and near-field technology, a large number of experimental and theoretical studies have investigated the nonlinear optical properties of LSP resonances in isolated NPs and near-field coupled NPs. Figure 2.10a–c presents the near-field two-photon excitation images of the Au nanosphere dimer [37]. From the dark field images of Au NP dimers under the action of incident light in different directions, it can be seen that the electric field at the gap is significantly enhanced when the dimer's arrangement direction is consistent with the polarization direction of light.

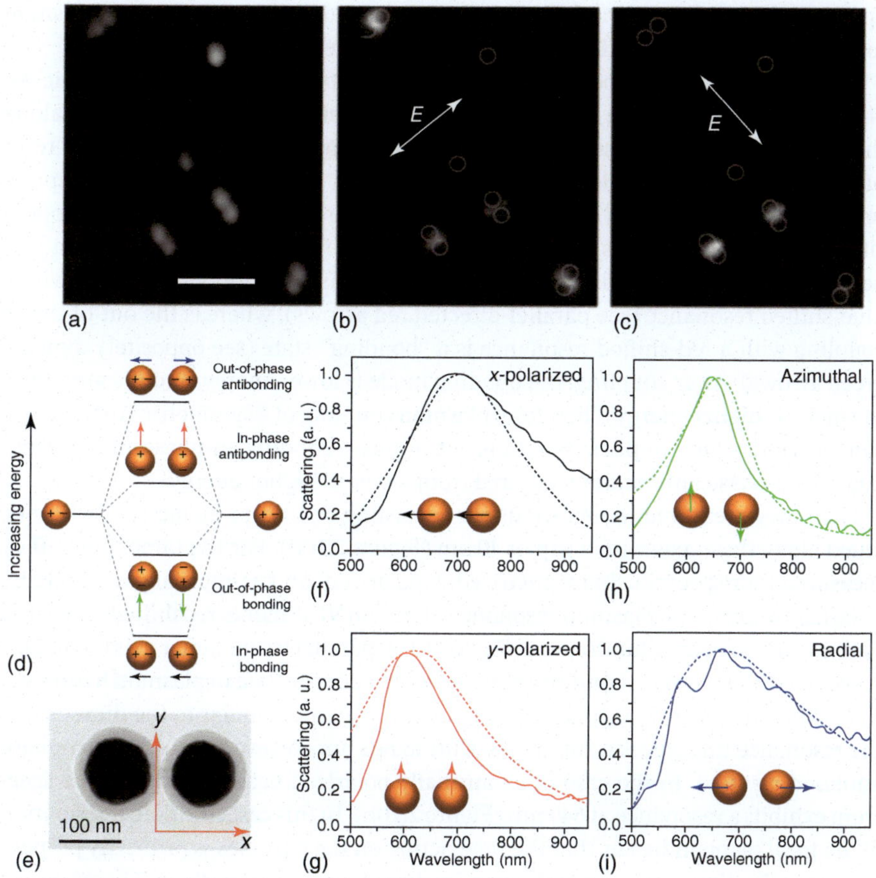

Figure 2.10 Near-field two-photon excitation images of Au nanosphere dimers. (a) Topography of the sample. (b, c) Two-photon excitation images. The excitation wavelength is 780 nm. Polarization of the excitation field is indicated by arrows in each panel [37]. Source: Okamoto et al. [37]/Reproduced with permission from © 2009, ELSEVIER. (d) Scheme of the plasmon hybridization model of a symmetric dimer, depicting in-phase/out-of-phase bonding/antibonding combinations of dipole moments. The colors are matched to the types of beams used, as shown in (f–i). (e) TEM image of one AuNP dimer of 100 nm diameter coated and separated by silica shells. The arrows in (f–i) indicate the instantaneous direction of the electric field acting on the Au NP's center at the focal plane (x–y plane) and the dipoles thereby induced. The two red crosses in (i) indicate the electric field is in the z-direction (x–z plane), which is the beam propagation direction. Source: Reproduced with permission from Deng et al. [38]; © 2018, American Chemical Society.

With the characterization development for nanostructures, study on single-particle optical property can be performed by setup for single-particle spectroscopy and beam profiles, as reported by N.F. Scherer group [38]. With the excitation of linearly polarized light, the hybridization picture predicts that in-phase coupling of the dipole moments is optically allowed, whereas the out-of-phase coupling is dark due to the cancellation of the oppositely oriented dipole moments. Then they applied azimuthally and radially polarized beams, directly exciting dark plasmon modes in symmetric Au NPs dimers at normal incidence.

The plasmon modes expected from the hybridization model for a symmetric Au NPs dimer oriented along the *x*-axis. When the incident beam is polarized along the dimer axis, the in-phase coupling forms a bonding state with a red-shifted resonance (Figure 2.10d, parallel-directed black arrows), and the out-of-phase coupling represents an antibonding state with a blue-shifted resonance (Figure 2.10d, oppositely directed blue arrows). The scenario is just the opposite for polarization perpendicular to the dimer axis. The in-phase coupling is an "antibonding" state with a blue-shifted resonance (see parallel-directed red arrows), whereas the out-of-phase coupling with a red-shifted resonance is a "bonding" state (see oppositely directed green arrows). After coupling, the bonding mode is lower energy, and the antibonding mode is higher energy. Therefore, plasmon coupling of a symmetric AuNP dimer can form four states: in-phase bonding, out-of-phase bonding, in-phase antibonding, and out-of-phase antibonding, ordered from lower to higher energy.

Then they measured the AuNP dimer consisting of 100 nm diameter AuNPs oriented along the *x*-axis with a gap of 40 nm (Figure 2.10d). For the same dimer, they measured four spectra using *x*-polarized, *y*-polarized, and azimuthally and radially polarized beams. The plasmon resonance of the AuNP dimer is redshifted to 699 nm for *x*-polarized beam illumination (Figure 2.10f, parallel to the dimer axis). This corresponds to the in-phase bonding state, the lowest energy state in plasmon hybridization. When the beam is *y*-polarized (Figure 2.10g, perpendicular to the dimer axis), the resonance undergoes a small blue shift to 603 nm corresponding to an in-phase antibonding state. In the case of azimuthally polarized beam excitation, the spectrum exhibits a resonance at 649 nm (Figure 2.10h). In this case, the field distribution at the focal plane indicates that the field acting on the particles is primarily perpendicular to the dimer axis, but with opposite directions in each particle. Therefore, we associate this plasmon resonance with the out-of-phase bonding state with the two dipole moments perpendicular to the dimer axis (and mutually antiparallel). The plasmon resonance is narrower than the bright modes, consistent with this being a dark mode. Finally, when the AuNP dimer is illuminated by the radially polarized beams, the spectrum shows two main resonances at 586 and 669 nm (Figure 2.10i). At the focal plane of the focused radially polarized beam, the field acting on each particle is now parallel to the dimer axis, but with opposite directions (antiparallel) in each particle. Therefore, the resonance at 586 nm is assigned to be the out-of-phase antibonding state, i.e. the highest energy in the model.

In addition to the great field enhancement effect at the nanogap, the spectrum of NPs assembly, such as chain and core–satellite structure, can also directly reflect the number of assembled particles or the length of linker molecule [40]. Using this feature of SPR, we can realize highly sensitive biological and chemical sensors because the assemblies depend sensitively on the absorbed molecular. Ruan et al. assembled different sized Au nanospheres into core/satellite nanostructures with 4-aminothiopheno (ATP) and 1,8-Octanedithiol (ODT) linkers, as presented in Figure 2.11a. The plasmon resonance wavelength undergoes a redshift from 648 to 723 nm after the attachment of the satellite Au nanospheres onto the

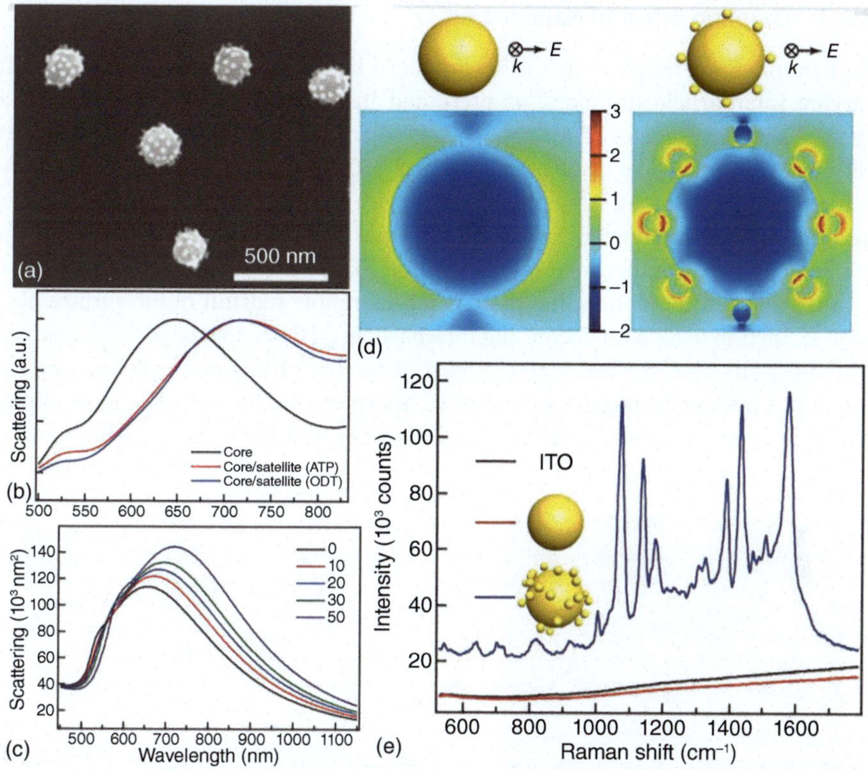

Figure 2.11 (a) SEM images of the core/satellite nanostructures on the Si substrate and at high magnification. (b) Normalized dark-field scattering spectra of the Au core NSs and core/satellite nanostructures deposited on ITO glass substrates. (c) Calculated scattering spectra of the core/satellite nanostructures with the number of the satellite NSs to be 0, 10, 20, 30, and 50. (d, e) Local electric field intensity enhancements and SERS spectra of the ATP-assembled core/satellite nanostructures. Source: Reproduced with permission from Ruan et al. [39]; © 2014, John Wiley and Sons Inc.

core nanosphere with ATP molecules. Similarly, the scattering peak redshifts to 716 nm for the ODT-assembled core/satellite nanostructures, as compared in Figure 2.11b. The redshifts are attributed to the plasmon coupling between the Au core nanospheres and satellite nanospheres. They also found that the ATP-linked core/satellite nanostructures with 18 satellites exhibit a larger redshift than the ODT-linked nanostructures with 24 satellites. This discrepancy between the experiments and calculations can be ascribed to the fact that the gap distance brought by ATP molecules with a shorter chain is slightly smaller than that caused by ODT molecules with a longer chain (Figure 2.11c). When the satellite number is gradually increased from 0 to 50, the scattering peak redshifts gradually. The assembled core/satellite nanostructures exhibit strongly enhanced Raman signals. Compared with the bare Au nanosphere cores, the large SERS enhancement factor of the core/satellite nanostructures and the richness of hot spots, as predicted by FDTD simulations (see Figure 2.11d,e).

2.6.2 Coupling Effect of Nanoprisms

Au nanoprisms arrangements can be produced by electron beam lithography with varying interparticle distances, as presented in Figure 2.12 [40]. Figure 2.12a,b showed the SEM images of Au nanoprisms (diameter 150 nm, height 17 nm) with varying 1D interparticle distances of 300 and 150 nm. Figure 2.12c,d presented the extinction spectra of Au nanoprisms pairs with the interparticle center-to-center distances as the parameter. Figure 2.12c shows the polarization direction of the incident light is parallel to the long particle pair axis, whereas Figure 2.12d is orthogonal to it. It is found that there is a remarkable redshift of the surface plasmon extinction peak with decreasing interparticle distance for the parallel case and a smaller, but nevertheless distinct blueshift for the orthogonal polarization. The resonance peak shift suggests a qualitative interpretation by a simple dipole–dipole interaction model known from spring oscillator or molecular systems, as shown in

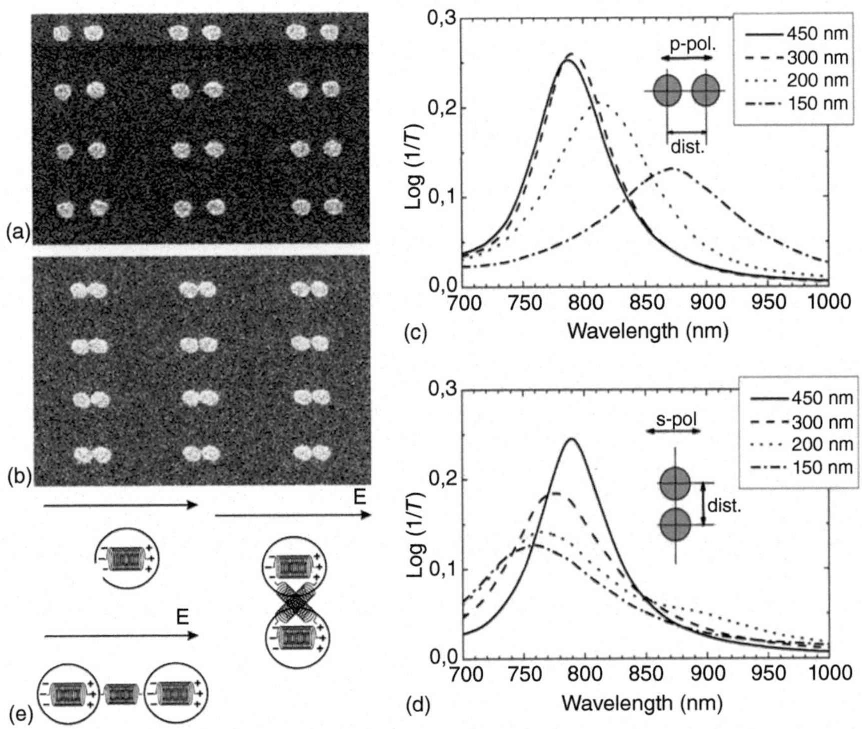

Figure 2.12 (a, b) SEM images of Au nanoprisms (diameter 150 nm, height 17 nm) with varying 1D interparticle distances of 300 and 150 nm. Extinction spectra of Au nanoprisms pairs with the interparticle center-to-center distances as the parameter. The orthogonal particle separation is kept constant (as shown in (a)). The polarization direction of the exciting light is (c) parallel to the long particle pair axis and (d) orthogonal to it. (e) Sketch to illustrate the electromagnetic interaction between closely spaced nanoparticles, an isolated particle, and a pair of close particles with the polarization of the exciting field parallel to the long particle pair axis and orthogonal to the long particle pair axis. Source: Ruan et al. [40]/Reproduced with permission from © 2014, ELSEVIER.

Figure 2.12e. Due to the confinement of the plasma electrons in the nanoparticle, the electric field of the irradiated light creates surface charges, which feel repulsive forces. It matches the condition for the system to become resonant.

However, when another particle is nearby, upon polarization additional forces act on both particles. If the driving field is parallel to the long-particle pair axis, this effect results in a weakening of the repulsive forces for the surface charges. The positive charge of the left particle faces the corresponding negative charge of the right one. Due to the attractive forces between these two charges, the repulsive forces within each particle are weakened, leading to a correspondingly lower resonance frequency. In contrast, when the driving field is normal to the long particle pair axis the charge distributions of both particles act cooperatively to enhance the repulsive action in both particles, thus increasing the resonance frequency. With increasing the particle size, the restoring force of the electronic vibration becomes weak, and the resonance frequency decreases, resulting in the SPR red shift in the spectra. This spring harmonic oscillator model can be effectively applied to qualitatively describe the SPR frequency shifting for both monomer NPs and many kinds of nanostructure assemblies.

Walsh and Negro [41] designed periodic planar arrays of Au nanocylinders and systematically investigated the effects of Fano-type coupling between long-range photonic resonances and localized surface plasmons on the second harmonic generation from periodic arrays of Au monomer and dimer geometries. Figure 2.13 demonstrated that the second harmonic generation (SHG) is enhanced when coupled photonic–plasmonic resonances of the arrays are excited at the fundamental pump wavelength, thus boosting the intensity of the electromagnetic near fields. With the excitation polarized along (red) and perpendicular to (blue) the dimer axis, and the polarization along the dimer axis, the array produces more than four times the second harmonic generation signal compared to perpendicular polarization at their respective peaks. The enhanced electromagnetic near fields in the gap were supported by FDTD numerical simulations, as shown in Figure 2.13b. And the intensity enhancement is concentrated in the dimer gap region for polarization along the axis.

2.6.3 Coupling Effect of Nanocubes

As it is known that the radii of the corners and edges have a significant effect on the optical extinction, plasmonic Au NPs with sharp ends and vertices, such as Au NBPs, have attracted much attention. Another nanostructure with rich, sharp corners is NC and nanocuboid. Compared with Au, Ag is more readily crystalized as cube-shaped nanostructures. The successful achievement was made by Cortie et al. [19]. They presented dipolar and higher-order plasmon resonances of Ag NCs and Ag nanocuboids with dipolar LSPR charges accumulated at corner sites.

Compared with Ag NCs, only a few literature was reported about the Au NCs, especially for the experimental work [42, 43]. According to Jeon et al. [42], it was evaluated the influence that the shape and vertices of Au NCs had on the sensitivity of homogeneous LSPR inflection points to changes in the RI. This was done by comparing Au NCs to Au nanospheres (Au NSs) that were of comparable size.

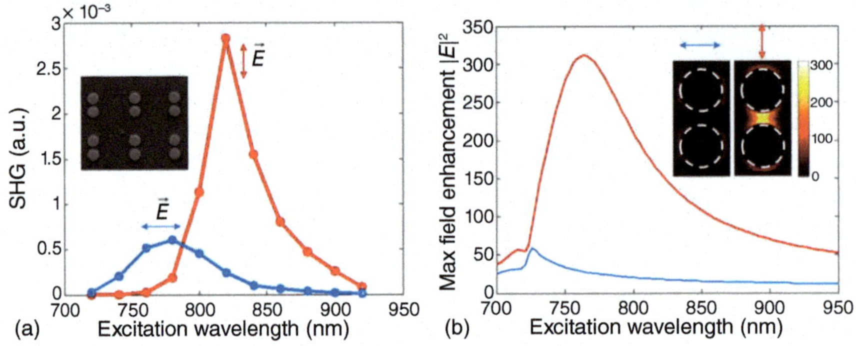

Figure 2.13 (a) Enhanced SHG spectra of diffractively coupled periodic array of Au nanocylinder dimers (SEM shown in the inset) with $D = 150$ nm and $\Lambda = 500$ nm for incident polarization along (red) and perpendicular to (blue) the dimer axis. (b) FDTD calculation of the maximum near-field intensity enhancement for the same array as in panel (a) for polarization along (red) and perpendicular to (blue) the dimer axis. Near-field maps of the intensity are shown in the inset for the two polarizations (indicated by the arrows) at the respective spectral maxima. Source: Reproduced with permission from Walsh and Negro [41]; © 2013, American Chemical Society.

According to the findings, watching the frequency shifts of the SPR peaks does not provide a better level of RI sensitivity than tracking the homogenous SPR inflection points. This is the case for both Au NCs and Au NSs, as shown in Figure 2.14a–c. In addition, in the homogeneous SPR inflection points, the RI sensitivity of single Au NCs with vertices was shown to be greater than that of single Au NSs of a size

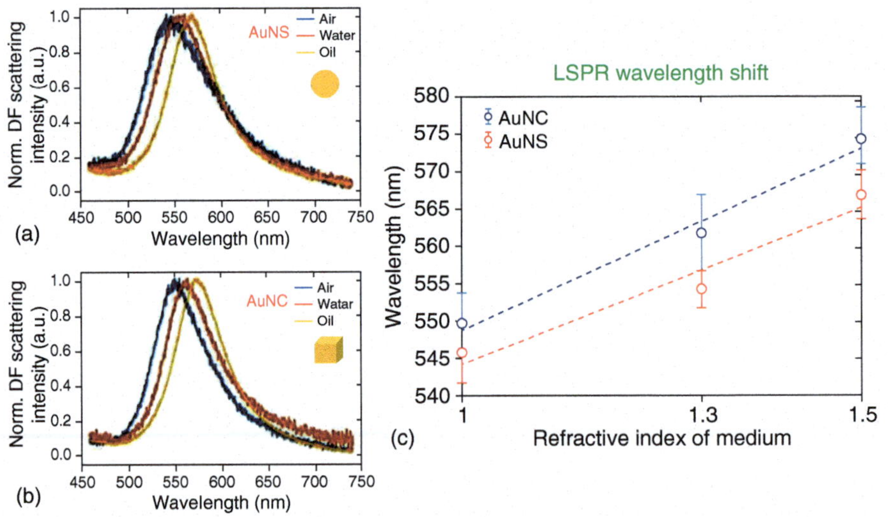

Figure 2.14 Change in the LSPR scattering spectra of single AuNS (a) and AuNC (b) in the different local RI media: air, water, and oil. (c) LSPR wavelength shifts for AuNS (red) and AuNC (blue) as a function of the local RI of medium. Source: Reproduced with permission from Jeon et al. [42]; © 2019, Springer Nature / Public Domain CC BY 4.0.

comparable to their own. As a result, they offered a more in-depth understanding of the RI sensitivity of homogeneous SPR inflection points of Au NCs with vertices for the purpose of their use in SPR-based biosensors.

They employed dark-field microscopy in their prior work to find a novel plasmon resonance effect for a single Ag NC. In this effect, the plasmon line form has two separate peaks when the particles are positioned on a glass substrate. This phenomenon can be seen for a single Ag NC. The dependence of the resonance on the size and shape of the NC was characterized, and it was discovered that the blue peak has a higher FOM for chemical sensing applications than other particle shapes that have been studied in the past. The correctness of spectrum assignments was validated by a comparison of the observed data with FDTD electrodynamics simulations, which will be present in the following part.

M. Pellarin et al. reported Fano transparency in rounded Au and Ag NC dimers induced by gap plasmon coupling [43]. The agreement between experiments and simulations is quite satisfactory both qualitatively (especially for the appearance of a Fano dip) and quantitatively. With decreasing the interparticle distance or increasing values of the edge rounding, the Fano dip is pronounced in the extinction spectra. They suggested that dimers of Au and Ag NCs form model plasmonic systems where the localized plasmon resonances sustained by each particle not only hybridize but also coexist with excitations of a different nature: surface plasmon polaritons confined within the Fabry–Perot cavity delimited by facing cube surfaces. The strong coupling between one of these highly localized modes and the highly radiating longitudinal dipolar plasmon of the dimer is responsible for the formation of a Fano resonance profile and the opening of a spectral window of anomalous transparency for the exciting light.

2.6.4 Coupling Effect of Nanorods

Among the 1D nanostructures, Au NRs have become one of the burgeoning nanostructures in the past two decades due to their anisotropic shape and adjustable plasmonic properties. More and more functional structures related to Au NRs also achieved unexpected plasmonic properties and improved performance. Au NRs-based porous nanostructures will be described in Chapter 4. Here, the coupling effect of Au NRs is described briefly.

Figure 2.15 shows two typical assemblies of Au NRs with the presence of Cys and HS-PEG-SH. The optical spectra indicated the SPR redshift and blueshift of Au NRs. A more detailed description can be obtained in Ref. [44]. Usually, TEM images of Au NRs assembly with end-to-end or side-by-side modes. It may be difficult to determine whether a particular assembly truly occurred in solution or it arose from the solvent evaporation during the TEM sample preparation, since TEM technique generally requires a dried sample. Ke et al. found the red color of Au NRs dispersed in water changed immediately with the addition of strong polar ether tetrahydrofuran (THF), as shown in Figure 2.16. They presented the color change evolution of the Au NRs colloids with different amounts of THF. Compared with the experimental and simulated optical spectra of Au NRs, it is indicated that the side-by-side assembly

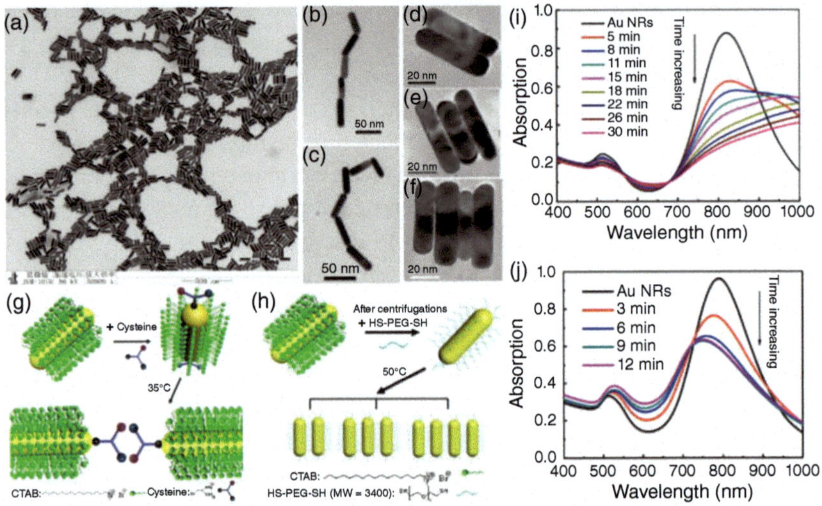

Figure 2.15 TEM images of mono-dispersed Au NRs (a) for end-to-end assembly (b, c) and side-by-side assembly of dimer (d), trimer (e), tetramer (f). (g, h) Schematic diagrams for the Cys-induced end-to-end assembly and HS-PEG-SH-induced side-by-side assembly of Au NRs. (i, j) Absorption spectra with time for the assembly process of Au NRs. Source: Liu et al. [44]/Reproduced with permission from © 2014, American Institute of Physics.

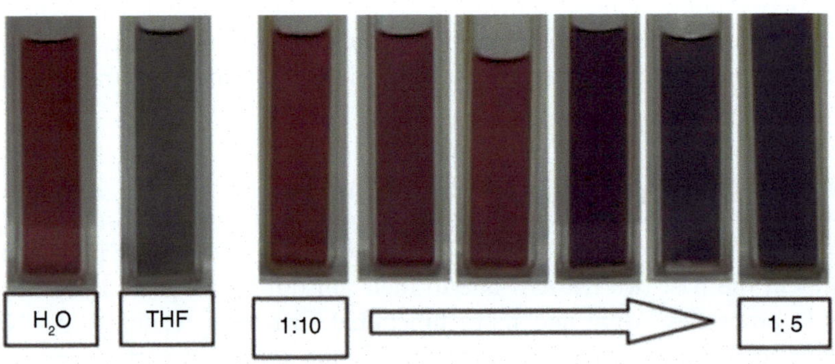

Figure 2.16 Pictures for Au NRs dispersed in H_2O and tetrahydrofuran (THF), and Au NRs dispersed in THF/H_2O mixture with increasing THF. Source: Ke et al. [45]/Reproduced with permission from © 2013, Royal Society of Chemistry.

process of the Au NRs with THF is driven by van der Waals attraction between the Au NRs and the steric repulsion at the ends of nanorods [45].

The beautiful colors of Au colloids have attracted tremendous attention due to their historical applications in art. As shown in Figure 2.17, it can be seen that there are different colors even for a certain nanostructure with different assembled configurations. The reversible assembly and disassembly of Au NRs in both end-to-end and side-by-side fashions were systematically studied by Wang's group [46, 47]. They demonstrated pH-controlled reversible assembly and disassembly of Au NRs in aqueous solutions using various assembling agents and found that the

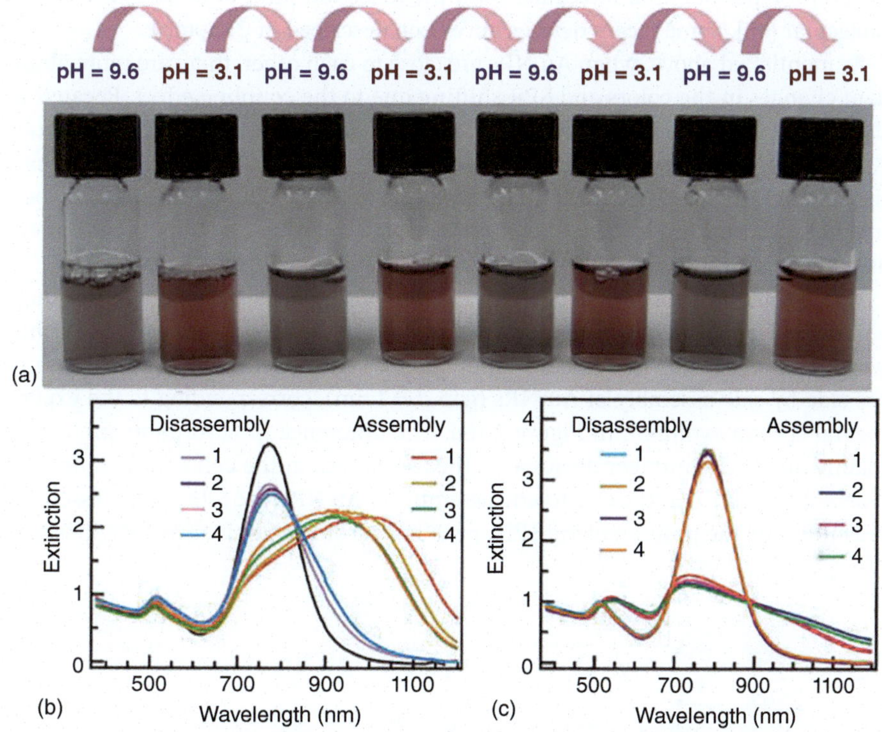

Figure 2.17 (a) Digital picture of four cycles of end-to-end assembly and disassembly of Au NRs induced by MPA. (b) Extinction spectra of Au NRs colloids recorded after four cycles of assembly and disassembly. The assembly was induced by adjusting the pH of the solution to 9.6 and the disassembly was obtained by adjusting the pH to 3.1. The black spectrum is the original Au NRs colloid. (c) Extinction spectra of Au NRs after four cycles of side-by-side assembly and disassembly. Source: Sun et al. [46]/Reproduced with permission from © 2008, JOHN WILEY & SONS, INC.

assembly occurs at higher pH and disassembly occurs at lower pH in the presence of 3-mercaptopropionic acid (MPA). The assembly is induced at lower pH and the disassembly is induced at higher pH with the agent of cysteine. The assembled and disassembled nanorods exhibit clearly different colors in a highly reproducible manner, which further indicates the robustness of this pH-controlled reversible assembly process.

In addition to the SPR sensitivities of Au NRs to aspect ratio and RI, many research studies proved that Au NRs can couple firmly with the adjacent units in different configurations, depending on the modified molecular. Using the SPR blue or red shifting of Au NRs dispersion, the tendency of surfactants coverage on Au NRs surface or ends can be inferred. More importantly, single perfect Au NR has both mirror and inversion symmetry and therefore does not bear any chiral optical properties. However, different Au NRs-based heterostructures and assemblies have been proposed to feature strong chiroptical activities. This effect has been used in the sensing of certain types of chiral molecules with the detection limit to be as low as single molecular [48]. Some recent devices have been described in Chapter 1.

As mentioned above, when Au NRs are close to each other, there are more abundant changes in the colors and SPR shifting due to the coupling effect. Because of the complex self-assembly process of Au NRs in a liquid phase, there will be various Au NRs assembly structures. For this case, FDTD simulation can be used to forecast the assembly mode of nanorods in the experiment. Results in Figure 2.18 are experimental spectra, TEM images, and FDTD simulated spectra for Au NRs of different assemblies with increasing the number of nanorods [32].

For the side-by-side assembly of Au NRs, the SPR_L peak blueshifts to the short wavelength region, and the transverse resonant peak appears to slightly redshift with the increase of the number of monomers. Figure 2.18b shows the spectra of the side-by-side assembly of Au NRs (gap $d = 1$ nm), corresponding to the experimental spectra evolution in Figure 2.18a. The difference in absorption strength is mainly due to the number decrease in the assemblies in the colloidal sample with assembly process. For the end-to-end assembly of Au NRs, the SPR_L of the assembly redshifts, and the position of the SPR_T peak has no obvious shift. The SPR_L position

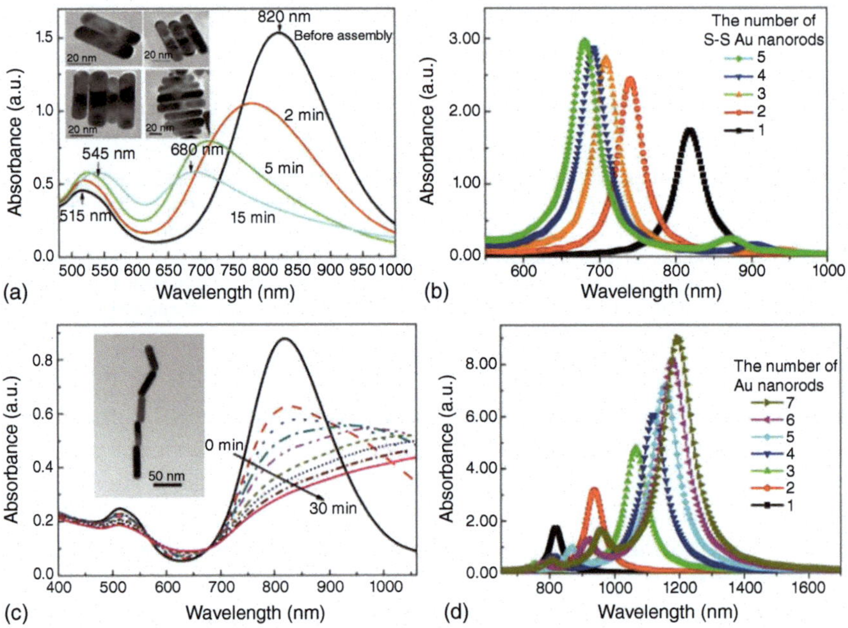

Figure 2.18 (a) Experimental spectra for side-by-side assembly of Au NRs with increasing time. Insets show the transmission electron microscopy (TEM) images of the different numbers of side-by-side oriented Au NRs. (b) The calculated SPR_L absorption spectra for side-by-side assembled Au NRs with increasing rod number. (c) Experimental spectra for end-to-end assembly of Au NRs with increasing time, inset shows the TEM images of oriented Au NRs. (d) The calculated SPR_L absorption spectra for the incident polarized light parallel to the end-to-end oriented Au NRs with different numbers. Source: Figure modified with permission from Ref. [32]; © 2014, Editorial office of Acta Physico-Chimica Sinica.

and shift evolution of the FDTD simulated spectra agree well with the experimental results. Additionally, FDTD results show that the spacing between nanorods also affects the position and intensity of SPR_L. For these two cases, the coupling effect decreases with the increase of the spacing between Au NRs, and the optical spectra are eventually similar to those of highly dispersed Au nanosystems.

2.6.5 Coupling Effect of Complex Structure

In addition to the nanorod trimers with different configurations, complex structures of rod-sphere-rod multimers were studied by Kumar et al., through experimental synthesis and calculation [49]. The modeled coupled plasmon resonance spectra are in accordance with the experimental results, which demonstrate that horizontal polarization can induce an antibonding mode in the high-energy region. For the dimer, excitation of the antibonding mode leads to electric field reduction due to repulsive coupling, and the magnitude of electric field is weaker than that for the corresponding bonding mode. However, when a sphere is introduced between the two rods, the magnitude of the electric field enhancement for the antibonding mode remains close to that of the bonding mode, as shown in Figure 2.19. More significantly, the "hot-spots" for the trimer are stronger than those for the dimer in the antibonding mode case. Such strong surface plasmon coupling was also observed from core–satellite nanoassemblies, which appear significantly red shifted from the SPR of the constituting NPs [50, 51]. More recently, a new type of coupling structure is nanoframe (or nanocage) with nanocavity, which will be described in Chapter 7.

For understanding the bonding and antibonding mode, Naomi J. Halas and Peter Nordlander proposed the concept of plasmon hybridization, which greatly simplifies the understanding of local SPR coupling [52]. The end-to-end model is similar to the σ hybridization of p orbital electrons in atom. While the side-by-side model is similar to the π hybridization of p orbital electrons in the outer layer of atoms. For each coupled nanostructure, it will give two new energy level orbits, namely, broad bright bonding state and dark narrow antibonding state. Compared with the bonding orbital, the antibonding orbitals with high energy are difficult to be excited by the light due to zero total dipole moment. The modes with zero total moment or weak dipole coupling are called sub-radiation, and these modes are usually narrow bands with long lifetime. When the coupled systems are more complex, the hybrid orbital model is still applicable.

Lorentzian formula was regarded as the fundamental line shape of a resonance for many years. Usually, the presence of several independent resonances is simply considered to be the sum of the intensities of the individual resonances. Thus, the interference effects owing to interacting resonances are not present in most typical resonance phenomena. When a certain asymmetry is introduced into the coupled systems, photonic crystals, and metamaterials, the interference effect between different plasmonic resonance modes can also be observed. Under this case, Fano resonance effect caused by destructive interference may be observed. This Fano effect can strongly reduce the absorption and scattering of the nanostructures, leading to the well-known phenomenon of electromagnetic transparency.

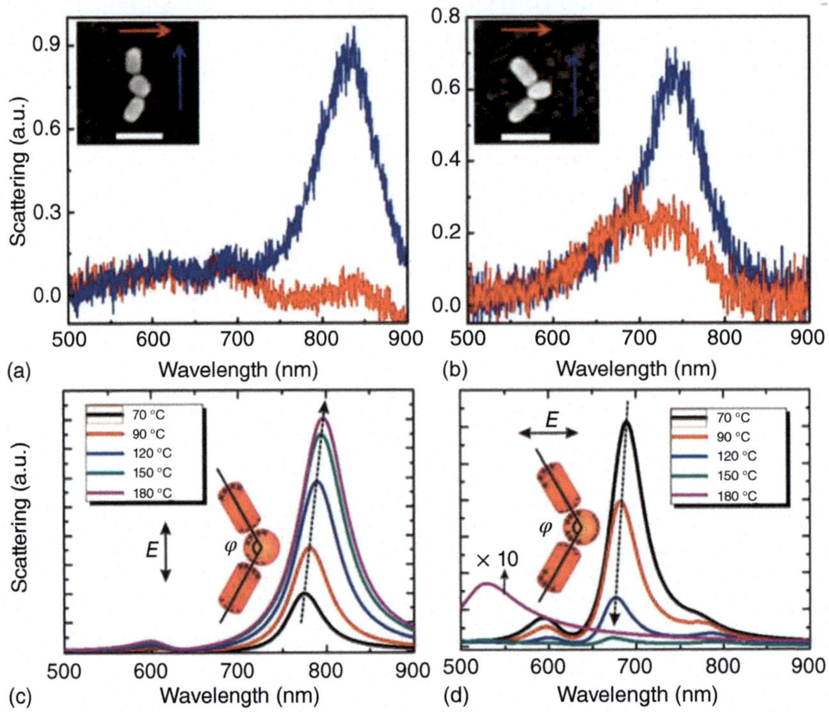

Figure 2.19 (a) Scattering spectra of nanorod-nanoparticle bend trimers having an angle of (a) ~165°, (b) ~90°. Inset shows the relevant SEM images of the nano-assemblies. Scale bar = 100 nm. The vertical and the horizontal polarization directions, and the spectra corresponding to these polarization directions are plotted in the respective colors. (c) Calculated scattering spectra (using finite element methods) and schematic views of the surface charge distributions associated with bonding modes with trimer angles. The gaps between the particles are fixed at 0.5 nm for modeling. (d) Peak wavelength of the bonding mode plotted as a function of the angle. Source: Reproduced with permission from Kumar et al. [49]; © 2013, Royal Society of Chemistry.

For more complex structures with broken symmetry, Fano resonances can arise due to the interaction of narrow dark modes with broad bright modes. A rich resonance spectrum emerges due to the hybridization of the plasmons of individual nanostructure components into bonding and antibonding combinations of plasmon modes. Verellen and coworkers presented an experimental investigation of Fano interference phenomena in individual plasmonic nanocomposites [53], including dolmens and disk-near-ring. Fano resonances are induced by the coupling of dark quadrupolar and higher-order modes with bright dipolar mode. From the large dolmen, the Fano interferences are clearly visible, respectively, at wavelengths of 850 and 1100 nm in the experimentally obtained extinction spectrum for polarization perpendicular to the pair axis. This result is in good agreement with the FDTD simulated spectra. From the surface charge distributions, it is apparent that the monomer dipole and quadrupole modes are mixed with each other, which makes the identification of the individual modes somewhat more difficult. Furthermore, it is found that the Fano line shapes depend sensitively on the dielectric gap between the individual resonators.

Since the Fano resonance was realized in plasmonic materials and metamaterials, this effect has been of great interest, as summarized by N.J. Halas [54]. The unique properties associated with this lineshape are highly promising for potential applications. The design of Fano resonances into optical systems may give rise to ultrasensitive media for chemical or biosensing, and new types of optical switches, interferometers, nanoantennas, displays, and other devices.

In addition to the nanorods multimers with side-by-side and end-to-end configurations, complex structures of rod-sphere-rod multimers were studied by Kumar et al. through experimental synthesis and calculation [49]. The modeled coupling plasmon resonance spectra are in accordance with the experimental results, which demonstrate that horizontal polarization can induce an antibonding mode in the high-energy region. For the dimer, excitation of the antibonding mode leads to electric field reduction due to repulsive coupling, and the magnitude of electric field is weaker than that for the corresponding bonding mode.

For the plasmon coupling between one Au NR and a small Au nanosphere heterodimer, the spatial perturbation of the nanosphere leads to distinct spectral changes in the heterodimer, as shown in Figure 2.20 [55]. When the nanosphere is located at the end of the nanorod, two scattering peaks are detected, with the higher-energy one much weaker than the lower-energy one. The two scattering peaks of the heterodimer arise from the plasmon coupling between the nanorod and nanosphere. When the nanosphere is moved from the end through the shoulder to the side of the nanorod at a fixed gap distance, the intensity of the higher-energy peak gets relatively stronger, and the two peaks become closer spectrally. When the nanosphere is at the exact central side of the nanorod, the two scattering peaks coalesce into one.

Furthermore, when a sphere is introduced between the two rods, the magnitude of the electric field enhancement for the antibonding mode remains close to that of the bonding mode. More significantly, the "hot-spots" for the trimer are stronger than those for the dimer in the antibonding mode case. Such strong surface plasmon coupling was also observed from core–satellite nanoassemblies, which appear significantly red shifted from the SPR of the constituting NPs [51].

Similar to the dimer of Au NR and a small Au NS, Au NC, and Au NS dimers are also fascinating for their controlled and enhanced plasmonic properties, as reported by D. Lee and S. Yoon [56]. It is found that plasmon coupling and SERS from nanogaps can be defined by different coupling shapes (with gap around the vertex or edge of Au NC). Upon resonant excitation wavelengths, the electric field is strongly localized in the nanogaps between the NPs. Strong plasmonic interactions between the Au NCs and Au NSs produce a new longitudinal plasmon coupling band at 790 nm, significantly red shifted from the SPR band of the isolated Au NCs or Au NSs in the range of 530–580 nm. The resonance plasmon coupling wavelength for the Au NC–Au NS dimers is also in the near-IR region, which is perfectly suited for many biomedical SERS applications.

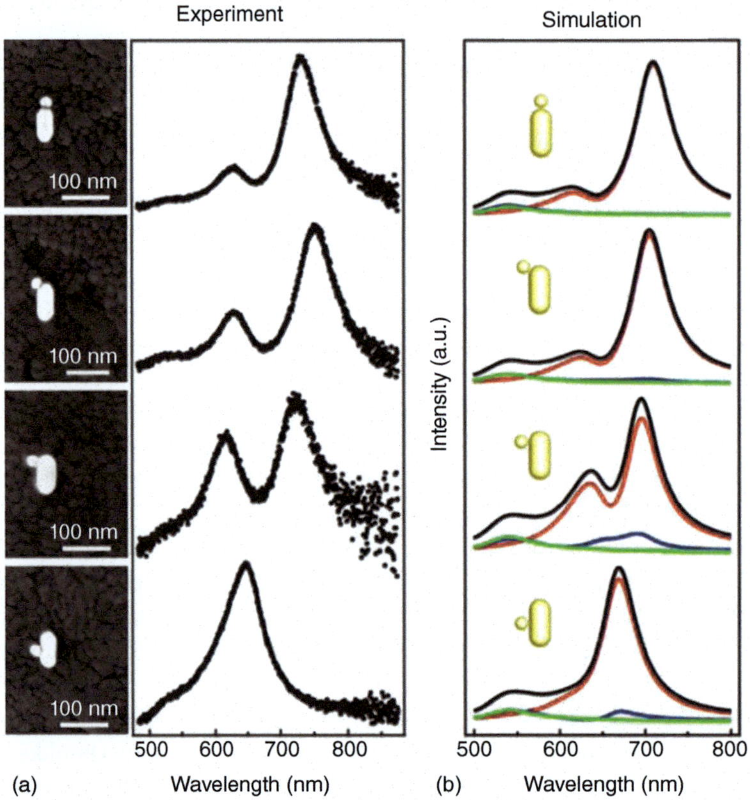

Figure 2.20 (a) Experimental and (b) simulated results for the plasmon coupling between one Au NR and a small Au NS located at different positions around the nanorods. Source: Reproduced with permission from Shao et al. [55]; © 2012, American Chemical Society.

For the combination of Au and Ag nanostructures, Ag NC and Au NSs core@satellites are fascinating for their controlled and enhanced plasmonic properties, as reported by G.W. Meng group [57]. Based on the fabricated Ag NCs (edge length ~100 nm) and Au NSs (diameter ~13 nm), they used C3-thiol linkers and electrostatic repulsion to achieve Ag NCs–Au NSs core@satellites, as indicated by Figure 2.21a. Upon resonant excitation wavelengths, the electric field is strongly localized in the nanogaps between the NPs. For a core@satellites, hot spots are generated at the bounding sites between the Au NSs satellites and the Ag-NCs core, as schematically illustrated by Figure 2.21b. The electric field enhancement at the "hottest" regions of the satellite structure of this model was found to be nearly two orders larger compared with that at the surface of Ag-NC core. This electric field enhancement depends largely on the distribution of the Au nanospheres and the packed amount, as shown in Figure 2.21c.

Then AFM-Raman correlated measurement was used to image the hot spots and study the enhancement factor of a singular nanostructure. For singular Ag NC (point 3 in Figure 2.21d), the signal is very weak but observable. Once the Ag NCs

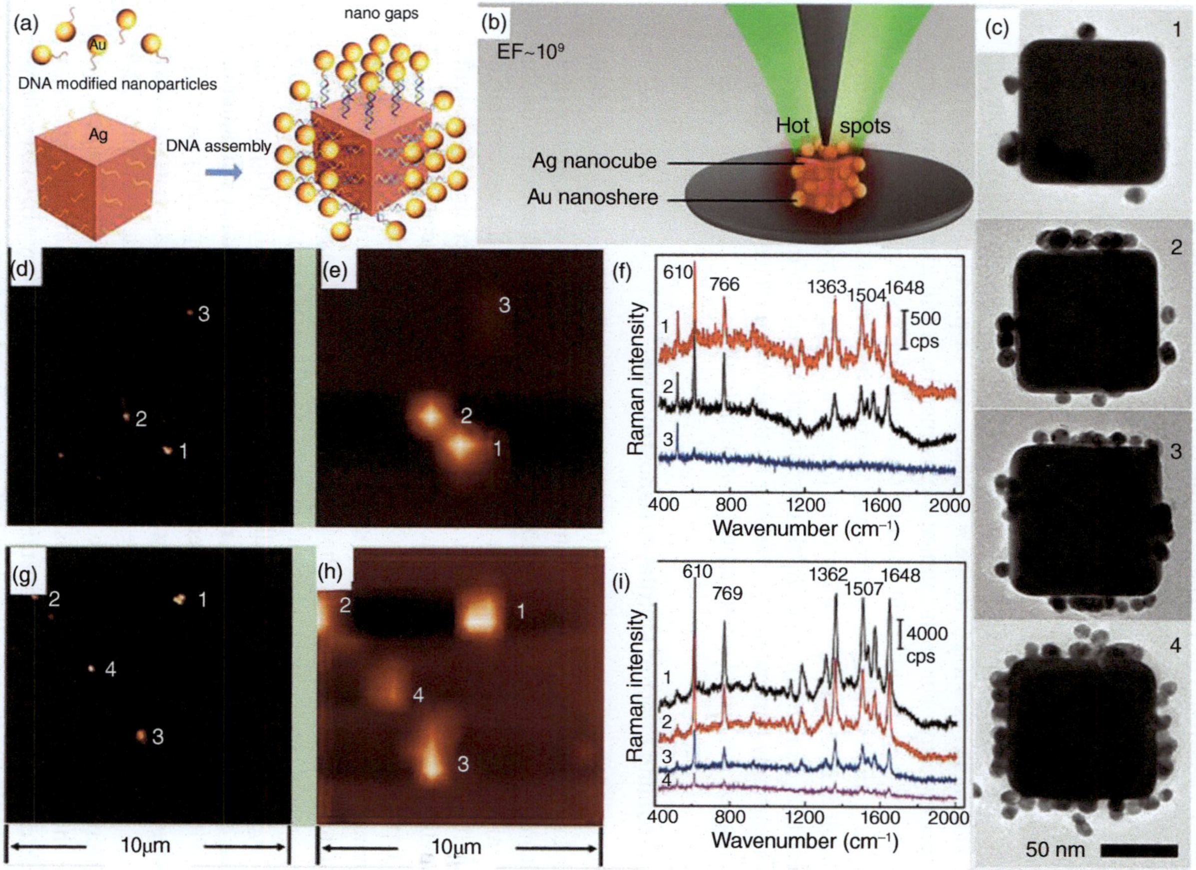

Figure 2.21 (a) Schematic illustration of DNA-assisted assembly of satellites using Au NSs and Ag NCs for the creation of abundant hot spots. (b) Schematic overview of the atomic force microscope-Raman experimental setup for the measurement of a single Ag NC@Au NSs surface-enhanced Raman scattering platform. (c) Typical TEM images of 13 nm Au NSs hybridized onto Ag NCs. AFM-correlated microscopic Raman measurements and *in-situ* Raman imaging (using rhodamine 6G as probe molecules) of Ag NCs (d, e) and Ag NC@Au NSs core@satellites (g–h). Snatched Raman spectra of Ag NCs (f) and core@satellites (i) at the marked hot spots from the corresponding Raman imaging. The laser line was 532 nm. Source: Huang et al. [57]/Reproduced with permission from © 2016, John Wiley & Sons.

are coupled as trimer (point 1 in Figure 2.21g), strong electromagnetic coupling is generated between neighboring Ag NCs. For the Ag-NC@Au-NSs, the size of the core@satellites increases slightly, as observed from the AFM images. Almost every hot area from the Raman imaging (Figure 2.21h,i) can yield high signal-to-noise ratio Raman spectrum. This can be attributed to the high-density hot spots induced by the strong plasmonic coupling between the neighboring Ag NCs, and more importantly, between the small Au NS satellites and Ag NC core.

For understanding the coupling effect in the dimmer and core–satellite structures, N.J. Halas and P. Nordlander proposed the concept of plasmon hybridization [58, 59], which greatly simplifies the understanding of local SPR coupling. The end-to-end model is similar to the σ hybridization of p orbital electrons in atom. While the side-by-side model is similar to the π hybridization of p orbital electrons in the outer layer of atoms. For each coupled nanostructure, it will give two new energy level orbits, namely, broad bright bonding state and dark narrow antibonding state. Compared with the bonding orbital, the antibonding orbitals with high energy are difficult to be excited by the light due to zero total dipole moment. The modes with zero total moment or weak dipole coupling are called sub-radiation, and these modes are usually narrow bands with long lifetime. When the coupled systems are more complex, the hybrid orbital model is still applicable.

2.6.6 Coupling Effect of Multilayer Nanostructure

Different from the above coupling nanostructures with external nanogap, core–shell, or multilayer nanostructures with inner cavity are also significant in plasmonics. The resonant frequencies of the multilayer nanostructure are determined by the hybridization of the plasmons of the inner and outer nanoshells. The coupling of core–shell and cavity nanostructures also optimize the resonance line shape with high-quality factor of optical resonance and localized field enhancement.

One typical example is the Au BRC consisting of Au NR enveloped by nanobox with nanocavity [36]. This Au BRC can be fabricated through a two-step wet-chemical method, including the overgrowth of Ag on the surface of Au NR and subsequent replacement reaction between $HAuCl_4$ and Au@Ag NR. When the excited light is perpendicular or parallel to the long axis of the Au BRC, plasmonic coupling of the core and shell would result in two splitting resonances. In addition to the SPR splitting, the strong field enhancement in the nanogap, nanocavity, and sharp edges of this coupling nanostructure will be of great importance, as simulated in Figure 2.22 for the surface charge distribution and electric-field enhancement of different parts. Experimentally, the morphological shape and field enhancement within the gap region were established by electron tomography in high-angle annular dark-field scanning transmission electron microscopy (HAADF-STEM) mode. Mapping of the near-field by EELS allows quantification of the spatial field enhancement of plasmonic modes (Note: the STEM and the EELS will be present in Chapter 3). At low energy losses (1.5–2.0 eV, 620–827 nm), the longitudinal dipolar mode (L1) of the nanorattle is excited. Energy losses ranging from 2.5 to 3.0 eV (413–496 nm) identify the first-order transversal cavity mode (T1). Here, the electric

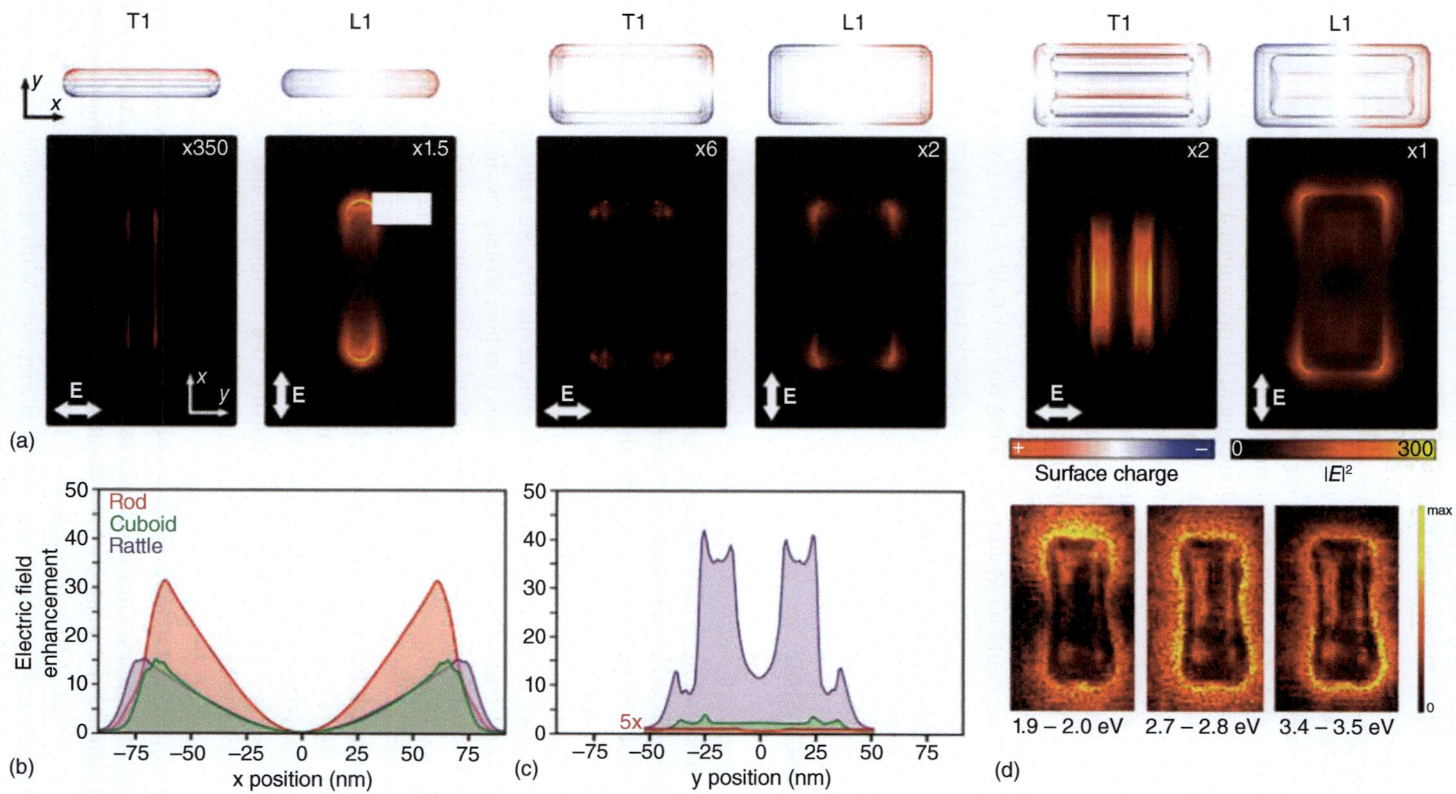

Figure 2.22 (a) Surface charge distribution and electric-field enhancement of the different plasmonic nanoparticles. The plasmonic modes were excited perpendicular and parallel to the long axes of the nanoparticles at the wavelength of the corresponding plasmonic resonance. Integrated electric field enhancement for longitudinal (b) and transversal excitation (c) (integrated in y–z and x–z, respectively, and normalized to the integrated distance and area). (d) STEM-EELS mappings at distinct energies show electric field enhancement inside of the cavity. Source: Reproduced with permission from Schnepf et al. [36]; © 2017, Royal Society of Chemistry.

field is homogeneously distributed inside the cavity. At energy losses between 3.0 and 3.5 eV (354–413 nm), the higher-order transversal mode is excited, showing a weaker field enhancement inside the cavity. The strong field enhancement is very useful for single-molecule fluorescence under one and two photoexcitation.

In this Yolk–shell nanostructure, the local fields coherent enhancement in the nanogap or interior cavity. Au NRs, for example, have a rather weak electric field enhancement along the cylindrical-shaped rod for transversal excitation. For longitudinal excitation, the field enhancement is pronounced at the tips, as can be also identified from the corresponding distribution of surface charges. This field concentration is increased for Au@Ag cuboids as a result of the sharp edges and inclusion of active Ag as a shell material. For axisymmetric nanorattles, a nanocavity is present and a confined electric field is observed within the dielectric cavity volume (transversal excitation). Surface charge calculations reveal the nature of the nanocavity mode. In these cross-section plots, two properties of the plasmonic modes could be identified: (i) the highest field enhancement occurs at the metallic–dielectric interface and (ii) the electric field decays exponentially into the dielectric material and environment. The field enhancement is most significant at the sharp corners and gap regions.

2.7 Conclusion

The basic knowledge of complex dielectric function of plasmonic NPs was introduced. Corresponding to experimental results, the complex dielectric functions of Au–Ag alloy nanostructures were modeled using an improved DCPM method. It is confirmed that DCPM can be used to obtain the reliable dielectric functions of other alloys (such as Au–Cu, Cu–Ag, and Cu–Al) for the prediction of optical properties. Then Mie theory, Gans function, DDA, and FDTD were applied to simulate the optical spectra of different NPs. When the plasmonic NPs are aggregated to form a dimer, chains, core–shell, cavity, and other complex structures, more and more unique physical phenomena appeared, such as enhancement of electromagnetic field in the nanogap, new SPR modes associated with coupling effect. In addition to the SNOM technique, such phenomena are generally difficult to be detected experimentally through far-field spectroscopy. However, as one vigorous and adaptable technique, the FDTD can be used to calculate the optical property of the coupled metal nanostructures. Optical properties of different assemblies of plasmonic nanostructures were described by comparing the simulated and experimental results, verifying the effectiveness and correctness of the FDTD method.

Metallic nanostructures are excellent materials for plasmonic applications, particularly noble metals of Au and Ag, because of their large negative values of the real part of permittivity. They have been used frequently in plasmonics due to their chemical stability and relatively low Ohmic loss in the Vis–NIR wavelength region. Colloids of various nanostructures with defined shapes and compositions can be synthesized through improved wet chemical methods. With the assistance of surfactants, colloidal methods have also demonstrated their flexibility of incorporating different plasmonic NPs into defined hybrid systems with controlled geometries. Therefore, colloidal chemical methods allow researchers to fully examine how the factors affect the performance of the hybrid structure, such as gap distance, the scattering cross-section, and also excitation wavelength. Comparing the simulation and experimental results, it is clear that single-NP imaging and theoretical modeling can guide the synthesis strategies, and forecast the performance of the NP.

References

1 Faraday, M. (1857). The Bakerian Lecture: experimental relations of gold (and other metals) to light. *Philos. Trans. R. Soc. Lond.* 147: 145–181.

2 Mie, G. (1908). Beiträge Zur Optik Trüber Medien, Speziell Kolloidaler Metallösungen. *Ann. Phys.* 330 (3): 377–445.

3 Quinten, M. (2011). *Optical Properties of Nanoparticle Systems: Mie and Beyond.* Wiley-VCH.

4 Kreibig, U. and Vollmer, M. (1995). Optical properties of metal clusters. In: *Springer Series in Materials Science*, vol. 25. Berlin: Springer.

5 Sharma, N., Joy, A., Mishra, A.K., and Verma, R.M. (2015). Fuchs Sondheimer-Drude Lorentz model and Drude model in the study of SPR based optical sensors: a theoretical study. *Opt. Commun.* 357: 120–126.

6 Bohren, C.F. and Huffman, D.R. (1998). *Absorption and Scattering of Light by Small Particles.* New York: Wiley.

7 Shahbazyan, T.V. and Stockman, M.I. (ed.) (2013). *Plasmonics: Theory and Applications.* Dordrecht, Heidelberg, New York, London: Springer, ISBN 978-94-007-7804-7 (eBook).

8 Garnett, M.J.C. (1904). Colours in metal glasses and metal films. *Phil. Trans. R. Soc. A* 203: 385–420.

9 Ghosh, S.K. and Pal, T. (2007). Interparticle coupling effect on SPR of Au nanoparticles: from theory to applications. *Chem. Rev.* 107 (11): 4797–4862.

10 Etchegoin, P.G., Ru, E.C., and Meyer, M. (2006). An analytic model for the optical properties of gold. *J. Chem. Phys.* 125 (16): 164705.

11 Lu, J.Y. and Chang, Y.H. (2010). Implementation of an efficient dielectric function into the finite difference time domain method for simulating the coupling between localized surface plasmons of nanostructures. *Superlattices Microstruct.* 47 (1): 60–65.

12 Rioux, D., Vallières, S., and Besner, S. (2014). An analytic model for the dielectric function of Au, Ag, and their alloys. *Adv. Opt. Mater.* 2 (2): 176–182.

13 Wang, X., Kan, C.X., Xu, J. et al. (2020). Dielectric function modelling and sensitivity forecast for Au-Ag alloy nanostructures. *Phys. Chem. Chem. Phys.* 22: 14932–14940.

14 Johnson, P.B. and Christy, R.W. (1972). Optical constants of noble metals. *Phys. Rev. B* 6: 4370–4379.

15 Lee, K.S. and El-Sayed, M.A. (2006). Gold and silver nanoparticles in sensing and imaging: sensitivity of plasmon response to size, shape, and metal composition. *J. Phys. Chem. B* 110: 19220–19225.

16 Fedrigo, S., Harbich, W., and Buttet, J. (1993). Collective dipole oscillations in small silver clusters embedded in rare-gas matrices. *Phys. Rev. B* 47: 10706–10715.

17 Mayer, K.M. and Hafner, J.H. (2011). Localized surface plasmon resonance sensors. *Chem. Rev.* 111: 3828–3857.

18 Huang, W.C. and Lue, J.T. (1994). Quantum size effect on the optical properties of small metallic particles. *Phys. Rev. B* 49: 17279–17285.

19 Cortie, M.B., Liu, F., Arnold, M.D., and Niidome, Y. (2012). Multimode resonances in silver nanocuboids. *Langmuir* 28: 9103–9112.

20 Kulkarni, S.K. (2015). *Nanotechnology: Principles and Practices*. Springer.

21 Gans, R. (1912). Über die form ultramikroskopischer goldteilchen. *Ann. Phys.* 342: 881–900.

22 Hu, M., Chen, J.Y., Li, Z.Y. et al. (2006). Gold nanostructures: engineering their plasmonic properties for biomedical applications. *Chem. Soc. Rev.* 35: 1084–1094.

23 Li, Z.S., Kan, C.X., and Cai, W.P. (2003). Tunable optical properties of nanostructured-gold mesoporous-silica assembly. *Appl. Phys. Lett.* 82: 1392–1394.

24 Draine, B.T. and Flatau, P.J. (1994). Discrete-dipole approximation for scattering calculations. *J. Opt. Soc. Am. A* 11: 1491–1499.

25 Chen, H., Shao, L., Li, Q., and Wang, J. (2013). Gold nanorods and their plasmonic properties. *Chem. Soc. Rev.* 42: 2679–2724.

26 Singham, S.B. and Bohren, C.F. (1988). Light scattering by an arbitrary particle: the scattering-order formulation of the coupled-dipole method. *J. Opt. Soc. Am. A* 5: 1867–1872.

27 Mulholland, G.W., Bohren, C.F., and Fuller, K.A. (1994). Light scattering by agglomerates: coupled electric and magnetic dipole method. *Langmuir* 10: 2533–2546.

28 Piller, N.B. (1997). Influence of the edge meshes on the accuracy of the coupled-dipole approximation. *Opt. Lett.* 22: 1674–1676.

29 Scheider, J.B. and Peden, I.C. (1988). Differential cross section of a dielectric ellipsoid by the T-matrix extended boundary condition method. *IEEE Trans. Antennas Propag.* 36: 1317–1321.

30 Yee, K.S. (1966). Numerical solution of initial boundary value problems involving maxwell's equations in isotropic media. *IEEE Trans. Antennas Propag.* 14 (5): 302–307.

31 Yin, G., Wang, S.-Y., Xu, M., and Chen, L.Y. (2006). Theoretical calculation of the optical properties of gold nanoparticles. *J. Korean Phys. Soc* 49: 2108–2111.

32 Li, Y.L., Kan, C.X., Wang, C.S. et al. (2014). Surface plasmon resonance coupling effect of assembled gold nanorods based on the FDTD simulation. *Acta Phys. Chim. Sin.* 30: 1827–1836.

33 Vigderman, L., Khanal, B.P., and Zubarev, E.R. (2012). Functional gold nanorods: synthesis, self-assembly, and sensing applications. *Adv. Mater.* 24: 4811–4841.

34 Wiley, B.J., Im, S.H., Li, Z.Y. et al. (2006). Maneuvering the surface plasmon resonance of silver nanostructures through shape-controlled synthesis. *J. Phys. Chem. B* 110: 15666–15675.

35 Ding, X.L., Kan, C.X., Mo, B. et al. (2012). Synthesis of polyhedral Ag nanostructures by a PVP-assisted hydrothermal method. *J. Nanopart. Res.* 14: 1000.

36 Schnepf, M.J., Mayer, M., Kuttner, C. et al. (2017). Nanorattles with tailored electric field enhancement. *Nanoscale* 9: 9376–9385.

37 Okamoto, H. and Imura, K. (2009). Near-field optical imaging of enhanced electric fields and plasmon waves in metal nanostructures. *Prog. Surf. Sci.* 84: 199–229.

38 Deng, T., Parker, J., Yifat, Y. et al. (2018). Dark plasmon modes in symmetric gold nanoparticle dimers illuminated by focused cylindrical vector beams. *J. Phys. Chem. C* 122 (48): 27662–27672.

39 Ruan, Q., Shao, L., Shu, Y. et al. (2014). Growth of monodisperse gold nanospheres with diameters from 20 nm to 220 nm and their core/satellite nanostructures. *Adv. Opt. Mater.* 2 (1): 65–73.

40 Rechberger, W., Hohenau, A., Leitner, A. et al. (2003). Optical properties of two interacting gold nanoparticles. *Opt. Commun.* 220: 137–141.

41 Walsh, G.F. and Negro, L.D. (2013). Enhanced second harmonic generation by photonic–plasmonic Fano-type coupling in nanoplasmonic arrays. *Nano Lett.* 13: 3111–3117.

42 Jeon, H.B., Tsalu, P.V., and Ha, J.W. (2019). Shape effect on the refractive index sensitivity at localized surface plasmon resonance inflection points of single gold nanocubes with vertices. *Sci. Rep.* 9: 1–8.

43 Pellarin, M., Ramade, J., Rye, J.M. et al. (2016). Fano transparency in rounded nanocube dimers induced by gap plasmon coupling. *ACS Nano* 10: 11266–11279.

44 Liu, J.S., Kan, C.X., Li, Y. et al. (2014). End-to-end and side-by-side assemblies of gold nanorods induced by dithiol poly(ethylene glycol). *Appl. Phys. Lett.* 104: 253105.

45 Ke, S.L., Kan, C.X., Liu, J.S., and Cong, B. (2013). Controlled assembly of gold nanorods using tetrahydrofuran. *RSC Adv.* 3: 2690–2696.

46 Sun, Z.H., Ni, W.H., Yang, Z. et al. (2008). pH-controlled reversible assembly and disassembly of gold nanorods. *Small* 4 (9): 1287–1292.

47 Zheng, J.P., Cheng, X.Z., Zhang, H. et al. (2021). Gold nanorods: the most versatile plasmonic nanoparticles. *Chem. Rev.* 121 (21): 13342–13453.

48 Ma, W., Kuang, H., Xu, L.G. et al. (2013). Attomolar DNA detection with chiral nanorod, assemblies. *Nat. Commun.* 4: 2689.

49 Kumar, J., Wei, X.Z., Barrow, S. et al. (2013). Surface plasmon coupling in end-to-end linked gold nanorod dimers and trimers. *Phys. Chem. Chem. Phys.* 15 (12): 4258–4264.

50 Gandra, N., Abbas, A., Tian, L.M., and Singamaneni, S. (2012). Plasmonic planet-satellite analogues: hierarchical self-assembly of gold nanostructures. *Nano Lett.* 12 (5): 2645–2651.

51 Yoon, J.H., Lim, J.H., and Yoon, S. (2012). Controlled assembly and plasmonic properties of asymmetric core-satellite nanoassemblies. *ACS Nano* 6 (8): 7199–7208.

52 Prodan, E.R., Halas, N.J., and Nordlander, P.J. (2003). A hybridization model for the plasmon response of complex nano-structures. *Science* 302: 419–422.

53 Verellen, N., Sonnefraud, Y., Sobhani, H. et al. (2009). Fano resonances in individual coherent plasmonic nanocavities. *Nano Lett.* 9: 1663–1667.

54 Lukyanchuk, B., Zheludev, N.I., Maier, S.A. et al. (2010). The Fano resonance in plasmonic nanostructures and metamaterials. *Nat. Mater.* 9: 707–715.

55 Shao, L., Fang, C., Chen, H. et al. (2012). Distinct plasmonic manifestation on gold nanorods induced by the spatial perturbation of small gold nanospheres. *Nano Lett.* 12: 1424–1430.

56 Lee, D. and Yoon, S. (2015). Gold nanocube–nanosphere dimers: preparation, plasmon coupling, and surface-enhanced Raman scattering. *J. Phys. Chem. C* 119: 7873–7882.

57 Huang, Z., Meng, G., Huang, Q. et al. (2017). Surface-enhanced Raman scattering from plasmonic Ag-nanocube@Au-nanospheres core@satellites. *J. Raman Spectrosc.* 48 (2): 217–223.

58 Willingham, B.A., Brandl, D.W., and Nordlander, P. (2008). Plasmon hybridization in nanorod dimers. *Appl. Phys. B: Lasers Opt.* 93 (1): 209–216.

59 Halas, N.J. (2005). Playing with plasmons: tuning the optical resonant properties of metallic nanoshells. *MRS Bull.* 30 (5): 362–367.

3

Advanced Characterizations for Plasmonic Nanostructures

Telescope and microscopy are two directions for studying matter. We are more interested in the latter one.

3.1 Introduction

Although the concept of plasmon was initially proposed in the 1950s by David Pines (1924–2018) and David Bohm [1, 2], rapid research progress on plasmon was made over the past two decades. Especially for the past few years, there are many starting journals, numerous papers, worldwide known research groups, and tremendous applications studies related to plasmonics (www.acemap.info).

Surface plasmon-related optical property is one of the important issues of plasmonic noble metal nanostructures. And the optical property is highly dependent on the morphologies, size, composition, and surrounding medium of plasmonic nanostructures. With the development of nanoscience and nanotechnology, many fabrication equipments and methods for plasmonics have been constructed and developed for continuous research demand in controlling nanostructures, enhancing performance, and expanding the application fields. Simultaneously, detailed characterizations of plasmonic nanostructures are the basis for a better understanding of the scientific problems in physiochemical processes and insight into the structure–property relations.

Nowadays, plasmon is not only a concept in fundamental research but also one source of technical category. For example, the plasmon photothermal effect can catalyze water decomposition and carbon dioxide reduction [3–5]. Using this catalysis reaction, photosynthetic new energy can be obtained just from water and carbon dioxide. This will be a breakthrough of 0–1 when the social burden of carbon dioxide was changed into an energy resource, which is extremely important in achieving the goal of carbon neutrality. The exploration of catalytic reaction mechanism and the detection of reaction process need advanced characterization technology.

Plasmonic Metal Nanostructures: Preparation, Characterization, and Applications, First Edition. Caixia Kan.

In addition to the normal optical spectrometer and electron microscope measurements, some state-of-the-art photoelectric characterizations for the metal nanostructures will be delivered, such as *in-situ* TEM, SERS, and scanning near-field optical microscopy (SNOM). Among these advanced characterizations, catalytic activity of metal nanostructures can be detected through SERS measurements; *in-situ* TEM can help us to trace the growth process of nanostructures in the atomic level, and plasmon excitation can be done well by SNOM.

3.2 Optical Property Characterization Technology

The optical properties of materials are related to the basic principles of solid-state physics, including the energy band structure, inter-band transition, and intra-band transition. In essence, the optical properties of solids are the absorption and scattering of light caused by the interaction between electromagnetic (EM) waves and electrons, excitons, lattice vibrations, and defects in solids. For insulator, ionic crystal, and semiconductor, there is a certain band gap between the conduction band and valence band, and there is no electronic energy level in the band gap. Solid will not absorb light if the radiated light is lower than band gap energy. Silicon crystal with band gap of 1.1 eV is black for the absorption of visible light and transparent to most infrared regions. Additionally, semiconductors also have excitons and defects that absorb light and then luminescent light, which is an important part of semiconductor physics.

There are a large number of electrons that can move freely in metal, forming an electron sea. This electron collective is also called bulk plasmon. As discussed in Chapter 1, the frequency of metal bulk plasmon is larger than the visible light, reflecting visible light to form metallic luster. From the perspective of the energy band, the valence band, and conduction band of the metal overlap, the empty energy level above the Fermi surface can accept excitation. The electrons are excited to a higher unoccupied energy level produce photocurrent. Instantaneously, the current is rapidly discharged to emit photons of the same wavelength to the surface of metal. That is why thin metal coatings have high reflectivity. Due to the relativistic contraction of the inner electron orbit and the expansion effect of the outer electron orbit, the band gap between outermost orbit and inner orbit is different for Au (2.3 eV) and Ag (4.8 eV).

Under the action of light, there are three motion forms within a material molecule. They are the motion of electrons relative to nuclei, the vibration of nuclei near equilibrium, and the rotation of molecules around the center of gravity. Therefore, molecules have three different energy levels, namely, electronic energy level, vibrational energy level, and rotational energy level. Using energy state transition for analysis is called spectral analysis. The energy difference of the electron energy level is in the range of 1–20 eV, and the absorption spectrum generated by the electron transition is in the ultraviolet–visible light. The vibration energy level difference of 0.05–1 eV results in absorption located in the infrared region, which is the infrared spectrum and molecular vibration spectrum. Table 3.1 summarized the interaction of a light beam with matter and the corresponding optical spectral technologies.

Table 3.1 Interaction of a light beam with matter and corresponding spectral technologies.

Spectra	Wavelength	Transition/motion	Spectral technology
X-ray	0.1–10 nm	K/L-layer electron	X-ray spectrometer
Ultraviolet	10–400 nm	Middle/valence electron	UV photometer
Visible	400–750 nm	Valence electron	vis spectrophotometry
Near-infrared	0.75–2.5 μm	Molecular vibration	NIR spectrophotometry
Mid-infrared	2.5–5 μm	Molecular vibration	Mid/IR spectrophotometry
Far-infrared	5–1000 μm	Molecular vibration/ rotation	Far/IR spectrophotometry
Microwave	1–1000 mm	Molecular rotation	Microwave spectroscopy

Spectral analysis can be divided into absorption spectrum, such as ultraviolet–visible–near-infrared (UV–vis–NIR) absorption spectrum, emission spectrum (such as fluorescence spectrum), and scattering spectrum (such as Raman spectrum). In this part, we mainly introduced UV–vis–NIR spectrometer and Raman spectrometer that commonly used in the characterization of optical properties of plasmonic nanostructures.

3.2.1 Absorption Spectrometer

3.2.1.1 Lambert's Law

When a beam of monochromatic light passes through a medium with a certain thickness, the absorption of the medium reduces the intensity of the transmitted light. Lambert's law was applied to express the intensity change of transmitted light. The monochromatic evaluation beam with intensity I_0 enters a uniform medium and propagates for a certain distance x. The light intensity I became $I - dI$ after passing through a thin layer dx (Figure 3.1).

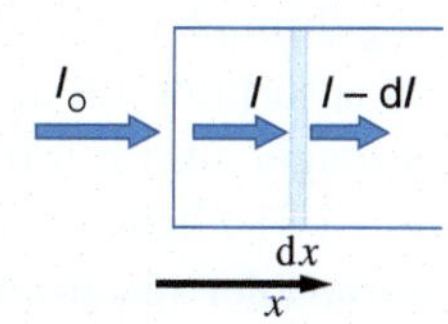

Figure 3.1 Schematic diagram of the optical path for Lambert's law.

The complex permittivity of a medium consists of two parts: the real part n and the imaginary part k. The real part represents the permittivity. While the imaginary part represents the attenuation loss of the material caused by various relaxation polarization caused by the internal steering polarization of the material that cannot keep up with the change of the external high-frequency electric field. The imaginary part is related to the conductivity (σ) of the medium.

When light enters the medium, the light vector decays exponentially $e^{-\omega kx/c}$ and light intensity accordingly decays in the form of $I = I_0 e^{-2\omega kx/c}$. After introducing one absorption coefficient (a) independent of the intensity of light, the propagation distance is $1/a$ when the energy decays to $1/e$ of its original intensity, i.e. $-aI = \frac{dI}{dx}$ or $I = I_0 e^{-ax}$. Then a can be obtained.

$$a = \frac{2\omega k}{c} = \frac{4\pi k}{\lambda} \tag{3.1}$$

As it is known that $\tilde{N} = n + ik$, we can describe the dielectric constant $\varepsilon_r = n^2 - k^2$, $2nk = \sigma/\omega\varepsilon_0$.

When light passes through the medium $\frac{1}{v^2} = \frac{(n+ik)^2}{c^2} = \frac{n^2-k^2}{c^2} + i\frac{2nk}{c^2}$, n and k can also be expressed as:

$$n^2 = \frac{\varepsilon_r}{2}\left[1 + \left(1 + \frac{\sigma^2}{\omega^2\,\varepsilon_r^2\varepsilon_0^2}\right)^{1/2}\right] \tag{3.2}$$

$$k^2 = \frac{\varepsilon_r}{2}\left[1 - \left(1 + \frac{\sigma^2}{\omega^2\,\varepsilon_r^2\varepsilon_0^2}\right)^{1/2}\right] \tag{3.3}$$

For nonconductive materials, there is no absorption, and the material is transparent. For semiconductors, absorption is related to the material and varies with the wavelength of light. For materials with large absorption coefficients, absorption of light is mainly concentrated in the very thin surface layer of the crystal. For atoms, the energy levels are discontinuous, and the difference between the two energy levels is a constant value. Therefore, the electron transition can only absorb photons of certain energy, leading to the line absorption spectrum, such as the well-known hydrogen atom line spectra. In the bulk crystal, the optical absorption spectrum is a continuous absorption band since many energy levels equivalent to the atomic energy level form the continuous energy band.

The absorption spectrum includes general absorption and selective absorption for certain wavelength regions. The former one refers to the independent weak absorption of light by a medium within a given wavelength range, such as visible light absorption by quartz and water. Transmitted light does not change color but only intensity after this general absorption. While selective absorption means that the medium has strong light absorption in a specific wave band. For example, protein molecules have a strong absorption for ultraviolet light of 280 nm. In the visible (vis) range, selective absorption means that light changes color and intensity after passing through the medium. In fact, absorption of light is composed of these two kinds of absorption for any materials. Although glass absorbs weakly visible light, it can strongly absorb infrared light with a wavelength greater than 750 nm and ultraviolet (UV) light with a wavelength less than 380 nm. The prism and cuvette in the ultraviolet spectrometer are made of quartz. While prisms in near-infrared (NIR) spectrometers are usually made of crystals, such as NaCl or NaF_2.

3.2.1.2 Spectrometer Device

Figure 3.2 is the optical path diagram of a UV–vis–NIR spectrometer. It is composed of light source, monochromator, absorption cell (sample rack), detector, data processing, and recording in computer.

The light sources include deuterium lamp, Xenon lamp, and tungsten of halogen lamp, providing light in the wavelength region of 150–2500 nm. Monochromatic light and required wave band beam can be obtained after the light from the source passes through the monochromator (dispersion prism or grating) that is a key component of spectrophotometer. Then by rotating the grating, the monochromatic light

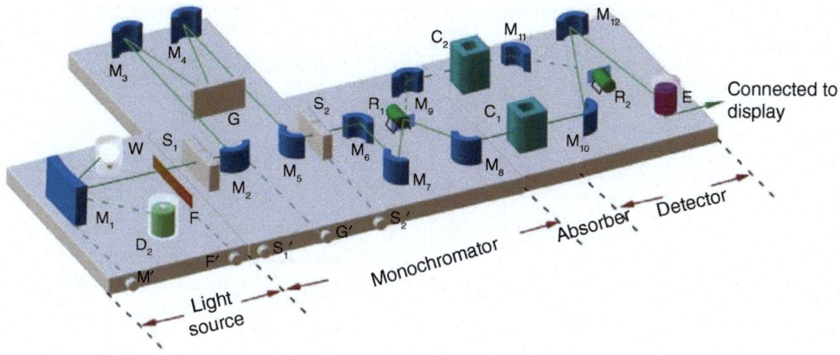

Figure 3.2 Diagram of dual beam UV–vis–NIR spectrophotometer.

of different wavelengths is irradiated to the sample cell (optical glass or plastic for vis and quartz glass or sapphire for UV region). The transmitted light is received by the detection system (photomultiplier tube). After transforming the optical signals into electrical signals and amplification, absorption of substances to various light wavelengths can be obtained.

3.2.1.3 Application of This Optical Spectrum

By using different modes of absorption, transmission, reflection, diffuse reflection, and other measurement functions of this UV–vis–NIR spectrophotometer, the corresponding spectra of liquid, film, and solid powder will be applied to qualitative identification of unknown samples and quantitative analysis on the physical parameters of absorbers. For instance, energy gaps can be calculated from the absorption edge for semiconductors, concentration from absorption coefficient, and purification; size analysis can be carried out from absorption/scattering band position in different wavelength regions and shapes; or morphological evolution of the nanostructures in the sample can be inferred from the peak position and spectrum shifting.

Generally, the response of semiconductor to incident light is very complex, including intrinsic absorption for the formation of photocurrent, discreted exciton absorption peaks (no photocurrent formation), free carrier absorption from the transition between different energy states of conduction band or valence band in longer wavelength, and broadband transition between impurity level and the whole energy band. Among them, the most commonly used is the accurate band gap calculation by choosing the appropriate band edge. Additionally, the direct band gap and the indirect band gap can also be distinguished from the shape of the absorption spectrum.

From the Tauc function $\alpha h\upsilon = C(h\upsilon - E_g)^n$, it is accepted that semiconductor is an indirect band gap for $n = 2$ and direct band gap for $n = 0.5$. For the homogeneous and transparent samples, because the absorption coefficient (α) is proportional to the absorbance, the constant (C) has little effect on the band gap calculation. And absorbance can be directly used as absorption coefficient for calculation. Here, we take the direct band gap semiconductor as an example. From the measured absorption spectrum, we can get a curve according to $(\alpha h\upsilon)^2 - h\upsilon$ and take the linear

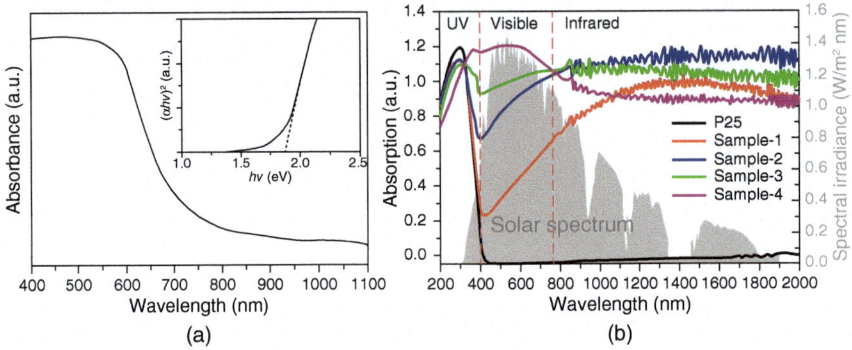

Figure 3.3 (a) UV–Vis–NIR spectrum of the typical $AgInS_2$ flowerlike nanoarchitectures [6]. Source: Adapted from Ref. [6]. (b) Solar spectrum and UV–Vis–NIR absorption spectra of TiO_x nanoparticles with different colors [7]. Source: Reproduced with permission from Ref. [7]; © 2002, American Association for the Advancement of Science.

part. The position where the extended line intersects with the abscissa is the band gap $E_g(eV): h\upsilon = 1240/\lambda$ (nm).

Figure 3.3 illustrates two typical spectra of the $AgInS_2$ flowerlike nanoarchitectures and TiO_2 obtained by different methods. From the $(\alpha h\upsilon)^2 - h\upsilon$ curve (energy mode) of inset in Figure 3.3a, the E_g (1.88 eV) agrees well with this material [6]. However, the band gap for TiO_2 is quite different due to the effect of size, phase structures and dopping from different methods. Different colors (white, gray, blue, and black) TiO_x can be obtained through hydrogenation process based on the pristine P25 TiO_2 nanocrystals, as shown in Figure 3.3b reduced TiO_x nanoparticle samples extend the photoresponse from the ultraviolet to the visible and near-infrared spectral regions [7].

When the size of the measured objects decreases, they will bring new size effects and quantum effects with the interaction between light and matter. And the effective refractive index of materials and band gap of semiconductors can be regulated. For plasmonic nanostructures, unique plasmon resonant light absorption at desired wavelength can be achieved by structure designing and controlling the subsequent synthesis process.

Results shown in Figure 3.4 are some typical colloid photos of Ag and Au nanostructures and their corresponding absorption spectra [8]. For the Ag nanoplates, as the concentration of PSS was increased, the Ag nanoplates became rounder, and finally transformed into oblate spheroidal shape (not shown here). These results indicated that Ag nanoplates could be easily degraded with the presence of PSS in the aging process. Since the color changes were related to the SPR properties of the Ag nanoplates, the extinction spectra were acquired to monitor the SPR evolutions of Ag nanoplates during the reaction. It is obviously that the in-plane dipole resonance peak gradually blue shifted, suggesting the decrease of edge/thickness ratio. In addition to the rich color originated from colloidal particles response to visible light, the thermal electrons from the localized field effect of strong plasmon absorption can be used to enhance the performance of different optoelectronic

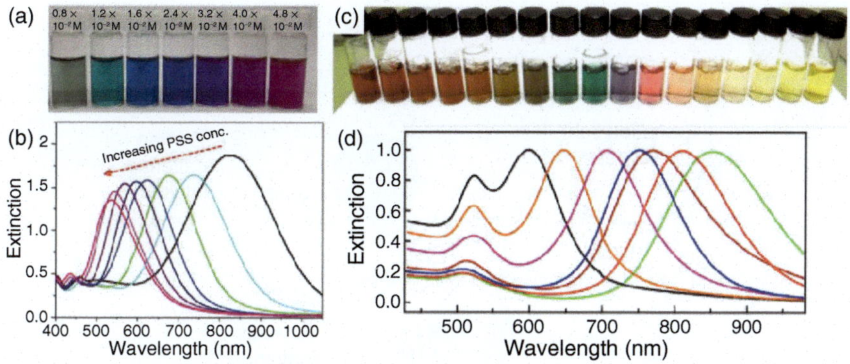

Figure 3.4 Some typical colloid photos of Ag and Au nanostructures, and the corresponding optical spectra. (a, b) Photograph and extinction spectra of Ag nanoplates after the addition of PSS at room temperature. The concentration of PSS in the systems was marked on the container. (c, d) Photograph and extinction spectra of Au NRs with increasing aspect ratios. Source: Reproduced with permission from Ref. [8]; © 2017, Royal Society of Chemistry.

devices and catalyze reaction processes. In addition to the SPR sensitivity to the absorbed molecule, medium, and aspect ratios, the photothermal effect of NIR light absorption has been used in thermotherapy and imaging in biomedicine. Among the burgeoning nanostructures with NIR photothermal effect, Au NRs were defined as the most versatile plasmonic nanoparticles. Therefore, it is of great significance and practical value to explore the application potential of optical properties of materials by using optical spectral characterization techniques.

3.2.2 Surface-Enhancement Raman Scattering

3.2.2.1 History and Principle of Raman Spectroscopy

Vibrational spectroscopy is an excellent method for identifying substances by providing the unique fingerprint spectra [9]. Raman spectroscopy is considered as one of the fastest and nondestructive analytical techniques that give the vibrational spectra and thus, it finds applications in various areas such as biotechnology, environmental monitoring, material science, medical and clinical chemistry, diagnostics, and surface analysis [10, 11].

Raman spectroscopy is specialized in measuring the energy difference between the scattered photon from the sample and the incident photon when light strikes a molecule. Such phenomenon of inelastic scattering of light was first discovered experimentally in 1928 by an Indian physicist Chandrashekhara Venkata Raman. Since then, this phenomenon has been referred to as Raman spectroscopy [12]. In his original experiment, by a telescope, sunlight was focused onto a sample that was either a purified liquid or a dust-free vapor. Another lens was placed by the sample to collect the scattered radiation. A system of optical filters was used to show the existence of scattered radiation with an altered frequency from the incident light the basic characteristic of Raman spectroscopy.

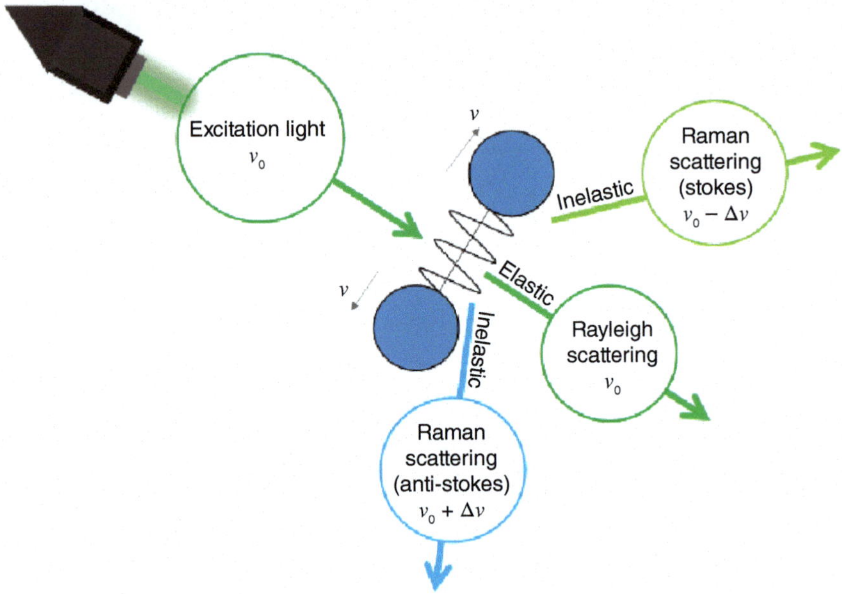

Figure 3.5 Molecular vibrations and scattering.

As mentioned above, Raman spectroscopy is a scattering form of molecular spectroscopy. As illustrated in Figure 3.5, the irradiation of a molecule with a monochromatic light (ν_0) could always result to two types of light scattering, elastic and inelastic. In elastic scattering, there occurs no change in photon frequency or any change in its wavelength and energy. Conversely, the other is inelastic scattering which is accompanied by the shift in photon frequency due to excitation or deactivation of molecular vibrations in which either the photon may lose or gain some amount of energy. Therefore, three types of phenomena can occur: (i) when light is incident on a molecule, it can interact with the molecule but the net exchange of energy (E) is zero. As a result, the frequency of the scattered light is the same as that of the incident light ($\nu = \nu_0$), which is well known as Rayleigh scattering. (ii) Light can interact with the molecule and the net exchange of energy is the energy of one molecular vibration. If the interaction causes the light photon to gain vibrational energy from the molecule, then the frequency of the scattered light will be higher than that of the incident light ($\nu = \nu_0 + \Delta\nu$), termed as anti-Stokes Raman scattering. (iii) If the interaction causes the molecule to gain energy from the photon, then the frequency of the scattered light will be lower than that of the incident light ($\nu = \nu_0 - \Delta\nu$), and this process is defined as Stokes-Raman scattering.

In terms of the view of quantum mechanics, as illustrated in Figure 3.6, if the energy of the incident photons is not larger enough to excite the molecule from the ground state to the lowest electronic state, the molecule will be excited to some virtual state between the two states. But since the electron of that excited molecule cannot stay long in the virtual state and needs to immediately go back to the ground

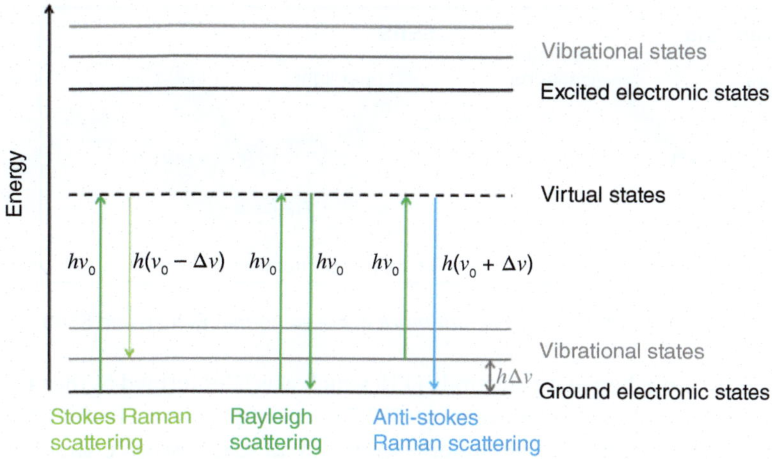

Figure 3.6 Energy level diagram of Rayleigh and Raman scattering.

state. Now, if the electron goes back to its original ground state, the wavelength of the emitted photon is then the same as that of the incident light photon, which is the case of Rayleigh scattering mentioned above. On the other hand, it is also possible for the electron to go to a virtual state that is different from where it is excited, giving rise to an energy difference between the emitted photon and the incident photon, and therefore a shift in wavelength (or an energy shift) for the emitted photon as compared to the incident photon. If the emitted photon has energy smaller than that of the incident photon, this scattering process is known as Stokes Raman scattering. And when the incident photon has a higher energy compared to the emitted photon, it is known as anti-Stokes Raman scattering. Besides, the relative intensities of Stokes and anti-Stokes processes depend on the population of the various states of the molecule. The population can be worked out from the Boltzmann equation but at room temperature, the number of molecules expected to be in an excited vibrational state other than any really low-energy ones will be small. Thus, compared to Stokes scattering, anti-scattering will be weak and will become weaker as the frequency of the vibration increase, due to decreased population of the excited vibrational states.

Overall, as mentioned above, Raman spectroscopy is based on the shift in the energy between the incident and scattered photons associated with the Stokes and anti-Stokes transitions, which is always measured as the change in the wavenumber (cm^{-1}). Since the shift in wavenumber highly depends upon the chemical composition of the molecules which are responsible for scattering. As a result, Raman spectroscopy is especially popular for its specificity in chemical and biological analysis and offers some distinct advantages over other spectroscopic methods. However, with respect to the conventional Raman spectroscopy, due to the low scattering cross sections (10^{-31}), Raman signal is intrinsically very weak, approximately six to 10 orders of magnitude less efficient than fluorescence and this greatly limited its broad applications. In 1974, Fleischmann and co-workers first reported the unexpected increase of the Raman signal from adsorbed pyridine on an

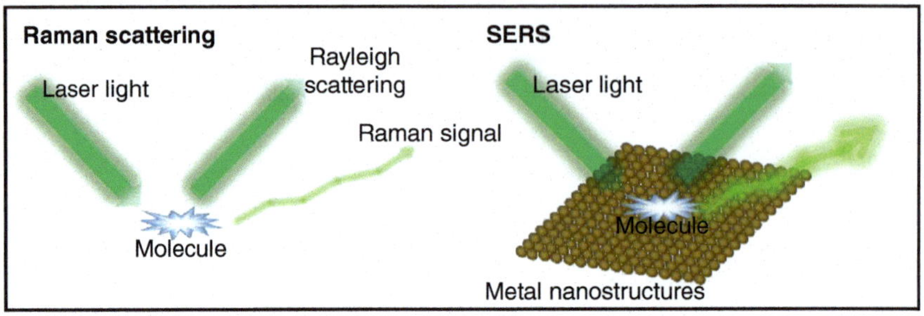

Figure 3.7 Schematic illustration of the difference between normal Raman and SERS.

electrochemically roughened Ag electrode [13], which was later termed as the effect of surface enhanced Raman spectroscopy (SERS), as compared in Figure 3.7. In addition to its high sensitivity, SERS inherits from Raman spectroscopy characteristics such as fingerprint, nondestructive, and label-free analytical nature. Therefore, SERS revived the interest of the research community in Raman spectroscopy as an analytical method and the past 40 years have also witnessed a dramatic increase in its applicability.

3.2.2.2 Research Progress of SERS

The discovery of SERS has a relatively short history. The first published paper reporting the strong Raman signal from the interface was done in 1974 by Fleischmann et al. [13]. The authors reported the unexpected increase of the Raman signal from adsorbed pyridine on an electrochemically roughened silver electrode, and they attributed the enhancement to a surface-area effect. But later in 1977, Jeanmaire and van Duyne and Albrecht and Creighton groups suggested enhancement factor of 10^5–10^6 [14, 15]. The connection with plasmon excitation was suggested by Albrecht and Creighton as a resonant Raman effect involving plasmon excitation, as proposed earlier by Phipott [16]. Subsequently, the connection of SERS intensities to enhanced fields arising from localized surface plasmons in rough metals was noted by Moskovits theoretically [17]. In 1979, Creighton and coworkers published a comprehensive experimental article on SERS from Ag/Au colloidal surfaces [18], in which they first correlated the sizes and materials of metallic colloids and the wavelength of SPR with SERS. The authors pointed that it was the surface plasma resonance of metallic colloids or Mie scattering that leads to the giant SERS effect, which the authors called “plasma resonance-enhanced Raman.” They also concluded that the dielectric functions of the metal colloids determined which materials were effective for SERS. This pioneering work expands the range of SERS substrates beyond electrochemically roughened electrodes or cold deposited surfaces to include colloidal particles, and at the same time, augmenting interest in the use of SERS to a broad scientific and engineering community embracing such disciplines as electrochemistry, heterogeneous catalysis, and biochemistry.

Within a few years, strongly enhanced Raman signals were verified for many different molecules, as illustrated in Figure 3.8, which had been attached to the SERS-active substrates. These SERS-active substrates are various metallic structures

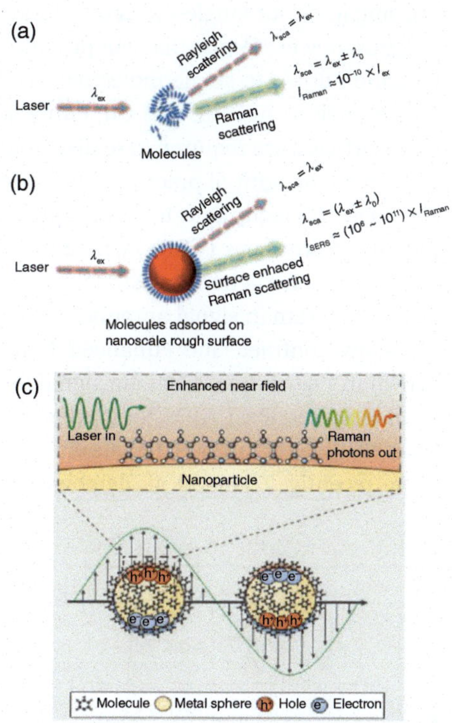

Figure 3.8 (a) Raman signal of molecules; (b) Raman signal of molecules adsorbed on rough metal nanoparticles; (c) the principle of localized plasmon effect on Raman signal.

with sizes on the order of tens of nanometers. The most common types of SERS-active substrates exhibiting the largest effects are colloidal Ag (in most cases) nanoparticles with size in the range of 10–150 nm, Ag electrodes, or evaporated Ag films.

Despite the enhancement, early SERS systems still immensely suffered from irreproducibility owing to ill-defined substrates fabricated by electrochemical roughening. Recent advances in nanofabrication and the 1997 discovery of single-molecule SERS, the ultimate limit of detection, have caused an extension of SERS from an interesting physical phenomenon to an effective analytical technique. In 1997, the Kneipp and Nie groups independently used distinct geometries of nanoparticles to achieve a single molecule-SERS signal [19, 20]. For some optically selected nanoparticles, enormous enhancement factor of 10^{14}–10^{15} had been estimated, which were, by many orders of magnitude, larger than the ensemble-averaged enhancement factors. Of importance is also that the single-molecule SERS spectra were both more intense and more stable than single-molecule fluorescence spectra. Furthermore, in 1999, Xu demonstrated the detection of molecular vibrations in single hemoglobin (Hb) protein molecules attached to isolated or immobilized Ag nanoparticles by SERS [21]. Combined with scanning electron microscopy (SEM) and atomic force microscope (AFM) analysis, they found that no Raman signal was detected on a single Ag nanoparticle; however, the majority of the dimers/trimers were active in Raman. Moreover, among the "hot" dimers, those with their dimer axis are oriented parallel to the incident electric field.

Comparable to single-molecule fluorescence spectroscopy, SERS is a powerful spectroscopy technique that can provide nondestructive and ultra-sensitive characterization down to single molecular level. However, generally substrates based on metals such as Ag, Au, and Cu, either with roughened surfaces or in the form of nanoparticles, are required to realize a substantial SERS effect, and this has severely limited the breadth of practical application of SERS. A number of approaches have extended the technique to nontraditional substrates, most notably tip-enhanced Raman spectroscopy (TERS) where the probe substance (molecule or material surface) can be on a generic substrate and where a nanoscale Au tip above the substrate acts as the Raman signal amplifier. It was not until 2000 that experiment evidence of locally confined and enhanced Raman scattering was demonstrated [22]. As shown in Figure 3.9a, a Raman signal increase of more than 30 times was obtained when the metalized AFM-tip was brought into contact with a thin brilliant cresyl blue (BCB) layer compared to measurements with the tip retracted from the sample. More recently, Zhang et al. demonstrated single molecule-TERS mapping which

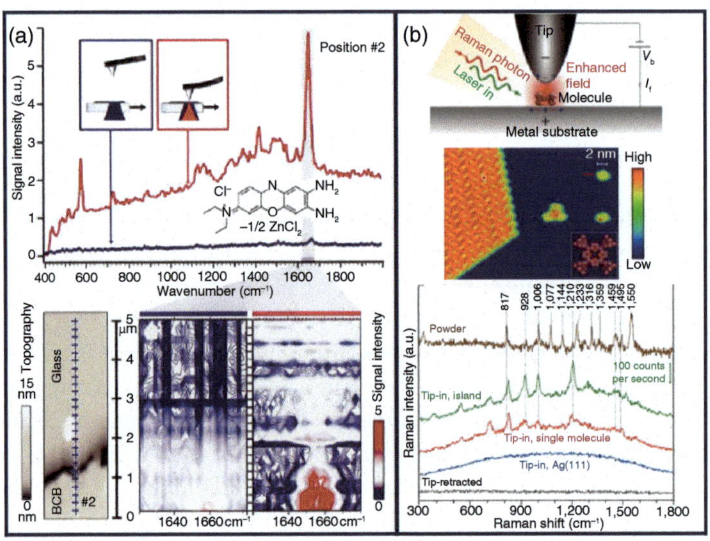

Figure 3.9 (a) Tip-enhanced Raman spectra of brilliant cresyl blue (BCB) dispersed on a glass support measured with Ag-coated AFM probe. The two Raman spectra in the upper part were measured with the tip retracted from the sample (in blue) and with the tip in contact with the sample (in red) at the position marked #2 in the topography image (left). Contour plots of 25 Raman spectra measured across sample. Red denotes high intensity; blue denotes low intensity. Middle: Corresponds to the Raman signal with tip retracted from the sample (in blue). Small background signals from glass, immersion oil led to many contour lines, resulting in stripes in the contour plot. Right: Raman signal with the tip in contact with the sample (in red). Source: Nie et al. [22]/Reproduced with permission from © 1997 American Association for the Advancement of Science. (b) Clean TERS spectra using well-defined tip and sample. Top: Schematic tunneling-controlled TERS in a confocal-type side-illumination configuration. Middle: STM topography of sub-monolayer H_2TBPP molecules on Ag (111). Bottom: TERS spectra for different conditions. Source: Stöckle et al. [23]/Reproduced with permission from ©2000, ELSEVIER.

achieved a spatial resolution of 0.5 nm [23], shown in Figure 3.9b. Thus, TERS has attracted attention for its potential application to *in-situ* measurements on solid surfaces and as a promising tool for exploring the frontiers of light-matter interactions including exotic photochemical reactions.

Besides TERS technique, in 2010, Tian's group developed a new generation of Raman technique, known as "shell-isolated nanoparticle-enhanced Raman spectroscopy" (SHINERS) to overcome the limitation of materials and morphology generalities in SERS [24]. They employed Au nanoparticles coated with an ultrathin silica or alumina shell (Figure 3.10a,b). The core–shell nanoparticles are directly deposited onto the surface under study and induce a large local electric field contact to the plasmonic nanoparticles with the analyte surface, which is very important for analyzing biological samples. A very recent work also from Tian's group demonstrated the graphene-coated Au nanoparticles for SHINERS (Figure 3.10c) [25]. The thickness of the graphene shell can be precisely controlled in the chemical vapor deposition (Figure 3.10c,d). In general, for the technique SHINERS, Au or Ag nanoparticles are employed with a layer of ultrathin and pinhole-free silica shell. Each Au core–inert shell nanoparticle can be considered as a TERS tip. Under a laser spot, the shell-isolated nanoparticles monolayer is equal to thousands of tips, thus the acquired Raman signal is very strong. Besides, in principle, this technique can be employed for the surface with any material and morphology, thus high-quality Raman spectrum can be obtained from single crystals of Pt, Au, or Si. Thus, SHINERS is becoming a promising and powerful tool in various fields, such

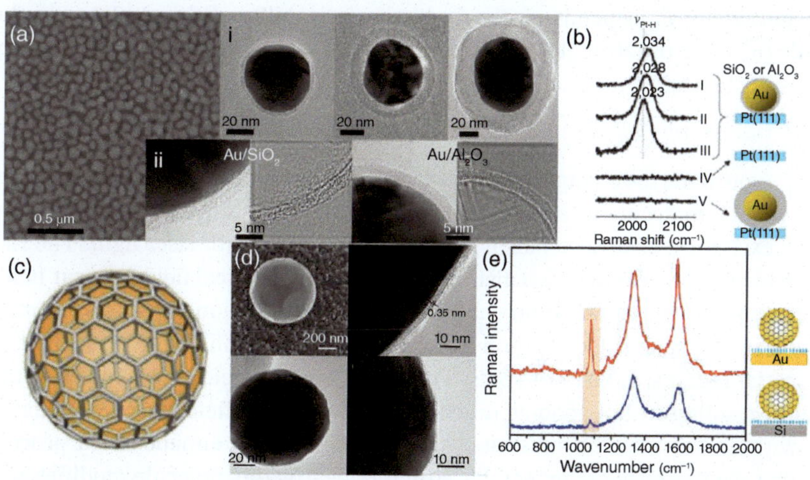

Figure 3.10 (a) SEM and high-resolution TEM images of Au@SiO_2 and Au@Al_2O_3 shell-isolated nanoparticles with different shell thicknesses. (b) Potential dependent SHINERS spectra for hydrogen adsorbed on a Pt (111) surface captured with/without shell-isolated nanoparticles. Source: Reproduced with permission from Ref. [24]; @2010 Springer Nature. (c) Scheme of a graphene-coated Au nanocrystal (Au@G). (d) SEM and high-resolution TEM images of Au@G. (e) SERS spectra of Au@G nanoparticles assembled on an Au substrate and a silicon substrate with mercaptobenzoic acid. Source: Zhang et al. [25]/Reproduced with permission from @ 2021 JOHN WILEY & SONS, INC.

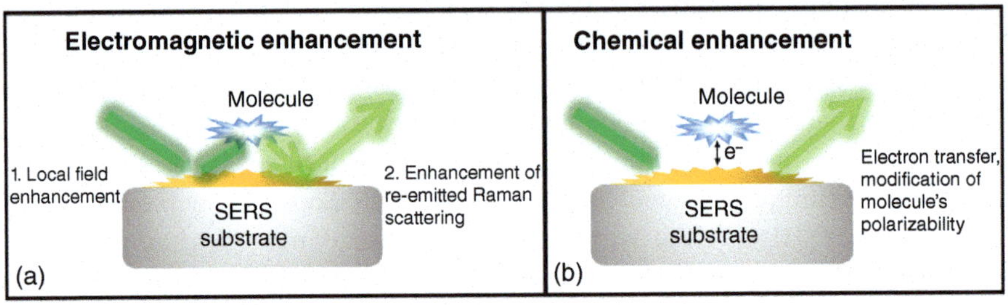

Figure 3.11 Two possible Raman enhancement possesses.

as surface science, material science, biological analysis, pharmaceutical analysis, and environmental protection.

3.2.2.3 The Enhancement Mechanism of SERS

As mentioned before, Moskovits proposed that the large signal was originated by the optical excitation of collective oscillations. However, additional studies proved that chemical mechanism could also contribute to the Raman enhancement in 1977 [15]. Now, it is accepted that there are two possible Raman enhancement possesses, as shown in Figure 3.11. To better explain this, it is necessary to describe the basic aspects of the Raman scattering effect. How the excitation and scattering process can be modified by the above enhancement mechanisms, which lead to the SERS effect.

When a molecule was oscillated under an incident electric field, it may induce polarization (P_0) and emit scattered light at a Raman-shifted frequency (ω_R). The magnitude of P_0 depends on the Raman polarizability of the electrons in the molecule (α_0^R) and the strength of the incoming EM radiation E_0 (with a frequency of ω_0), as Eq. (3.4) [26]:

$$P_0\left(\omega_R\right) = \alpha_0^R\left(\omega_0, \omega_R\right) E_0\left(\omega_0\right) \tag{3.4}$$

where α_0^R represents the change in the electron cloud during the molecular vibration.

Then when a molecule was placed at the surface of a metallic substrate that supports the excitation of surface plasmon when illuminated with laser light. The EM field enhancement possesses two distinct contributors: the local field (or near field) enhancement and the re-radiation enhancement. The excitation of surface plasmon induces a strong spatial localization and hence amplification of the laser light in small spatial regions, called hot spots. Therefore, the EM field experienced by the molecules residing in hot spots is much stronger than the field they would experience without the metallic substrate. And the re-radiation enhancement process means that the presence of the metallic structure nearby the molecule modifies also the efficiency with which the molecule radiates Raman power; this occurs because the power by a dipole (i.e. the molecule oscillating at the Raman frequency) depends on the environment in which it is embedded.

When a metal nanoparticle is irradiated with an incoming light E_0, the conductive electrons are delocalized into collective oscillations, which generate an EM field, which is referred as localized surface plasmons resonance (LSPR). Generally, the local EM field (E_{loc}) associated with the LSPR is stronger than the incident EM field, which results in the enhancement of the EM field in SERS by a factor of $G_{ex} = [E_{loc}(\omega_0)/E_0(\omega_0)]^2$.

Similarly, other sources of oscillation, such as dipoles or quadrupoles, lead to the re-emitted of the LSPR. Typically, the Raman polarizability of molecules interacting with metal nanostructures is about one or three orders of magnitude larger than that of free molecules. Therefore, the interaction of molecules with the vicinal metal nanostructures leads to the mutual excitation of the Raman polarizability by the local EM field. In such a process, the EM field in SERS is enhanced by a factor of $G_R = [E_{loc}(\omega_R)/E_0(\omega_R)]^2$. For Raman modes with low vibrational frequency, ω_0 can be considered roughly equal to ω_R, and the EM field enhancement factor $G_{ex}(\omega_0)$ and $G_R(\omega_R)$ are comparable. Therefore, the overall enhancement of the EM field in SERS (G) scales with the fourth power of the local EM field enhancement, as defined in Eq. (3.5) [19]:

$$G = \frac{\left|E_{Loc}(\omega_R)\right|^4}{\left|E_0(\omega_0)\right|^4} \tag{3.5}$$

Therefore, a tiny change in the local field enhancement can lead to a significant variation in SERS enhancement. This expression is referred to as the zero Stokes shift limit of the $|\boldsymbol{E}|^4$ approximation; it is more accurate at small Raman shifts and if $\boldsymbol{E}_{Loc}(\boldsymbol{\omega}_R)$ is not too sharp near $\boldsymbol{\omega}_0$. This $|\boldsymbol{E}|^4$approximation is widely applied in SERS. However, there are some specific situations in which its use is not appropriate and a more sophisticated approach is required. Typically, these cases involve the interpretation of polarization measurements. An average enhancement factor can be worked out by integrating the single molecule enhancement in the above equations, over the possible orientations of the molecule and over the surface of the substrate.

Although the EM enhancement considerably exceeds the chemical enhancement in terms of magnitude; chemical mechanism also plays an important role because it determines the spectral pattern of the SERS spectra (i.e. the Raman shifts and the band intensity ratios). The chemical effect has also been studied by many authors [27, 28]. The adsorption of a molecule on a substrate can be classified on the basis of the strength of the interaction. Physical adsorption refers to an adsorption process in which the interaction enthalpy is less negative than −25 kJ/mol; chemical adsorption, instead, refers to an adsorption process in which the enthalpy is more negative than −40 kJ/mol. In the former case, Van der Waals forces drive the adsorption process; therefore, the structure of the molecule is only slightly modified. A stronger perturbation is expected in the latter case, in which a chemical bond is formed between the molecule and the surface. In both situations, although at a different extent, the electronic and geometrical structure of the molecule is altered by the interaction with the surface and, hence, the Raman cross-sections of its vibrational modes will be, in general, different with respect to those of the free molecule. The chemical enhancement can be defined as:

$$\boldsymbol{G}_{\mathbf{SERS}}^{\mathbf{Chem}} = \frac{\boldsymbol{\sigma}_{\boldsymbol{k}}^{\mathbf{ads}}}{\boldsymbol{\sigma}_{\boldsymbol{k}}^{\mathbf{free}}} \tag{3.6}$$

where $\sigma_{\boldsymbol{k}}^{\mathrm{ads}}$ and $\sigma_{\boldsymbol{k}}^{\mathrm{free}}$ are the Raman cross-sections of the $\boldsymbol{k}$-th vibrational mode of the adsorbed and of the free molecule, respectively. Two different mechanisms can contribute to the chemical enhancement: (i) Nonresonant chemical effect. The interaction between the molecule and the metal does not lead to the formation of a new electronic state (the molecular orbitals lay at energies not close enough to the Fermi level of the metal); however, such interaction may induce an appreciable change in the geometrical and electronic structure of the molecule, that reveals as a mild modification of the Raman shifts and of the intensity of the vibrational modes. (ii) Resonant charge transfer chemical effect. The interaction between the molecule and the metal brings about the creation of a metal-molecule charge transfer state. If the Raman scattering is excited with a laser source in resonance or pre-resonance with this state, some Raman modes can be strongly enhanced, in particular those coupled to the allowed electronic transitions (resonant Raman scattering).

3.2.2.4 Advantages and Applications of SERS

The advantages of SERS over conventional Raman were briefly mentioned before but are worth re-iterating since these benefits in conjunction with the shortcomings of Raman are responsible for the widespread of this technique. SERS proves a more sensitive and selective detection and quantification platform. Noninvasive, chemically specific and spatially resolved, analysis facilitates the exploration of SERS-based nanoprobes in diagnostic and theranostic applications with improved clinical outcomes compared to conventional technology. Other key advantages of SERS enhancement are that a reduced laser power and acquisition time can be incorporated for the analysis of fragile samples with minimal loss of spectral sensitivity whereas with conventional Raman some compromise must be made between the laser power and measurement time to obtain sufficiently distinct spectra without incurring sample damage.

Besides, the sharp molecularly specific spectra produced by SERS also promote the technique for use in multiplexed scenarios including multiple disease detection and in many instances, differences can be determined by eye without the need for the incorporation of chemometric methods. Conventional Raman is unlikely to be employed in such a manner since the spectral differences between healthy and diseased samples are extremely subtle, let alone between multiple diseases in a single sample, and generally require the implementation of statistical methods for their differentiation. Due to these advantages, researchers have applied SERS to a very wide range of applications in various areas, including biomedical, environmental analysis, food safety, forensic science, chemical warfare, and so on, as illustrated in Figure 3.12. Some interesting applications of SERS were highlighted briefly.

SERS in disease diagnosis Exciting advancements were noticed in the field of disease diagnosis by the implementation of SERS which is credited for the noninvasive nature and narrow spectral bands. SERS is applied to investigate from the building blocks (DNA and RNA) to functional complex structures (peptides and proteins) to even whole cells and tissues. Several studies have been published with respect to successful single- and double-stranded DNA detection with and without Raman labels and also interactions between DNA molecules and

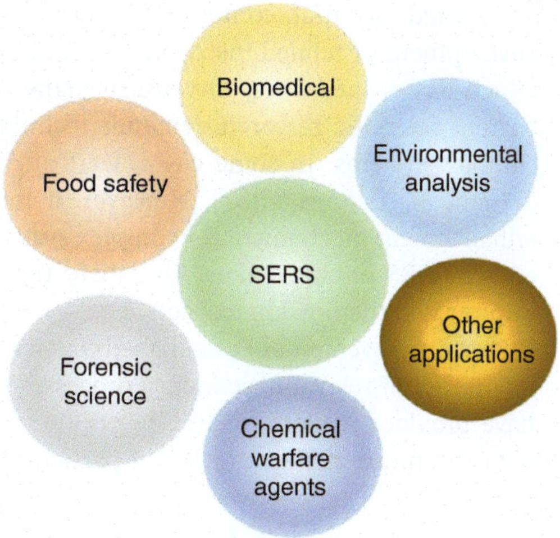

Figure 3.12 SERS applications in different areas.

small molecules [29]. Conformation changes induced by target molecules can also be monitored. Of particular interest in medical diagnostic is the development of SERS tags for cancer detection and monitoring cytotoxic effects of chemotherapeutic agents to cancel cells.

Food safety SERS can be employed to detect banned substances, including illegal dyes and foreign proteins in food. The detection of drugs and explosives is another potential application. SERS can also be applied as biosensor for glucose sensing or as pH sensor. Another fascinating field where SERS could be applied is the investigation of catalytic systems, such as molecular adsorbates at catalytically relevant interfaces, temperature control, or studies during galvanostatic electrodeposition.

Chemical warfare agents Gas phases chemical detection is of critical importance for the sensing of highly toxic molecules, such as chemical warfare agents and toxic industrial chemicals. The key challenge in detecting such analytes with SERS is to overcome the typical lack of interaction of the molecules with the substrate, allowing for the detection of a SERS signal. Despite this difficulty, explosives such as the half-mustard agent and dinitrobenzenethiol have been successfully detected by SERS. One way to circumvent the surface interaction challenge is to use water-soluble analytes that are converted from the gas phase to the liquid phase and can be detected using a device such as a combined microfluidics-SERS sensor, which has allowed for the detection of most explosives. Beyond the ability to detect chemical warfare agents in a laboratory, i.e. in the context of large, technical, and relatively expensive apparatus, significant interest is the detection of chemical warfare with SERS in the field. The size of small and portable Raman spectrometers is close to that of a smartphone. Overall, SERS is a promising method for the ultrasensitive detection of chemical species that are relevant to security.

Overall, since its discovery of SERS more than 40 years ago, it has been confirmed as a powerful technique to probe simple as well as complex molecules. A broad spectrum of physical, chemical, and analytical applications has been proposed, ranging from materials and environmental science to biology and medicine.

Such huge potential, combined with extraordinary technological progress in the development of related instrumentation, has resulted in an explosion of research, which has pushed SERS forward in many different directions [30]. Nevertheless, current scientific research demands techniques to be able to analyze even more complex systems. In some cases, featuring smaller and more precisely designed 2D and 3D nanogaps [31]. However, it remains challenging to understand the enhancement mechanism for complex systems, and it requires rigorous evaluation of the SERS enhancement. To satisfy this requirement, SERS must be further improved in the flowing respects.

First, improvements of established classical EM and chemical mechanisms, implementation of relevant secondary processes, and consideration of quantum effects have provided a more precise description and prediction of SERS enhancement, with the future demonstration of particularly high values within nanometer and sub-nano scale gaps between metal nanoparticles or between nanoparticles and flat metal substrates. One of the major driving forces behind the recent progress of SERS has been the practical achievement of nanostructures with excellent quality and tailored morphologies. However, it remains challenging to understand the enhancement mechanism for complex systems, and it requires rigorous evaluation of the SERS enhancement.

Second, high sensitivity, efficiency, and reproducibility of SERS substrates are recurrently discussed as key factors. In this context, specific platforms have been proposed and synthesized toward a variety of sensing strategies based on SERS nanotags, chiral-selective systems, SHINERS, intragap core-shell particles, and remote SERS, often achieving SM detection. The development and implementation of statistical data analysis currently offer quantification and identification of multiple molecules/components within complex mixtures, which are crucial advancements for biomedical and environmental applications. However, with the continuous advancement of nanofabrication technology, the current focus is on the development of SERS substrates with low signal fluctuation, low fluorescence or photoluminescence backgrounds, high uniformity and stability, and less photo-induced reaction or desorption.

In addition, applications of SERS in biosensing have a rapid increase during the past decade, related to the detection of cancer biomarkers. However, progress toward becoming a widespread clinical diagnostics technology has still lagged behind possibly due to the use of highly diluted samples in complex matrixes and the need for sophisticated instrumentation. Some of these issues could be solved by using brighter probes, such as SERS tags combined with microfluidic systems, using portable devices. In the case of *in vivo* applications, limited tissue penetration depth has hindered implementation in full-body or deep tissues. Therefore, applications where the laser beam can access the diseased area have a better chance to succeed.

3.3 Electron Microscopy

The word microscope is derived from the Greek *mikros* (means small) and *skopeo* (means look at). From science and life, there has been an interest in being able to look at smaller and smaller details of the world around us. Biologists wanted to examine the structure of cells, bacteria, and even viruses. Materials scientists wanted to see imperfections in metals and ceramics. Detailed study of rocks, minerals, and fossils on a microscopic scale provides insight into the origins of our planet.

It is accepted that one of the earliest magnifying instruments for observing small objects was made by Dutchman Antony van Leeuwenhoek (1632–1723) from the Netherlands. It consisted of a powerful convex lens and an adjustable holder for the object. The limiting factor in Van Leeuwenhoek's microscope was the single convex lens, which was remedied later by the addition of a second lens, micro focusing, and lighting systems, giving us the basis of optical microscopes today for revealing the microcosm. Optical microscopes are the ones most familiar to us. They use visible light and transparent lenses to see objects as small as about 1 μm. In the process of perfection and development of the optical microscope, it is found that the magnification and separation resolution are always limited to the levels of ~1000× and ~200 nm no matter how to improve the lens and structure of the optical microscope. Scientists tried to achieve better resolution, but they soon realized that the resolving power of the microscope was limited not only by the number and quality of the lenses but also by the wavelength of the light of illumination. This limitation is recognized from the diffraction of visible light source induced by optical microscope.

Since the light source limits the magnification and resolution of the microscope, the light source with a shorter wavelength should be replaced to improve the performance of microscope. Using ultraviolet light gave a small improvement on the distribution of optical microscope. However, most of the observed objects strongly absorb ultraviolet light, and ultraviolet optical microscope is rarely used. If an X-ray is used as the light source, the resolution can be greatly improved, but no optical lens is suitable for converging X-ray.

In the 1920s, Louis de Broglie proposed that accelerated electrons should exhibit wavelike properties with a wavelength substantially shorter than light [32]. Due to wave-particle duality, the wavelength of an electron beam after a certain acceleration voltage is shorter than that of an X-ray. Thus, using the electron beam as the light source, the resolution of the microscope may be improved by several orders of magnitude. Furthermore, it was discovered that electric and magnetic fields could be conceived of as a lens to bend and focus the electron beam, like a glass lens for visible light. Hence, it was feasible to build an instrument that uses electrons to illuminate a specimen and surpass the resolution limit of optical microscope. Thus, electron microscopes use a beam of charged particles instead of light and use EM lenses to focus the particles. They can see features as small as an angstrom (about the size of hydrogen atom). Table 3.2 summarized signals and technologies generated by an electron beam passing through a particle matter.

In fact, theoretical and experimental basis for the electron microscope has been established prior to this equipment, including vacuum technology in the 1640s, electron discovery by J.J. Thomson in 1897, proposal of wave-particle duality in 1924, discoverage of magnetic fields focusing electron beams by Busch in 1926 [33], and observation of low energy electron diffraction on the (111) plane of Ni single crystal by Davidson.

Table 3.2 Interaction of an electron beam with a matter and corresponding electronic techniques (Signal depth from sample mainly depends on the accelerating voltage and atomic number of matter).

Electron-sample	Signals	Depth in sample	Electronic technique
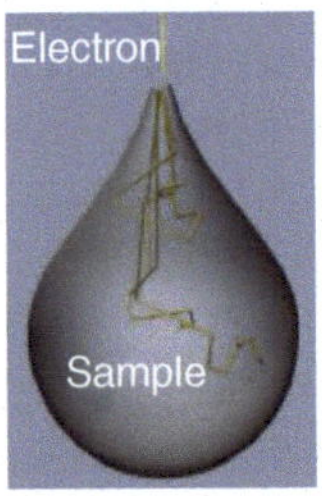	Backscattered electron	<1 μm	SEM/EELS
	Secondary electron	5–50 nm	
	Transmission electron	/	TEM
	Absorbed electron	/	Absorbed electron image
	Characteristic X-ray	0.5–5 μm	EDS
	Auger electron	<1 nm	Auger spectrum/EM

The wavelength of electron can be calculated from the function of wave-particle duality $\lambda = {}^{h}/_{mv} = {}^{h}/_{\sqrt{2meV}} = \frac{12.25}{\sqrt{V}}\,\text{Å}$. Therefore, the wavelength of electron decreases from 0.037 Å to 0.014 Å with increasing the voltage from 100 to 200 kV (The acceleration voltage of transmission electron microscope [TEM]). Theoretically, the resolution of electron microscope is about 0.02 Å under 100 kV accelerating voltage. However, the actual resolution of the electron microscope is ~2 Å, which is far from the theoretical prediction. This is due to the imperfection and various aberrations of the magnetic lens used for focusing electron in the microscope, such as spherical aberration, chromatic aberration, astigmatism, and distortion. Nevertheless, the resolution of electron microscope is more than 1000 times higher than that of optical microscope.

The EM "lens" in the microscope is defined as the effect of EM field on electrons that is very similar to that of light wave through optical lens. As an electronic optical device, electron microscopes can be used to observe the morphology of object and identify the crystal structure and chemical composition. There are two electron microscopes, TEM and SEM that are widely applied for the characterization of plasmonic nanostructures.

3.3.1 Transmission Electron Microscope

In the early 1930s, Ernst Ruska at the University of Berlin, along with Max Knoll, developed the idea into a practical reality. The first TEM was built in Germany [34], for which Ruska was awarded the Nobel Prize for Physics in 1986. In 1934, a third lens was added, demonstrating a resolution of 50 nm. The first commercial TEM equipment with the resolution of 10 nm was constructed by Siemens AG in 1939. In the 1970s, with the establishment of theory and technique for high-resolution TEM, atoms arrangement in the crystals can be observed. Then study on the crystalline structure, composition, and element distribution was carried out by means of EELS, SAED, and electron spectrum. Along with the rapid development of nanoscience and

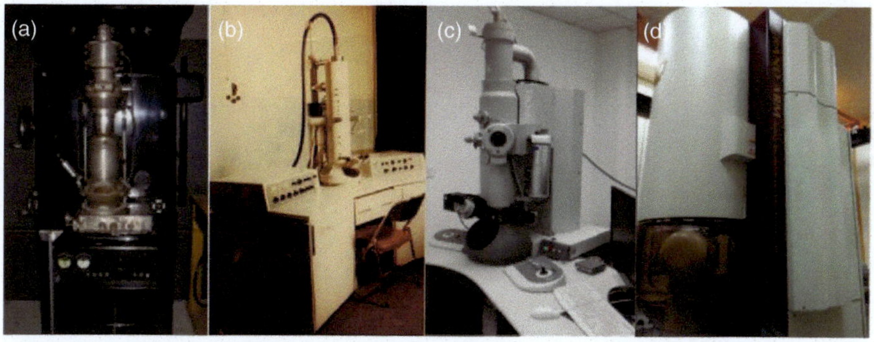

Figure 3.13 Typical TEM of different ages for observation of Au, Ag nanoparticles. (a) The first commercial TEM was built in Siemens Laboratory in 1938. (b) DX-2 electron microscope was developed in 1965 at Changchun Institute of Optics, Fine Mechanics and Physics, Chinese Academy of Sciences. (c) Tecnai G2 F30 microscope with functions of HRTEM and EDS Mapping. (d) Double spherical aberration-corrected STEM (JEM-ARM300F) in 2019 with spatial resolution of 80 pm.

nanotechnology in the 1990s, AC-TEM and cell-TEM were built based on TEM for the demand of ultra-high resolution of atom distribution in one protocell and *in-situ* observing the crystal formation process. Some typical TEM equipments of different generations were indicated in Figure 3.13. Today, electron microscopes have reached resolutions of better than 0.05 nm, more than 4000 times better than a typical light microscope and 4 000 000 times better than the human eyes.

There are four main components to a TEM, including electronics, an electron optical column, a vacuum system, control panels, and recording system, as shown in Figure 3.14. From the electronics, electron beam from the high-voltage generator was focused and deflected by lens. In the electron optical column, the elements analogous to those of a light microscope and the glass lenses are replaced by EM lenses. Unlike glass lenses, the power (focal length) of magnetic lenses can be tuned by changing the current through the lens coil. The electron beam emerges from the electron gun and passes through a thin specimen. The transmitting electrons are collected, focused, and projected onto the viewing device at the bottom of the column. The entire electrons path from gun to camera must be under vacuum condition. The images and diffraction patterns can be observed on the fluorescent screen, which were usually recorded by film, digital camera, or computers.

When the electron beam passes through the optical column, the electrons change the direction after collision with the atoms in the sample, resulting in solid angular scattering. This is the signal collected by STEM to be introduced later. The transmitting electron numbers are different corresponding to the thickness of the sample. The transmitting electron beam carries the information of the sample, which is primarily amplified by objective lens and comprehensively amplified by the subsequent intermediate lens and projection projector lens. In this duration of amplification, the diffraction image can also be obtained through changing the current of the intermediate lens. Therefore, TEM can present us with both an inverted space diffraction pattern and the real space two-dimensional image of samples.

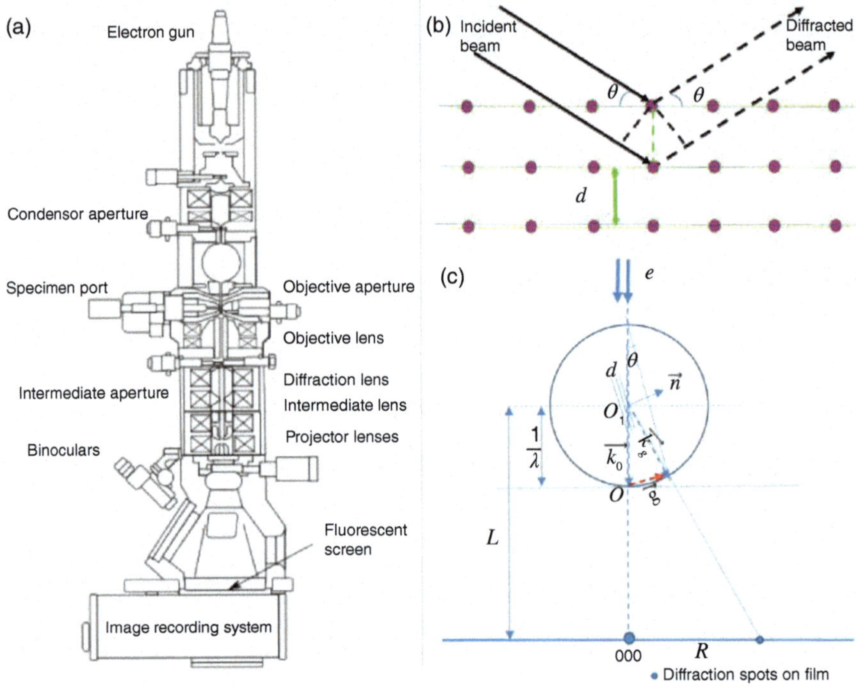

Figure 3.14 (a) Diagram of TEM equipment. (b) Bragg diffraction of incident light beam or electron beam on crystal plane. (c) Diagram of Bragg's law in Ewald sphere.

With the wide usage of computers in the control systems and data storage, the TEM working environment is no longer a dark room, and the film is gradually withdrawn from the recording images. Morphology image with magnification and bars is the main recorded information of two-dimensional morphology image of nanomaterials. In the following image processing, the bar is an important reference. However, different from the Fourier transform of lattice to diffraction spots, the inverse Fourier transform can transform the diffraction spots to lattice. TEM parameters are not included in the film of digitized diffraction, and some important calculations cannot be carried out on the digitized diffraction image that can be arbitrarily enlarged and reduced.

From the Bragg's law ($2d_{\mathrm{kl}}\sin\theta = n\lambda$), θ is the angle between incident electron beam and crystal plane. For practical characterization technology, $n = 1$, $\sin\theta = \lambda/2d$. The accelerating voltage of TEM is usually in the range of 100–200 kV, and the electron wavelength is ~10^{-3} nm far less than the crystal plane spacing (10^{-1} nm). Therefore, the diffraction angle (θ) is about 10^{-2} rad. And the crystal plane that can produce Bragg diffraction is almost parallel to the incident electron beam. In 1913, Bragg and his son successfully measured the crystal structure of diamond with the help of X-ray and put forward the "Bragg formula," which laid a foundation for the establishment of modern crystallography.

Combining Bragg's law and Ewald sphere, diffraction image in the inverted space can be obtained, as shown in Figure 3.14c. Here, AO is incident electron beam direction, part of which passes through the sample along $\overrightarrow{k_0}$ and diffracts in the direction of $\overrightarrow{k_g}$. With reflection point as the center and $\frac{1}{\lambda}$ as radius, one can draw the Ewald sphere. G is inverted lattice point, OG is reciprocal vector of a certain (hkl) crystal plane. Therefore, the reciprocal diffraction points are all on the surface of Ewald sphere. As it is known that the diffraction crystal plane is almost parallel to the incident electron beam. The wavelength of electron λ is much smaller than the crystal plane spacing d (note: the radius $\frac{1}{\lambda}$ of Ewald sphere is not proportional to the camera constant [L] in the figure). Then the relationship of these parameters can be obtained.

$$\sin\theta = \frac{OG}{AO} = \frac{1/d}{2/\lambda} \tag{3.7}$$

$$\sin 2\theta = \frac{R}{L} = \frac{2/d}{2/\lambda} = \frac{\lambda}{d} \tag{3.8}$$

$$L\lambda = Rd \tag{3.9}$$

This is one important relation in the TEM analysis of selected area electron diffraction (SAED) for spots recorded on the film. For single crystals grown along a preferred direction, the diffraction spots are in equidistant symmetrical about the center (000). L has a specific value for different electron microscopes (generally tens of mm). R is the distance (unit of mm) from the spot to the (000) center on the specific-sized film equipped with TEM. The wavelength λ can be calculated from the function of wave-particle duality. Therefore, the crystal plane spacings and some (hkl) crystal planes can be calculated accurately from the diffraction pattern, including single crystal spots and polycrystalline rings. Additionally, one can get the crystal axis [uvw], since the crystal planes induced diffraction are almost parallel to the incident electron beam.

In fact, Bragg's law only discusses the diffraction of electrons by crystals from a view of geometry without considering the position and density of atoms that are reflected in the structural factor F_{hkl} (not described here). This structure factor determines the (hkl) of different crystal structures, such as (111), (200), (220), (311), and other modes of all odd or all even for *fcc* structures. There is no odd–even mixed plane for the diffraction spots, such as (100, 110), (210). Therefore, Bragg's law determines the diffraction angle, and the diffraction intensity depends on F_{hkl} structure factor. For thin film samples, especially for *fcc* structures with stacking fault, there will be dynamically forbidden fractional diffraction [35]. The regular displacement of the atoms in the crystal will cause the change of electron diffraction, resulting in the appearance of original extinction spots, such as fractional diffraction $^{(422)}/_3$ and $^{2(422)}/_3$, as shown in Figure 3.15 [36]. This anomalous diffraction will be described in Chapter 5 for Au nanoplates.

3.3.2 Scanning Electron Microscope (SEM)

The images of matter and structure by TEM is a two-dimensional transmission diagram. Similar to the shadow of a tree in the sun, it cannot reflect the real information on the tree (branch structure, leaf morphology, whether there are fruits

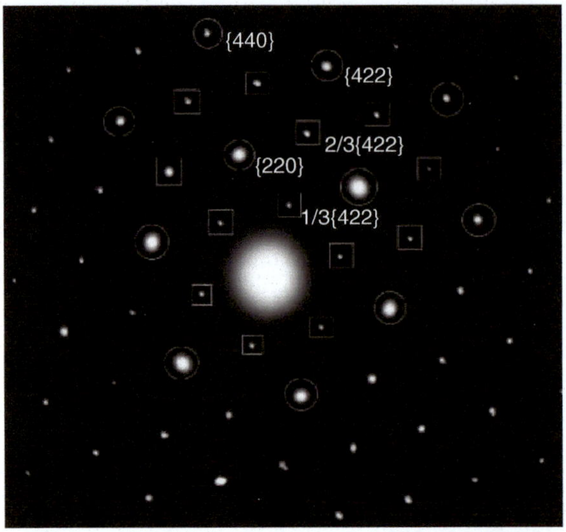

Figure 3.15 One typical SAED pattern of Au nanoplate, showing the fractional diffraction. Source: Kan et al. [36]/Reproduced with permission from © 2010 JOHN WILEY & SONS, INC.

Figure 3.16 (a) An elephant shadow formed in the sun by three children. (b) SEM image of Ag nanocubes modified with small particles. Source: Zhu et al. [37]/Reproduced with permission from © 2009, Springer Nature.

and animals). As shown in Figure 3.16a, the elephant shadow formed in the sun is a specific combination of three children. Therefore, TEM can not present the real morphology of the structure, especially the surface information. The real structures of some square nanostructures in Figure 3.13f are nanocubes with small nanoparticles modification on the surfaces, as indicated by SEM characterization in Figure 3.16b [37].

The working principle of SEM was proposed by German scientist Knoll and the original SEM model was constructed in 1935. The first laboratory-level SEM was made in the United Kingdom in 1942, and produced as a commodity in Cambridge Scientific Instruments Limited in 1965. Then with the successful development of a field emission electron gun, FE-SEM was obtained with high resolution.

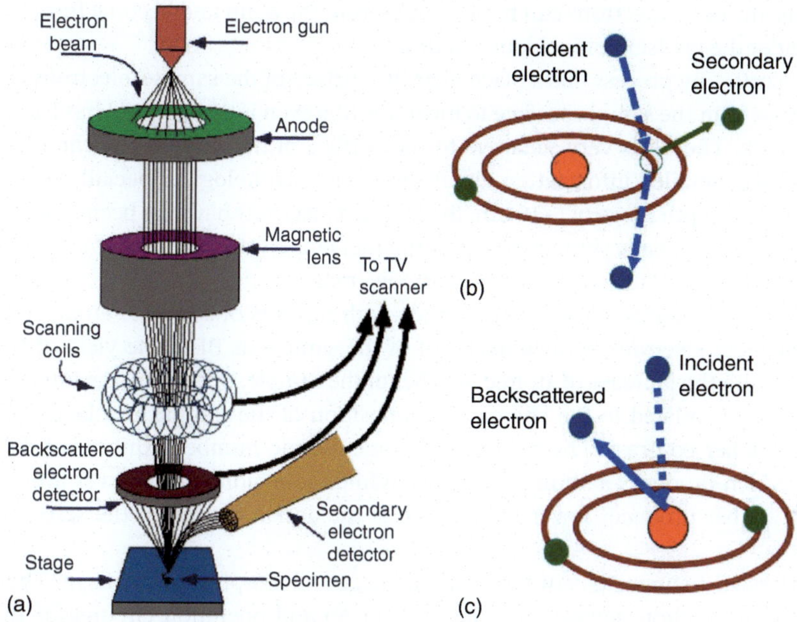

Figure 3.17 (a) Diagram of SEM instrument, Formation of secondary electrons (SEs) (b) and backscattering electrons (BEs) (c) under incident electron.

Similar to TEM, SEM consists of an electron optical column, a vacuum system, electronics, and software, as shown in Figure 3.17. However, the SEM technique does not impose any restriction on specimen and the specimen chamber can be larger and thick. In addition to the large sample size and low accelerating voltages, the most important difference between TEM and SEM is the imaging process. Rather than the broad static beam used in TEM, the SEM beam is focused to a fine point and scans line by line over the sample. This beam is scanned in a rectangular raster over the specimen and the intensities of various signals created by interactions between the beam electrons and the specimen are measured and stored in computer memory. The resolution of the new generation SEM can reach 1 nm, with large field depth of, large view field, and good stereoscopic sense. With the combination of SEM and other analytical instruments, it can be used to observe the micromorphology and analyze the composition of the material. In the imaging process, there are two image modes collected, respectively, from secondary electrons (SEs) and backscattering electrons (BEs), as illustrated in Figure 3.17b,c. The SE signal is frequently applied in imaging. Through photoelectric signal conversion, the stored values are then mapped as variations in brightness. The image varies with the topography of the sample surface, in which the edges are bright and the recesses are dark.

The SEs are the outer electrons of an atom bombarded by the incident electrons. Because of the small binding energy between the nucleus and the outer valence electrons, the outer electrons can be separated from the atom and become free electrons under the collision of incident electrons.

If this scattering process takes place near the surface of the sample, electrons can also escape from the sample surface to form free SEs even if the material has a large escape work. The SE is very sensitive to the surface of the sample and can effectively display detailed information about the micromorphology. Especially for the surface with a depth range of 5–10 nm, the incident electrons have not been reflected yet. Thus, the SE area is similar to that of the incident electron beam, which is one important reason for the high resolution of SEs image.

When the incident electron beams collide with nucleus of a sample atom, some of the incident electrons bounce back out of the sample as BEs. The yield of BEs increases with the increase of atomic number of the sample. Therefore, the intensity of the signal is related to the chemical composition of the sample, displaying the atomic number contrast. The region with large atomic number exhibits a bright region, which produces a strong SE signal. While the region with a lower average atomic number produces fewer BEs, which is the darker region on the screen or photos.

Then how to distinguish between element image and morphological image? There are two p–n junction device detectors (M and N) and operation circuits for this function, as arranged in Figure 3.18. For samples with flat surfaces and different elements, the signal from M is similar to that from N detector. The subtracted signal is 0, while the superimposed signal multiplied, which is the electronic image only containing the information of element components. For samples with uniform composition but rough surface, the signal from M is contrary to that from N detector. The subtracted signal is multiplied, while the superimposed signal is 0. For this case, the multiplied signal is the information of morphology. Figure 3.19 presented some BEs

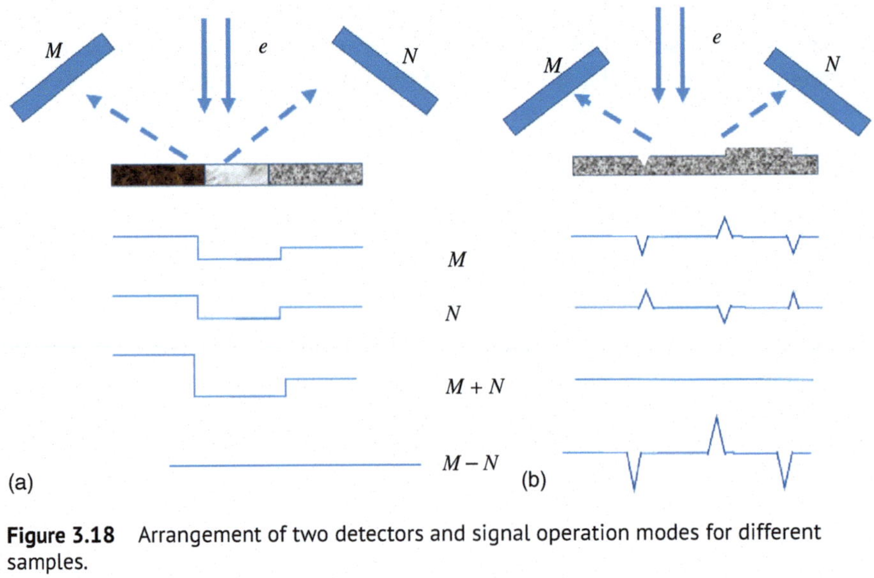

Figure 3.18 Arrangement of two detectors and signal operation modes for different samples.

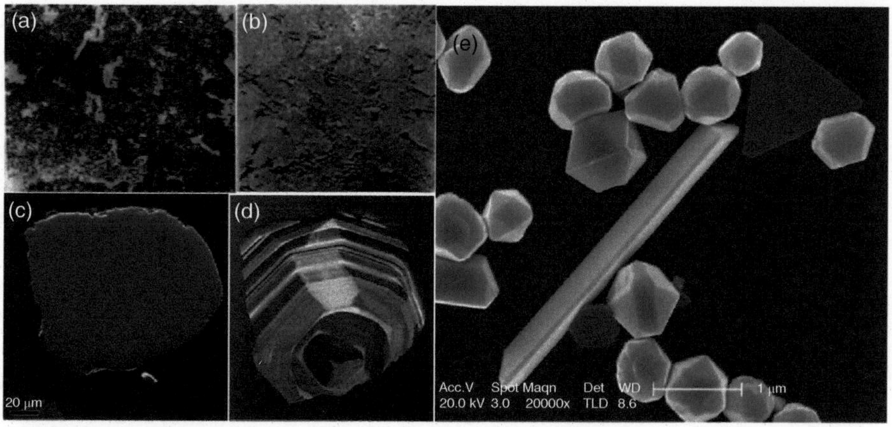

Figure 3.19 Element and morphology diagrams of polished aluminum alloy (a, b) and single-element crystal (c, d). (e) Au nanostructures with different dimensional morphologies.

and SEs images of polished aluminum alloy and single-element crystal, showing the element image and morphology image.

3.3.3 Combination TEM and SEM

3.3.3.1 Aberration Correction

As mentioned above, the resolutions of both TEM and SEM do not achieve the theoretical expectation under the high accelerating voltage. In fact, there are similarities between magnetic lenses and optical lenses, in addition to the influence of diffraction effect on resolution, there is also the influence of aberration on resolution, including special aberration, astigmatism, and chromatic aberration. The astigmatism is caused by the asymmetry of the magnetic field, which can be eliminated by introducing an astigmatist. Chromatic aberration correctors have been successfully incorporated into the TEM platform. The spherical aberration is caused by the different electron deflecting abilities in the near-axis region and the far-axis region of the magnetic field. The spherical aberration of EM lens is the main factor of aberration, and it cannot be supplemented or corrected by the combination design of convex lens and concave lens.

The TEM contains a number of magnetic lenses, including objective lens, condensing lens, intermediate lens, and projection lens. In the imaging process, the aberration of the objective lens will be amplified during the electron passing through subsequent lens. In most cases, TEM resolution is limited primarily by spherical aberration, resulting in a general blurring of the image and a phenomenon called delocalization, in which periodic structures appear to extend beyond their actual physical boundaries.

The point resolution of electron microscope is defined as $r = A\lambda^{3/4}Cs^{1/4}$. Cs is the special aberration coefficient with unit of mm, such as Cs is 0.64 mm for HRTEM, and 0.43 mm for STEM. You can achieve a smaller r by decreasing Cs and/or λ. In practice, there is a practical limit on λ due to the beam damage. Therefore, additional hardware to compensate *Cs* is developed. And this hardware is called correctors similar to the glasses for eyes.

In 1936, Scherzer first proposed the inevitable existence of spherical aberration in electron microscopy [38]. He also designed hardware in the optical path to reduce the spherical aberration. This hardware did not come into being until the 1970s when Harald Rose designed an EM corrector to correct spherical aberration and chromatic aberration [39]. Then in the 1990s, Harald Rose, Maximilian Haider, and Knut Urban successfully corrected the spherical aberration and chromatic aberration of objective lens by using EM 4-pole and 8-pole fields [40]. These three German physicists overcame this difficult problem that has puzzled the electron microscope industry for more than half a century. In 1998, the first AC-TEM in the world was successfully developed and achieved sub-angstrom resolution. After more than 80 years of development, the resolution of analytical TEM has reached 0.2 nm, and the spatial resolution of the current AC-TEM can reach 50 pm, far beyond the classical optical diffraction limitation.

3.3.3.2 STEM

There are two main TEM operating modes: conventional (HR)TEM mode and scanning transmission electron microscopy (STEM) mode. Correspondingly, there are two types of spherical aberration correctors, involving image Cs corrector (AC-TEM) and Probe Cs corrector (AC-STEM). The photoelectron intensities of two modes are quite different, like flash light source and laser light source. The transmitted electrons that have been scattered through a relatively large angle are detected using a high-angle annular dark field (HAADF) detector. In the TEM mode, the incident beam is broad and unfocused. While the incident beam focused and most relevant subtype is HAADF-STEM imaging in the STEM mode. STEM combines the principles of TEM and SEM and can be performed on either type of instrument. STEM enables the use of many signals produced from the interaction of electron and matter, including SEs, scattered beam electrons, characteristic X-rays, and electron energy loss. Compared with the conventional SEM, the primary advantage of STEM is the improvement in spatial resolution under the requirement of very thin samples.

Images in Figure 3.20a–c are the catalytically active icosahedral Pd nanoparticles by coating a few atomic Pt layers [40]. The atomic-resolution HAADF-STEM image indicates the presence of an average of 2.7 Pt layers in the shell. This structures was denoted as core-shell nanocrystals shown Pd@$Pt_{2.7L}$ icosahedra. The authors further analyzed the bimetallic nanocrystals by energy-dispersive X-ray spectroscopy (EDX) mapping to confirm the formation of a Pd–Pt core–shell structure (Figure 3.20d). From the atomic-resolution HAADF-STEM images shown in Figure 3.20(e-f), it was found that the packing of Pt atoms in the overlayers did not exactly follow the underlying Pd atoms of the core. Instead, the Pt atoms underwent relaxation to take a corrugated structure due to the existence of both tensile strain and twin boundaries on the surface of a Pd icosahedral seed. Therefore, the spherical aberration correction electron microscope can be used to detect the structure and atomic arrangement of matter in the atomic scale, as shown in Figure 3.21 for GaAs and Si crystals [42] (https://www.fei.com/products/tem/themis-z-for-materials-science). Now the Themis Z delivers impressive STEM resolution down to ~39 pm at 300 kV.

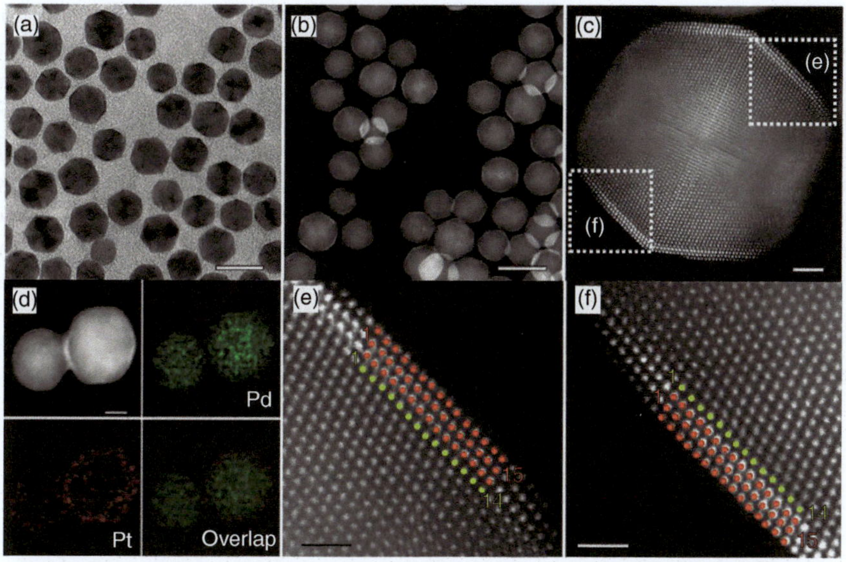

Figure 3.20 Structural and compositional analyses of the Pd@Pt2.7L icosahedra. (a) TEM and (b) HAADF-STEM images. Scale bar, 20 nm. (c) Atomicresolution HAADF-STEM image taken from a single particle along a two-fold symmetry axis, showing three atomic layers of Pt in the shell (with a brighter contrast) and the Pd atoms in the core. Scale bar, 2 nm. (d) HAADF-STEM image of two $Pd@Pt_{2.7L}$ icosahedra and the corresponding energy-dispersive X-ray spectroscopy mapping of Pd and Pt, confirming a core-shell structure. Scale bar, 2 nm. (e,f) Atomic-resolution HAADF-STEM images taken from the edges marked by boxes in (c), revealing the detailed arrangements of Pd and Pt atoms (green dots: Pd atoms; red dots: Pt atoms). Scale bar, 1 nm. Source: Reproduced with permission from Ref. [40].

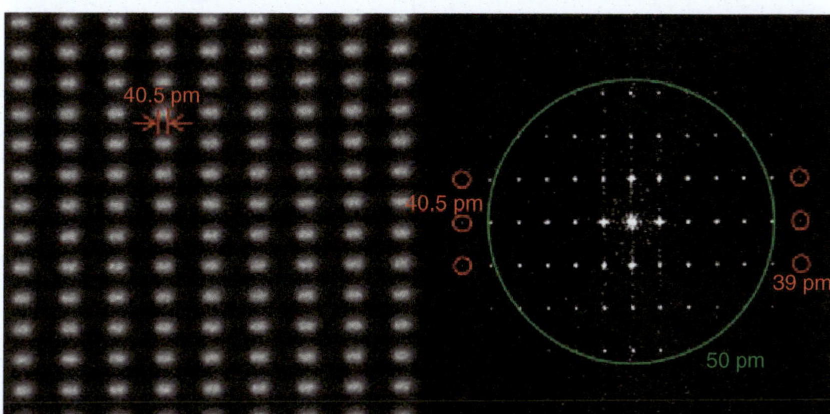

Figure 3.21 39 pm STEM resolution demonstrated on GaN[212]; Taken at 300 kV using a Thermo Fisher Scientific Spectra 300 Transmission Electron Microscope. Source: Fu et al. [41] Thermo Fisher Scientific/from the website of (https://www.fei.com/products/tem/themis-z-for-materials-science/).

With the ultrahigh resolution of STEM for atomic arrangement, it is possible for scientists to study phase transformation, monatomic catalysts, atomic segregation, and structural evolution in alloys [41]. Since the discovery of high-temperature superconductivity in copper oxide materials, sustained efforts were made to understand the origins of this phase and discover new cuprate-like superconducting materials. Due to the proximity of nickel and copper, nickelates have been reported as close cuprate analogs. More currently, nickelate superconductors were achieved without chemical doping by Pan et al. Figure 3.22a,b showed the layered crystal structure description of the layered nickelates of $Nd_6Ni_5O_{16}$ Ruddlesden-Popper

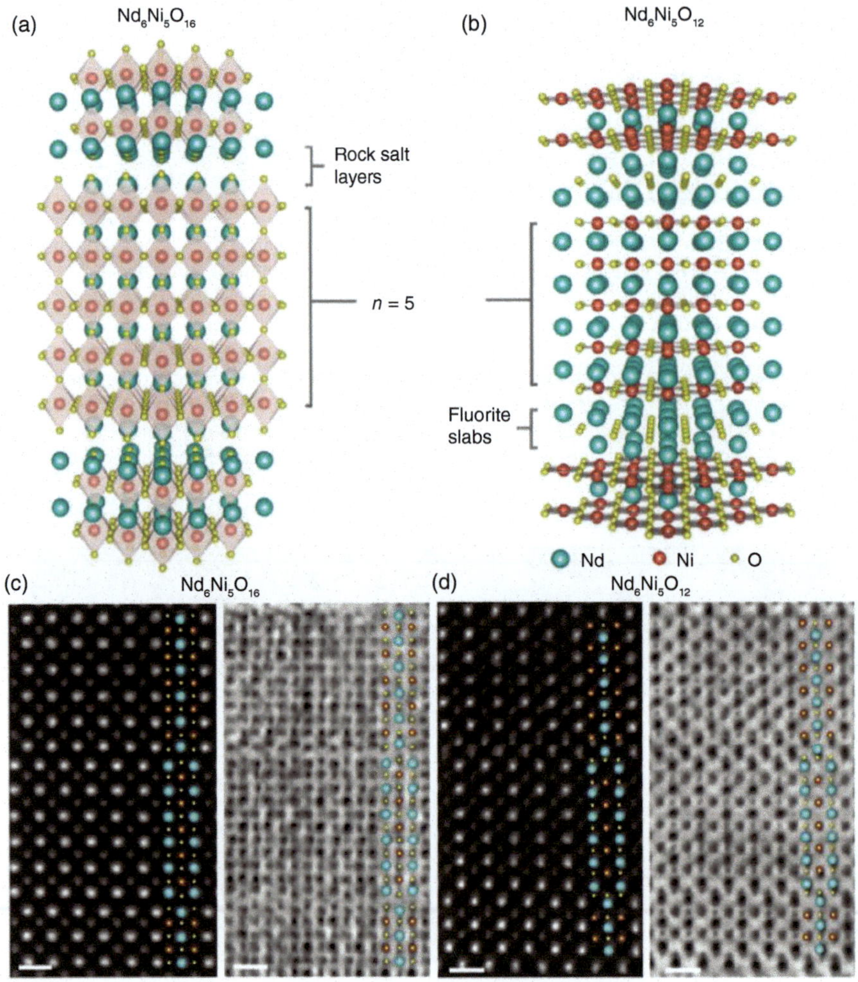

Figure 3.22 Crystal structures of the layered nickelates in the $Nd_6Ni_5O_{16}$ Ruddlesden–Popper phase (a) and $Nd_6Ni_5O_{12}$ reduced square-planar phase (b). HAADF and ABF STEM images of the $Nd_6Ni_5O_{12}$ (c) and $Nd_6Ni_5O_{12}$ (d). Nd, Ni, and O are represented by turquoise, red, and yellow dots, respectively. Scale bars, 5 Å. Source: Reproduced with permission from Ref. [43]; © 2022, Springer Nature.

phase and $Nd_6Ni_5O_{12}$ reduced square-planar phase [43]. Atomic-resolution HAADF images of well-ordered regions confirm both the shortening of the *C*-axis lattice constant and the removal of apical oxygen in the neodymium planes from the $Nd_6Ni_5O_{16}$ (Figure 3.22c) to the $Nd_6Ni_5O_{12}$ phase (Figure 3.22d). The presence of different superlattice was proved by the peaks shift in the X-ray diffraction spectra. The temperature-dependent resistivities indicated a metal-to-insulator transition for $Nd_6Ni_5O_{16}$ Ruddlesden–Popper film and superconductivity for $Nd_6Ni_5O_{12}$ film.

Thanks to the advances in aberration correctors, another important measurement is electron energy loss spectrometry (EELS) carried out in STEM, as shown in Figure 3.23. One particular energy was selected for examination and the spectral intensity was plotted as a function of position, resulting in an energy-filtered spectrum image, which is commonly known as an EELS map. In most STEM/EELS experiments, the structural information is collected by a HAADF detector, which registers only the electrons that are scattered to high angles. EELS can analyze elements 1–99 and is more sensitive to light elements. There are usually three parts in the EELS (Figure 3.23b): zero-loss peak from unscattered or fully elastic collision electrons, low loss area less than 50 eV (such as plasmon peak excited in this region), and high ionization loss peak of excited inner electrons. As presented in Chapter 2, plasmon can be excited by a high electron beam with the EELS mapping [45]. SP modes for the individual Au NR or coupled Au NRs dimer can be observed through experimental EELS maps, agreeing well with the FDTD simulations.

3.3.4 *In-Situ* Electron Microscope

The invention of the TEM was driven, in part, by the increasing interest in structures with sizes beyond the resolving power of the optical microscope. With decreasing

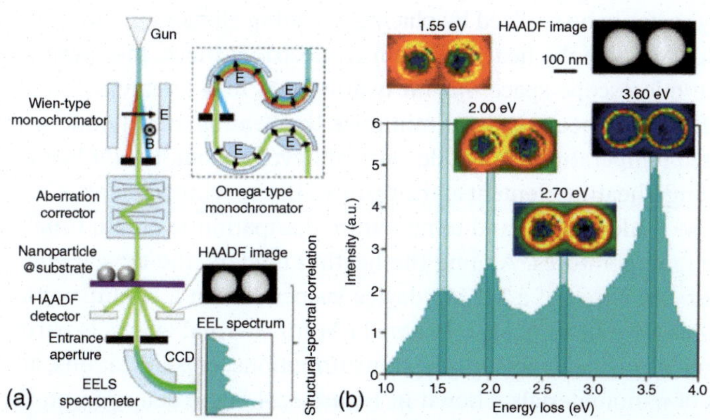

Figure 3.23 Schematic of EELS acquisition in STEM equipped with a Wien-type monochromator (inset in a) and an aberration corrector. (b) An EELS of one Ag dimer revealing different LSPR modes supported by the system. Also presented are the HAADF image and LSPR mode maps generated at loss energies of 1.55, 2.00, 2.70, and 3.60 eV. The green dot in the HAADF image is the beam position used for the spectrum acquisition. Source: Reproduced with permission from Ref. [44]. © 2018, American Chemical Society.

the characteristic size of microelectronic devices to the nanometer size, controllable preparation of various hyperfine structures is the important foundation. For micromachining and failure analysis of nano-devices, it is necessary to study how to control the structural evolution of metal nanocrystals at the atomic scale. In most physical deposition and all of the liquid phase chemical synthesis of nanomaterials, the formation of the structure depends on the method summarized by repeated experiments. With the increase in reaction complexity, the traditional repeated testing methods become inefficient. Due to the lack of direct means to characterize the nucleation and growth of nanocrystals in liquid environment at the atomic scale, many scientific problems about the growth mechanism of nanocrystals have not yet formed theories, in which some views about the nucleation and growth are usually controversial. For plasmonic nanostructures, they are generally prepared through chemical methods. The chemical synthesis of metal nanocrystals is still an art, not a science at present, as outlined by Y. Xia et al. [46]

As mentioned above in this chapter, UV–vis–NIR spectroscopy is a simple means to characterize the structural evolution of metal nanocrystals in liquid environment. For example, the etching process of Au NRs by $FeCl_3$ can be detected by UV–vis–NIR spectra, in which the blue shifting of the longitudinal SPR of Au NRs indicated the decrease of aspect ratio in the etching process. And the SPR shifting is coincident with the *in-situ* TEM images. The measurement of UV–vis–NIR spectra is simple and low cost. It can quickly reflect the possible structural and property changes of the sample during the reaction process. However, both the UV–vis–NIR spectra and ordinary TEM cannot directly reflect the detailed change information of the sample morphology. Therefore, it is necessary to develop the equipment for *in-situ* observation on the growth and interaction process under electron microscope.

The key unit for the *in-situ* observation is the TEM specimen holder. Depending on what stimulus is to be applied, a corresponding specimen holder can be designed. Heating is normally easier to be realized by placing a heating element at the tip of a TEM specimen holder. Unlike the heat treatment in electric furnace, the specimen was confined in the microscope space. *In-situ* heating was first researched in the 1950s, and many efforts had been made to minimize the heating area atomic with resolution at elevated temperatures. After decades of development, *in-situ* heating was realized by placing a heating element at the tip of a TEM specimen holder, which normally uses precise Joule heating and temperature dissipation to obtain tunable and stable specimen temperatures. Among the heating holders, the typical one is MEMS-based holder composed of a MEMS device suspended on a Si chip, which produces localized resistive heating [47]. *In-situ* TEM for thermal effects is particularly interesting in melting behaviors, phase transformations, crystallizations, and chemical reactions of nanomaterials. Shown in Figure 3.24 is the detection on the alloying and dealloying behaviors in Au@Ag core–shell nanorods by thermal stimulus by L.T. Sun and C.X. Kan groups [48].

Here, Au@Ag core–shell nanorods were used as an example to investigate the structure and phase behaviors in bimetallic nanocrystals. A combinatorial method by combining a chip-based heating holder with aberration-corrected TEMs was used to apply thermal pulse treatments on the Au@Ag nanorods. Figure 3.24a shows the

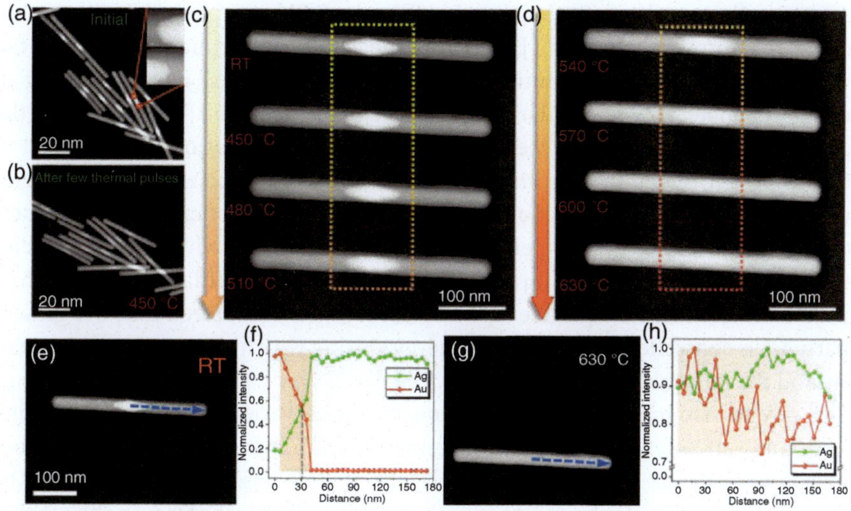

Figure 3.24 (a) Dark-field STEM image of the initial Au@Ag core–shell nanorods at room temperature. The inset shows the interfaces of a typical nanorod. (b) Dark-field STEM image of the Au@Ag nanorods after exerting a few thermal pulses. The thermal pulse is set as 450 °C with a duration time of one second. (c, d) STEM image sequences showing the alloying process of a typical Au@Ag nanorod. (e–h) Dark-field STEM images and corresponding EDS line scans of the Au@Ag core-shell nanorod before and after *in-situ* heating process. Source: He et al. [48]/Reproduced with permission from ©2022 Royal Society of Chemistry.

STEM image of the initial Au@Ag heterostructures with sharp Au–Ag interfaces illustrated in the inset. Figure 3.24b shows the structure change after exerting a few thermal pulses. As can be seen, the boundary in the *Z* contrast between Au and Ag becomes blurry after this treatment, indicating that inter-diffusion should have taken place in the nanorods under the stimulus of thermal pulses. Figure 3.24c,d show the structure/phase evolution tracked after each thermal pulse. It can be seen that the Au core disperses significantly at elevated temperatures. Together with the dispersion of the Au core, the brightness of the rod ends was also enhanced, implying the occurrence of alloying. EDS line scans were also performed on the Au@Ag nanorod during each of the thermal pulses. In the initial stage, the EDS line scan (Figure 3.24) exhibited a slope corresponding to the bipyramid Au core and a sharp boundary at the Au–Ag interface. The signal of Au decreases to zero from this boundary to the far end. The extensions of the Au signal could be distinctly observed. Although it is still unable to track the diffusion channels at the atomic scale, the diffusion trend of the Au atoms suggests that an even alloying phase can be obtained with sufficient thermal pulses.

In addition to the heating treatment, *in-situ* TEM technology can apply other external excitations, such as force and electricity, to the prepared sample through improving the electron microscope and its sample holder. In this way, the dynamic change process of the sample under these external excitations can be observed in real time. In addition to the application of physical field to sample in the TEM, cell-TEM with one placement of liquid cell in the TEM sample holder is another successful improvement [49–51]. Most of the *in-situ* experiments focus on the structural and properties evolution under external stimulus, the monitoring of chemical changes is a particular challenge at present. *In-situ* TEM requires the capability of recording what is seen at an adequate time resolution.

Before the introduction of *in-situ* cell TEM, the growth process of nanoparticles could not be observed directly. Some processes and theories of crystal growth have also been proposed. Ostwald ripening is a typical process, in which small particles are preferentially dissolved and precipitated on the surface of large particles [52]. On the contrary, for the digestive ripening theory, it is considered that the growth of particles is controlled by the surface energy, during which the small-size particles grow and the large-size particles dissolve [53].

The physiochemical properties of nanostructures varied with the controlling growth and self-assembly in the wet chemical system, which affects their further application in device construction. Therefore, if it is possible to investigate the formation process on atomic scale, one can establish the mechanism in the etching and growth procession of nanostructures in liquid phase.

The formation of metal nanocrystals involves three steps of atoms nucleation, seeds formation, and seeds growth into nanocrystals with different shapes depending on the conditions. According to the Wulff principle, the fcc metals, such as Au Ag Cu and Pt group metals, preferred the formation of thermodynamically stable Wulff polyhedron in vacuum condition, inducing the minimum free energy. However, in the real wet chemical system, the crystal growth is far away from the thermodynamic stability, but controlled by dynamics, especially for the highly anisotropic growth of metal nanostructures with the assistance of surfactant. Additionally, the atom deposition and scattering on the crystal surface also depend on the temperature, pH, solvents, and other parameters, which will be discussed in Chapter 7 for the study of nanoframe.

With the development of TEM in resolution and sample holder, it is possible to observe the atomic formation process of sample enclosed in a micro liquid chamber [54]. Cell-TEM can provide spatiotemporal resolution for the evolution of nanostructures in the liquid system. First, the spatial resolution of cell-TEM has reached the atomic level, which can characterize the growth and etching process of nanoparticles. Second, in terms of time resolution, direct electron detection camera can record sub-millisecond level with ultra-high time resolution [55–57]. For example, the observation of Au NRs formation process can be performed by enveloping reaction chemicals within the graphene liquid cell and the etching process of Au NRs with ligands of CTAB and Cys [58]. The different etching processes indicated the different coverage of ligand on the surface of Au NRs. Additionally, liquid chamber and environmental conditions are suitable for the nucleation and growth of nanoparticles in solution. The circulating type chamber can directly promote the reaction by adding a reductant. And in the closed chamber, the nucleation and growth can usually be stimulated by heating, photoelectric excitation, and electron beam excitation.

Liquid-cell AC-STEM has undergone a rapid development for direct observations of structural evolution and the dynamic processes in liquid phases and liquid–solid

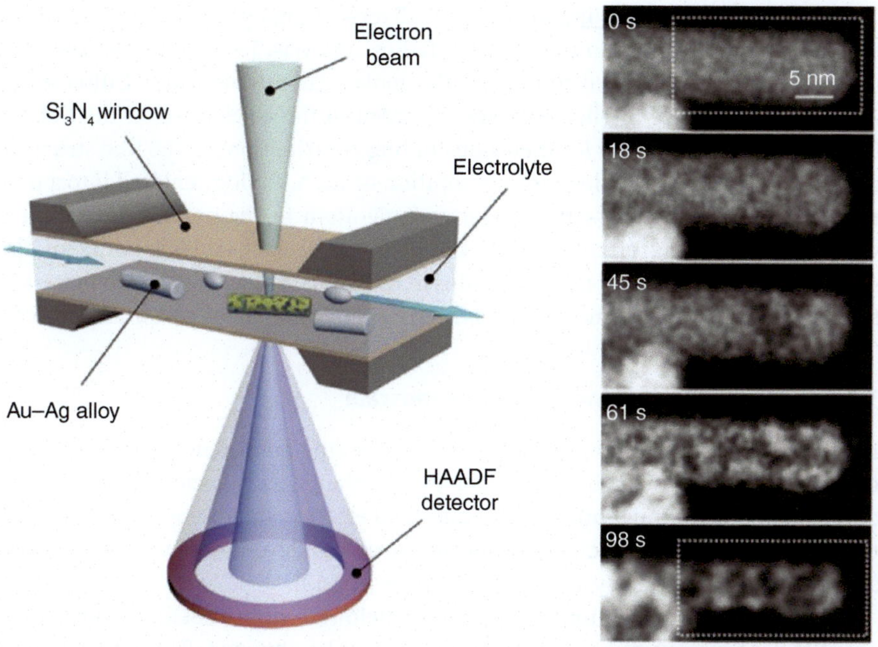

Figure 3.25 Schematic view of the liquid-cell setup for *in-situ* AC-STEM characterization. Time sequential *in-situ* observation on the dealloying of AgAu NRs captured within ~98 seconds, showing volume shrinkage. Source: Reproduced with permission from Ref. [59]; © 2020, American Chemical Society.

interfaces. Since the liquid-cell AC-STEM can provide time-resolved real space information on structure and chemistry in a length scale of sub-nanometer, this technique enabled the investigations of dynamic processes of many important physical phenomena and chemical reactions that are difficult to realize with any other techniques. P. Liu et al. provide comprehensive experimental observations on the dealloying process from initial surface etching to nanopore coarsening and offer microscopic insights of dealloying kinetics [59]. Surprisingly, they found a large volume change in the nanorods before and after de-alloying, as shown in Figure 3.25. Then they compared the sizes of other nanoparticles before and after dealloying without drying. It is found that the volume change apparently increases with a decrease in the initial size, attributed to the formation of a dense shell with a thickness of about one ligament. The surface ligaments are analogous to surface atoms, whose low coordinates bring them closer, which was used to explain the volume shrinkage. With the help of *in-situ* liquid-cell TEM, researchers can directly visualize the coalescence process of Au nanocrystals. It is found that the Au atomic nanobridge forms between adjacent nanocrystals that are separated by a ~0.5 nm hydration layer. The nanobridge structure first induces initial contact of Au nanocrystals over their hydration layers and then surface diffusion and grain boundary migration to rearrange into a single nanocrystal [60].

Electron microscopes are both amazing and incredible scientific instruments used by research laboratories and universities for nanoscience and nanotechnology.

The applications for these instruments are diverse, ranging from particle analysis to material characterization to industrial failure analysis. In life sciences, electron microscopes are being used to explore the molecular mechanisms of disease. In 2017, the Nobel Prize in chemistry was awarded to three scientists in the field of freeze electron microscopy for their outstanding contributions to the fine structure of biological macromolecules with a resolution of 1.2 Å. Biology entered into a new era with the successful observation on a single atom in the protein. Researchers worldwide are using electron microscopes for a deeper understanding of the structure–property–function relationships in materials and newly created materials for the next generation of fuel cells, solar cells, catalysis, and optoelectronic devices.

3.3.5 STM Induce Luminescence at Nanoscale

For the perspective of improving image resolution, the predicted resolution limit of conventional optical imaging methodology was the primary factor motivating the development of scanning probe techniques and electron microscopy. These related techniques have enabled scientists to gain the atomic level of the world, and even atomic micromachining by atom manipulation [61, 62].

From the perspective of improving electric field resolution, detailed information about the interaction of light with objects should be obtained. The interaction of light with a microscope specimen results in the generation of both far-field and near-field light components. The former one propagates through space in an unconfined manner and is utilized in conventional microscopy. The latter one is consisted of a non-propagating field that carries more high-frequency information and has the greatest amplitude near the surface of objects. As discussed in Chapter 1, the near-field light usually goes undetected because it decays exponentially within a distance less than the wavelength of the light.

Scanning tunneling microscopy (STM), invented by Binnig and Roher [63], offers atomic resolution for imaging due to the highly localized nature of the tunneling current across the tunnel junction. It not only enables directly observing and manipulating individual atoms and molecules. Besides, STM can induce luminescence (STML), which combines highly localized electron excitations with optical detection, and offers a unique platform to investigate optical properties of matters with atomic-scale precision. This technique can excite an electrically excited light source (~10 nm) [64–66]. Besides, it also enables the molecular spectroscopy in the tunnel junction, which greatly expands the measurement limit and provides new opportunities for the study of electroluminescence down to sub-nanometer scale.

Taking the advantage of STML with ultrahigh spatial resolution and the optical resolution, it is possible to investigate the EM local density of states of single-plasmonic nanostructure, from which the near field optical information is obtained. We also did such kind of work to associate the plasmonic nanostructures with their near-field optical information [67]. As illustrated in Figure 3.26a, the local excitation of the EM modes at specific locations on the individual triangular sample with a size around 600 nm is carried out with a commercial STM head mounted on an inverted optical microscope in air. The emission of light is associated with the

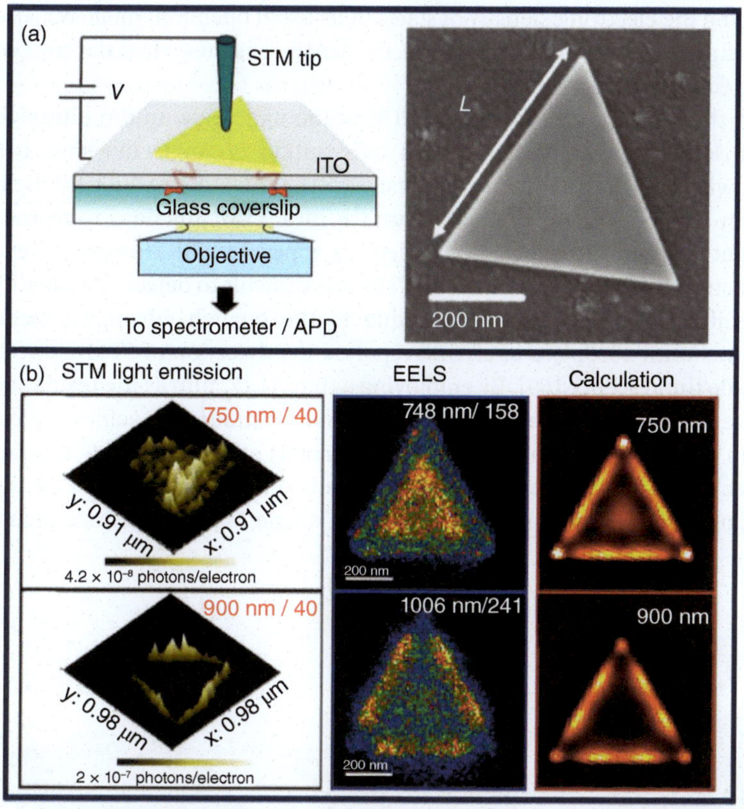

Figure 3.26 (a) Schematic of the experimental setup and SEM image of one Au platelet. The inelastic tunneling current between the STM tip and metal sample excites the plasmonic modes of the triangle platelet, which then decay radiatively. (b) STM-excitation photon yield maps, EELS measurements in the range of 1.5–1.8 eV (top) and 1.1–1.4 eV (bottom) and calculated intensity of the light emitted by the tunneling STM nanosource. Source: Cao et al. [67]/Reproduced with permission from © 2020, American Chemical Society.

radiative decay of the near-field EM modes of the nanostructure. In Figure 3.26b, the number of photons emitted per electron in a wavelength range of 750 ± 20 nm as a function of excitation position is shown, while, for a wavelength range of 900 ± 20 nm was different. For the higher energy range, more light is emitted per electron when the STM tip is located in the center of the upper face of the platelet, while for the lower energy range, the most efficient excitation of light occurs when the STM tip is located on the edge of the nanoparticle. Such striking contrast would suggest that different types of EM modes are excited. Furthermore, as we mentioned before, EELS is a loss spectroscopy and it reflects the total EM local density of states projected along the axis of the incoming electron beam, which is an advanced technique to map the dark and bright modes of the plasmonic nanostructure. The consistent results from EELS and theoretical calculation with COMSOL in Figure 3.26b demonstrated that STML may be considered not only a tool for mapping the electronic density of states in its usual operation mode but also for investigating the near-field radiative EM local density of states when the emitted light is collected far field.

Apart from the light source induced by STM at the nanoscale, in the ultrahigh vacuum with low temperature, the optical information of single molecule can be obtained with STML. Following the first experimental observation of light emission from metal and semiconductor surface in a STM junction, numerous reports on light emission from adsorbed molecules appeared [68]. However, single molecular electroluminescence induced by STM was difficult to detect and identify molecule-specific features from emission due to the quench effects on metal substrate. It was until 2003, Qiu et al. first reported the detection of STM-induced single-molecule fluorescence [69]. By employing a thin Al_2O_3 film as the insulating layer to reduce the interaction between molecules and substrate, they achieved electrically driven optical signal coming from a single Zn(II)-etioporphyrinI (Zn EtioI) molecule with vibrational features which are sensitive to the tip location on the molecule (Figure 3.27a). Later, in the same group, single Mg-porphyrin molecule

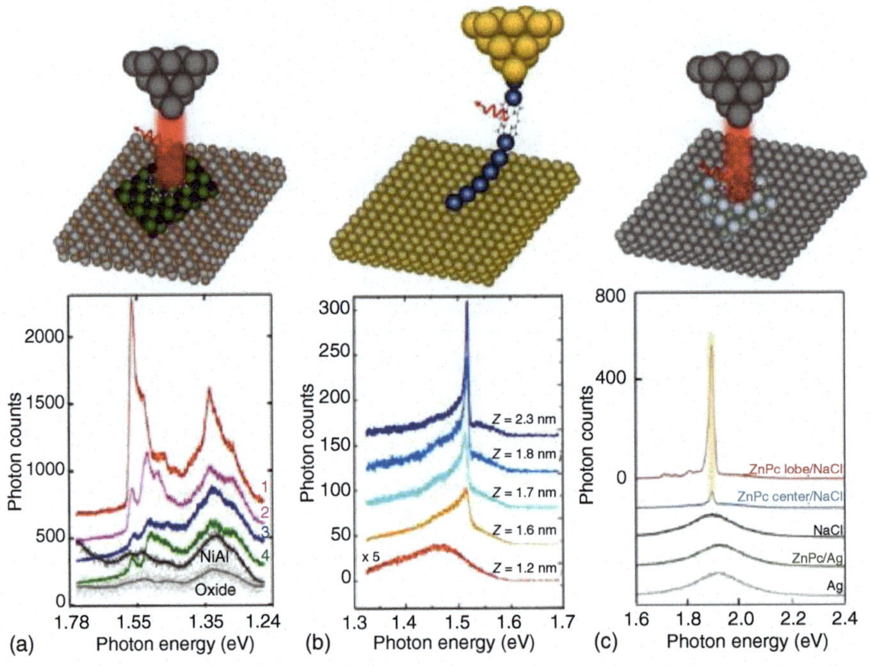

Figure 3.27 Sketch of the experiments and light spectra for a single-molecule transducer between electronic circuits and surface plasmons. (a) Light emission spectra from individual porphyrin molecules adsorbed on an ultrathin Al film grown on a NiAl (220) surface. Source: Adapted from Ref. [69]. (b) Light emission spectra as a function of the detachment of the molecular emitter from the Au (111) surface. Source: Adapted from Ref. [70]. (c) STML emission spectra ZnPc molecule on NaCl/Ag(100) with a much greater intensity over the yellow shaded energy region 1.89–1.91 eV. Source: Adapted from Ref. [71].

decoupled on the same substrate of NiAl (220) was also reported. However, in these above-mentioned achievements, the luminescence cannot be definitely attributed to the fluorescence of the Zn EtioI molecule, since the main emission line energy did not match the fluorescence energy obtained with conventional optical methods. One of the possibilities could be attributed to the strong interaction between the molecule and the insulating Al_2O_3 film used to decouple the molecule.

Later, further efforts have been devoted to developing innovative methods to decouple the STML of a single molecule from the metallic substrate. In 2016, Chong et al. obtained the electroluminescence from a porphyrin molecule which is suspended between the STM tip and metallic surface by an oligothiophene wire (Figure 3.27b) [70]. By attaching the polymeric molecular wire and progressively lifting the complex, and thus varying the decoupling between the porphyrin molecule and the substrate, they showed a continuous variation from a broadband emission to a narrow line emission, which is corresponding to the fluorescence of the porphyrin molecule. It is also important to note that it was the first time that the STM-induced light emission could be unambiguously attributed to the single molecular fluorescence. Unfortunately, this suspended configuration seriously reduced the spatial resolution, making the investigation impossible for the direct molecular environment on the optical properties.

Shortly after this achievement, a more ideal system was proposed by Zhang et al. in which sodium chloride (NaCl) was employed to decouple a single zinc-phthalocyanine (ZnPc) molecule from an Ag (200) substrate (Figure 3.27c) [71]. The STML signal acquired on a single ZnPc presents a sharp fluorescence emission peak at 1.9 eV which is corresponding to the photoluminescence measurement on the same kind of molecule in a frozen matrix. Furthermore, as indicated in Figure 3.27c, they also identified that excitation over the lobe generates emission with a much greater intensity than that above the ZnPc center, indicating the enhanced coupling between the tip plasmons and the molecular exciton compared in this case. The NaCl decoupling method, so far, was proven to be the most efficient way to preserve fluorescence signals from single molecule.

STML provides an opportunity to probe the optical properties of individual molecules with high spatial, spectral and temporal resolution. In multichromophoric molecular assemblies, the STML technique is also allowed to investigate more complex energy-transfer process. Thanks to the manipulation capability of the STM, one can arrange the chromophores with atomic-scale precision so as to transfer energy between distant donor and acceptor centers. For example, Cao et al. have dealt with the possibility of using a STM to study and control the transfer of energy between distant chromophores [72]. They developed an extremely controlled approach to mimic energy funneling occurring in light-harvesting complex by using the basic unit, i.e. molecular chromophores and focused on characterizing the detailed role of intermediate molecules between donor and acceptors. As presented in Figure 3.28a, a trimer in which a large gap molecule (PdPc) is separated from a small gap molecule (H_2Pc) by an intermediated gap molecule (ZnPc). In this configuration, the central chromophore may act as an ancillary molecule, accepting

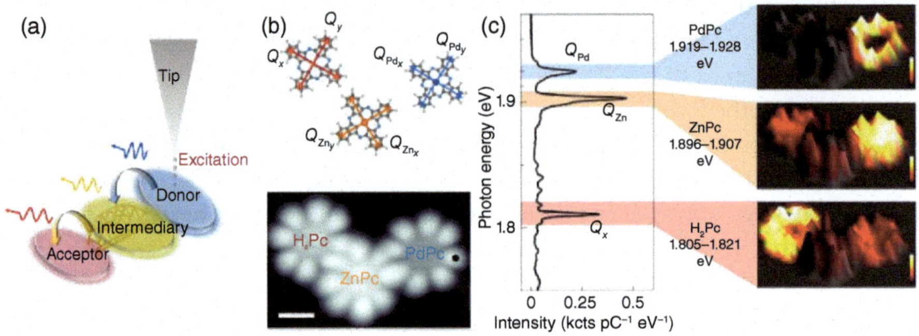

Figure 3.28 (a) Sketch of the experiments in which acceptor and Donor pairs were separated by energy-gap intermediate chromophores. (b) STM image (bottom) and model of PdPc–ZnPc–H_2Pc trimer (top) obtained by controlled STM-tip manipulations of individual chromophores on the NaCl substrate. Scale bar, 1 nm. (c) The black dot in the STM image of b indicates the STM tip position used to acquire the STML spectrum HRFM maps of the emission lines of PdPc, ZnPc, and H_2Pc. Source: Reproduced with permission from Ref. [72].

energy from PdPc before transferring it to H_2Pc. The fluorescence spectrum in Figure 3.28c, obtained for the tip located on PdPc, confirms this picture. It reveals emission lines characteristic of the three chromophores, which include the distant H_2Pc acceptor. Moreover, hyper-resolved fluorescence microscopy (HRFM) maps for the three fluorescence lines also demonstrate that sequential energy transfer can be used to funnel energy between distant centers with a high efficiency and show that sequential resonant energy transfer (RET) process with small energy steps may be more advantages than single RET events with large energy gaps. This general result illustrates, at the level of three chromophores, the efficient strategy used by photosynthetic systems to convey energy from collection centers to distant reaction centers, a concept also plays in artificial multichromophoric components.

3.4 Conclusion

Nanotechnology is the foundation of future high-tech development. The relationships among the chemical composition, phase structure, and microstructure are key factors in determining their performance and applications. Characterization technology and its development have become an essential part in the field of nanotechnology. In this chapter, some characterization techniques, including mechanism, typical results, and their development, were presented.

On a spatial scale, the understanding of the materials world develops toward two infinite directions of telescope and microscopy. For the later one, scientists and researchers believe that there's as much to learn through the scientific exploration of "inner space". Although people can track the process of atoms nuclei, atoms deposition, and crystal formation with the assistance of optical spectrometer and electron microscopy, complete theory and full understanding of nanomaterials have not yet been established. There are still some contradictory experimental phenomena and conflicting opinions on the characteristic information of nanomaterials. Anyway, plasmonics itself has brought tremendous progress to many fields. In the following chapters, a study on some typical plasmonic nanostructures will be delivered.

References

1 Pines, D. and Bohm, D. (1952). A collective description of electron interactions: II. Collective vs individual particle aspects of the interactions. *Phys. Rev.* 85: 338.

2 Bohm, D. and Pine, D. (1953). Coulomb interactions in a degenerate electron gas. Collective description of electron interactions: III. *Phys. Rev. A* 92: 609–625.

3 Tang, H., Chen, C.-J., Huang, Z. et al. (2020). Plasmonic hot electrons for sensing, photodetection, and solar energy applications: a perspective. *J. Chem. Phys.* 152: 220901.

4 Chen, C., Zhou, L., Yu, J. et al. (2018). Dual functional asymmetric plasmonic structures for solar water purification and pollution detection. *Nano Energy* 51: 451–456.

5 Gao, C., Lyu, F., and Yi, Y. (2021). Encapsulated metal nanoparticles for catalysis. *Chem. Rev.* 121 (2): 834–881.

6 Liu, Z., Tang, K., Wang, D. et al. (2013). Facile synthesis of $AgInS_2$ hierarchical flowerlike nanoarchitectures composed of ultrathin nanowires. *Nanoscale* 5: 1570–1575.

7 Khan, S.U.M., Al-Shahry, M., and Ingler, W.B. Jr., (2002). Efficient photochemical water splitting by a chemically modified n-TiO_2. *Science* 297: 243–245.

8 Wijaya, Y.N., Kim, J., Choi, W.M. et al. (2017). A systematic study of triangular silver nanoplates: one-pot green synthesis, chemical stability, and sensing application. *Nanoscale* 9: 11705–11712.

9 Stipe, B.C., Rezaei, M.A., and Ho, W. (1998). Single-molecule vibrational spectroscopy and microscopy. *Science* 280: 1732–1735.

10 Das, R.S. and Agrawal, Y.K. (2011). Raman spectroscopy: recent advancements, techniques and applications. *Vib. Spectrosc.* 57: 163–176.

11 Mulvaney, S.P. and Keating, C.D. (2000). Raman spectroscopy. *Anal. Chem.* 72: 145–158.

12 Raman, C.V. and Krishnan, K.S. (1928). A new type of secondary radiation. *Nature* 121: 501–502.

13 Fleischmann, M., Hendra, P.J., and Mc Quillan, A.J. (1974). Raman spectra of pyridine adsorbed at a silver electrode. *Chem. Phys. Lett.* 26: 163–166.

14 Jeanmaire, D.L. and VAN Duyne, R.P. (1977). Surface Raman spectroelectrochemistry: part 1. Heterocyclic, aromatic, and aliphatic amines adsorbed on the anodized silver electrode. *J. Electroanal. Chem. Interfacial Electrochem.* 84: 1–20.

15 Albrechet, M.G. and Creighton, J.A. (1977). Anomalously intense Raman spectra of pyridine at a silver electrode. *J. Am. Chem. Soc.* 99: 5215–5217.

16 Philpott, M.R. (1975). Effect of surface Plasmons on transitions in molecules. *J. Chem. Phys.* 62: 1812–1817.

17 Moskovits, M. (1978). Surface roughness and the enhanced intensity of Raman scattering by molecules adsorbed on metals. *J. Chem. Phys.* 69: 4159–4161.

18 Creighton, J.A., Blatchford, C.G., and Albrecht, M.G. (1979). Plasma resonance enhancement of Raman scattering by pyridine adsorbed on silver or gold sol particles of size comparable to the excitation wavelength. *J. Chem. Soc. Faraday Trans.* 75: 790–798.

19 Nie, S. and Emory, S.R. (1997). Probing single molecules and single nanoparticles by surface-enhanced Raman scattering. *Science* 275: 1102–1106.

20 Kneipp, K., Wang, Y., Kneipp, H. et al. (1997). Single molecule detection using surface-enhanced Raman scattering (SERS). *Phys. Rev. Lett.* 78: 1667–1670.

21 Xu, H., Bjerneld, E.J., Käll, M., and Börjesson, L. (1999). Spectroscopy of single hemoglobin molecules by surface enhanced Raman scattering. *Phys. Rev. Lett.* 83: 4357–4360.

22 Stöckle, R.M., Suh, Y.D., Deckert, V., and Zenobi, R. (2000). Nanoscale chemical analysis by tip-enhanced Raman spectroscopy. *Chem. Phys. Lett.* 318: 131–136.

23 Zhang, R., Zhang, Y., Dong, Z.C. et al. (2013). Chemical mapping of a single molecule by plasmon-enhanced Raman scattering. *Nature* 498: 82–86.

24 Li, J.F., Tian, X.D., Li, S.B. et al. (2010). Shell-isolated nanoparticle-enhanced Raman spectroscopy. *Nature* 464: 392–395.

25 Zhang, Y.J., Chen, Q.Q., Chen, X. et al. (2021). Graphene-coated Au nanoparticle-enhanced Raman spectroscopy. *J. Raman Spectrosc.* 52: 439–445.

26 Pérez-Jiménez, A.I., Lyu, D., Lu, Z. et al. (2020). Surface-enhanced Raman spectroscopy: benefits, trade-offs and future developments. *Chem. Sci.* 11: 4563–4577.

27 Wu, D.Y., Liu, X.X., Duan, S. et al. (2008). Chemical enhancement effects in SERS spectra: a quantum chemical study of pyridine interacting with copper, silver, gold and platinum metals. *J. Phys. Chem. C* 112: 4195–4204.

28 Valley, N., Greeneltch, N., Van Duyne, R.P., and Schatz, G.C. (2013). A look at the origin and magnitude of the chemical contribution to the enhancement mechanism of surface-enhanced Raman spectroscopy (SERS): theory and experiment. *J. Phys. Chem. Lett.* 4: 2599–2604.

29 Cialla, D., März, A., Böhme, R. et al. (2012). Surface-enhanced Raman spectroscopy (SERS): Progress and trends. *Anal. Bioanal. Chem.* 403: 27–54.

30 Langer, J., Aberasturi, D.J.D., Aizpurua, J. et al. (2020). Present and future of surface-enhanced Raman scattering. *ACS Nano* 14: 28–117.

31 Pan, R., Yang, Y., Wang, Y. et al. (2018). Nanocracking and metallization doubly defined large-scale 3D plasmonic sub-10 nm-gap arrays as extremely sensitive SERS substrates. *Nanoscale* 10: 3171–3180.

32 De Broglie, L. (1923). Waves and quanta. *Nature* 112: 540–540.

33 Busch, H. (1927). Über die Wirkungsweise der Konzentrierungsspule bei der Braunschen Röhre. *Archiv für Elektrotechnik* 18 (6): 583–594.

34 Knoll, M. and Ruska, E. (1932). Das elektronenmikroskop. *Zeitschrift für physik* 78: 318–339.

35 Morriss, R.H., Bottoms, W.F.R., and Peacock, G. (1968). Growth and defect structure of lamellar gold microcrystals. *J. Appl. Phys.* 39: 3016.

36 Kan, C.X., Wang, C.S., Li, H.C. et al. (2010). Gold microplates with eell-defined shapes. *Small* 6: 1768–1775.

37 Zhu, J., Kan, C.X., Zhu, X.G. et al. (2007). Synthesis of perfect silver nanocubes by a simple polyol process. *J. Mater. Res.* 22: 1479–1485.

38 Scherzer, O. (1936). Some defects of electron lenses. *Z. Phys.* 101: 593–603.

39 Rose, H. (1990). Outline of a spherically corrected semiaplanatic medium-voltage transmission electron microscope. *Optik* 85: 19–24.

40 Zhang, L., Roling, L.T., Wang, X. et al. (2015). Platinum-based nanocages with subnanometer-thick walls and well-defined, controllable facets. *Science* 349: 412–416.

41 Fu, X., Wang, X.D., Zhao, B. et al. (2022). Atomic-scale observation of non-classical nucleation-mediated phase transformation in a titanium alloy. *Nat. Mater.* 21: 290–296.

42 Haider, M., Rose, H., Uhlemann, S. et al. (1998). A spherical-aberration-corrected 200 kV transmission electron microscope. *Ultramicroscopy* 75: 53–60.

43 Pan, G.A., Segedin, D.F., LaBollita, H. et al. (2022). Superconductivity in a quintuple-layer squareplanar nickelate. *Nat. Mater.* 21: 160–164.

44 Wu, Y., Li, G., and Camden, J.P. (2018). Probing nanoparticle plasmons with electron energy loss spectroscopy. *Chem. Rev.* 118: 2994–3031.

45 Chu, M.W., Myroshnychenko, V., Chen, C.H. et al. (2009). Probing bright and dark surface plasmon modes in individual and coupled noble metal nanoparticles using an electron beam. *Nano Lett.* 9: 399–404.

46 Xia, Y., Xiong, Y., Lim, B., and Skrabalak, S.E. (2009). Shape-controlled synthesis of metal nanocrystals: simple chemistry meets complex physics? *Angew. Chem. Int. Ed.* 48 (1): 60–103.

47 Zhang, M., Olson, E., Twesten, R. et al. (2005). In-situ transmission electron microscopy studies enabled by microelectromechanical system technology. *J. Mater. Res.* 2: 1802–1807.

48 He, L.B., Lei, S., Ran, Y.T. et al. (2023). Revealing the alloying and dealloying behaviours in AuAg nanorods by thermal stimulus. *Nanoscale Adv.* 5: 685–692.

49 Sutter, E.A. and Sutter, P.W. (2017). In-situ liquid cell electron microscopy of Ag-Au galvanic replacement reactions. *Nanoscale* 9 (3): 1271–1278.

50 Jonge, N., Houben, L., Dunin-Borkowski, R.E., and Ross, F.M. (2018). Resolution and aberration correction in liquid cell transmission electron microscopy. *Nat. Rev. Mater.* 4 (1): 61–78.

51 Callaway, E. (2020). Revolutionary microscopy technique technique sees individual atoms. *Nature* 582: 156–157.

52 Voorhees, P.W. (1985). The theory of ostwald ripening. *J. Stat. Phys.* 38: 231–252.

53 Banfield, J.F., Welch, S.A., Zhang, H.Z. et al. (2000). Aggregation-based crystal growth and microstructure development in natural iron oxyhydroxide biomineralization products. *Science* 289: 751–754.

54 Liao, H.G., Zherebetskyy, D., Xin, H. et al. (2014). Facet development during platinum nanocube growth. *Science* 345: 916–919.

55 Liao, H. and Zheng, H. (2016). Liquid cell transmission electron microscopy. *Annu. Rev. Phys. Chem.* 67: 719–747.

56 Jiang, Y., Zhu, G., Dong, G. et al. (2017). Probing the oxidative etching induced dissolution of palladium nanocrystals in solution by liquid cell transmission electron microscopy. *Micron* 97: 22–28.

57 Zhu, C., Liang, S.X., Song, E. et al. (2018). In-situ liquid cell transmission electron microscopy investigation on oriented attachment of gold nanoparticles. *Nat. Commun.* 9: 421.

58 Hauwiller, M.R., Ye, X.C., Jones, M.R. et al. (2020). Tracking the effects of ligands on oxidative etching of gold nanorods in graphene liquid cell electron microscopy. *ACS Nano* 14: 10239–10250.

59 Liu, P., Chen, Q., Ito, Y. et al. (2020). Dealloying kinetics of AgAu nanoparticles by in-situ liquid-cell scanning transmission electron microscopy. *Nano Lett.* 20 (3): 1944–1951.

60 Jin, B., Sushko, M.L., Liu, Z. et al. (2018). In-situ liquid cell TEM reveals bridge-induced contact and fusion of Au nanocrystals in aqueous solution. *Nano Lett.* 18: 6551–6556.

61 Pham, V.D., Kanisawa, K., and Fölsch, S. (2019). Quantum rings engineered by atom manipulation. *Phys. Rev. Lett.* 123: 66801.

62 Theobald, J.A., Oxtoby, N.S., Phillips, M.A. et al. (2003). Controlling molecular deposition and layer structure with supramolecular surface assemblies. *Nature* 424: 1029–1031.

63 Binnig, G., Rohrer, H., Gerber, C., and Weibel, E. (1982). Surface studies by scanning tunneling microscopy. *Phys. Rev. Lett.* 49: 57–61.

64 Coombs, J.H., Gimzewski, J.K., Reihl, B. et al. (1988). Photon emission experiments with the scanning tunnelling microscope. *J. Microsc.* 152: 325–336.

65 Gimzewski, J.K., Reihl, B., Coombs, J.H., and Schlittler, R.R. (1988). Photon emission with the scanning tunneling microscope. *Zeitschrift für Phys. B Condens. Matter.* 72: 497–501.

66 Lyo, I.W. and Avouris, P. (1991). Field-induced nanometer-to atomic-scale manipulation of silicon surfaces with the STM. *Science* 253: 173–176.

67 Cao, S., Zapata-Herrera, M., Campos, A. et al. (2020). Probing the radiative electromagnetic local density of states in nanostructures with a scanning tunneling microscope. *ACS Photonics* 7: 1280–1289.

68 Flaxer, E., Sneh, O., and Cheshnovsky, O. (1993). Molecular light emission induced by inelastic electron tunneling. *Science* 262: 2012–2014.

69 Qiu, X.H., Nazin, G.V., and Hot, W. (2003). Vibrationally resolved fluorescence excited with submolecular precision. *Science* 299: 542–546.

70 Chong, M.C., Reecht, G., Bulou, H. et al. (2016). Narrow-line single-molecule transducer between electronic circuits and surface plasmons. *Phys. Rev. Lett.* 116: 1–5.

71 Zhang, Y., Luo, Y., Zhang, Y. et al. (2016). Visualizing coherent intermolecular dipole-dipole coupling in real space. *Nature* 531: 623–627.

72 Cao, S., Rosławska, A., Doppagne, B. et al. (2021). Energy funnelling within multichromophore architectures monitored with subnanometre resolution. *Nat. Chem.* 13: 766–770.

Part II

Precise Preparation of Plasmonic Nanostructures

4

Core–Shell and Porous Nanorods with Hot Spots

Why gold nanoparticles are more precious than pretty gold? Mostafa A. El-Sayed, 2006.

4.1 Introduction

Noble metals nanostructures, especially Au, Ag, Pd, and Pt, have been widely applied in the fields of photoelectronics, catalysis, energy, and biomedical science, due to their particular plasmonic properties and excellent catalytic activities [1–5]. Different from spherical nanoparticles (NPs), anisotropic nanocrystals present unique surface plasmon resonance (SPR) properties highly dependent on the sizes, composition, and surroundings. It was not until the early 2000s that multiple groups reported high-yielding colloidal syntheses of anisotropic metal nanostructures with high uniformity: Xia and co-workers for Ag nanocubes, Murphy and El-Sayed and co-workers for Au NRs, and Mirkin and co-workers for Ag nanoplates. For the case of 1D nanostructures, Au NRs (Au NRs), Au nanobipyramids (Au NBPs), and Ag nanorods (Ag NRs) are of particular interest due to their longitudinal and transverse SPR modes [6]. The transverse SPR (SPR_T) of Au nanostructures is located at ~530 nm, while the longitudinal SPR (SPR_L) can be regulated by adjusting the aspect ratios. In general, anisotropic elongated metal nanostructures in the size of tens nanometers show tunable SPR properties. When the size is reduced to 5 nm, nanocrystals would possess remarkable catalytic activity. Inspired by the feasible combining SPR and catalytic activity performance, great efforts have been devoted to developing the synthetic methods to fabricate complex metallic nanostructures based on nanorods with desired applications in solar energy, surface-enhanced Raman spectroscopy, plasmon sensing, optoelectronic, and so on [7, 8].

To optimize the physical and chemical properties of nanocrystals, multimetallic nanocrystals are of great importance due to synergetic effects of metal elements. Typically, the sedimentary of a second metal on the surface of as-prepared nanocrystals can alter the interface energy and electron transfer process, further improving the structural performances. For example, the reaction rate of 4-nitrophenol in the present of Au@Ag core–shell NPs can be accelerated compared to monometallic

Plasmonic Metal Nanostructures: Preparation, Characterization, and Applications, First Edition. Caixia Kan.

NPs [9]. The reason is that the catalytic activity is highly dependent on the surface structure of the catalyst. It has been illustrated that plasmon-excited hot spots play an important role in the photocatalytic performance. For photocatalysis, the conversion of reactants relies on the intensity of light excitation. Plasmonic resonance of metallic nanocrystals matched with incident light can increase the light absorption efficiency. Therefore, the mixture of nanostructures with different plasmonic resonance wavelengths can effectively absorb sunlight. It is obtained that anisotropic nanocrystals are promising candidates to serve as plasmonic catalysts.

A large number of research studies have been carried out to fabricate bimetallic nanocrystals with desired plasmonic and catalytic properties. 1D Au nanostructures, such as Au NRs and Au NBPs, were usually used as inner cores to synthesize bimetallic nanostructures due to their tunable SPR in the Vis–NIR region, high chemical stability, and well-developed preparation methods. Wang and coworker fabricated nanorice and nanodumbbell with tunable plasmonic properties and effectively catalytic activities via site-selective surface etching of Au@Ag nanocuboids [10]. Majima et al. [11] developed Pd-modified Au NRs, which served as the light absorber and catalytically active site and can drive hydrogen generation. In addition, the site-selective Pd deposition on Au NBPs was also applied for photocatalysis. This experimental study displayed the contribution of the spatial distribution of plasmonic hot electron to photocatalytic activity.

Bimetallic nanocrystals with core–shell and alloyed structures are emerging as more important materials than monometallic nanostructures. They are expected to display not only a combination of the properties associated with two distinct metals but also new properties and capabilities due to a synergy between the two metals. More importantly, bimetallic nanostructures usually show composition-dependent surface structure and atomic segregation behavior. Therefore, they appealed more interest in science and various technical fields.

Recent work on bimetallic core–shell and porous nanorods was presented in this chapter. The structural features and tunable SPR properties of 1D nanostructures were first introduced, including Au NRs and Au NBPs. Then based on the as-prepared Au nanostructures, study on the anisotropic nanostructures was carried out, such as core–shell, alloy, Yolk–shell, and porous nanostructures. These novel nanostructures exhibited high stability, excellent plasmonic properties, and catalytic performance, in which the catalytic performance was tested through surface enhanced Raman spectroscopy (SERS) measurement.

4.2 One-Dimensional Au Nanostructures

Au nanostructures have attracted wide attention due to their unique photo- electric and catalytic properties attributed to large surface area to volume ratio and spatial confinement effect of free-electron. Typically, larger specific surface area will induce the appearance of more surface sites with high curvature and more dangling bonds. Thus, the chemical reactivity and surface binding energy are improved. The spatial confinement effect of free-electron can give rise to many promising properties, such as the transfer of energy level, nonlocal dielectric responses, and enhanced optical transition. Among various Au nanostructures, ID nanostructures (including Au NRs and Au NBPs) are high profile due to two special SPR modes. The SPR_L is derived from the collective oscillation of electron along the longitudinal axis under the excitation of electric field. The SPR modes are related to the size, aspect ratio, and purity. Therefore, it is significant to prepare Au nanostructures with controllable size, tunable aspect ratio, and high purity for application development. Seed-mediated growth containing the nucleation and growth of seeds is one of the most popular methods to synthesize well-defined Au NRs and Au NBPs. It has been reported that the seed structure plays an important role in the products [12–15]. Seeds are prepared by reducing metal salt via a strong reductant. The prepared seeds are further added to the growth solution in the presence of metal salt, weak reductant, and surfactant. The seed structure plays a critical role in the morphology of final products. Take Au seeds for example, Au seeds usually involve single-crystal, single-twinned crystal, and five-fold twinned crystal structures. When single crystalline Au seeds were used, Au NRs can be produced. When single-twinned Au seeds were employed, Au nanospheres can be obtained. When the five-fold twinned seeds were exploited, bipyramid-shaped Au nanostructures with twinned boundaries around the long axis can be obtained, as shown in Figure 4.1 [16].

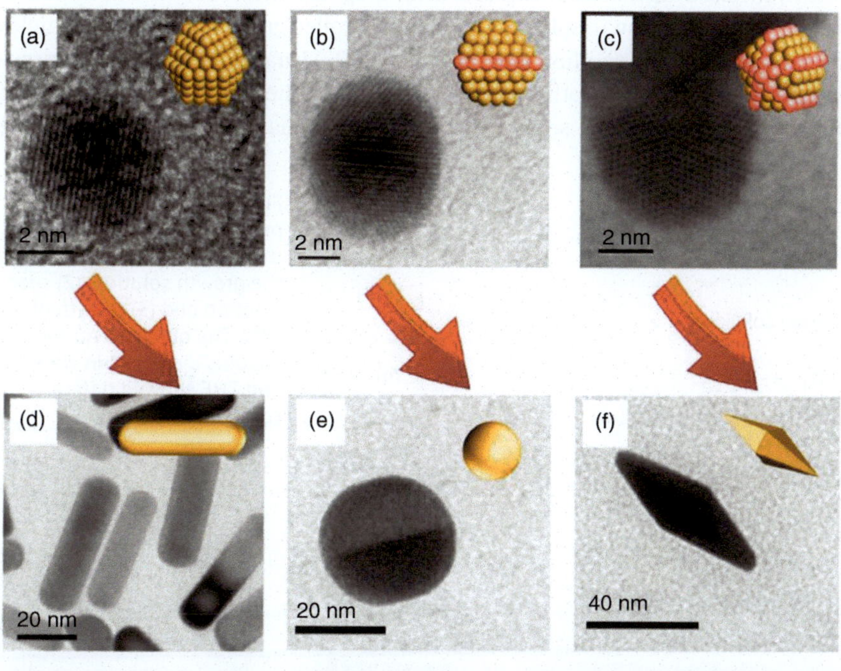

Figure 4.1 HRTEM images and schematic diagrams of ultrafine Au seeds with (a) single-crystal, (b) single-twinned crystal, and (c) 5-fold twinned crystal structures. (d–f) TEM images of Au nanostructures obtained with different seeds. Source: Xu et al. [16]/Reproduced with permission from © 2020 Society of Photo-Optical Instrumentation Engineers.

4.2.1 Precise Growth of Au NRs

Au NRs have attracted enormous attention due to their highly adjustable morphology, plasmonic properties, and applications in wide technical fields. With the improvement in synthetic methods, a variety of materials have been achieved on the base of Au NRs with improved plasmonic properties. They all possess numerous unique functionalities and potential applications in a broad range of scientific as well as technological scenarios. Therefore, significant progress has been made in the synthesis of Au NRs and Au NR-based heterostructures. With the development of advanced characterization techniques and powerful simulation systems, growth mechanism of Au NRs has been qualitatively understood through controllable synthesis of Au NRs covered by specific surfaces. Some important issues have been focused on the assembly of Au NRs into functional units for device, highly sensitive detection of coupling target molecules, plasmonic-related properties, and relevant application of Au NRs.

As one of the chemical methods, seed-mediated technique has become a classic and the most widely used method for the synthesis of Au NRs. The work describing the seed-mediated protocols was pioneered by Murphy, Liz-Marzán, and El-Sayed groups [17–19]. Then many researchers continuously optimized this seeded growth of Au NRs with a variety of additives as colloidal stabilizers and shape-directing agents. In the seed-mediated procedure, three steps are involved, the nucleation of seeds, preparation of growth solution, and the controlled growth of the seed particles in the growth solution, as illustrated by Figure 4.2. Usually, cetyltrimethylammonium bromide (CTAB) and other surfactants were applied to direct the growth of the anisotropic crystals from small Au NPs (~2 nm in diameter) seeds with a trace

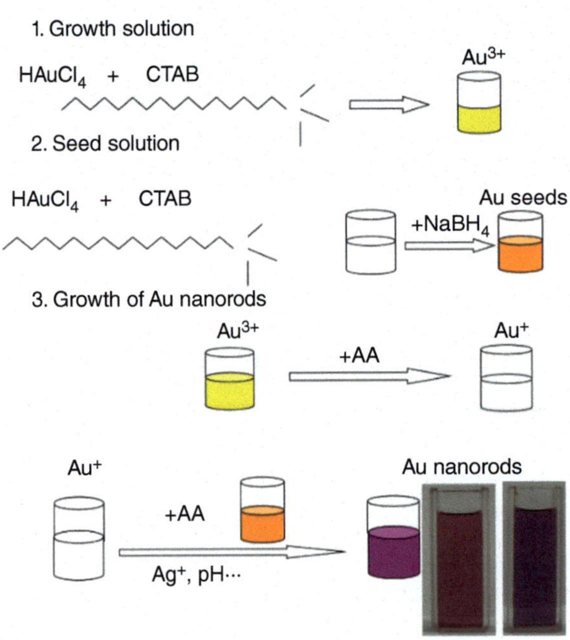

Figure 4.2 The preparation process of Au NRs, including three steps: (1) preparation of the growth solution, (2) seed solution and (3) growth of Au NRs. The obtained Au NRs colloids dispersed in water and THF showed different colors.

of Ag^+ ions [20]. Au NRs with desired aspect ratios can be obtained by adjusting the parameters in the reaction system, comprising Au precursor ($HAuCl_4$), Ag^+ ions, reducing agent, surfactant, halide ions, and the environmental conditions. Each parameter in the reaction system will influence the final production, even the existence of trace Ag^+ ions.

For example, the seeds produced using CTAB have a single-crystalline cuboctahedral structure, which typically gives single-crystalline of final Au NRs. It has been shown that nanorods grow longitudinally along the [100] direction and have an octagonal cross-section. After the addition of the Au seeds into the growth solution, the Au atoms initially grow at almost the same rate along the symmetry-equivalent directions on the seed. Another key parameter is the addition of a trace of Ag^+ ions, which play an essential role in breaking the growth symmetry. The role of the Ag^+ ions in the seed-mediated procedures was improved to induce the formation of single crystals through direct adsorption on the CTA^+ in the form of CTA^+-$AgBr^{2-}$ complex. The concentration of Ag^+ ions during growth can determine the particle size at which the symmetry breaks and the aspect ratio of the resultant Au NRs [21–23].

As shown in Figure 4.3, overgrown of Au seeds remains spherical and multiply twinned particles in the absence of Ag^+ ions, while Au seeds overgrown in the

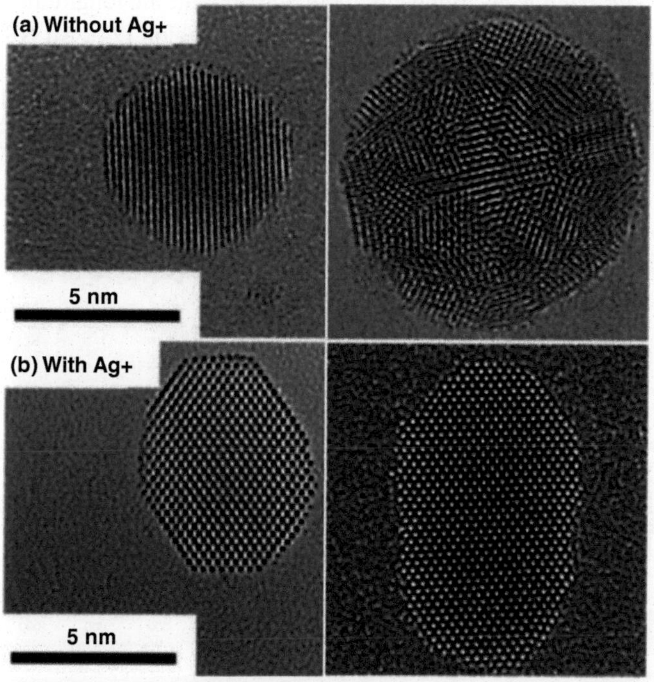

Figure 4.3 Schematic representation of the proposed key steps of the symmetry-breaking process in single-crystal Au NR growth. (a) Au seed particles overgrown in the absence of Ag^+ ions remain spherical and multiply twinned. (b) Au seeds overgrown in the presence of Ag^+ ions into anisotropic structure. Source: Reproduced with permission from Walsh et al. [22]; © 2015, American Chemical Society.

presence of Ag^+ ions are anisotropic in single crystalline nature. Then the Au NRs will undergo isotropic overgrowth, leading to the exposure of the facets and hence the creation of cuboidal Au nanocrystals. It is worthy of noting that one may create Au nanostructures of various sizes and form using the same Au NRs as the seeds by carefully altering the CTAB content in the growth fluid, as well as the pH. In contrast, the Au seeds nucleated in citrate or in a silver-free environment typically have a multiply twinned structure, which usually gives Au NRs with a pentagonal cross-section. The multiply twinned Au NRs are bound by (200) or (220) facets, while the tips are bound by five (111) faces and five (111) twin boundaries run longitudinally along the nanorods. Furthermore, different head shapes for Au NRs can be produced by subsequent overgrowth using the growth solution made of the reactants at varying concentrations. In particular, their end shapes are highly dependent on synthetic conditions.

Figure 4.4a–d presents TEM images of monodisperse Au NRs using different additives and the corresponding photographs of aqueous dispersion of Au NRs with different aspect ratios grown with the direction agent CTAB. The beautiful colors of Au colloids have attracted tremendous attention due to their historical applications in art, as shown in Figure 4.4e–f for some Au NRs with SPR in the visible region. Localized surface plasmon resonances in Au NRs produce two extinction bands, in which the transverse SPR band is positioned at about 520 nm and the longitudinal one is located at a longer wavelength tunable in the Vis–NIR region [24]. Au NRs are also promising candidates in the optics and other plasmon-related fields because they have strong resonant absorption on incident light with a controllable tunability spanning in the Vis–NIR region by simply varying the aspect ratio and assembly process [25].

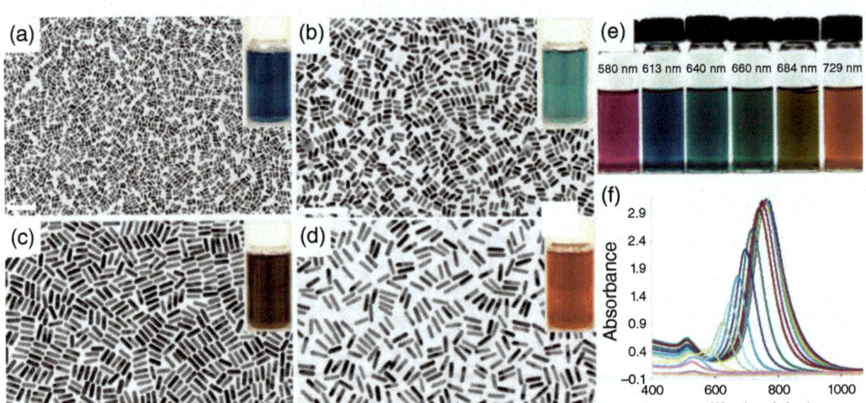

Figure 4.4 TEM images and photograph of the corresponding aqueous dispersion of monodisperse Au NRs using additives of salicylate-based sodium salt (a, b), aromatic acid (c), and $AgNO_3$ (d). Scale bars: (a–c) 200 nm, (d) 100 nm. Source: Adapted from ye et al. [26]. (e–f) colloids photograph and UV–Vis–NIR absorbance spectra of Au NRs with SPR peaks labeled on the bottle. Source: Vigderman et al. [27]/Reproduced with permission from © 2012, JOHN WILEY & SONS, INC.

Au NRs with the anisotropy in geometry have two SPR modes derived from free electrons' collective oscillation along the short (SPR_T) and long axis direction (SPR_L), which makes excellent SPR optical properties, such as extremely large electric field enhancements, largely tunable resonance wavelengths, huge optical extinction cross sections, generation of energetic charge carriers, and photothermal conversion [28–30]. Compared with the isolated Au NRs, the assemblies of Au NRs are also interesting as they exhibit some distinctive properties and offer new functionalities [31–33]. In the assembly, the spacing and arrangement of NPs are closely related to the properties of the assembly. When Au NRs are assembled closely, the coupled interaction will occur in the SPR of the monomers, which can lead to changes in the scattering/absorption properties of the Au NRs, including the shift of the resonance peak, variation in the absorption, and change in the shape of the scattering spectrum. Especially, the densely populated electromagnetic hotspots formed in the gap regions between the adjacent Au NRs. The optical and electrical properties of the assembly, depending on the coupling interaction of the coupled Au NRs, show more complex synergistic properties than that of the Au NRs monomers, which plays an important role in the development of nanomaterials in the future [34, 35].

4.2.2 Precise Growth of Au Nanobipyramids

Au NBP typically consists of two pentagonal pyramids with five-fold twinned planes aligned along the length axis. The unique structural features, compared to Au NRs, enable Au NBPs more fascinating SPR optical properties, especially the narrow full width at half maximum (FWHM). The preparation procedure for Au NBPs is similar to the Au NRs via the seed-mediated method, including the preparation of the seed solution, growth solution, and the subsequent growth of the seed particles. The main difference being that seeds solution derives from the seeds of five-fold twinned crystal structures as opposed to the seeds of single-crystal or single-twinned crystal. It was found that citrate is benefit to the formation of twinned crystal seeds, while high temperature is also conductive to the secondary ripening of seeds to form five-fold twinned crystal structures [36]. High-purity Au NBPs can be synthesized by using high-purity penta-twinned crystal seeds under the action of surfactant and Ag^+ ions [16]. In detail, initial Au seeds were prepared through a fast reduction reaction of $HAuCl_4$ by $NaBH_4$ in cetyltrimethylammonium chloride (CTAC) solution with the presence of the citrate under vigorous stirring room temperature. And then, the seeds solution was heated at 80 °C for 90 minutes. The thermally treated seeds were mainly consisted of five-fold twinned crystals. Finally, the seeds were added to the growth solution containing CTAB, $HAuCl_4$, $AgNO_3$, HCl, and ascorbic acid (AA). The high yield of Au NBPs was obtained after stirring at 30 °C for 2 hours.

The size, aspect ratio, and purity of Au NBPs were determined by the surfactants, precursor concentration, and reduction rate. In the growth solution, CTAB was selected as a surfactant. If CTAC with a weaker affinity for Au was used to replace CTAB, the final product was dominated by elongated and javelin-like nanostructures [37]. Moreover, the foreign metallic ions (Ag^+) also play an important role in the morphology of nanostructures. With the amount of crystal seed decreasing,

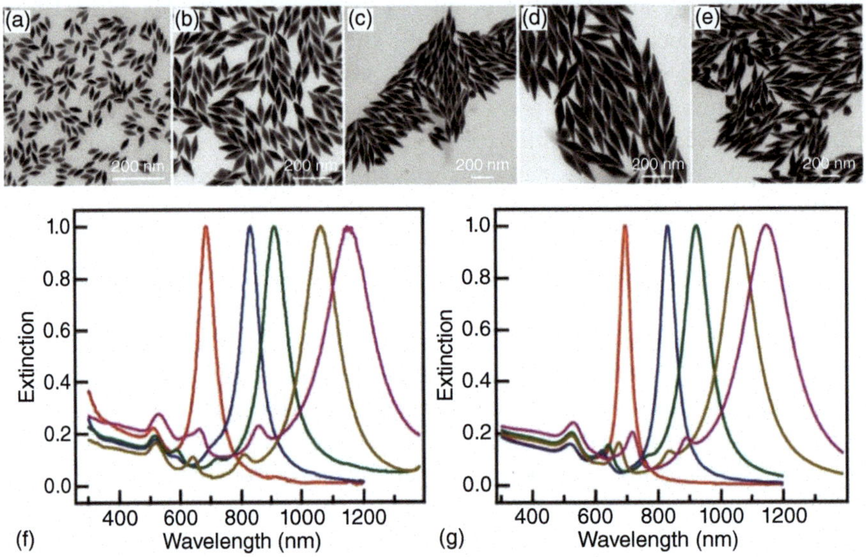

Figure 4.5 (a–e) TEM images of Au NBPs with different aspect ratios. (f) The experiment and (g) FDTD simulation results of Au NBPs with different aspect ratios, corresponding to the TEM of (a–e). Source: Li et al. [38]/Reproduced with permission from © 2015 JOHN WILEY & SONS, INC.

the aspect ratios of Au NBPs would gradually increase [38–40], as shown in Figure 4.5a–e [41, 42].

Due to the anisotropy characteristic, Au NBPs exhibit significant longitudinal surface plasmon resonance band, shifting in the range of Vis–NIR spectral region when Au NBPs interact with external electromagnetic waves. Figure 4.5f,g shows the experiment and FDTD simulated optical spectra of Au NBPs with different aspect ratios. For Au NBPs with small size and aspect ratio, the optical spectrum shows a longitudinal resonance absorption peak, corresponding to the dipole resonance mode. With the size and aspect ratio increasing, the surface plasmon will exhibit two or more resonance absorption peaks, corresponding to the longitudinal quadrupole resonance mode and multipole resonance modes, in which the dipole resonance mode is the strongest among them. Many research results have shown that Au NBPs have a narrowed FWHM and high environmental sensitivities compared with that Au NRs. Additionally, the localized field enhancement of Au NBPs is also obviously higher than that of Au NRs possibly due to two sharp ends.

The yield of Au NBPs is related to the seeds' crystal structure, surfactant, and the purification process. For example, Liu and Guyot-Sionnest [41] presented that the Au nanostructures synthesized by seed-mediated method are strongly dependent on the crystal structure of Au seeds. They prepared two types of seeds, containing single crystalline and twinned crystalline using CTAB and sodium citrate as the surfactants, respectively. Au NBPs as the by-products of Au NRs formed under the same growing condition. And the yield of Au NBPs is only proximate to be

~30% at the present of single crystalline Au seeds. Liz-Marzan group regarded that thermal treatment of small Au seeds can lead to a subsequent drastic improvement in extensive twinning, and greatly increase the Au NBPs yield (>85%) of five-fold twinned NPs [15]. This modified seeded growth method can greatly improve the yield and reproducibility of Au NBPs. And the seeds can also be stored for long periods of time. The crystal structure of the seeds thus defines the yield of the obtained NPs, which is closely related to reaction kinetics including temperature, salt concentration, and the presence of ligand ions. By adjusting the parameters, the yield of Au NBPs can be further improved to 90% [43].

On the other hand, the subsequent purification is also very important. Due to the weight of a single Au NP being close to that of a single Au NBPs, it is difficult to separate and purify the Au NPs efficiently by conventional centrifugation. By introducing strong electrolyte and suitable surfactant, such as CTAB, CTAC, and benzyldimethylhexadecylammonium chloride (BDAC), and utilizing the purification mechanism induced by depletion interaction, the effective separation and high purified Au NBPs can be realized. Combined with seed ripening and self-precipitation separation, the preparation purity of Au NBPs can reach 100% [15, 38]. The controllable preparation of high purity of Au NBPs provides a good-based material for further preparation, construction, and development of Au NBPs-based nanostructures. Due to the excellent surface plasmon resonance characteristics, Au NBPs have wide application prospects in biosensing, immunoassay, dark field optical imaging, fluorescence enhancement, drug carrier photothermal therapy, and surface Raman enhancement detection [14, 44–46].

Based on the as-prepared Au NRs and Au NBPs, complex nanostructures, such as core–shell, porous, and Yolk–shell, have also been explored through precise control on the deposition and over-growth of other materials on the surface of Au NRs and Au NBPs.

4.3 Core–Shell Nanostructures

As illustrated above, 1D Au nanostructures present excellent monodispersity and tunable SPR properties. And they have been widely applied in the fields of optical information storage, photocatalysis, sensing, and thermal therapy. In the construction of bimetallic nanostructures, the second noble metal Ag was commonly considered for the same crystalline structure and lattice matching as those of Au. When the thickness of Ag shell was more than 5 nm, the properties of Au@Ag core–shell nanostructure are similar with those of monometallic Ag nanostructures.

In the case of Ag nanostructures, they are also promising in the fundamental research and applications of photoelectric field since they possess smaller dielectric function and higher interband transition energy. The strong SPR_L of Ag NRs is similar to that of Au NRs, which can be regulated in a large spectral region. Ag NRs also present remarkable SPR_T mode and multipolar resonance in the UV spectral region, while the SPR_T of Au NRs is weak and located at 530 nm. In addition, Ag NR shows better performances than those of Au NR with the same size, including the electric field enhancement, Raman-enhanced effect, and refractive index sensitivity. Unfortunately, poor stability of mono-metallic Ag NRs impedes the development of these applications.

With the rapid development of nanoscience and nanotechnology, nanomaterials composed of a single element possessing monotonous properties cannot satisfy the requirements. Core–shell nanostructures consist of an inner core and outer shell from different materials that have attracted wide attention, which combines several novel properties of different materials. For example, the stability of Au@Ag core–shell nanostructures can be improved, and the optical properties can be dramatically regulated by changing the shell thickness. The biocompatibility can also be improved by coating an outer SiO_2 shell on the surface of nanostructures.

Due to the highly purified 1D Au nanostructures, the overgrowth of shell (metal or oxide materials) on Au nanostructures provides one useful method for the fabrication of multicomponent nanostructures. In this section, 1D Au nanocrystals including single-crystalline Au NRs and five-fold twinned crystalline Au NBPs were used as the inner core. Figure 4.6 presents the schematic illustration for the synthesis of well-defined core–shell Au@Ag bimetal nanostructures of different shapes through heteroepitaxial growth, such as nanocuboids, nanorods, nano dumbbell. For further annealing treatment and application, mesoporous SiO_2 shell was introduced on the surface of as-prepared core–shell nanostructures.

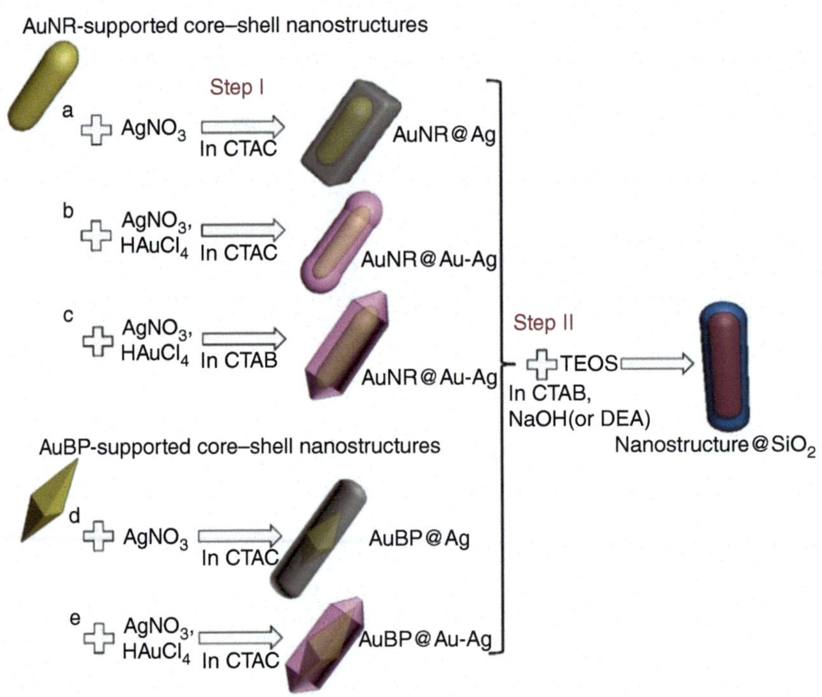

Figure 4.6 Schematic illustration of the synthetic route of Au NR- and Au NBP-supported core–shell nanostructures.

4.3.1 Core–Shell Nanostructures Based on Au NRs

4.3.1.1 Au@Ag Core–Shell Nanocuboids

Uniformed Au@Ag core-shell nanocuboids were fabricated by one-step of Ag overgrowth on the surface of Au NRs with the presence of surfactant [9]. The precursor $AgNO_3$ was reduced by AA and Ag atoms deposited on the surface of Au NRs. Ag source can also be provided by the addition of AgCl particles suspended in water instead of the free Ag^+ ions from $AgNO_3$. AgCl particles were expected to form $AgCl_2^-$ ions in the CTAC solution.

Here, for understanding the growth mechanism of Ag nanocuboid, it is necessary to analyze the structure of the Au NRs first, as shown in Figure 4.7. Typically, the cross-section of Au NRs is a regular octagon, consisting of four (200) and four (220) crystal faces on the lateral sides. Two tips are (200) crystal faces and truncated faces are composed of eight alternating (200) and (220) faces with axial growth direction [001]. The surface energy of the crystal surfaces is in the order of $\gamma_{(220)} > \gamma_{(200)} > \gamma_{(111)}$ [47]. Therefore, Au NRs with rich high-energy (220) crystal facets are generally dispersed in solution with the presence of CTAB or CTAC surfactants. And Au NRs are still not stable in the liquid phase for a long time.

During the growth of the Ag nanocuboids, due to the more CTAC surfactant molecular were adsorbed on the ends of the Au NRs, the reduced Ag atoms are preferentially deposited on the lateral and truncated (220) crystal faces, leading to these crystal facets to be gradually disappeared and eventually form (200) planes, as shown in Figure 4.7b. The cross-section of the Au NRs changes from regular octagon to square. As a result, the Ag nanostructure maintains anisotropic overgrowth along the lateral direction and finally becomes nanocuboid in shape. Throughout the process, the Au NR remains at the center of the entire nanostructure. It is worth stating that the surface energy on the sharp edges of this nanocuboid is high, and the edge atoms are extremely easy to the trend of surface reconstruction, resulting in a

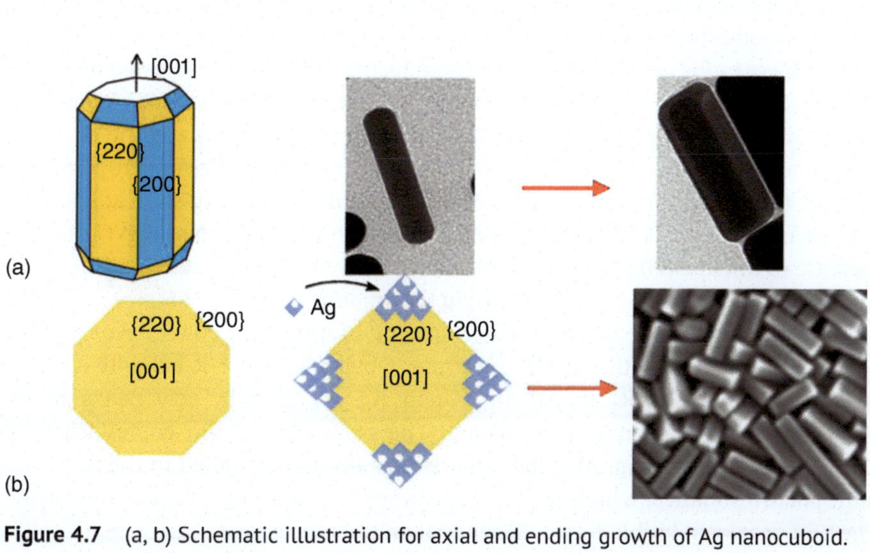

Figure 4.7 (a, b) Schematic illustration for axial and ending growth of Ag nanocuboid.

surface-passivated nanocuboid shape with a very narrow-edged surface, illustrated by the SEM image of Ag nanocuboids.

It was known that the lattices of Au and Ag are well-matched, which was beneficial for the formation of a uniform Ag shell on the surface of Au crystal. The growth condition, especially surfactant, was also significant for the final products. CTAC is an ideal surfactant to prepare uniform Ag shell. Although CTAB has the same aliphatic tail with CTAC, the products are dumb-bell and symmetric boat-like Ag shell, respectively [17, 48–50]. The reason was that CTAB preferred to bind more closely to the side (220) facets of Au NRs than the ending (111) facets. Thus, more Ag atoms are deposited on the surface of two ends of Au NRs, leading to the formation of dumb-bell Au@Ag nanostructures. The formation of boat-like nanostructures was ascribed to the competition between the strong stabilization of CTAB to the (220) facet and minimization of overall surface energy. Four (220) side facets show different growth rates. Especially, the growth rate for two of the neighboring (220) facets was more rapid than that for the remaining facets. As a result, Au core was adjacent to the Ag (220) facet with low growth rate, leading to the generation of asymmetrical core–shell nanostructures.

In order to adjust the growth rate of crystal faces, Niidome et al. applied the mixed micellar solution of CTAC and CTAB as surfactants to prepare Au@Ag core–shell NRs [51]. In this case, $AgNO_3$ was reduced by AA at room temperature, and NaOH was added to regulate the pH value at 5.7. They found that CTAC loosely packed on the surface of Au NR, inducing more rapid deposition than CTAB. Rapid nucleation of Ag shell was beneficial for the formation of uniform shell. However, the subsequent rapid growth would induce the formation of an inhomogeneous Ag shell. That is, the growth rate of shell should be suppressed to obtain uniform shell. In the absence of NaOH, the reduction power of AA was decreased, accompanying with the low growth rate [52]. It should be pointed out that the appropriate growth temperature and surfactant are very important for the formation of uniform shell. For example, the presence of CTAC was advantageous for the uniform core–shell NRs under the condition of 60 °C.

The aspect ratios of Au@Ag core–shell nanostructures can be adjusted by changing the additional amount of Ag precursors, as illustrated in Figure 4.8. It has been known that single crystal Au NRs possess the side face of (220) and the end faces terminated by (111) and (200). Because the growth rate along the Au [110] direction is faster than that along the [100] direction. The thickness of Ag-shell at both side surface and two ends simultaneously increases with the addition of $AgNO_3$, but the increase of thickness on the side surface is faster than that on the ends. Therefore, the aspect ratios decrease with increasing the amount of $AgNO_3$.

Figure 4.8f presents the extinction spectra of the four cuboid Au NR@Ag core–shell, together with Au NR substrate. Obviously, Au NR exhibits a strong SPR_L and a weak SPR_T at 510 nm. Upon Ag overgrowth, new peaks in the short-wavelength region were observed. The peak at ~350 nm comes from the characteristic peak of Ag. And the other two peaks corresponded to the SPR_T and multipolar resonance, which slightly red shifted with the aspect ratios increasing. The SPR_L peaks in visible region dependent on the aspect ratios can be regulated in a large wavelength region and gradually blue shifted with the addition of Ag.

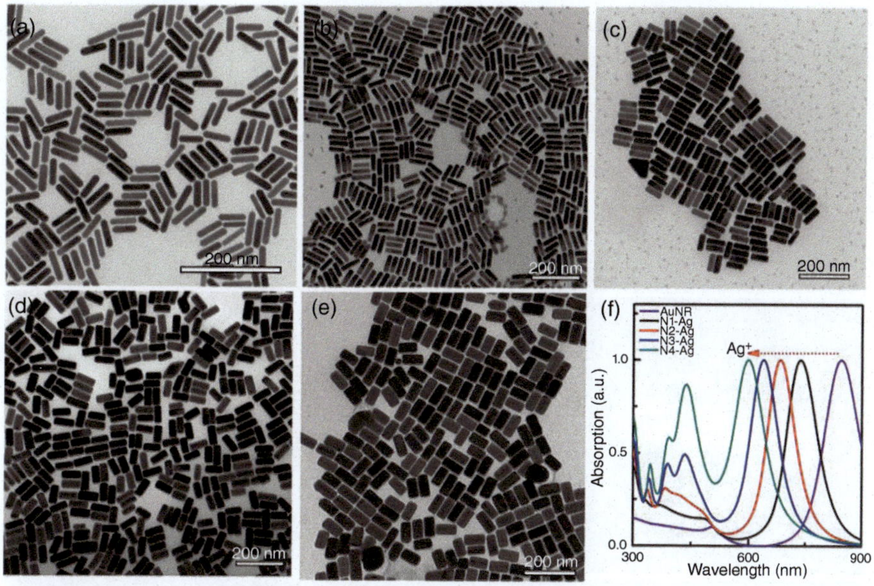

Figure 4.8 TEM images and UV–Vis–NIR spectra of Au NR-supported core–shell NRs with increasing the Ag precursor addition. Source: Ke et al. [9]/Reproduced with permission from © 2019 ELSEVIER.

4.3.1.2 Au@AuAg Dumbbell Nanostructures

Similarly, an effective approach was developed to fabricate Au core@AuAg alloyed shell NRs based on co-reduction of Au and Ag salts via the guidance of 1D Au nanostructures. In this case, Au NR-supported AuAg alloy shell was obtained by simultaneous reduction of Au and Ag precursors, where the injection pump with dual channel was used to control the addition rate of precursors. The capping agents influence the morphology of final products. In the presence of CTAB, the reduced metal atoms deposited on the surface of Au NRs to form fusiform-shaped NRs with two pointed ends. If the surfactant (CTAB) was replaced by CTAC, the products would be dumbbell-shaped NRs with two rounded ends. Figure 4.6b,c expresses the synthesis process of AuAg alloy shell. TEM images in Figure 4.9 showed the effect of surfactants on the morphology of products. EDX mapping displayed the distribution of Au and Ag in the nanostructure.

In the formation process of AuAg alloy nanoshell, the synergy of Ag^+ ions and capping agent influenced the morphologies of final products. Because of the abundant reductant, the dripping precursors would be reduced instantaneously. It has been known that the binding energy of Br^- with Au is stronger than Cl^-. And the solubility of AgBr is lower than AgCl. Therefore, Ag^- ions tend to bind with Br^-. In the presence of CTAB, partially insoluble AgBr would preferentially occupy high-energy crystalline planes on the ends. The co-reduced Au–Ag atoms would deposit on the side facets, leading to the formation of fusiform-shaped nanostructures. When CTAC was introduced as a capping agent, the reduced metal atoms would prefer to occupy the ends of nanostructures due to the weak competitiveness of Cl^- with Ag. And thus, the final products were dumbbell-shaped nanostructures. It is found that the

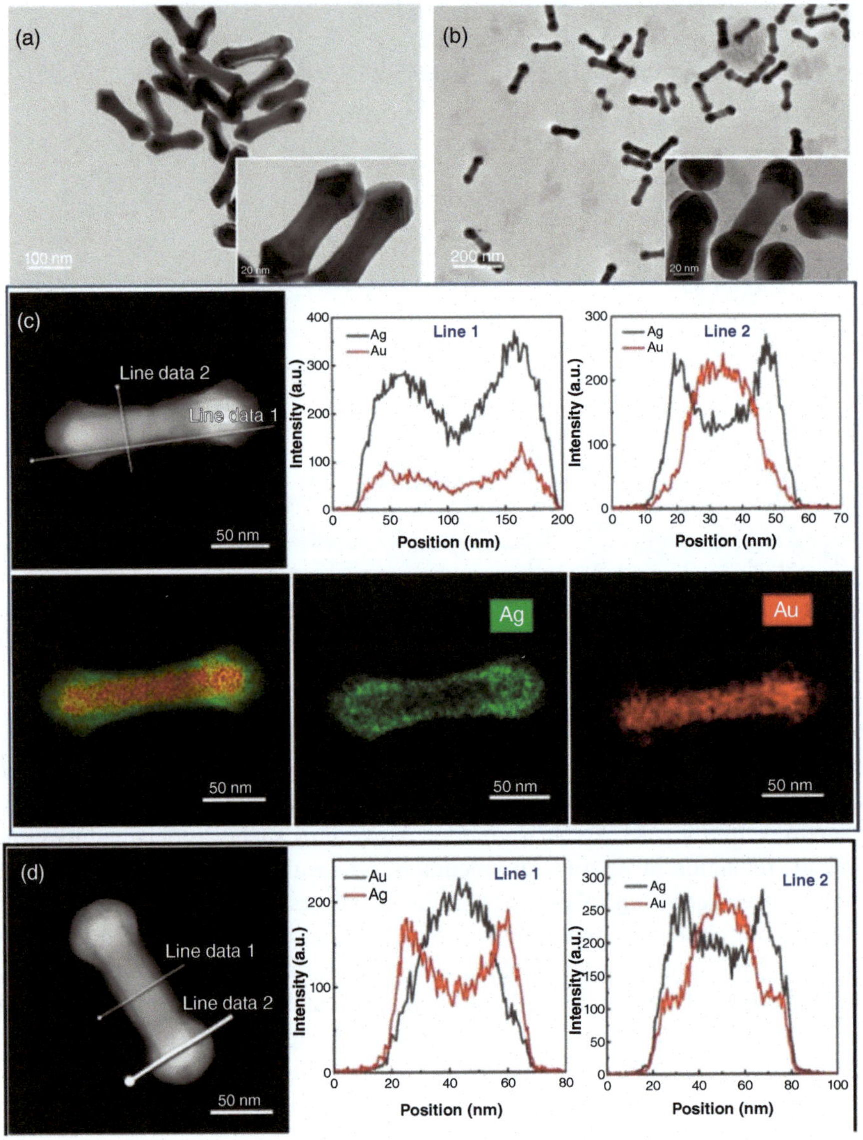

Figure 4.9 TEM images and EDX mapping of fusiform and dumbbell-shaped Au NR@AuAg alloyed NRs.

morphology of the product can be controlled by the injection rate of precursors and concentration of reducing agent. It has been reported by Han and coworkers that the synthesis of alloyed Au–Ag shell on single-crystalline Au NRs [53]. Due to the remarkably different redox potential between $HAuCl_4$ and $AgNO_3$, the products tend to generate core–shell nanostructures.

Remarkably, the reduction rate plays a key role in the formation of alloy shell. AA as reductant is a popular choice for simultaneous reduction of $HAuCl_4$ and

$AgNO_3$. However, the reducibility of AA was influenced by the pH value of the reaction solution. It had been found that NaOH was usually introduced to neutralize the solution and adjust the reduction rate. When the metallic precursors were added fast, the products presented an Au-enriched inside and a pure-Ag outside on the surface of Au NR. When the addition was slow, the reduced Au and Ag atoms simultaneously deposited on the sides and ends of Au NRs [54]. The injection pump with dual-channel can be used to slowly inject precursor into the Au NRs colloid. Therefore, the reduced Au and Ag atoms simultaneously deposit on the surface of Au NRs. Therefore, the resulting alloyed nanoshell had uniform Au and Ag atoms distribution without phase segregation.

The aspect ratios of products can be regulated by changing the concentration of Ag and Au precursors. Notably, the regulation process also influenced the aspect ratio of the final products. Taking fusiform-shaped NRs with two pointed ends, for example, the reduced metal atoms gradually grow along the side faces from two ends with increasing the precursor volume with the same molar concentrations. The continual deposition of Au and Ag atoms on the surface leads to a decrease in aspect ratios (Figure 4.10). Similar to the extinction peaks of Au NRs, Au @AuAg core–shell NRs also possess two resonance modes of SPR_L and SPR_T. Obviously, the aspect ratio of core–shell NRs decreased compared with that of Au NRs. Therefore, the SPR_L peak blue shifted. The addition of Ag induced the SPR_T blueshift slightly.

4.3.1.3 (Au@Ag NR)@SiO_2

Although Au@Ag core-shell NPs present promising for catalytic application, two limitations restrict their applications. One is that Ag nanoshell can be easily oxidized in harsh environment. The other is that Au@Ag NRs cannot be recycled, since organic solution during the catalytic reaction can dissolve the surfactant attached to

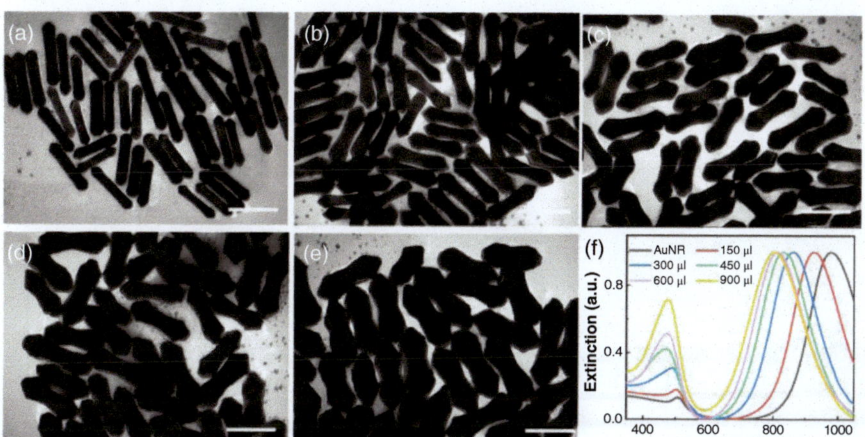

Figure 4.10 TEM images and corresponding UV–vis–NIR spectra of fusiform-shaped Au NR@AuAg alloyed NRs by changing the addition amounts of Ag and Au precursors (The authors' not published work).

the surface of nanostructures, leading to aggregation of nanostructures. Therefore, nanostructures with rich hotspots provide a feasible approach to optimize catalytic performances. Various metallic oxides, such as Cu_2O, TiO_2, ZnO, CeO_2, and so on, were applied as a surface coating to improve the stability and reusability of nanostructures.

In addition to the metal oxides, it has been found that mesoporous silica (SiO_2) can be achieved at room temperature as promising coating layer for reaction channel [55]. Previous studies show that the nanostructures with SiO_2 coating possess excellent thermal and chemistry stability [56]. Moreover, SiO_2 layer as an inert gain material is transparent at the visible region and does not destruct the SPR band of nanostructures. Different nanostructures with mesoporous SiO_2 ($mSiO_2$) coating layer were fabricated via a modified Stöber method, which was illustrated by step II in Figure 4.6. First, the as-prepared metal nanostructures were dispersed into CTAB (1 mM) solution. And then the pH value was adjusted to 11 by adding NaOH or diethanol amine. Subsequently, the mixed solution of TEOS/EtOH (the volume ratio of 1 : 9) was added. Finally, the reaction mixture was vibrated for 20 hours at room temperature. The thickness of SiO_2 layer increased by increasing TEOS.

Figure 4.11 shows the TEM images and extinction spectra of Au NR@$mSiO_2$ with different thicknesses. Obviously, the mesoporous SiO_2 coated nanostructures remained the same two resonance modes of SPR_L and SPR_T. The SPR_L peaks gradually red shifted as the thickness of SiO_2 layer increased due to the change of the surrounding medium of Au NR, while the SPR_T was located at ~530 nm without obvious shifting. It was known that the SPR_L is sensitive to the dielectric

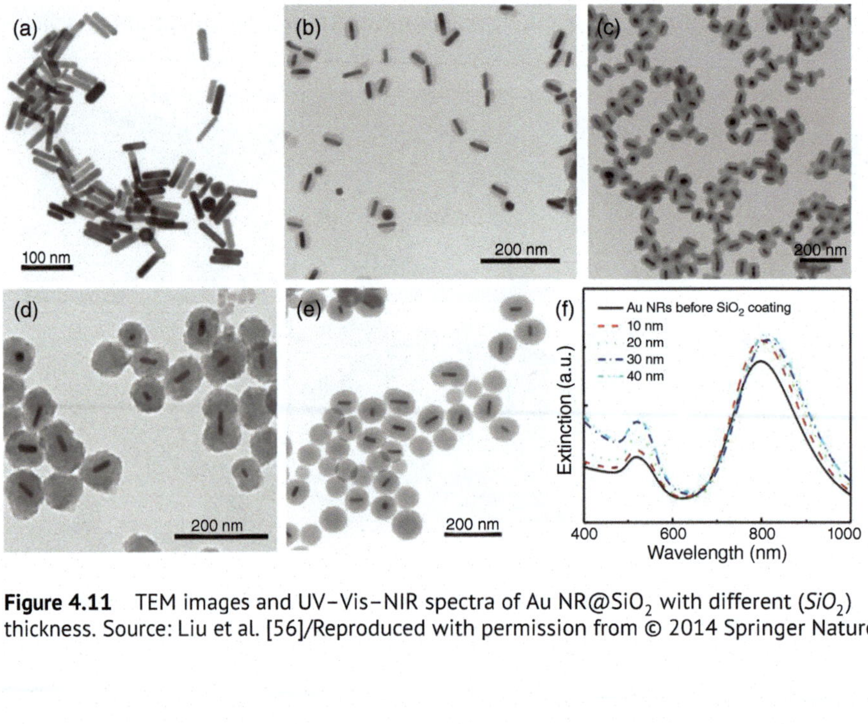

Figure 4.11 TEM images and UV-Vis-NIR spectra of Au NR@SiO_2 with different (SiO_2) thickness. Source: Liu et al. [56]/Reproduced with permission from © 2014 Springer Nature.

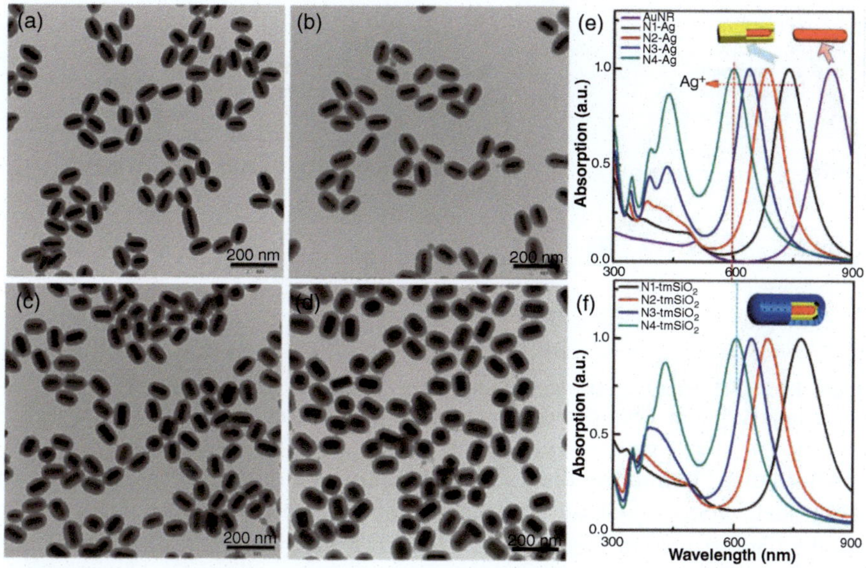

Figure 4.12 TEM images (a–d) and UV–Vis–NIR spectra (e, f) of different Au NR@Ag before and after modification of uniform $mSiO_2$. Source: ke et al. [9]/Reproduced with permission from © 2019 ELSEVIER.

constant of surrounding. With further increasing TEOS, the mesoporous SiO_2 shell was in quasi-spherical shape and some SiO_2 nanospheres without Au NRs core can be observed under TEM measurement. There was no obvious SPR_L shift for Au NR when the thickness of mesoporous SiO_2 shell is more than 30 nm. The stability of Au NRs with mesoporous SiO_2 coating was improved. Under the laser irradiation, Au NR@$mSiO_2$ was stable, while Au NRs would be deformed into spherical particles.

For Au @Ag NR with SiO_2 coating, the fabrication process was similar to that of Au NR@$mSiO_2$. Usually, NaOH was a popular reagent to regulate pH value. Due to the etching of OH^- for Ag, diethanolamine (DEA) was selected instead of NaOH to adjust pH value. Inevitably, the alkaline growth condition would induce the formation of Ag_2O. The above section introduced the preparation of Au @Ag NR (Figure 4.8). Figure 4.12 illustrates TEM images and absorption spectra of Au NRs with increasing Ag layer thickness and further coating uniform SiO_2 shell, which exhibited the same variation trend. The uniform SiO_2 shell would lead to the redshift of SPR_L, as indicated by red and blue lines in spectra.

Although the modification of SiO_2 layer improves the structural stability, the SiO_2 capping layer hydrolysis would appear in an aqueous solution, which limited the application in water. Moreover, the remanent surfactant from the process of forming SiO_2 layer would occupy the active sites, reducing the catalytic efficiency. An effective approach was proposed to open the mesoporous channel and remove the surfactant CTAB by ultrasonic treatments in the presence of HCl.

In detail, the process can be divided into four steps as illustrated in Figure 4.13. The first step was the synthesis of well-defined Au@Ag core–shell NRs dispersed in

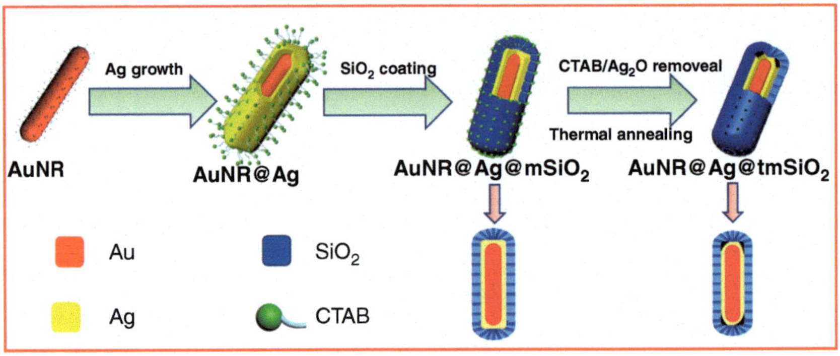

Figure 4.13 Schematic illustration of the synthetic route of stable Au NR-based nanocatalysts. Source: Reproduced with permission from Xia et al. [9]; © 2019 Elsevier.

CTAB solution. To improve the stability of nanostructures or prevent the oxidation of Ag shell, $mSiO_2$ shell was formed on the surface of nanostructures via the modified Stöber growth method. The growth of SiO_2 was in an opening alkaline environment at room temperature. Therefore, the ultrasonic treatment in the mixed solution of HCl and ethanol was carried out to remove the surfactant around the nanocrystals and in the mesoporous channels, forming nanostructures with rich active sites. Finally, the prepared nanostructure coated with $mSiO_2$ would proceed twice thermal treatments at 60 °C and 100 °C to improve the resistivity for the hydrolysis of SiO_2, which was defined as nanostructure@$tmSiO_2$. Thus, the residual surfactant can be removed.

Figure 4.14 presents TEM and optical spectra of Au NR@$tmSiO_2$ and (Au @Ag NR) @$tmSiO_2$ before and after ultrasonic. It can be found that under the thermal treatment, SiO_2 layer remained mesoporous structure. As it was known, the $mSiO_2$ capping would lead to the redshift of SPR peaks due to an increase in local dielectric constant of the surrounding medium. Figure 4.11d,e illustrates that when the SiO_2 layer thickness was about 25 nm, SPR_L peak red shifted to ~30 nm. After a fast-cleaning route, the SPR_L blue shifted about 20 nm. It was ascribed to the changes in the surrounding medium from the removal of surfactants and the morphology from surface etching.

For the case of (Au @Ag NR) @$tmSiO_2$, the residual Ag_2O can be removed and the surface of metallic nanocrystals would generate some surface defects manifested by the atomic size of terraces and steps. The formed Au@Ag core–shell NRs present an improved catalytic property, since these defects can serve as active catalytic sites. The structural characteristics by dark-field STEM imaging and complementary energy dispersive spectroscopy (EDS) elemental mapping in Figure 4.15 revealed the morphology change, especially at the corner sites. After the ultrasonic treatment and HCl etching process, some concaved sites and cavities appeared at the edges and ends. The rough Ag shell brought out abundant high-index surfaces, atomic steps, and terraces, which can serve as active sites for further increasing catalytic activity of the nanocrystals.

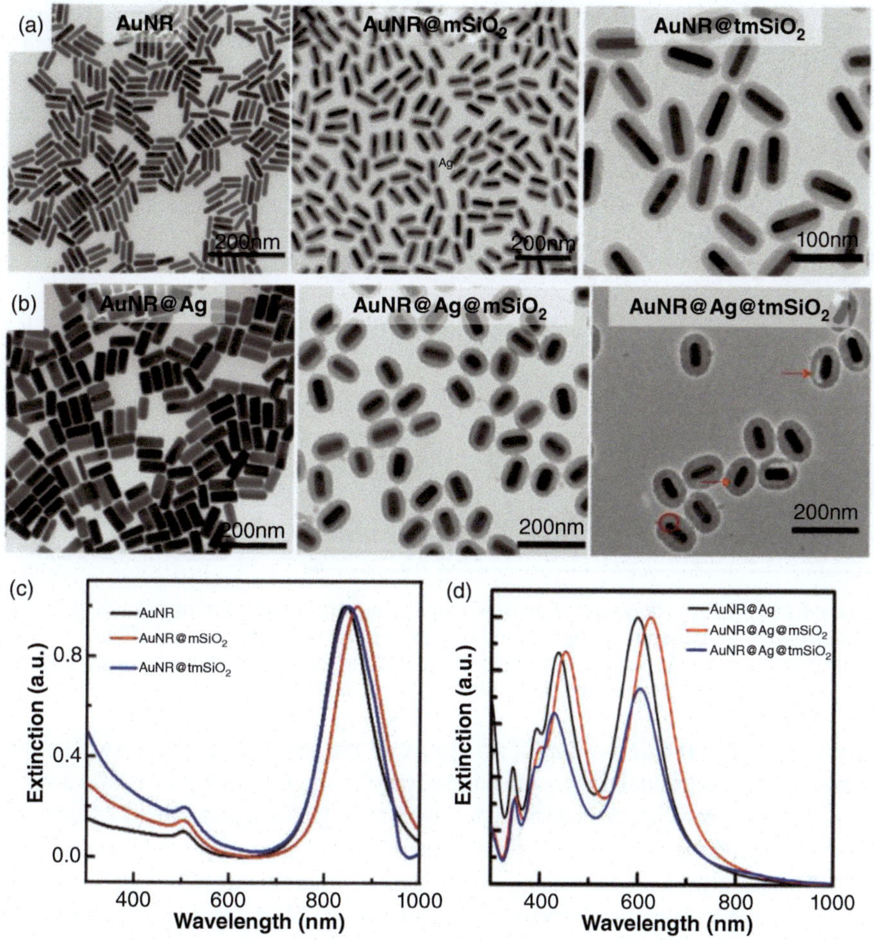

Figure 4.14 TEM images of Au NRs (a) and Au@Ag core–shell nanocuboids (b) with mesoporous SiO_2, (c and d) the corresponding extinction spectra of the samples (a and b). Source: Ke et al. [9]/Reproduced with permission from © 2019 ELSEVIER.

4.3.2 Core–Shell Nanorods Based on Au NBPs

4.3.2.1 Au NBP@Ag Core–Shell Nanostructures

Due to the excellent size and morphology uniformities, Au NBP with five-fold twinned crystal also is a promising substrate to direct the fabrication of anisotropic composite nanostructures. The synthesis approaches of Au NBP-supported nanostructures are similar to that of Au NR-based nanostructures [57–59], including two steps of the preparation of Au inner-core and overgrowth of Ag (Figure 4.6d). In this case, the as-prepared Au NBPs were dispersed, respectively, in solution with the presence of CTAC and BDAC, followed by adding the mixture of $AgNO_3$ and AA into the growth solution for the growth of Au NBP@Ag NRs.

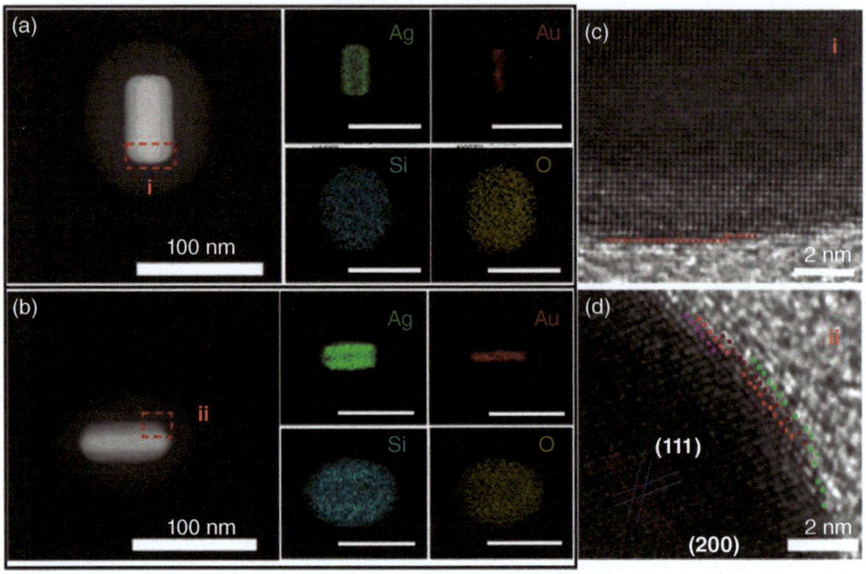

Figure 4.15 STEM images and elemental maps of Au, Ag, Si, and O of one (Au@Ag NR)@mSiO_2 nanostructure (a) before and (b) after HCl etching and ultrasonication treatment, and the corresponding HRTEM images in panels c-i (c) and d-ii (d). Source: Ke et al. [9]/Reproduced with permission from © 2019 ELSEVIER.

Different from the deposition of Ag atoms on the single crystalline Au NRs into nanocuboids, the reduced Ag atoms deposit along the twinned planes of Au NBPs into Au@Ag core–shell NRs with high aspect ratios. The aspect ratios can be regulated by varying the addition of Ag precursor. Figure 4.16a–j displayed the core-shell Au-NBP@Ag NRs with increasing aspect ratios, wherein Au NBPs are located in the center of NRs (see the elemental mapping image). With increasing aspect ratios of the nanorods, the SPR_L extends to the middle infrared region, as indicated by Figures 4.16l. The formed Au-NBP@Ag NRs also presented a five-fold twinned structure bounded by five (200) facets on the side surfaces, ten (111) facets at two ends, and small (220) facets at the edges.

The effect of surfactants, such as CTAB, CTAC, and BDAC, on products was investigated. When CTAB was applied as surfactant, the morphology of products was invariable and remained Au NBPs, which was illustrated by the TEM and optical spectrum, as shown in Figure 4.17a. When CTAC (or BDAC) was served as surfactant, Ag atoms deposited on the surface of Au NBPs, forming rod-like structures, as indicated by the inset in Figure 4.17b,c. It can be obtained that the aspect ratio of products for the case in CTAC was longer than that in BDAC. It is well known that surfactants acting as soft templates form 1D polymeric chain to guide the Ag overgrowth on the surface of Au NBPs. The above discussion has acquired that Br^- ions from CTAB have strong bound action to the side surface of Au, thus blocking the deposition of Ag. As a result, the products in the presence of CTAB were still dominated by Au NBPs. For CTAC and BDAC with the same aliphatic tails and different head groups, the difference on products sizes mainly resulted from two factors

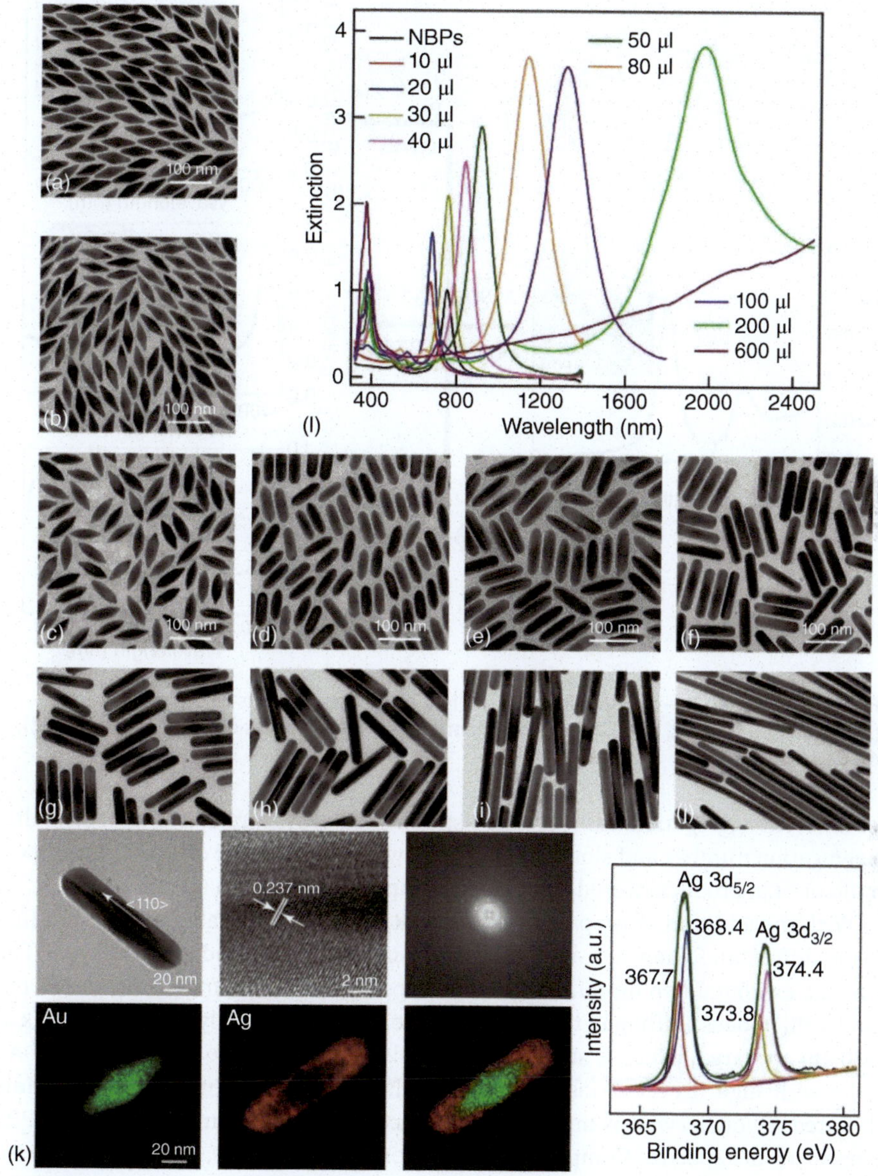

Figure 4.16 (a–j) TEM images of Au NBP@Ag NRs with different aspect ratios. Source: Reproduced with permission from Xu et al. [59]; © 2023 Royal Society of Chemistry. (k) Characterization of a single Au NBP@Ag NR, including HRTEM image of a single Ag nanorod, the fast Fourier transform spectra of area, elemental mapping images of Au NBP@Ag NR, and XPS spectra of Ag 3d of Ag nanorod. (l) UV–Vis–NIR spectra of different aspect ratios. Source: Xu et al. [60]/Reproduced with permission from © 2021 Royal Society of Chemistry.

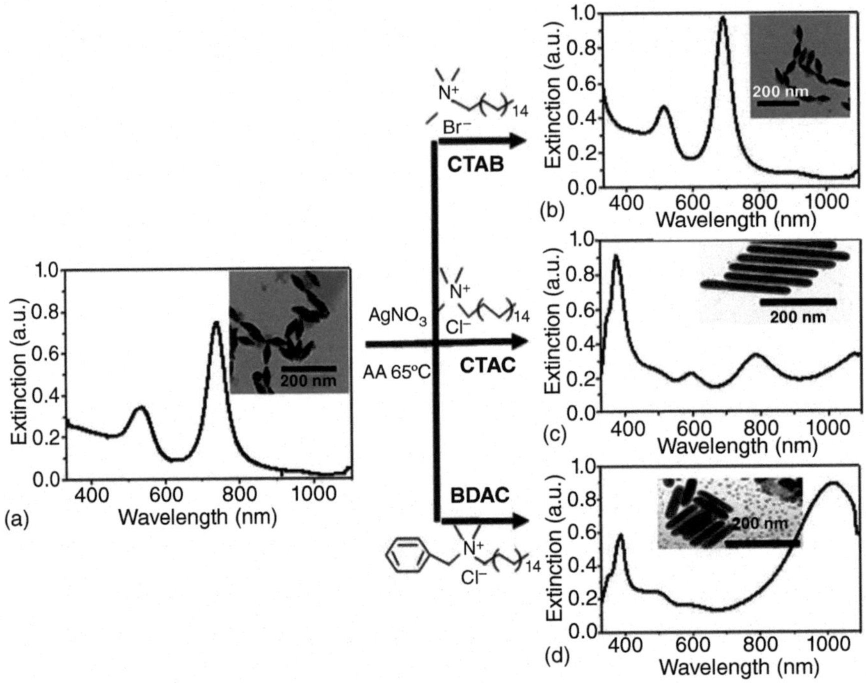

Figure 4.17 UV–Vis–NIR spectra of (a) Au NBPs and the products in the presence of (b) CTAB, (c) CTAC, and (d) BDAC. Source: Reproduced with permission from Wei et al. [58]; ©2016 Elsevier.

of different aggregation number and distinguished growth rate. Due to the smaller aggregation number and slower growth rate with the presence of BDAC, the aspect ratio of final products would decrease in comparison of that in CTAC.

With the addition of Ag precursor, the SPR_L of Au NBP located at ~780 nm first blue shifted and then red -shifted. The SPR_T at 530 nm gradually decayed. The spectra exhibited two prominent resonance peaks originated from SPR_T at 390 nm and SPR_L modes shifting in the Vis–NIR region. Simultaneously, a shoulder peak at 350 nm appeared, which is attributed to the plasmon resonance of bulk Ag. Au@Ag NRs with high aspect ratios will regulate the SPR_L mode into near and middle infrared region, as well as multipole resonance modes between the SPR_T and SPR_L appear in the spectra. Compared with monometallic Ag NRs, Au NBP-supported Ag NRs not only possess high uniformity in size and morphology but also exhibit an improved chemical stability and excellent tunable optical properties.

4.3.2.2 Au NBP@AuAg Core–Shell Nanostructures

For the preparation of Au NBP-supported AuAg alloy nanostructures [61, 62], the process was similar to that of core–shell Au@AuAg NRs. The metal precursors ($HAuCl_4$ and $AgNO_3$) were simultaneously added dropwise into Au NBPs seed solution with the presence of surfactant of CTAC and reducing agent (sodium ascorbate). The synthetic route was illustrated in Figure 4.6. For this case, the

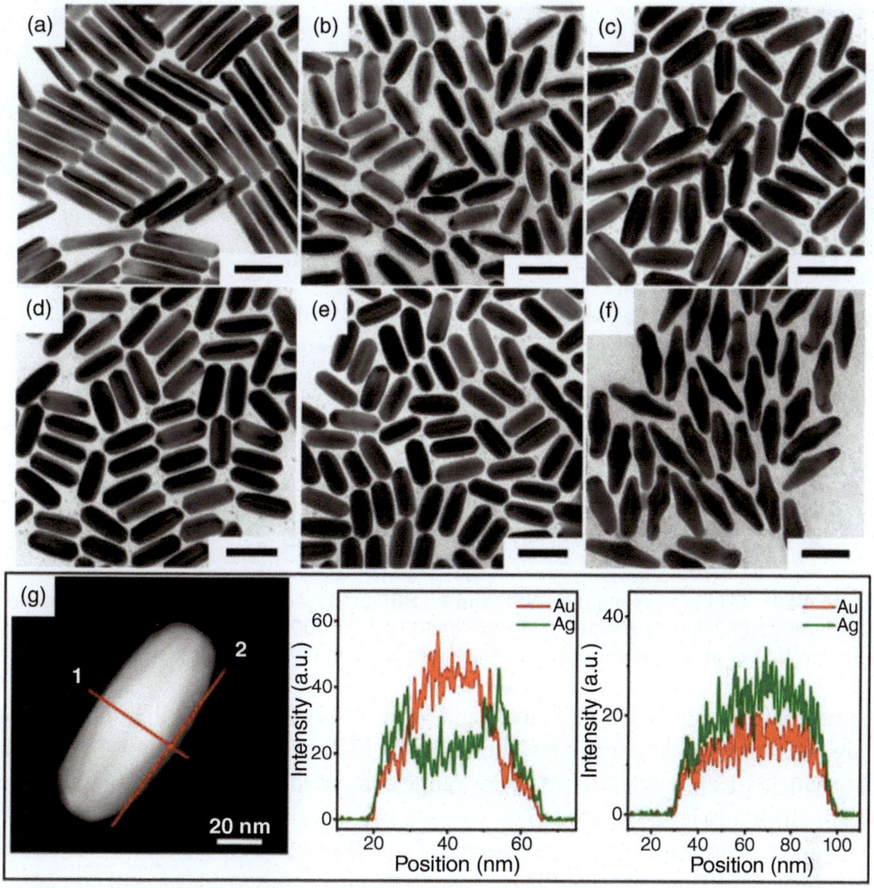

Figure 4.18 TEM images of Au NBP@Au_xAg_{1-x}NRs with different component ratios: (a) $x = 0$, (b) $x = 0.2$, (c) $x = 0.4$, (d) $x = 0.6$, (e) $x = 0.8$, and (f) $x = 1$. (g) HAADF-STEM images of Au NBP@AuAg alloyed NRs and corresponding the line scanning results. (Scale bars: 100 nm). Source: Wang et al. [61]/Reproduced with permission from © 2020 IOP Publishing.

component ratios of alloy shell were controlled by changing the concentration of precursors on the condition that the added volume was constant. According to the section above, the reduced Ag atoms would deposit along the crystal axis of Au NBPs. However, the epitaxy growth of Au–Ag alloyed shell is different from the overgrowth of pure Ag based on Au NBPs. Figure 4.18 shows TEM images of Au NBP@Au_xAg_{1-x} NRs with different component ratios. It can be found that the bimetallic products were rice-like nanostructures. The HAADF-STEM images and line-scanning results demonstrated the formation of Au–Ag alloyed shell on the surface of Au NBPs. The growth process was explored by changing the concentrations and additional volumes of two metal precursors.

For Au NBP@$Au_{0.4}Ag_{0.6}$ NRs ($x = 0.4$), the products changed from nanorice to pointy ended nanorod with increasing the precursors, as shown in Figure 4.19. For the bimetallic Au–Ag deposition, the reduced metal atoms would preferentially

Figure 4.19 TEM images of (a) Au NBP, and Au NBP@$Au_{0.4}Ag_{0.6}$NRs with different volumes of precursor: (b) 50 μl, (c) 200 μl, (d) 300 μl, (e) 400 μl, (f) 1000 μl. (Scale bars: 100 nm). Source: Wang et al. [61]/Reproduced with permission from © 2020 IOP Publishing.

deposit on the high-index side facets and grow slowly along [110] direction, while pure Ag atoms would deposit fast along the [110] direction. As a result, the length of products was restricted to a smaller range compared with the overgrowth of Ag on Au NBPs into long nanorods.

Figure 4.20 presents the optical spectra of Au NBP@Au_xAg_{1-x} NRs with different component ratios and Au NBP@$Au_{0.4}Ag_{0.6}$ NRs with different addition of precursor. For the samples of Au NBP@Ag NRs and Au NBP@Au NRs ($x = 0$ and $x = 1$), the SPR_L peaks are located at 1000 and 895 nm. When x changed in the range of 0.2 to 0.8, the SPR_T peaks can be regulated in the spectral region of 380–520 nm. Interestingly, the SPR_L peaks blueshift obviously from 1000 to ~685 nm. Moreover, the incorporation of alloyed shell brought narrower FWHM for the SPR_L band compared with that of Au NBP@Ag NRs and Au NBP@Au NRs. For Au NBP@$Au_{0.4}Ag_{0.6}$ NRs, the SPR_L first blue shifted and then red shifted with increasing the precursors. The intensity of SPR_T increased remarkable simultaneously with the SPR_L shifting. The inset of Figure 4.20b indicated that both the length and width of nanostructures increased linearly with the increasing addition of precursors.

Figure 4.21 expresses the optical spectra of Au NBP@Au_xAg_{1-x} NRs with different Au–Ag components before and after the etching treatment with $Fe(NO_3)_3$. The variation of SPR modes demonstrated that Au_xAg_{1-x} alloyed shell ($x = 0.2$) was unstable due to the high Ag content that can be etched with the addition of $Fe(NO_3)_3$ solution. With increasing the Au content ($x = 0.4$), Au NBP@Au_xAg_{1-x} NRs showed high stability. It had been reported that the binding energies of Au–Au and Au–Ag bonds were lower than that of Ag–Ag bonds. Au–Ag bond possessed the lowest binding energy and expressed excellent chemical and thermal stability.

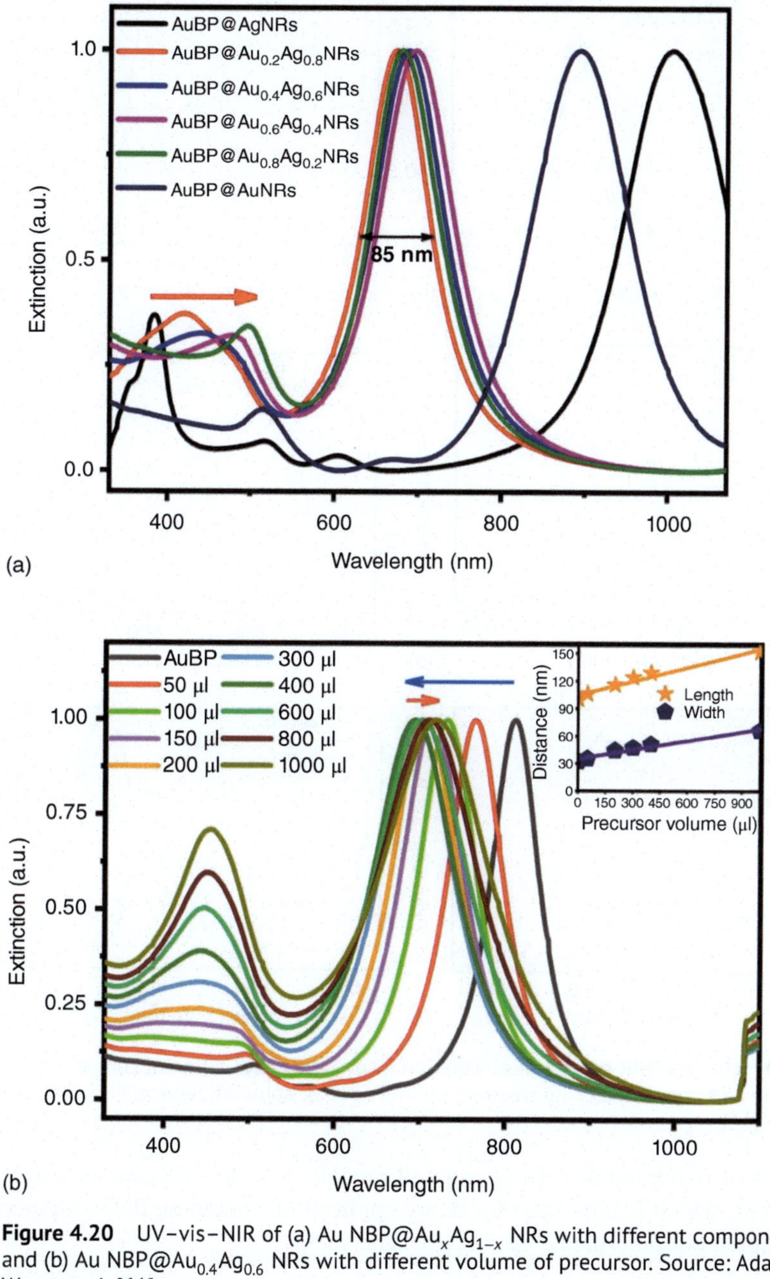

Figure 4.20 UV–vis–NIR of (a) Au NBP@Au_xAg_{1-x} NRs with different component ratios and (b) Au NBP@$Au_{0.4}Ag_{0.6}$ NRs with different volume of precursor. Source: Adapted from Wang et al. [61].

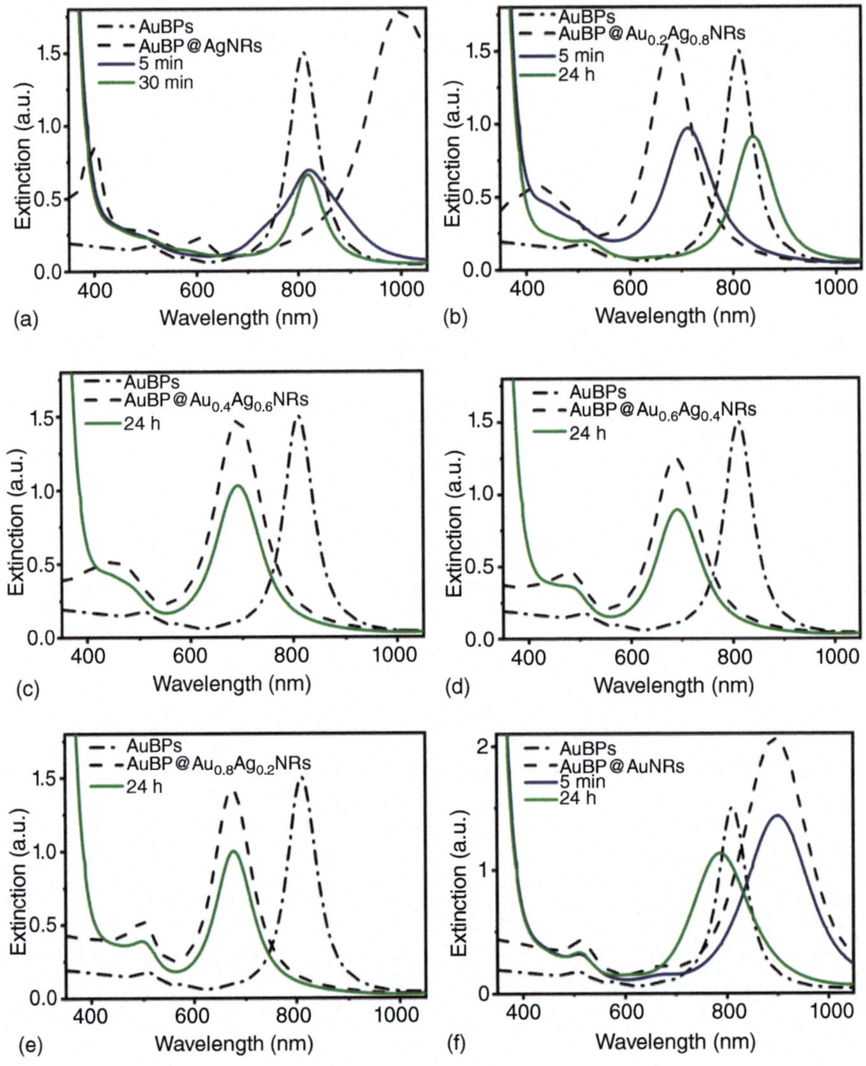

Figure 4.21 UV-Vis-NIR spectra of Au NBP@Au_xAg_{1-x} NRs with different component ratios before and after the etching treatment with $Fe(NO_3)_3$ solution. Source: Adapted from Wang et al. [61].

Alloyed metal nanocrystals with strong and tunable SPR can be stable in a harsh environment, which is important for many applications including field-enhanced spectroscopies, plasmon-mediated catalysis, and so on.

4.4 Alloy Au/Ag Nanorods

The above sections introduced the synthesis of Au-supported Ag NRs and alloyed Au–Ag nanoshells. Alloyed nanostructures show synergetic advantages combining excellent chemical stability of Au nanocrystals and plasmonic performances of Ag nanocrystals, which have been applied in the fields of molecular detection, catalysis, plasmonic sensing, and imaging. For example, Au–Ag alloyed nanospheres can be served as a better SERS sensor than monometallic Au nanospheres in quantitative detecting endonuclease [62, 63]. The chemical reactivities are associated with surface structure, component, and segregation [64, 65].

According to the atomic spatial arrangement, alloyed nanostructures can be classified into uniform alloys and intermetallic nanoclusters. Many efforts have been devoted to the synthesis of uniform alloyed nanostructures with controllable size, morphology, and composition [66–71]. The common synthetic approaches mainly include co-reduction, thermal treatment, galvanic replacement, and Kirkendall effect. Among the methods, thermal annealing is the most straight and effective strategy to facilitate the phase transition and rearrangement of metal atoms, in which the d–d orbital interactions were carried out. This method has been widely applied to synthesize bimetallic nanostructures and intermetallic semiconductors. In addition, liquid-phase-based method is also a powerful strategy, which involves hydrothermal method, electrochemical method, co-reduction, and seed-mediated process mentioned above. The synthesis procedure is governed by the thermodynamic and kinetic contributions.

In this section, the synthesis of anisotropic Au–Ag alloyed NRs was first introduced, which involves uniform alloy and intermediated nanostructures. Especially, uniform Au–Ag alloy NRs with tunable aspect ratios were fabricated by thermal annealing. The plasmonic performances can be widely regulated by the composition ratio and aspect ratios. The chemical stability was also investigated.

4.4.1 Co-Reduction Method

In consideration of atomic ordering, the bimetallic nanostructures can be divided into two patterns including uniform alloy and intermetallic cluster-in-cluster arrangement. Co-reduction provides a straightforward and effective approach for the synthesis of alloyed nanostructures, in which two or more metal precursors were simultaneously reduced. The nucleation and growth of nanocrystals are dependent on the reduction rate of metal ions. And the reduction rate is controlled by the synergistic contributions of reaction temperature, reducing agent, surfactants, and coordination ligand [72, 73].

It has been known that the reduction potential of the metal ions presents intrinsic discrepancies. Although the lattice structures of Au and Ag are matched, in which the metal–metal bond lengths for Au (2.88 Å) and Ag (2.90 Å) are almost identical, the reduction potentials of Au^{3+}/Au and Ag^{+}/Ag are 1.50 and 0.8 V, respectively. In consideration of the thermodynamic contributions, the surface energy of Au (97 meV $Å^{-2}$) is higher than that of Ag (78 meV $Å^{-2}$). And the cohesive energy of Au (3.81 eV per atom) is larger than that of Ag (2.95 eV per atom). These discrepancies decrease the metal-metal bond strength of Ag–Au, leading to the enrichment of Ag on the surface of Au and form the core–shell nanostructures. However, the reduction rate can be held in synchronization by regulating the molar ratio of precursors. Han et al. [74] achieved Au–Ag alloyed nanospheres by co-reduction of Au and Ag precursors. Sun et al. [75, 76] recommended a one-pot synthesis of monodisperse Au_xAg_{1-x} alloy NPs by simultaneously reducing $HAuCl_4$ and $AgNO_3$ in a solution of octadecene at 120 °C with oleylamine as the reducing agent and capping agent. Considering the thermodynamic contribution, the maximization of (111) planes can minimize the surface energy of NPs, as well as compensate the extra strain energy [77]. Although the reduction rate can be controlled by the relative concentration of metallic precursors, comprehensive parameters must be considered in the reactive system, including solvent, reducing agents, capping agents, surfactants, and temperature.

Zheng and co-workers reported an ultrafast synthesis of wavy Au–Ag alloy nanowires via a seed-mediated co-reduction [78]. In this case, Au seeds with tunable size in the range of 2–10 nm are capable to manipulate the diameter of final products. Metallic precursors ($HAuCl_4$ and $AgNO_3$) were reduced by AA at 60 °C in the present of CTAC. The detail structures and morphologies characterizations of Au–Ag wavy nanowires by HRTEM, HAADF-STEM, EDS, STEM-EDX, and XPS demonstrated the presence of the alloy phase of Au–Ag, as shown in Figure 4.22. Unfortunately, the ultrafine nanowires with uncontrollable and nonidentical length limited further applications.

4.4.2 Thermal Annealing Method

Thermal annealing is another powerful method for the synthesis of alloy nanocrystals. It has been proven that when two metal precursors possess different reduction potential, there is a greater probability for the synthesis of core–shell nanocrystals. Notably, core–shell nanocrystals at a higher temperature will present interface infusion, leading to the formation of alloy nanocrystals, as shown in Figure 4.23a. Moreover, random alloy nanocrystal annealed at an elevated temperature will transform into an intermetallic nanocrystal [78]. The homogeneity can be enhanced though the inter-infusion between Au and Ag atoms by means of annealing [76, 79, 80], laser irradiation [81, 82], or ultrasonic treatment [83, 84]. Sun et al. [76] reported an organic solvothermal method for the synthesis of multi-crystalline AuAg alloyed NPs though annealing Ag@Au core–shell NPs in oleylamine at 100 °C. Yin et al. [85] prepared fully alloyed NPs by annealing Au@Ag@SiO_2 core–shell structures at 1000 °C in a tube furnace in a nitrogen atmosphere. It was found that these alloyed NPs are single crystalline in nature. The transformation was attributed that the high temperature accelerated the diffusion of Au and Ag atoms and compensated for the crystallographic defects.

Using the annealing protocol, uniformly alloyed Au–Ag NRs were achieved by annealing Au@Ag@SiO_2 core–shell NRs in a furnace at a higher temperature [86]. The synthesis process shown in Figure 4.23b mainly involved four steps of the preparation of Au@Ag core–shell NRs, SiO_2-shell coating, thermal annealing, and etching. Tailoring the addition of Ag precursor can tune the optical properties in a wide wavelength region. The as-prepared core–shell Au@Ag NRs served as templates to fabricate alloyed Au–Ag NRs that remain the morphology of the original morphology.

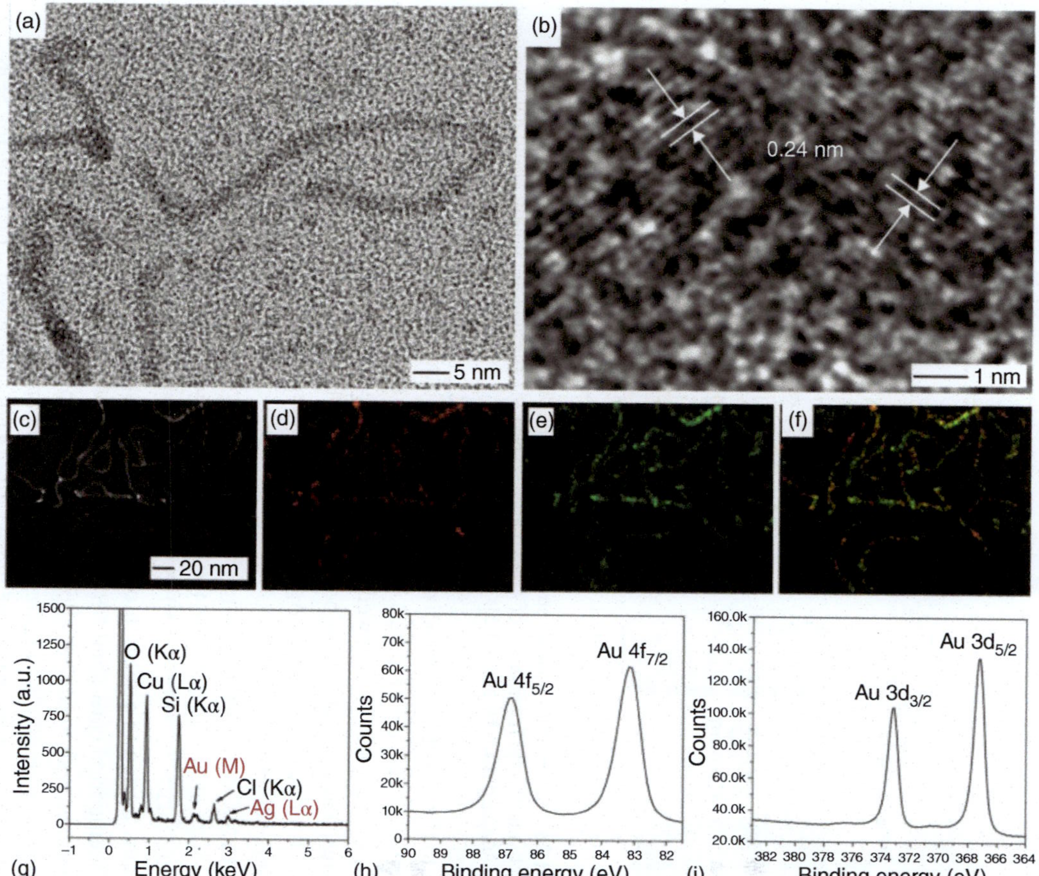

Figure 4.22 Morphology and structural characterization of ultrathin AuAg wavy nanowires: (a, b) HRTEM; (c) HADDF-STEM; (d–f) STEM mapping: (d) Au, (e) Ag and (f) overlap of Au and Ag; (g) EDS; (h, i) XPS. The scale bar in (c) applies to (d–f). Source: Wu et al. [78]/Reproduced with permission from © 2021 JOHN WILEY & SONS, INC.

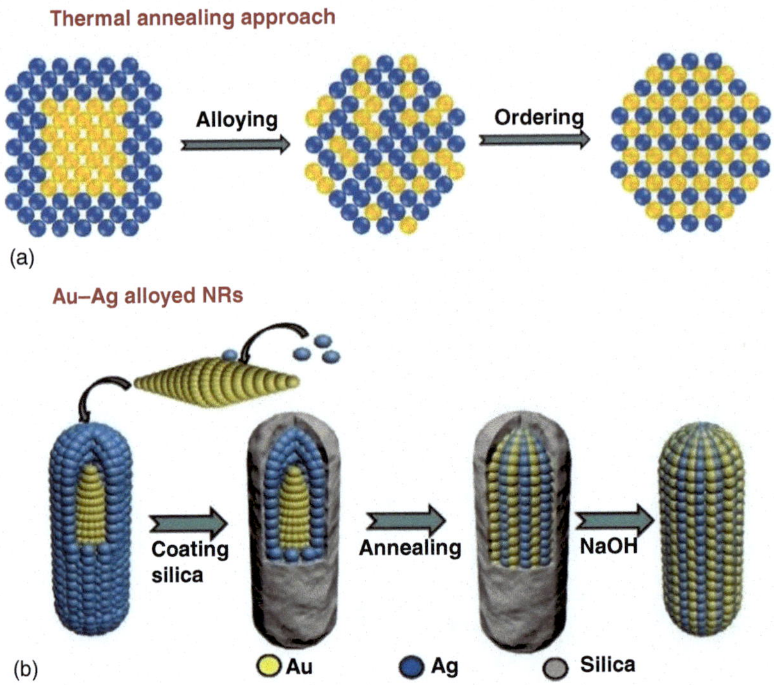

Figure 4.23 Schematic illustration showing (a) atoms distribution in thermal annealing approach and (b) the fabrication of Au–Ag alloyed NRs via thermal annealing approach. Source: Adapted from Ni et al. [86].

Due to the possible spheriodized transformation and sublimation of Ag atoms at high temperature, the prepared Au@Ag core–shell NRs (Figure 4.24a) were capped by SiO_2 (Figure 4.24b), followed by annealing in N_2 atmosphere at different temperatures. The SiO_2 layer can be removed by NaOH or HF solution. Through this process, the interface between Ag-shell and Au-core disappeared and Au–Ag alloyed NRs were achieved.

For Au NR-supported core–shell NRs, the interface between Ag-shell and Au-core can be seen after annealing at 450 °C, as signed by the red arrow in Figure 4.26a. Compared with Au NRs, Au NBPs possessed a high index of side facets, and thus the atoms tended to diffuse into Ag shell. As a result, the alloying temperature of Au NR-supported core–shell NRs was higher than Au NBP-supported one. A series of experiments showed that alloyed Au–Ag NRs formed when the annealing temperature is up to 600 °C (Figures 4.25c,d). Compared with the core–shell Au@Ag NRs template, the aspect ratio of alloyed Au–Ag NRs decreased. Therefore, the SPR_L peak blue shifted (Figure 4.26e). Simultaneously, multipolar resonance modes for the core–shell Au@Ag NRs in the short wavelength range disappeared, and only one SPR_T band can be observed for the alloyed Au–Ag NRs located at ~410 nm. Though annealing core–shell Au@Ag NRs, alloyed Au–Ag NRs with well-defined size can be prepared. The SPR_L can be tuned in the visible region (Figure 4.26f).

It has been known that Au NBPs-supported core–shell Au@Ag NRs exhibited high tunability in aspect ratios. Therefore, the aspect ratios of alloyed Au–Ag NRs also

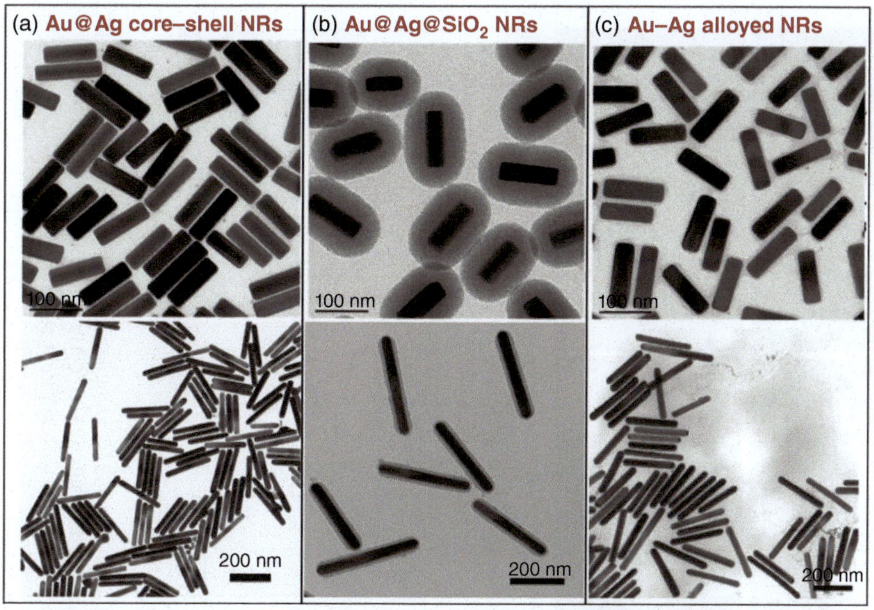

Figure 4.24 TEM images of (a) Au NRs- and Au NBPs- supported core–shell NRs, (b) mSiO_2-encapsulated core–shell NRs, and (c) Au–Ag alloyed NRs. (The authors' not published work)

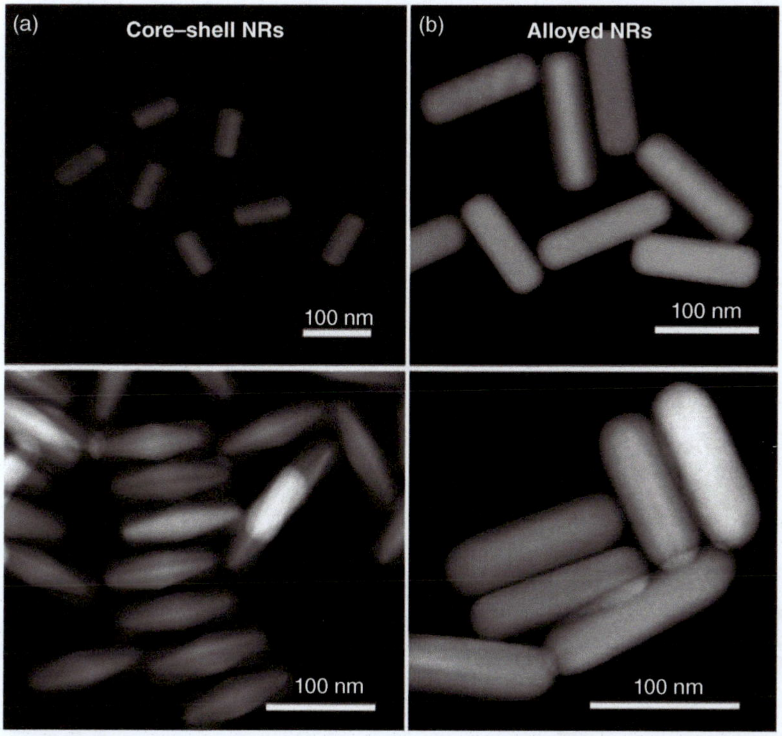

Figure 4.25 (a) HAADF-STEM images of Au NR- and Au BP-supported core–shell NRs. Source: Reproduced with permission from Gao et al. [85]; © 2019 Optical Society of America. (b) HAADF-STEM images of the corresponding Au-Ag alloyed NRs. Source: Chen et al. [77]/Reproduced with permission from © 2021 ELSEVIER.

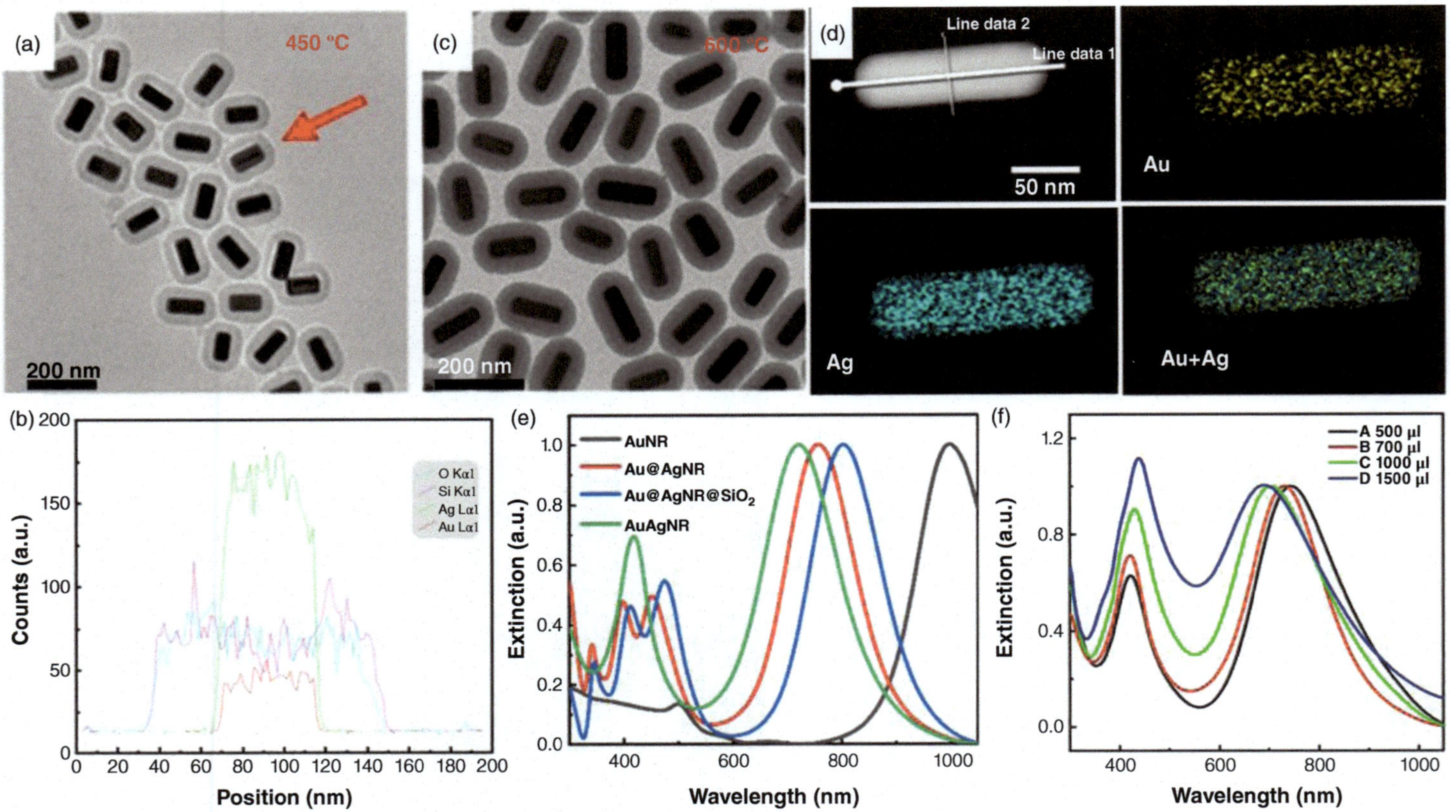

Figure 4.26 (a) TEM image of the products by annealing Au NR-supported core–shell Au@Ag NRs at 450 °C, and (b) corresponding line scanning of EDX. (c) TEM image and (d) EDX mapping of the products by annealing Au NR-supported core–shell Au@Ag NRs at 600 °C, UV–Vis–NIR spectra of (e) the samples during the synthetic process and (f) the products by annealing Au NR-supported core–shell NRs with different aspect ratios at 600 °C. Source: Ke et al. [9]/Reproduced with permission from © 2019 ELSEVIER.

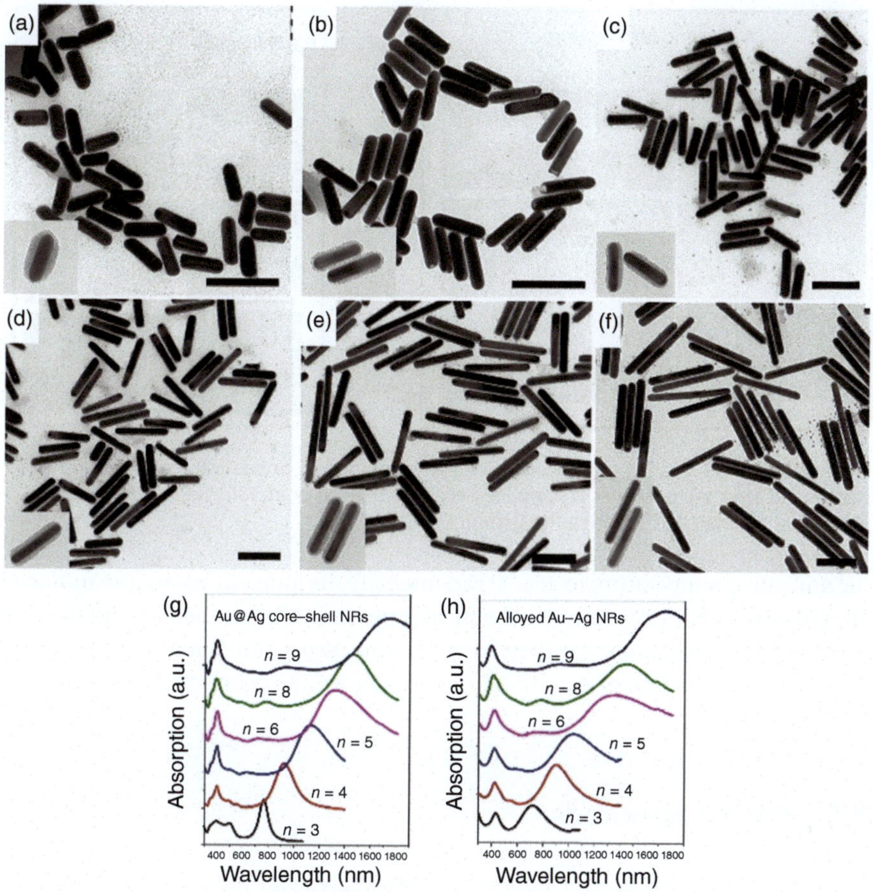

Figure 4.27 TEM images and corresponding UV-vis-NIR spectra of core–shell Au@Ag NRs and alloyed Au-Ag NRs with different aspect ratios. The insets are TEM images of core–shell Au@Ag NRs and alloyed Au–Ag NRs coating with thin mSiO_2 nanoshell. (Scale bars: 200 nm). Source: Ni et al. [86]/Reproduced with permission from © 2019 Optica Publishing Group.

can be tuned in a large range, as illustrated in Figure 4.27a–f. Correspondingly, the SPR modes can be regulated in a large wavelength region. Figure 4.27g,h presents optical spectra of different aspect ratios of core–shell Au@Ag NRs and alloys Au–Ag NRs. Alloyed Au–Ag NRs possessed two dominated SPR_L and SPR_T. The SPR_L peak changed from visible to infrared region, which was mainly related to the aspect ratio. A new dimension on composition ratio was provided to tune the plasmonic properties. It had been known that the characteristic resonance bands of Au and Ag are 520 nm and 400 nm, respectively. The SPR_T peak dependent on the composition ratios between Au and Ag would blueshift with an increase in Ag content.

For Au NBP-supported core–shell Au@Ag NRs, alloyed Au–Ag NRs formed when the annealing temperature was 400 °C, which can be illustrated by the EDX images in Figure 4.28. Au atoms were mainly distributed in the middle of NRs. Increasing

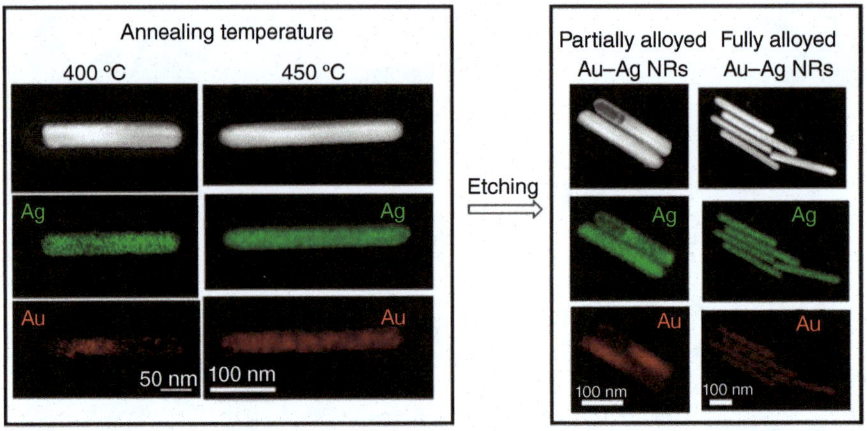

Figure 4.28 EDX images of alloyed Au–Ag NR obtained by annealing at 400 °C and 450 °C before and after etching in H_2O_2 and $NH_3 \cdot H_2O$. Source: Ni et al. [86]/Reproduced with permission from © 2019 Optica Publishing Group.

the annealing temperature to 450 °C can intensify the atom diffusion into uniformly alloyed Au–Ag NRs. The physicochemical properties of core–shell and alloyed NRs were distinguished, although they have the same aspect ratio and the composition ratio. The introduction of Au–Ag alloy can improve the chemical and thermal stability of NRs [85].

4.5 Porous Nanorods

Unlike solid NPs, porous, or hollow nanostructures can provide large surface-to-volume ratios and catalytically active surfaces. Although galvanic replacement and dealloying have been widely used to synthesize porous or hollow nanostructures, galvanic replacement is not suitable for AuAg NRs. Dealloying is a robust method to fabricate nanoporous materials, and this approach has been used to develop various nanoporous alloys for many different applications in the past decade [88–96]. In particular, open bi-continuous nanoporous structures constructed through dealloying process have demonstrated extraordinary catalytic activities, high stability, and improved selectivity [97–99].

4.5.1 Porous AuAg NRs

In general, porous nanostructures adopt a quasi-spherical geometry to attain a minimum surface free energy. Therefore, the fabrication of homogeneously dispersed and fully alloyed porous Au–Ag nanostructures exhibiting distinct edges or corners is restricted by the lack of a reliable synthesis strategy. On the other hand, the well-defined edges or corners and nanopores are the beneficial features that generate a localized electromagnetic field for SERS enhancement. Thus, a reliable synthesis strategy for porous Au–Ag nanostructures exhibiting distinct corners and edges is highly desired to utilize the valuable SERS properties of the nanostructures. Porous alloyed Au–Ag nanorods (P-AuAg NRs) represent a group of catalysts whose aspect ratios and composition can be tuned to achieve the optimal catalytic performance toward specific reactions. For example, longitudinal surface plasmon resonance peaks of the P-AuAg NRs can be regulated across visible to near-IR regions, which are beneficial for applications of solar energy in photocatalysis and photothermal energy conversion. In addition, the nanoscale porosity gives rise to inherent hotspots upon plasmonic excitation. Recently, Chen and co-workers reported the dealloying kinetics of Au–Ag NRs utilizing the in situ liquid-cell under AC-STEM equipment [100]. Nevertheless, there are few reports for the direct and precise control of nanoporous in rod-like alloy nanostructures. Also, the properties of P-AuAg NRs for catalytic applications have not been explored.

A schematic of the synthesis of P-AuAg NRs is shown in Figure 4.29. Au@Ag NRs were synthesized by retarding the epitaxial overgrowth of Ag shells on the surface of Au NRs using hexadecyltrimethylammonium chloride (CTAC) as a surfactant. Mesoporous SiO_2 shell acted as a protective layer that encapsulated the Au@Ag NR (Au@Ag NR@mSiO_2), followed by annealing the silica-capped NPs at 600 °C for one hour to ensure an atomically homogeneous mixture of Ag and Au. Anisotropic Au@Ag NRs, in contrast to isotropic nanospheres, tend to transform into spherical NPs and experience Ag sublimation at high annealing temperatures (e.g. 930 °C) which can enhance the mobility of the metal atoms [84]. It was also reported that Au@Ag NRs can be fully alloyed after treatment at 450 °C and N_2 protection for one hour [101], while the Au@Ag NRs formed inhomogeneous structures under the same annealing temperature and time. Consequently, annealing at 600 °C appeared to be suitable for the alloying process of Au@Ag NRs. Because silicate condensation is associated with the collapse of inherent mesopores during alloying, the SiO_2 shell was significantly densified. The literature shows that SiO_2 shells do not worsen the

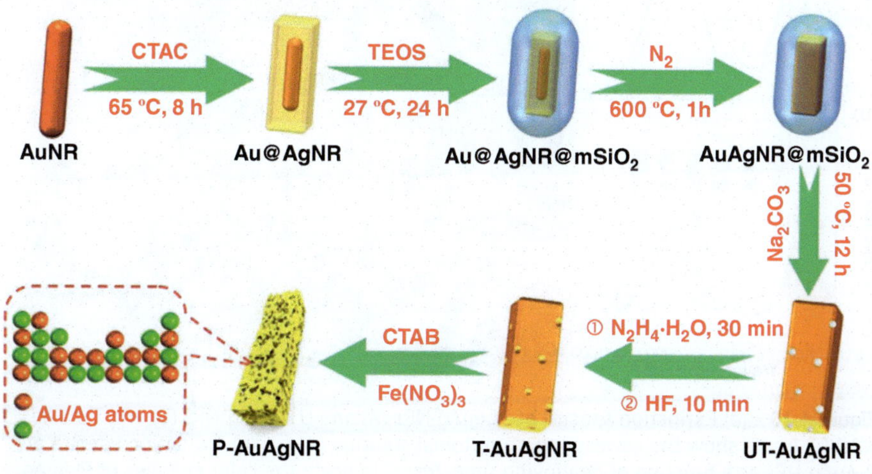

Figure 4.29 Schematic illustration of the synthetic route of P-AuAg NRs nanocatalysts. TEOS: tetraethyl orthosilicate. Source: Reproduced with permission from Ke et al. [87]; © 2021 Elsevier.

plasmonic properties of the core and provide protection against etching [9, 102]. However, the SiO_2 shells hinder many applications, especially in SERS-related chemical catalysis. They block the interfacial adsorption and exchange of thiolated ligands on the surfaces of nanocatalysts and also interrupt interactions between nanostructures and reactions catalyzed on the surface of NPs [103–105]. In this case, the catalytic efficiency is reduced, and alloyed P-AuAg NRs without the SiO_2 shell must be achieved.

Silica removal by alkaline etching is a reversible chemical reaction process, during which the breakage of SiO_2 under alkaline conditions is accompanied by the redeposition of dissolved silica species [106, 107]. Zheng et al. reported that the redeposition of dissolved silica species can be guided by cationic surfactants (CTA^+) [108]. Comparatively, HF etching is a very fast and irreversible process, but it has a strongly corrosive action. Since it was not possible to completely remove the mesoporous SiO_2 shell in a basic solution, SiO_2 coated AuAg NRs were dispersed in a neutral surfactant (1 wt% PVP) solution. To reduce HF consumption, Na_2CO_3 can be used as the etchant at 50 °C for 12 hours for the first step.

Significantly, apart from the disappearance of the two octupolar modes, a sharp peak associated with the transverse dipolar SPR mode for AuAg NRs appeared at 420 nm. Meanwhile, the SPR_L peak blue-shifted (Figure 4.30a). This was assigned to both the permittivity changes and a slight decrease in the aspect ratio after alloying. Because of the partial oxidation of the Ag shell during storage and experiments, it is

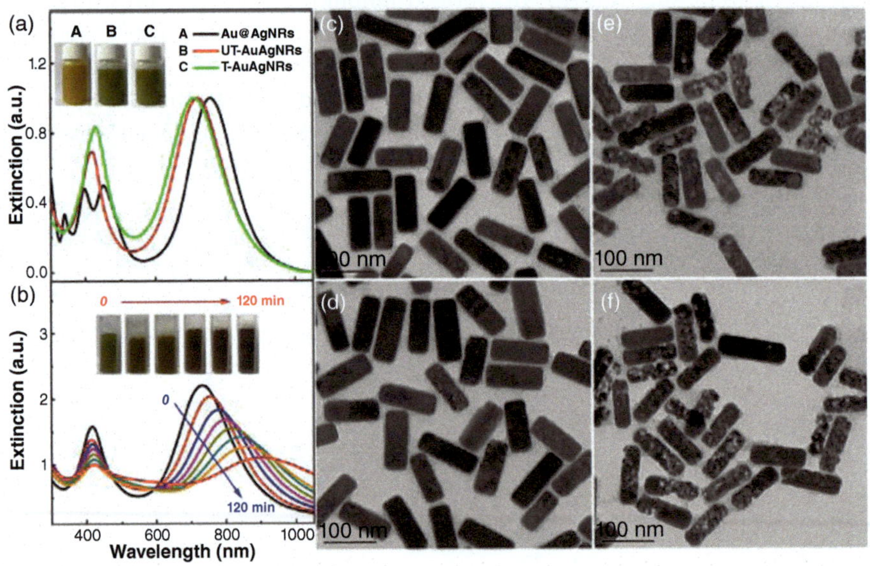

Figure 4.30 (a) Extinction spectra of Au@Ag NRs (A) and UT-AuAg NRs (B) and T-AuAg NRs (C). Insets show the corresponding colloidal solution colors. (b) Extinction spectra of P-AuAg NRs as a function of dealloying time. Insets indicate the color changes of P-AuAg NRs during dealloying. (c–f) Representative TEM images of P-AuAg NRs obtained after etching with $Fe(NO_3)_3$ for different times (c) 5 minutes, (d) 10 minutes, (e) 30 minutes, and (f) 120 minutes. Source: Ke et al. [87]/Reproduced with permission from © 2021 ELSEVIER.

necessary to minimize the loss of Ag before HF addition. Silver oxide (Ag_2O) on the AuAg NRs surface was treated with hydrazine ($N_2H_4 \cdot H_2O$) and in-situ reduced into Ag atoms. After $N_2H_4 \cdot H_2O$ treatment, the longitudinal plasmonic peak of alloyed AuAg NRs (T-AuAg NRs) exhibited a blueshift and the intensity of the SPR_T increased slightly compared with untreated AuAg NRs (UT-AuAg NRs), indicating that surficial Ag_2O reacted with $N_2H_4 \cdot H_2O$. Interestingly, as a comparison, the color of the NRs colloid changed from yellow to dramatic blue-green before and after treatment, as indicated by the inset of Figure 4.30a.

The resulting alloyed AuAg NRs were subsequently dealloyed by the addition of $Fe(NO_3)_3$ in an aqueous solution of 0.05 M CTAB, which partially dissolved Ag from the alloyed AuAg NRs. It is well-known that the SPR of nanostructures is strongly dependent on their size, shape, and composition; thus, changes in the plasmonic peaks of the alloyed P-AuAg NRs as a function of dealloying time can be monitored by the optical UV–Vis extinction spectra (Figure 4.30b). The longitudinal peak moved progressively toward the low-energy region, along with a declining peak intensity and broadening peak width, while the transverse peak position remained unchanged, with a slow decrease in its intensity. Finally, the longitudinal peak moved to >900 nm, and the transverse mode peak nearly disappeared from the extinction spectra.

In addition, the colloidal solution experienced a gradual color change from green to dark blue upon increasing the etching time (inset of Figure 4.30b). With increasing etching time, more and more nanopores formed and evenly dispersed throughout the nanostructure. Eventually, a highly nanoporous alloy network was reconstructed, as shown in Figure 4.29c–f. The alloyed P-AuAg NRs displayed completely irregular shapes for both solid metallic ligaments and empty pore channels. The size of nanopores increased with the erosion time, while the overall size of the alloyed P-AuAg NRs shrank remarkably. There is a large difference in the coordinating ability of acetylacetone with Fe^{3+} and Fe^{2+} ions, so the dealloying process can be terminated by the addition of acetylacetone into the suspension and separation by centrifugation. Acetylacetone strongly coordinated with Fe^{3+} ions, which quickly reduced the concentration of free Fe^{3+} in the colloidal suspension. As a consequence, the corresponding electrode potential decreased, and the reaction rate between Fe^{3+} and Ag^+ slowed down. Under this condition, it allowed us to obtain the desired intermediates for subsequent characterizations. The convenient tunability of the plasmonic peaks makes it possible for the alloyed P-AuAg NRs to achieve optimal plasmonic activity.

Based on the aforementioned procedure, the alloyed AuAg NRs were fully alloyed. The atomic-scale surface structural characterization of the alloyed P-AuAg NRs was resolved using AC-STEM. Figure 4.31a shows a representative HAADF-STEM image of alloyed P-AuAg NRs at low magnification. Electron diffraction patterns acquired of the alloyed P-AuAg NRs displayed a single set of sharp diffraction spots, which unambiguously showed that the alloyed P-AuAg NRs were single crystalline, as indicated by the inset of Figure 4.31a. Since the solid ligaments and empty channels were randomly oriented, five different regions, including surface facets alongside convex and concave curvatures, were arbitrarily selected to analyze HAADF-STEM images

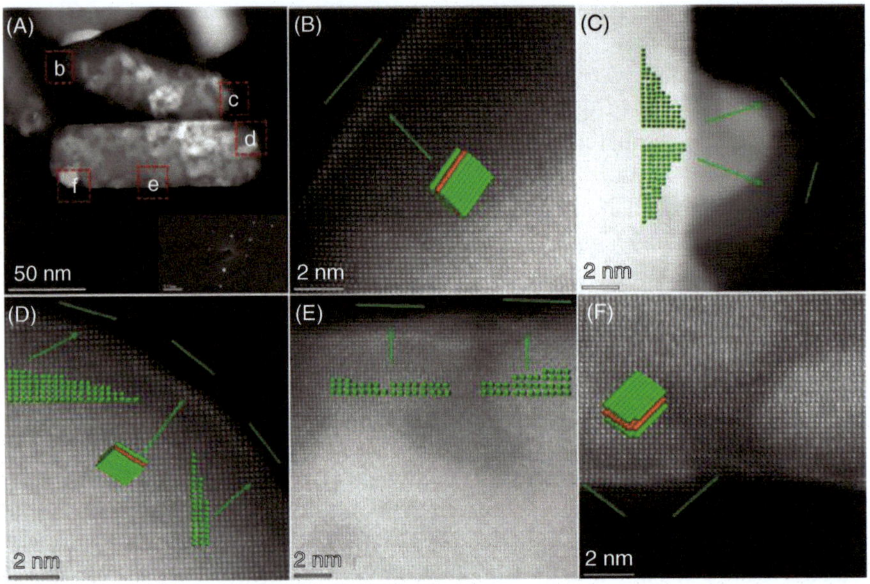

Figure 4.31 (A) Aberration-corrected HAADF-STEM image of alloyed P-AuAg NRs at low magnification. The diffraction pattern (inset) shows that alloyed P-AuAg NRs are single crystalline. The marked squares denote the areas resolved by HAADF-STEM in the following observation. (B–F) Surface planes with monatomic terraces, steps, kinks, and vacancies were observed. The schematics illustration atomic-scale structural characterization of the corresponding high-index or low-index faceting are exhibited in each graph inset. Source: Ke et al. [87]/Reproduced with permission from © 2021 ELSEVIER.

Figure 4.31b–f for different parts of Figure 4.31a. In the curved regions, the surfaces were terminated by high-index facets with a high density of monatomic terraces. Abundant steps, kinks, and some vacancies were also observed. It is believed that these various catalytic sites on the surface of alloyed P-AuAg NRs will play a critical role in improving the catalytic activity.

4.5.2 Chemical Stability of P-AuAg NRs

Stability is extremely critical for practical applications, and in general, alloyed NPs are more stable than single-component systems. However, it is unclear whether the alloyed NPs have significantly enhanced chemical stability against corrosive environments containing different surfactants and oxidants. Previously, a similar annealing procedure was used to prepare alloyed Au–Ag nanospheres and nanorods, which were added into an aqueous solution of polyvinylpyrrolidone (PVP, k30). They exhibited high chemical stability in the presence of hydrogen peroxide H_2O_2 and ammonia $(NH_3 \cdot H_2O)$ [85, 101, 109]. Correspondingly, when the alloyed AuAg NRs were dispersed in 0.05 M CTAC solution in the presence of H_2O_2 (1.5 vol%) and $NH_3 \cdot H_2O$ (1 vol.%), they showed strong etching resistance, as demonstrated by a negligible loss in their extinction intensity, even after exposure to the solution for 24 hours (Figure 4.32a). Using 0.05 M CTAB or 1 wt% PVP instead of CTAC as

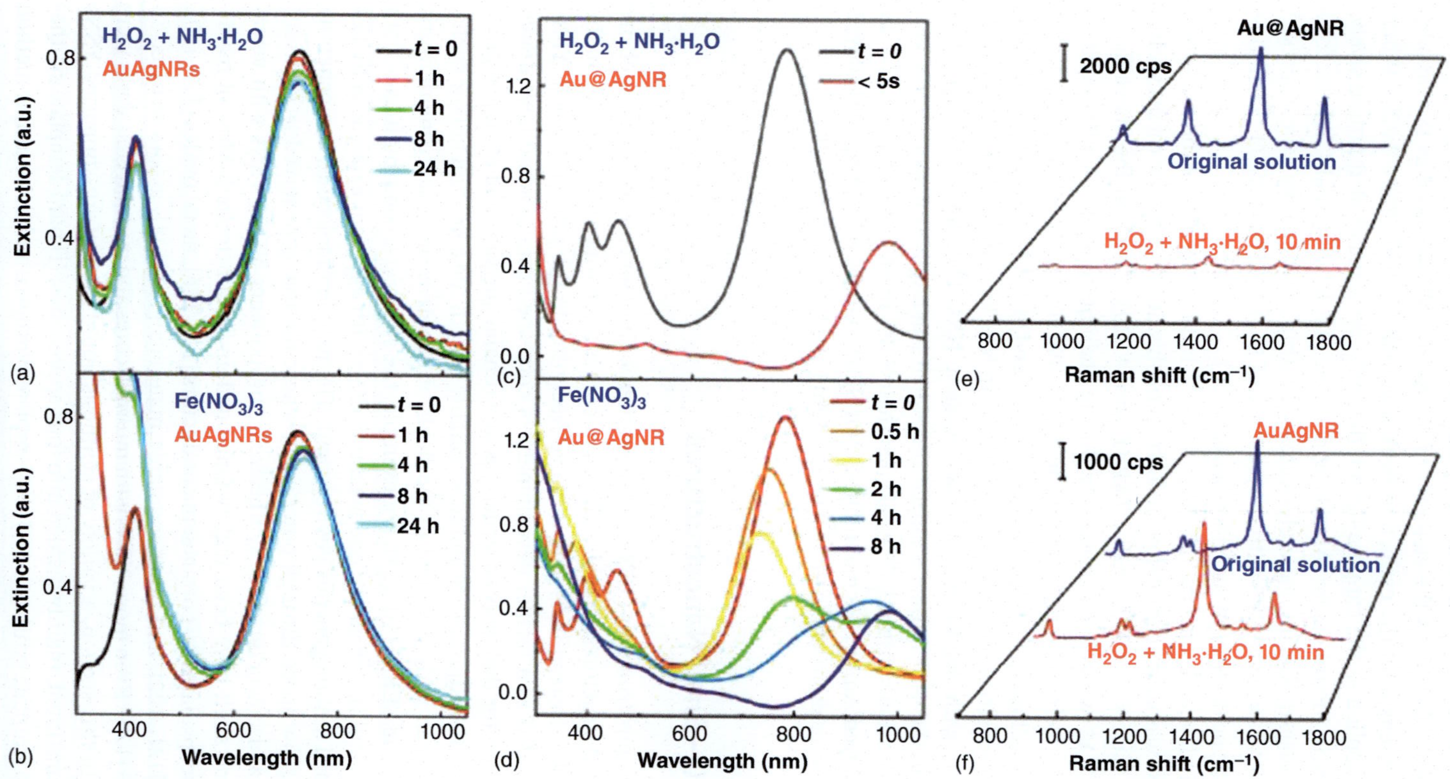

Figure 4.32 Extinction spectra of (a) alloyed AuAg NRs and (c) core–shell Au@Ag NRs exposed to an aqueous etching solution, H_2O_2 (1.5 vol%) and $NH_3 \cdot H_2O$ (1 vol%) in the presence of CTAC, and using 120 µl $Fe(NO_3)_3$ as the etchant for (b) alloyed AuAg NRs and (d) core–shell Au@Ag NRs. Colloidal SERS spectra of 4-NTP on (e) core–shell Au@Ag NRs and (f) alloyed AuAg NRs dispersed in an aqueous etching solution, H_2O_2 (1.5 vol%) and $NH_3 \cdot H_2O$ (1 vol%) in the presence of CTAC for 10 minutes. Source: Reproduced with permission from Ke et al. [87]; © 2021 Elsevier.

the capping ligand produced a similar phenomenon. If H_2O_2 and $NH_3 \cdot H_2O$ were replaced with $Fe(NO_3)_3$ as the etchant or if CTAC or PVP was used as the capping ligand, the intensity of the extinction peak was maintained (Figure 4.32b). When CTAB was applied as the capping ligand, the SPR_L peak red-shifted progressively, as described in Figure 4.30b. In striking contrast, the longitudinal LSPR peak of core–shell Au@Ag NRs immediately decreased within 5s when they were dispersed in 0.05 M CTAC or 1 wt% PVP with the presence of H_2O_2 and $NH_3 \cdot H_2O$, generating characteristic peaks only of Au NRs (Figure 4.32c).

When CTAB was replaced with CTAC, the Fe^{3+}-induced etching of the core–shell Au@Ag NRs was significantly slowed down, probably due to the more robust shelter of Ag (200) facets by CTAC than CTAB (Figure 4.32d). Based on the above facts, Fe^{3+} and CTA^+ worked synergistically during etching. SERS is an ultrasensitive spectroscopic technique that can provide exhaustive structural details of the target molecules attached to metallic surfaces. To this end, 4-NTP was used as a Raman probe to demonstrate the merits of alloyed AuAg NRs by comparing their colloidal SERS activity with that of the core–shell Au@Ag NRs. The addition of H_2O_2 and $NH_3 \cdot H_2O$ to the 4-NTP functionalized core–shell Au@Ag NRs dispersed in CTA^+ or PVP suspension substantially weakened their SERS intensities (Figure 4.32e). In contrast, alloyed AuAg NRs maintained their SERS signals, indicating that alloyed AuAg NRs were more stable than core–shell Au@Ag NRs (Figure 4.32f).

4.6 Yolk–Shell Nanostructures

Yolk–shell nanostructures (YSNs) or nanorattles with inner core and outer shell are relative novel structures, in which the inner core is not covalently connected to the shell material. Instead, it is able to move freely inside the protective shell. YSNs can be regarded as a further development of core–shell catalysts and have drawn great attention for their special structures and tunable physical and chemical properties. Due to the adjustability and functionality in both the cores and hollow shells, YSNs have been widely applied in the fields of nanoreactors, drug/gene delivery, biosensors, and SERS [110–114]. For example, the synergetic contributions of both metal cores and hollow shells endowed the YSNs with excellent catalytic activities that provided a homogeneous environment. The reaction molecules can penetrate into the hollow space through outer shell, where the free core served as catalysis. Since Xia and coworkers first synthesized Au@polymer YSNs, YSNs with versatile core and shell have been widely fabricated and characterized. Most of metal-core are encapsulated by porous semiconductors [115, 116], metal oxides [117–120], polymers [121, 122], and other inorganic materials [123–128], as summarized in Figure 4.33 [129]. Even more, YSNs with multiple cores and multilayered shells were prepared. In this section, a variety of synthesis approaches of Au-supported YSNs were demonstrated, mainly divided into four approaches: (i) hard-template synthesis, (ii) soft-template synthesis, (iii) self-template synthesis, and (iv) multiple method combination synthesis.

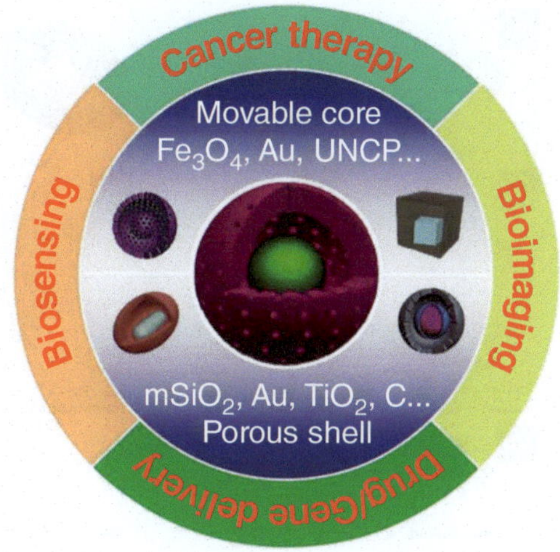

Figure 4.33 Yolk–shell nanostructures and their biomedical applications. Reproduced with permission from Lin et al. [129]; © 2018, Wiley-Blackwell.

4.6.1 Hard-Templating Synthesis

Hard template is the most popular method for fabricating YSNs. In general, the hard-template approaches mainly involved the selective etching process, which was proceeded from the core-supported epitaxial growth. The pre-synthesized Au nanostructures served as templates were coated with one or two layers to build the sacrificial layer and target layer. The material of target layer can be similar with that of sacrificial layer. And then in accordance to the differences of material in physical and chemical properties, appropriate methods were used to remove the sacrificial layer, the outer layer of core, or inner layer of target layer, leading to the formation of multilayered core@void@shell nanostructures.

During the selective etching process, the part of core (or shell) was selectively removed by the dissolution reaction of solvent. SiO_2 was one of the most common hard templates because of a simple synthesis via Stöber process and rapid movement by alkali or hydrogen fluoride (HF) treatment. SiO_2 layer was coated on the surface of Au nanocrystals by the hydrolysis of TEOS. Subsequently, the uniform polymer shell with tunable thickness was formed on the functionalized surface of Au@SiO_2 NPs. And then the interlayer of SiO_2 was dissolved by HF or NaOH aqueous solution.

To produce YSNs with a mesoporous SiO_2 shell, some mild etchants have been explored in the SiO_2-templating synthesis. Shi and coworkers reported the preparation of Au@mSiO_2 and ellipsoidal Fe_2O_3@mSiO_2 YSNs by using sodium carbonate or ammonia solution as the etching agent, which could selectively remove the silica interlayer while retaining the SiO_2 outer shell [128]. Usually, the synthesized Au@mSiO_2@mSiO_2 NPs were or hydrothermal treatment in ammonia solution, the sandwich-structured NPs were successfully converted into YSNs. Kamata and co-workers [130] first employed SiO_2 as sacrificial template to generate the Au@poly(benzyl methacrylate) (Au@PBzMA) YSNs. Similarly, Schüth coated two layers (middle layer of SiO_2 and outer layer of ZrO_2) on the surface of Au NPs [117]. The thin ZrO_2 shell was mesoporous.

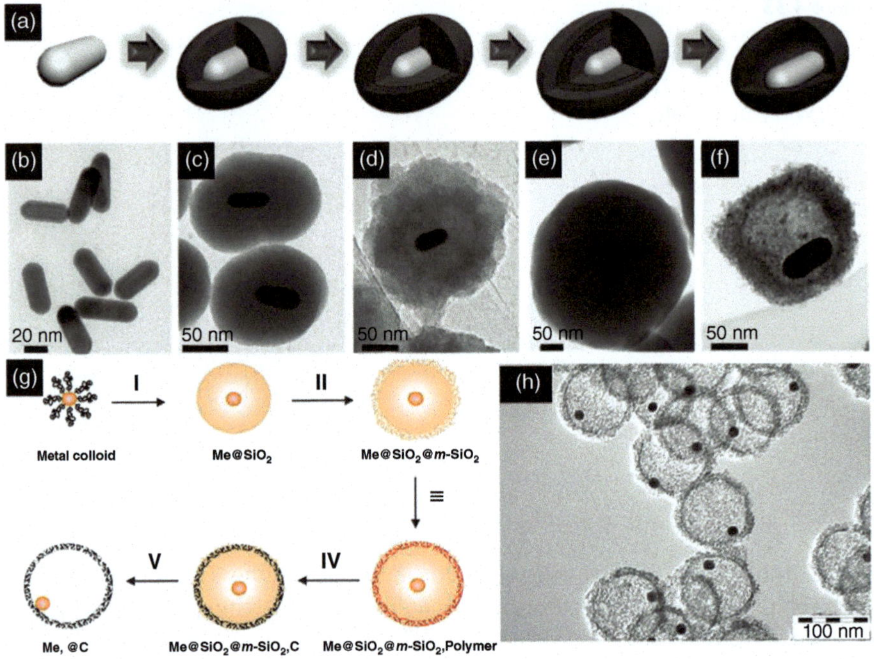

Figure 4.34 (a) Schematic illustration of the generation process of AuNR@TiO_2 YSNs. TEM images of (b) AuNRs, (c) AuNR@SiO_2 NPs, (d) AuNR@SiO_2@amorphous TiO_2 NPs, (e) AuNR@SiO_2@amorphous TiO_2@$SSiO_2$ NPs, and (f) AuNR@TiO_2 YSNs. (g) Strategy for the formation of encapsulated metal nanoparticles in porous carbon shells by nanocasting using the Me@SiO_2@m-SiO_2 NPs as an exotemplate. (h) TEM image of Au@C YSNs produced using mesophase pitch as carbon precursors. Reproduced with permission from Lin et al. [129]/Reproduced with permission from © 2018 JOHN WILEY & SONS, INC.

It has been known that SiO_2 can be dissolved in the excess NaOH solution. Therefore, NaOH can penetrate into ZrO_2 shell and reaction with SiO_2, leading to the formation of Au@ ZrO_2 YSNs by using SiO_2 as the sacrificial template. Lu and coworkers also fabricated Au@poly(*N*-isopropylacrylamide) (Au@PNIPA) YSNs by the removement of SiO_2 [121]. Furthermore, in addition to serving as a sacrificial template, SiO_2 can be applied as a protective shell to synthesize Au NR@TiO_2 YSNs [131, 132]. The synthesis process involved three layers coating including SiO_2-inner layer, TiO_2-interlayer, and SiO_2-outer layer. The presence of SiO_2 outer layer can prevent the destruction of the amorphous TiO_2 under the calcination. After the calcination, the SiO_2 layers were removed by NaOH to obtain Au NR@TiO_2 YSNs, as shown in Figure 4.34a–f.

Besides acting as a hard template for the synthesis of outer shell on the surface of NPs, SiO_2 shell can serve as the interlayer between Au NP and mesoporous SiO_2 [133]. Some mild etching agents (sodium carbonate or ammonia solution) were used to selectively remove the SiO_2 interlayer and Au@SiO_2@$mSiO_2$ nanostructures converted into Au@$mSiO_2$. The research found that Au@SiO_2@$mSiO_2$ nanostructures under calcinated at 500 °C would generate pores. The porous

SiO_2 layer as exo-template provided adequate spaces to accommodate the outer layer materials. Using this method, YSNs can be fabricated with the assistance of porous SiO_2. The porous shell was filled with polymer, leading to the formation of core–shell (Au@SiO_2@$mSiO_2$, polymer). The polymer shell would carry out carbonization at high temperature under vacuum or an inert atmosphere. YSNs were obtained with removing the SiO_2 layer with NaOH or HF.

Carbon has been applied as a hard template due to its low cost and ease of removal. Schüth *et al.* reported the SiO_2-templating fabrication of metal@C (Me@C) YSNs with the assistance of porous SiO_2 as the exotemplate through nanocasting process in polymer precursors. The carbonization of the polymer coating was carried out at high temperatures. After the removal of the SiO_2 with NaOH or HF, Me@C YSNs were obtained (Figure 4.34g,h). Two typical YSNs were shown in Figure 4.34 [129].

The above approaches were suitable for the synthesis of YSNs with the inner core of Au NPs. Ag also can be applied as a hard template to prepare YSNs. The strategy involved the fabrication of multi-layered core@shell nanostructures. Some researchers have reported on the fabrication of YSNs composed of Au NR in mesoporous SiO_2 capsule, where Au@Ag core–shell was an ideal intermediate [134, 135]. The above section has introduced the synthesis of Au@Ag core–shell NR and mesoporous SiO_2 coating. Ag shell was selectively removed by adding the excess hydrogen peroxide (H_2O_2) and generate Au NR@$mSiO_2$m YSNs (path two in Figure 4.34a). H_2O_2 can oxide Ag because the reduction potential of H_2O_2 (E_0 = 0.867 V) was higher than that of Ag^+/Ag (E_0 = 0.799 V) and lower than that of Au^{3+}/Au (E_0 = 1.51 V). Figure 4.34 presents the TEM images and absorption spectra of samples during the synthesis of Au NR@SiO_2 YSNs. The SPR modes of Au NR@$mSiO_2$ YSNs were almost identical with pure Au NRs. The formed YSNs exhibited tunable SPR properties due to the well-defined Au NR-core. And the $mSiO_2$ provided a nanocavity with the advantages of low cytotoxicity, high stability, and easy functionalization.

4.6.2 Soft-Template Synthesis

Vesicles as effective soft templates to prepare YSNs are composed by self-assembly of amphiphilic molecules with hydrophilic and hydrophobic components. The vesicles attached to the surface of inner-core nanostructures played the role of bridge between outer shell and inner core. Xu et al. applied anionic vesicle containing lauryl sulfonate betaine and sodium dodecyl benzenesulfonate as soft template [136] to synthesize YSNs in the presence of 3-aminopropyltriethoxysilane. Subsequently, Lu and coworkers [137] reported the preparation of Au@$mSiO_2$ YSNs via the vesicle template formed by a fluorocarbon surfactant, as shown in Figure 4.34. The core-vesicle complex was obtained by adding core material into a water-ethanol solution including a fluorocarbon surfactant and ammonia. And then mesoporous SiO_2 shell was coated on the surface of the core-vesicle complex with the addition of TEOS. After the hydrolysis and thermal treatment, YSNs were synthesized. Using the soft template approach, YSNs with various types of inner core, such as Au nanostructures, Fe_2O_3, and SiO_2, had been successfully synthesized.

4.6.3 Self-Template Synthesis

Core–shell nanostructures can be converted into YSNs by partially removing inner core. For example, Song and co-workers demonstrated the formation of Au@ SiO_2 YSNs by selective etching Au core in Au@ SiO_2 core–shell nanostructures [138]. When metallic NPs were coated by the mesoporous SiO_2, reaction agents can pass into the mesoporous channel and react with metalcore, as shown in Figure 4.33c. In this case, Au NPs were coated with mesoporous SiO_2 shell. And then Au cores were partially etched by KCN solution through the following reaction [139]:

$$4\mathrm{Au} + 8\mathrm{CN}^- + \mathrm{O}_2 + 4\mathrm{H}^+ \rightarrow 4\mathrm{Au(CN)}^{2-} + 2\mathrm{H}_2\mathrm{O}$$

$$4\mathrm{Au} + 8\mathrm{CN}^- + O_2 + 4\mathrm{H}^+ \rightarrow 4\mathrm{Au(CN)}^{2-} + 2\mathrm{H}_2\mathrm{O}$$

The size of Au core can be regulated by changing the amount of KCN. With increasing the addition of KCN, the diameters of Au core were reduced. Note that, the etching rate of Au-core surface was anisotropic, which induced a relatively broad size distribution of Au core. It was attributed to the different surface energies, and the stable lattice planes in a facet-centered-cubic crystal preferred to remain much more (111) facets to form faceted particles. Therefore, this KCN etching method cannot be applied into the fabrication of 1D Au nanostructures-embedded YSNs.

Different from the convenient etching route, Au-supported YSNs can be synthesized through chemical reaction between metal ions and $\mathrm{SiO_2}$ shell, which can form hollow metal-silicate shell. For example, Au@SiO_2 core–shell nanospheres can be converted into YSNs with an Au core and a nickel silicate shell via the reaction between nickel nitrate and SiO_2 [140]. Similarly, other metal shell, such as magnesium silicate and copper silicate, can be prepared. Ren and coworkers synthesized Au@X@ SiO_2 YSNs based on $\mathrm{NH_3 - NH_4Cl}$ buffer through a facile two-step route. The Au@ X@ SiO_2 YSNs showed excellent stability against high-temperature sintering and desirable excellent catalytic performances [141]. This approach can be developed to fabricate Au NR@ X@ SiO_2 YSNs.

4.6.4 Multiple Step Methods

The "ship-in-a-bottle" method is an alternative method to construct YSNs including three steps: (i) the preparation of hollow shell as nanoreactors, (ii) the accessing of core precursors into shell, and (iii) the formation of core inside the hollow shell. Usually, spherical Au NP was obtained within SiO_2 shell through this method. For example, Mou and coworkers [124] prepared Au@SiO_2 YSNs by using a water-in-oil microemulsion as template. The water-in-oil microemulsion was applied in the synthesis of monodispersed hollow SiO_2 nanospheres. And then, the Au precursor was added to the as-prepared hollow SiO_2 nanospheres dispersed in water. Shaik and co-workers reported the synthesis of metal@ SiO_2 YSNs by regulating pH of the reaction system. Monodispersed mesoporous SiO_2 shell dispersed in metal precursor aqueous solution can be applied as nanoreactors. The spherical and polyhedral metal cores were formed inside the SiO_2 shell, which was tuned by pH value. A critical pH > 4 was required to synthesize YSNs [142]. SiO_2 shell mainly consisted of hydroxyl species (≡Si – OH) and siloxane bridges (≡Si – O – Si≡). At the condition of pH > 4, the hydroxyl species were deprotonated (≡Si – O–) with a net negative charge, which possessed a strong reducing ability. As a result, metal precursors were reduced and nucleated within the hollow shell, leading to the formation of metal@SiO_2 YSNs. However, it was difficult for the "ship-in-a-bottle" method to prepare 1D Au nanostructures-embedded YSNs.

4.7 Conclusion and Remarks

Plasmonic nanorods-based bimetallic nanocrystals exhibit different properties and applications from their bulk counterparts. It has been well established that the remarkable properties are closely associated with their size, shape, composition, structure, and crystal phase. Therefore, precise control and optimization on the nanostructures are of critical importance. For the bimetallic nanostructures, core–shell, intermetallic, and uniformly alloyed nanocrystals have been intensively explored in recent years.

Plasmonic Au NRs and Au NPs-supported bimetallic nanostructures presented in this chapter and their related hybrid structures are very fascinating due to the unique structural feature and plasmonic properties. One of the distinct features of these structures as compared to monometallic systems is the plasmonic coupling between the core and porous shell, which results in an electromagnetic hotspot within the nanostructure. In these physical systems, near-field interactions between molecules and plasmons are especially strong. The advantages of Au@Ag NR-based nanocatalysis are also expressed in the remarkable plasmonic properties, since Ag possessed the highest plasmonic activity than other metals.

For the formation of stable bimetallic nanorods, direct growth of alloyed nanoshells on the surface of Au NRs is an easy and effective method. But the aspect ratios of the bimetallic nanorods usually decrease, leading to a blue shift and broadening of the SPR property. For the alloyed and porous Au–Ag nanorods, high-temperature annealing approach is commonly used for precise control over the composition and atomic ordering, it is still a big challenge to manipulate the size and shape of the nanocrystals, as annealing process at high temperature often destroy the original shape of as-synthesized nanocrystals in most cases and lead to severely nanocrystals aggregation/sintering.

To solve this issue, surface coating mesoporous SiO_2 is required to deposit on the surface of the as-synthesized nanocrystals before annealing. Mesoporous silica nanoshell through the well-established sol–gel process is an ideal inorganic material for the encapsulation of metal NPs. It is a chemically inert oxide and can retain its structural and morphological integrity under thermal conditions. These advantages make mesoporous silica an attractive coating material to stabilize metal NPs for highly durable catalysis in a broad range. When the core–shell Au@Ag nanorods were protected by mesoporous SiO_2 nanoshell, the existence of SiO_2 would maintain excellent optical properties and recyclable performances after multiple cycles. Unfortunately, this process involves multistep. And the removal of coating is necessary because the coating layers will dramatically impede nanoscale mass transport.

In terms of photocatalysis, many efforts have been devoted to enhancing the photocatalytic activity, but fewer papers focused on the fundamental aspects of plasmon-enhanced photocatalysis. Debates have existed for years on the dominant underlying mechanism of Au NR-based photocatalysis, be it plasmon-induced hot electron generation or the rise in the local temperature caused by the plasmonic photothermal conversion effect. The combined advantages of plasmon and nanocatalysis would promote the plasmon-catalytic applications for different wavelengths of light, which will be delivered in Chapter 11. For the preparation methods and practical applications, it can be predicted that there is a large room to extensively develop improved synthetic strategies to further improve the structure and performance for various practical applications.

References

1 Linic, S., Christopher, P., and Ingram, D.B. (2011). Plasmonic-metal nanostructures for efficient conversion of solar to chemical energy. *Nat. Mater.* 10: 911–921.

2 Rycenga, M., Cobley, C.M., Zeng, J. et al. (2011). Controlling the synthesis and assembly of dilver nanostructures for plasmonic applications. *Chem. Rev.* 111: 3669–3712.

3 Jiang, N.N., Zhuo, X.L., and Wang, J.F. (2018). Active plasmonics: principles, dtructures, and applications. *Chem. Rev.* 118: 3054–3099.

4 Kristensen, A., Yang, J.K.W., Bozhevolnyi, S.I. et al. (2017). Plasmonic colour generation. *Nat. Rev. Mater.* 2: 16088.

5 Shahbazyan, T.V. (2013). Theory of plasmon-enhanced metal photoluminescence. *Nano Lett.* 13: 194–198.

6 Chen, H.J., Shao, L., Li, Q., and Wang, J.F. (2013). Gold nanorods and their plasmonic properties. *Chem. Soc. Rev.* 42: 2679–2724.

7 Pescaglini, A., Martin, A., Cammi, D. et al. (2014). Hot-electron injection in Au NR-ZnO nanowire hybrid device for near-infrared photodetection. *Nano Lett.* 14: 6202–6209.

8 Zheng, J.P., Cheng, X.Z., Zhang, H. et al. (2021). Gold nanorods: the most versatile plasmonic nanoparticles. *Chem. Rev.* 121: 13342–13453.

9 Ke, S.L., Kan, C.X., Ni, Y. et al. (2019). Construction of silica-encapsulated gold-silver core-shell nanorod: Atomic facets enrichment and plasmon enhanced catalytic activity with high stability and reusability. *Mater. Design* 177: 107837.

10 Jing, H., Zhang, Q.F., Large, N. et al. (2014). Tunable plasmonic nanoparticles with catalytically active high-index facets. *Nano Lett.* 14: 3674–3682.

11 Zheng, Z.K., Tachikawa, T., and Majima, T. (2015). Plasmon-enhanced formic acid dehydrogenation using anisotropic Pd-Au NRs studied at the single-particle level. *J. Am. Chem. Soc.* 137: 948–957.

12 Langille, M.R., Personick, M.L., Zhang, J., and Mirkin, C.A. (2012). Defining rules for the shape evolution of gold nanoparticles. *J. Am. Chem. Soc.* 134: 14542–14554.

13 Xia, Y.N., Gilroy, K.D., Peng, H.C., and Xia, X.H. (2017). Seed-mediated growth of colloidal metal nanocrystals. *Angew. Chem. Int. Ed.* 56: 60–95.

14 Strobbia, P., Languirand, E., and Cullum, B.M. (2015). Recent advances in plasmonic nanostructures for sensing: a review. *Optim. Eng.* 54: 100902.

15 Sanchez-Iglesias, A., Winckelmans, N., Altantzis, T. et al. (2017). High-yield seeded growth of monodisperse pentatwinned gold nanoparticles through thermally induced seed twinning. *J. Am. Chem. Soc.* 139: 107–110.

16 Xu, J., Kan, C.X., Lou, Y.K. et al. (2020). Structural transformation of Au seeds: the influence of temperature and surfactants. *J. Nanophoton.* 14: 016004.

17 Nikoobakht, B. and El-Sayed, M.A. (2003). Preparation and growth mechanism of gold nanorods (NRs) using seed-mediated growth method. *Chem. Mater.* 15: 1957–1962.

18 Murphy, C.J. and Jana, N.R. (2002). Controlling the aspect ratio of inorganic nanorods and nanowires. *Adv. Mater.* 14: 80–82.

19 Jana, N.R., Gearheart, L., and Murphy, C.J. (2001). Wet chemical synthesis of high aspect ratio cylindrical gold nanorods. *J. Phys. Chem. B* 105: 4065–4067.

20 Ye, X.C., Zheng, C., Chen, J. et al. (2013). Using binary surfactant mixtures to simultaneously improve the dimensional tunability and monodispersity in the seeded growth of gold nanorods. *Nano Lett.* 13: 765–771.

21 Jackson, S.R., McBride, J.R., Rosenthal, S.J., and Wright, D.W. (2014). Where's the silver? imaging trace silver coverage on the surface of gold nanorods. *J. Am. Chem. Soc.* 136: 5261–5263.

22 Walsh, M.J., Barrow, S.J., Tong, W.M. et al. (2015). Symmetry breaking and silver in gold nanorod growth. *ACS Nano* 9: 715–724.

23 Tong, W.M., Walsh, M.J., Mulvaney, P. et al. (2017). Control of symmetry breaking size and aspect ratio in gold nanorods: underlying role of silver nitrate. *J. Phys. Chem. C* 121: 3549–3559.

24 Eustis, S. and El-Sayed, M. (2005). Aspect ratio dependence of the enhanced fluorescence intensity of gold nanorods: Experimental and simulation study. *J. Phys. Chem. B* 109: 16350–16356.

25 Sun, Z.H., Ni, W.H., Yang, Z. et al. (2008). pH-controlled reversible assembly and disassembly of gold nanorods. *Small* 4: 1287–1292.

26 Ye, X.C., Jin, L.H., Caglayan, H. et al. (2012). Improved size-tunable synthesis of monodisperse gold nanorods through the use of aromatic additives. *ACS Nano* 6: 2804–2817.

27 Vigderman, L., Khanal, B.P., and Zubarev, E.R. (2012). Functional gold nanorods: synthesis, self-assembly, and sensing applications. *Adv. Mater.* 24: 4811–4841.

28 Takahashi, Y., Ide, N., and Yamada, S. (2014). Characteristics of gold nanorods and their applications to analytical sciences. *Bunseki Kagaku* 63: 551–561.

29 Xie, H., Wang, K.Y., Li, Y. et al. (2021). Gold nanorod@ruthenium oxide core-shell heterostructures: Synthesis, single-particle characterizations, and enhanced hot electron generation. *Adv. Opt. Mater.* 9: 2002136.

30 Havel, F., Novotny, F., Jelinek, M. et al. (2014). Optimalization of optothermal properties of gold nanorods. In: *Nanocon 2013, 5th International Conference*, 209–214.

31 Zaibudeen, A.W., Khawas, S., and Srivastava, S. (2021). Understanding multi-scale assembly mechanism in evaporative droplet of gold nanorods. *Colloids Interface Sci. Commun.* 44: 100492.

32 Fan, Z.Y., Tebbe, M., Fery, A. et al. (2016). Assembly of gold nanoparticles on gold nanorods using functionalized poly(*N*-isopropylacrylamide) as polymeric "glue". *Part. Part. Syst. Char.* 33: 698–702.

33 Ferrier, R.C., Koski, J., Riggleman, R.A., and Composto, R.J. (2016). Engineering the assembly of gold nanorods in polymer matrices. *Macromolecules* 49: 1002–1015.

34 Liu, J.S., Kan, C.X., Li, Y.L. et al. (2014). End-to-end and side-by-side assemblies of gold nanorods induced by dithiol poly(ethylene glycol). *Appl. Phys. Lett.* 104: 253105.

35 Li, Y.L., Kan, C.X., Wang, C.S. et al. (2014). Surface plasmon resonance coupling effect of assembled gold nanorods based on the FDTD simulation. *Acta Physico-Chimica Sinica* 30: 1827–1836.

36 Barnard, A.S., Young, N.P., Kirkland, A.I. et al. (2009). Nanogold: a quantitative phase map. *ACS Nano* 3: 1431–1436.

37 Chateau, D., Liotta, A., Vadcard, F. et al. (2015). From gold nanobipyramids to nanojavelins for a precise tuning of the plasmon resonance to the infrared wavelengths: experimental and theoretical aspects. *Nanoscale* 7: 1934–1943.

38 Li, Q., Zhuo, X.L., Li, S. et al. (2015). Production of monodisperse gold nanobipyramids with number percentages approaching 100% and evaluation of their plasmonic properties. *Adv. Opt. Mater.* 3: 801–812.

39 Chow, T.H., Li, N.N., Bai, X.P. et al. (2019). Gold nanobipyramids: an emerging and versatile type of plasmonic nanoparticles. *Acc. Chem. Res.* 52: 2136–2146.

40 Zhu, X.Z., Xu, J., Zhang, H. et al. (2020). Gold nanobipyramid-embedded ultrathin metal nanoframes for in-situ monitoring catalytic reactions. *Chem. Sci.* 11: 3198–3207.

41 Liu, M.Z. and Guyot-Sionnest, P. (2005). Mechanism of silver(I)-assisted growth of gold nanorods and bipyramids. *J. Phys. Chem. B* 109: 22192–22200.

42 Kou, X.S., Zhang, S.Z., Tsung, C.K. et al. (2006). Growth of gold nanorods and bipyramids using CTEAB surfactant. *J. Phys. Chem. B* 110: 16377–16383.

43 Wang, Y., Fu, W., and Hu, X. (2015). Facile synthesis of high-purity single-twinned Au nanocrystals through manipulating reaction kinetics. *CrstEngComm* 17 (35): 6636–6640.

44 Lee, S., Mayer, K.M., and Hafner, J.H. (2009). Improved localized surface plasmon resonance immunoassay with gold bipyramid substrates. *Anal. Chem.* 81: 4450–4455.

45 Rao, W.Y., Li, Q., Wang, Y.Z. et al. (2015). Comparison of photoluminescence quantum yield of single gold nanobipyramids and gold nanorods. *ACS Nano* 9: 2783–2791.

46 Lee, C.H., Hankus, M.E., Tian, L. et al. (2011). Highly sensitive surface enhanced Raman scattering substrates based on filter paper loaded with plasmonic nanostructures. *Anal. Chem.* 83: 8953–8958.

47 Wang, Z., Gao, R., Nikoobakht, B., and El-Sayed, M. (2000). Surface reconstruction of the unstable (110) surface in gold nanorods. *J. Phys. Chem. B* 104: 5417–5420.

48 Ah, C.S., Do Hong, S., and Jang, D.J. (2001). Preparation of $Au_{core}Ag_{shell}$ nanorods and characterization of their surface plasmon resonances. *J. Phys. Chem. B* 105: 7871–7873.

49 Liu, M.Z. and Guyot-Sionnest, P. (2004). Synthesis and optical characterization of Au/Ag core/shell nanorods. *J. Phys. Chem. B* 108: 5882–5888.

50 Xiang, Y.U., Wu, X.C., Liu, D.F. et al. (2008). Gold nanorod-seeded growth of silver nanostructures: From homogeneous coating to anisotropic coating. *Langmuir* 24: 3465–3470.

51 Okuno, Y., Nishioka, K., Nakashima, N., and Niidome, Y. (2009). Rapid formation of silver shells on gold nanorods in a micellar solution of hexadecyltrimethylammonium chloride. *Chem. Lett.* 38: 60–61.

52 Okuno, Y., Nishioka, K., Kiya, A. et al. (2010). Uniform and controllable preparation of Au-Ag core-shell nanorods using anisotropic silver shell formation on gold nanorods. *Nanoscale* 2: 1489–1493.

53 Huang, J.F., Zhu, Y.H., Liu, C.X. et al. (2015). Fabricating a homogeneously alloyed AuAg shell on Au NRs to achieve strong, stable, and tunable surface plasmon resonances. *Small* 11: 5214–5221.

54 Liu, J.B., Feng, L.L., Hu, Z.J. et al. (2013). Au core/Au-Ag alloy shell nanorods: composition- and shape-tailored optical responses. *J. Nanosci. Nanotechnol.* 13: 1006–1010.

55 Lee, S., Kumari, N., Jeon, K.W. et al. (2018). Monofacet-selective cavitation within solid-state silica-nanoconfinement toward Janus iron oxide nanocube. *J. Am. Chem. Soc.* 140: 15176–15180.

56 Liu, J.S., Kan, C.X., Cong, B. et al. (2014). Plasmonic property and stability of core-shell Au@SiO_2 nanostructures. *Plasmonics* 9: 1007–1014.

57 Xu, H.Y., Kan, C.X., Miao, C.Z. et al. (2017). Synthesis of high-purity silver nanorods with tunable plasmonic properties and sensor behavior. *Photonics Res.* 5: 27–32.

58 Wei, J.J., Kan, C.X., Lou, Y.K. et al. (2016). Synthesis and stability of bimetallic Au@Ag nanorods. *Superlattices Microst.* 100: 315–323.

59 Xu, J., Zhu, X.Z., Xu, L.H. et al. (2023). Template-directed growth of Ag nanostructures: soft templates versus hard templates. *Nanoscale* 15: 1687–1694.

60 Xu, J., Yun, Q.R., Zhang, H. et al. (2021). A novel deposition mechanism of Au on Ag nanostructures involving galvanic replacement and reduction reactions. *Chem. Commun.* 57: 8332–8335.

61 Wang, X., Kan, C.X., Xu, J. et al. (2020). Gold nanobipyramid enveloped in alloyed nanoshell for stable plasmonic sensors. *J. Phys. D-Appl. Phys.* 53: 295303.

62 Wang, X., Kan, C.X., Xu, J. et al. (2020). Dielectric function modelling and sensitivity forecast for Au-Ag alloy nanostructures. *Phys. Chem. Chem. Phys.* 22: 14932–14940.

63 Ruban, A.V., Skriver, H.L., and Norskov, J.K. (1999). Surface segregation energies in transition-metal alloys. *Phys. Rev. B* 59: 15990–16000.

64 Si, Y.M., Bai, Y.C., Qin, X.J. et al. (2018). Alkyne-DNA-functionalized alloyed Au/Ag nanospheres for ratiometric surface-enhanced Raman scattering imaging assay of endonuclease activity in live cells. *Anal. Chem.* 90: 3898–3905.

65 Bozzolo, G., Ferrante, J., Noebe, R.D. et al. (1999). Surface segregation in multicomponent systems: Modeling of surface alloys and alloy surfaces. *Comput. Mater. Sci.* 15: 169–195.

66 Molenbroek, A.M., Haukka, S., and Clausen, B.S. (1998). Alloying in Cu/Pd nanoparticle catalysts. *J. Phys. Chem. B* 102: 10680–10689.

67 Alayoglu, S. and Eichhorn, B. (2008). Rh-Pt bimetallic catalysts: synthesis, characterization, and catalysis of core–shell, alloy, and monometallic nanoparticles. *J. Am. Chem. Soc.* 130: 17479–17486.

68 Chen, H.C., Chou, S.W., Tseng, W.H. et al. (2012). Large AuAg alloy nanoparticles synthesized in organic media using a one-pot reaction: their applications for high-performance bulk heterojunction solar cells. *Adv. Funct. Mater.* 22: 3975–3984.

69 Liu, Y.L. and Walker, A.R.H. (2010). Monodisperse gold copper bimetallic nanocubes: facile one-step synthesis with controllable size and composition. *Angew. Chem. Int. Ed.* 49: 6781–6785.

70 Choi, B.S., Lee, Y.W., Kang, S.W. et al. (2012). Multimetallic alloy nanotubes with nanoporous framework. *ACS Nano* 6: 5659–5667.

71 He, W.W., Wu, X.C., Liu, J.B. et al. (2010). Design of AgM bimetallic alloy nanostructures (M = Au, Pd, Pt) with runable morphology and peroxidase-like activity. *Chem. Mater.* 22: 2988–2994.

72 You, H.J. and Fang, J.X. (2016). Particle-mediated nucleation and growth of solution-synthesized metal nanocrystals: A new story beyond the LaMer curve. *Nano Today* 11: 145–167.

73 LaMer, V.K. and Dinegar, R.H. (1950). Theory, production and mechanism of formation of monodispersed hydrosols. *J. Am. Chem. Soc.* 72 (11): 4847–4854.

74 Han, S.W., Kim, Y., and Kim, K. (1998). Dodecanethiol-derivatized Au/Ag bimetallic nanoparticles: TEM, UV/VIS, XPS, and FTIR analysis. *J. Colloid Interface Sci.* 208: 272–278.

75 Wang, C., Yin, H.G., Chan, R. et al. (2009). One-pot synthesis of oleylamine coated AuAg alloy NPs and their catalysis for CO oxidation. *Chem. Mater.* 21: 433–435.

76 Wang, C., Peng, S., Chan, R., and Sun, S.H. (2009). Synthesis of AuAg alloy nanoparticles from core/shell-structured Ag/Au. *Small* 5: 567–570.

77 Chen, M., Feng, Y.G., Wang, X. et al. (2007). Silver nanoparticles capped by oleylamine: formation, growth, and self-organization. *Langmuir* 23: 5296–5304.

78 Wu, Q.S., Zhang, G.G., Ma, Y.Y. et al. (2021). Ultrafast synthesis of wavy gold-silver alloy nanowires with tunable diameters in the range of 2–10 nm via a seed-mediated co-reduction. *Eur. J. Inorg. Chem.* 2021: 1152–1159.

79 Yan, Y.C., Du, J.S.S., Gilroy, K.D. et al. (2017). Intermetallic nanocrystals: syntheses and catalytic applications. *Adv. Mater.* 29: 1605997.

80 Shore, M.S., Wang, J.W., Johnston-Peck, A.C. et al. (2011). Synthesis of Au(core)/Ag(shell) nanoparticles and their conversion to AuAg alloy nanoparticles. *Small* 7: 230–234.

81 Hodak, J.H., Henglein, A., Giersig, M., and Hartland, G.V. (2000). Laser-induced inter-diffusion in AuAg core-shell nanoparticles. *J. Phys. Chem. B* 104: 11708–11718.

82 Peng, Z.Q., Spliethoff, B., Tesche, B. et al. (2006). Laser-assisted synthesis of Au-Ag alloy nanoparticles in solution. *J. Phys. Chem. B* 110: 2549–2554.

83 Wang, X., Zhang, Z.Y., and Hartland, G.V. (2005). Electronic dephasing in bimetallic gold-silver nanoparticles examined by single particle spectroscopy. *J. Phys. Chem. B* 109: 20324–20330.

84 Radziuk, D.V., Zhang, W., Shchukin, D., and Mohwald, H. (2010). Ultrasonic alloying of preformed gold and silver nanoparticles. *Small* 6: 545–553.

85 Gao, C.B., Hu, Y.X., Wang, M.S. et al. (2014). Fully alloyed Ag/Au nanospheres: combining the plasmonic property of Ag with the stability of Au. *J. Am. Chem. Soc.* 136: 7474–7479.

86 Ni, Y., Kan, C.X., He, L.B. et al. (2019). Alloyed Au-Ag nanorods with desired plasmonic properties and stability in harsh environments. *Photonics Res.* 7: 558–565.

87 Ke, S.L., Kan, C.X., Zhu, X.Z. et al. (2021). Effective fabrication of porous Au-Ag alloy nanorods for in-situ Raman monitoring catalytic oxidation and reduction reactions. *J. Mater. Sci. Technol.* 91: 262–269.

88 Zugic, B., Wang, L.C., Heine, C. et al. (2017). Dynamic restructuring drives catalytic activity on nanoporous gold-silver alloy catalysts. *Nat. Mater.* 16: 558–559.

89 Biener, J., Wittstock, A., Zepeda-Ruiz, L.A. et al. (2009). Surface-chemistry-driven actuation in nanoporous gold. *Nat. Mater.* 8: 47–51.

90 Lang, X.Y., Hirata, A., Fujita, T., and Chen, M.W. (2011). Nanoporous metal/oxide hybrid electrodes for electrochemical supercapacitors. *Nat. Nanotechnol.* 6: 232–236.

91 Wittstock, A., Zielasek, V., Biener, J. et al. (2010). Nanoporous gold catalysts for selective gas-phase oxidative coupling of methanol at low temperature. *Science* 327: 319–322.

92 Zhong, J.H., Chimeh, A., Korte, A. et al. (2018). Strong spatial and spectral localization of surface plasmons in individual randomly disordered gold nanosponges. *Nano Lett.* 18: 4957–4964.

93 Ron, R., Haleva, E., and Salomon, A. (2018). Nanoporous metallic networks: fabrication, optical properties, and applications. *Adv. Mater.* 30: 201706755.

94 Lyu, X., Jia, Y., Mao, X. et al. (2020). Gradient-concentration design of stable core-shell nanostructure for acidic oxygen reduction electrocatalysis. *Adv. Mater.* 32: 2003493.

95 Lee, W.J., Park, E.Y., Choi, D. et al. (2020). Colloidal porous AuAg alloyed nanoparticles for enhanced photoacoustic imaging. *ACS Appl. Mater. Interfaces* 12: 32270–32277.

96 Zhang, T., Sun, Y.Q., Hang, L.F. et al. (2018). Periodic porous alloyed Au-Ag nanosphere arrays and their highly sensitive SERS performance with good reproducibility and high density of hotspots. *ACS Appl. Mater. Interfaces* 10: 9792–9801.

97 Fujita, T., Guan, P.F., McKenna, K. et al. (2012). Atomic origins of the high catalytic activity of nanoporous gold. *Nat. Mater.* 11: 775–780.

98 Erlebacher, J., Aziz, M.J., Karma, A. et al. (2001). Evolution of nanoporosity in dealloying. *Nature* 410: 450–453.

99 Kamiuchi, N., Sun, K., Aso, R. et al. (2018). Self-activated surface dynamics in gold catalysts under reaction environments. *Nat. Commun.* 9: 2060.

100 Liu, P., Chen, Q., Ito, Y. et al. (2020). Dealloying kinetics of AgAu nanoparticles by in-situ liquid-cell scanning transmission electron microscopy. *Nano Lett.* 20: 1944–1951.

101 Albrecht, W., van der Hoeven, J.E.S., Deng, T.S. et al. (2017). Fully alloyed metal nanorods with highly tunable properties. *Nanoscale* 9: 2845–2851.

102 Hanske, C., Sanz-Ortiz, M.N., and Liz-Marzan, L.M. (2018). Silica-coated plasmonic metal nanoparticles in action. *Adv. Mater.* 30: 170700.

103 Zhang, H., Wang, C., Sun, H.L. et al. (2017). In-situ dynamic tracking of heterogeneous nanocatalytic processes by shell-isolated nanoparticle-enhanced Raman spectroscopy. *Nat. Commun.* 8: 15447.

104 Li, J.F., Huang, Y.F., Ding, Y. et al. (2010). Shell-isolated nanoparticle-enhanced Raman spectroscopy. *Nature* 464: 392–395.

105 Li, J.F., Tian, X.D., Li, S.B. et al. (2013). Surface analysis using shell-isolated nanoparticle-enhanced Raman spectroscopy. *Nat. Protoc.* 8: 52–65.

106 Zhang, T.R., Ge, J.P., Hu, Y.X. et al. (2008). Formation of hollow silica colloids through a spontaneous dissolution-regrowth process. *Angew. Chem. Int. Ed.* 47: 5806–5811.

107 Chen, Y., Xu, P.F., Chen, H.R. et al. (2013). Colloidal HPMO nanoparticles: silica-etching chemistry tailoring, topological transformation, and nano-biomedical applications. *Adv. Mater.* 25: 3100–3105.

108 Fang, X.L., Chen, C., Liu, Z.H. et al. (2011). A cationic surfactant assisted selective etching strategy to hollow mesoporous silica spheres. *Nanoscale* 3: 1632–1639.

109 Bai, Y.C., Gao, C.B., and Yin, Y.D. (2017). Fully alloyed Ag/Au NRs with tunable surface plasmon resonance and high chemical stability. *Nanoscale* 9: 14875–14880.

110 Lou, X.W., Archer, L.A., and Yang, Z.C. (2008). Hollow micro-/nanostructures: synthesis and applications. *Adv. Mater.* 20: 3987–4019.

111 Zhao, Y. and Jiang, L. (2009). Hollow micro/nanomaterials with multilevel interior structures. *Adv. Mater.* 21: 3621–3638.

112 Liu, J., Qiao, S.Z., Hu, Q.H., and Lu, G.Q. (2011). Magnetic nanocomposites with mesoporous structures: synthesis and applications. *Small* 7: 425–443.

113 Park, J.C. and Song, H. (2011). Metal@silica Yolk-shell nanostructures as versatile bifunctional nanocatalysts. *Nano Res.* 4: 33–49.

114 Cao, A., Lu, R., and Veser, G. (2010). Stabilizing metal nanoparticles for heterogeneous catalysis. *Phys. Chem. Chem. Phys.* 12: 13499–13510.

115 Ikeda, S., Ishino, S., Harada, T. et al. (2006). Ligand-free platinum nanoparticles encapsulated in a hollow porous carbon shell as a highly active heterogeneous hydrogenation catalyst. *Angew. Chem. Int. Ed.* 45: 7063–7066.

116 Liu, R., Mahurin, S.M., Li, C. et al. (2011). Dopamine as a carbon source: the controlled synthesis of hollow carbon spheres and Yolk-structured carbon nanocomposites. *Angew. Chem. Int. Ed.* 50: 6799–6802.

117 Arnal, P.M., Comotti, M., and Schuth, F. (2006). High-temperature-stable catalysts by hollow sphere encapsulation. *Angew. Chem. Int. Ed.* 45: 8224–8227.

118 Lee, I., Joo, J.B., Yin, Y.D., and Zaera, F. (2011). A yolk@dhell nanoarchitecture for Au/TiO_2 catalysts. *Angew. Chem. Int. Ed.* 50: 10208–10211.

119 Galeano, C., Guttel, R., Paul, M. et al. (2011). Yolk-shell gold nanoparticles as model materials for support-effect studies in heterogeneous catalysis: Au, @C and Au, @ZrO_2 for CO oxidation as an example. *Chem. A Eur. J.* 17: 8434–8439.

120 Wang, Y.F., Biradar, A.V., and Asefa, T. (2012). Assembling nanostructures for effective catalysis: supported palladium nanoparticle multicores coated by a hollow and nanoporous zirconia shell. *ChemSusChem* 5: 132–139.

121 Wu, S., Dzubiella, J., Kaiser, J. et al. (2012). Thermosensitive Au-PNIPA yolk-shell nanoparticles with tunable selectivity for catalysis. *Angew. Chem. Int. Ed.* 51: 2229–2233.

122 Liu, B., Zhang, W., Feng, H.L., and Yang, X.L. (2011). Rattle-type microspheres as a support of tiny gold nanoparticles for highly efficient catalysis. *Chem. Commun.* 47: 11727–11729.

123 Huang, X.Q., Guo, C.Y., Zuo, L.Q. et al. (2009). An assembly route to inorganic catalytic nanoreactors containing sub-10-nm gold nanoparticles with anti-aggregation properties. *Small* 5: 361–365.

124 Wu, S.H., Tseng, C.T., Lin, Y.S. et al. (2011). Catalytic nano-rattle of Au@hollow silica: towards a poison-resistant nanocatalyst. *J. Mater. Chem.* 21: 789–794.

125 Sanles-Sobrido, M., Perez-Lorenzo, M., Rodriguez-Gonzalez, B. et al. (2012). Highly active nanoreactors: nanomaterial encapsulation based on confined catalysis. *Angew. Chem. Int. Ed.* 51: 3877–3882.

126 Fang, X.L., Liu, Z.H., Hsieh, M.F. et al. (2012). Hollow mesoporous aluminosilica spheres with perpendicular fore channels as catalytic nanoreactors. *ACS Nano* 6: 4434–4444.

127 Kim, M., Park, J.C., Kim, A. et al. (2012). Porosity control of Pd@SiO_2 yolk-shell nanocatalysts by the formation of nickel phyllosilicate and its influence on Suzuki coupling reactions. *Langmuir* 28: 6441–6447.

128 Shi, Y.L. and Asefa, T. (2007). Tailored core-shell-shell nanostructures: sandwiching gold nanoparticles between silica cores and tunable silica shells. *Langmuir* 23: 9455–9462.

129 Lin, L., Song, J., Yang, H., and Chen, X. (2018). Yolk-shell nanostructures: design, synthesis, and biomedical applications. *Adv. Mater.* 30: 1704639.

130 Kamata, K., Lu, Y., and Xia, Y.N. (2003). Synthesis and characterization of monodispersed core-shell spherical colloids with movable cores. *J. Am. Chem. Soc.* 125: 2384–2385.

131 Li, A., Zhang, P., Chang, X.X. et al. (2015). Gold nanorod@TiO_2 Yolk-shell nanostructures for visible-light-driven photocatalytic oxidation of benzyl alcohol. *Small* 11: 1892–1899.

132 Chen, Y., Chen, H.R., Guo, L.M. et al. (2010). Hollow/rattle-type mesoporous nanostructures by a structural difference-based selective etching strategy. *ACS Nano* 4: 529–539.

133 Galeano, C., Baldizzone, C., Bongard, H. et al. (2014). Carbon-based yolk-shell materials for fuel cell applications. *Adv. Funct. Mater.* 24: 220–232.

134 Choi, E., Kwak, M., Jang, B., and Piao, Y. (2013). Highly monodisperse rattle-structured nanomaterials with gold nanorod core-mesoporous silica shell as drug delivery vehicles and nanoreactors. *Nanoscale* 5: 151–154.

135 Zhang, Q., Cobley, C.M., Zeng, J. et al. (2010). Dissolving Ag from Au-Ag Alloy nanoboxes with H_2O_2: a method for both tailoring the optical properties and measuring the H_2O_2 concentration. *J. Phys. Chem. C* 114: 6396–6400.

136 Wu, X.J. and Xu, D.S. (2009). Formation of Yolk/SiO_2 shell structures using surfactant mixtures as template. *J. Am. Chem. Soc.* 131: 2774–2775.

137 Liu, J., Qiao, S.Z., Hartono, S.B., and Lu, G.Q. (2010). Monodisperse yolk-shell nanoparticles with a hierarchical porous structure for delivery vehicles and nanoreactors. *Angew. Chem. Int. Ed.* 49: 4981–4985.

138 Lee, J., Park, J.C., and Song, H. (2008). A nanoreactor framework of an Au@SiO_2 yolk/shell structure for catalytic reduction of p-nitrophenol. *Adv. Mater.* 20: 1523–1524.

139 Giersig, M., Ung, T., Liz-Marzan, L.M., and Mulvaney, P. (1997). Direct observation of chemical reactions in silica-coated gold and silver nanoparticles. *Adv. Mater.* 9: 570–571.

140 Wang, Y.Q., Tang, C.J., Deng, Q.A. et al. (2010). A versatile method for controlled synthesis of porous hollow spheres. *Langmuir* 26: 14830–14834.

141 Dong, K., Liu, Z., and Ren, J.S. (2013). A general and eco-friendly self-etching route to prepare highly active and stable Au@metal silicate yolk-shell nanoreactors for catalytic reduction of 4-nitrophenol. *nEngComm* 15: 6329–6334.

142 Shaik, F., Zhang, W.Q., and Niu, W.X. (2017). A generalized method for the synthesis of ligand-free M@SiO_2 (M = Ag, Au, Pd, Pt) yolk-shell nanoparticles. *Langmuir* 33: 3281–2386.

5

Nanowires for Conductive Films and Electromagnetic Shielding

Metal nanowires are promising in the ultraflexible, highly efficient EMI shielding, and self-healable triboelectric nanogenerator.

5.1 Introduction

One-dimensional (1D) nanostructures with high aspect ratio (AR) have been manifested as an ideal transport carrier for quantum particles (e.g. electrons, phonons, and photons), due to their radial dimension at or below some certain characteristic lengths, such as the Bohr radius, the wavelength of the light, phonon mean-free path, critical size of magnetic domains, and exciton diffusion length [1, 2]. At the time of the 2000s, 1D nanostructures could only be synthesized from two metals: Au and Ag. The synthesis required the use of appropriate capping agents to induce anisotropic growth. Since then, colloidal synthesis of nanorods (NRs) and nanowires (NWs) have been reported for different metals, such as Cu, Ni, Pd, and Al, most of which still rely on capping agents to achieve anisotropic growth. Researchers can now not only control the AR of various 1D metal nanostructures but also produce them with different cross-sections. In addition, colloidal syntheses have been further developed to produce bimetallic and trimetallic nanostructures, as well as heterostructural systems consisting of a metal and a dielectric or semiconductor.

Recently, the number of NWs-based publications has experienced an exponential increase, as shown in Figure 5.1a for recent 20 years [3]. Since the seminal review on the semiconductor NWs in 2010 by the Yang group [4], the research field of NWs has attracted significant attention, and their applications have been extended to numerous fields, as statistically classified in Figure 5.1b, such as electronics, energy storage materials, optoelectronics, catalysis, sensors, and biology [5, 6]. These exciting developments were mainly achieved by several world-famous groups, including the Lieber group at Harvard University, the Yang group at UC Berkeley, the Xia group at Georgia Tech, and so on, as summarized in Figure 5.1c for development from 2010 to 2019 [3] and some work from the author's group in recent years. Particularly, due to the diversity of components and structure, NWs find applications

Plasmonic Metal Nanostructures: Preparation, Characterization, and Applications, First Edition. Caixia Kan.

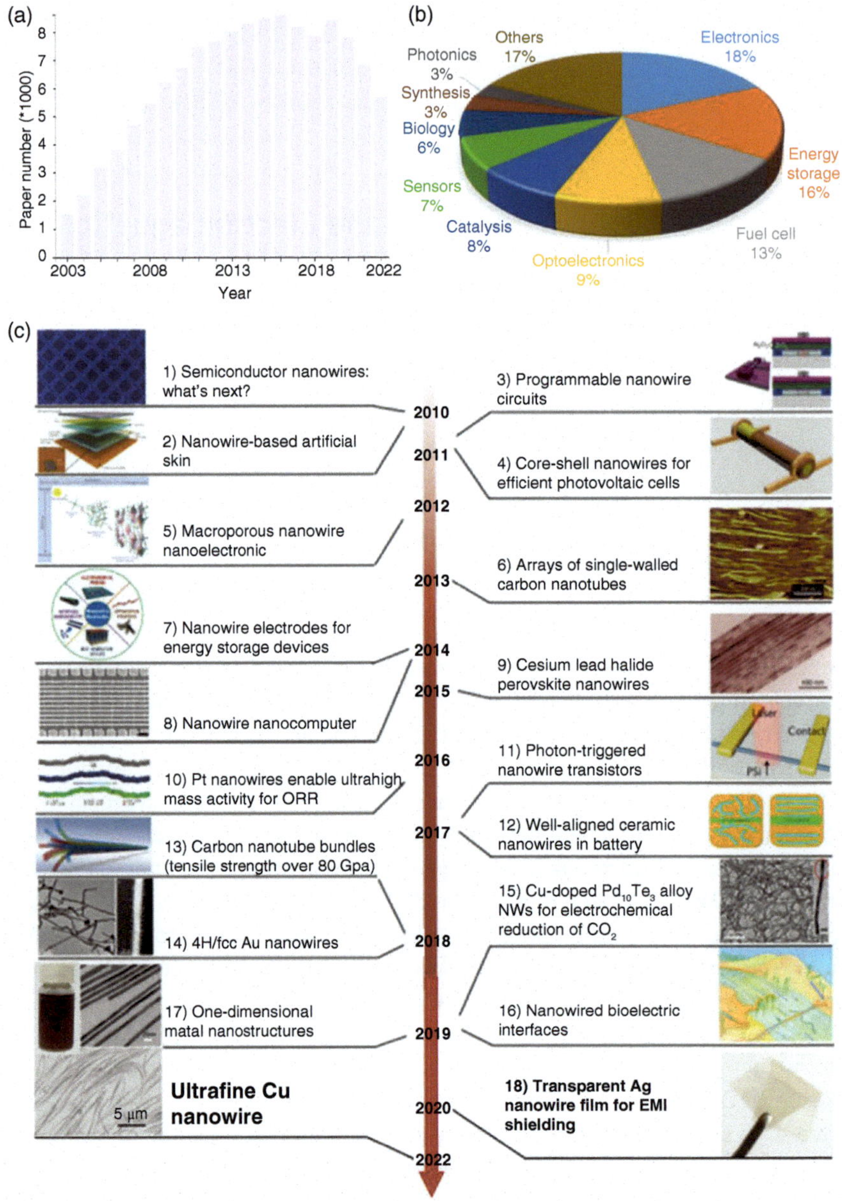

Figure 5.1 (a) The number of publications containing the keyword "nanowire" during the past two decades (from web of science). (b) Relative distributions of publications about the application of NWs. (c) The timeline about the key developments of NWs since 2010. Source: He et al. [3]/Reproduced with permission from @2019, John Wiley & Sons.

in newly emerging areas, such as artificial skin, nanoelectronics, photovoltaic cells, electrodes for energy storage, transistors, electromagnetic shielding, and even bioelectrics [7]. Moreover, the arrangement of NWs can also bring new properties. For example, the programmed NWs can act as circuits and even be used in an NW computer [2, 8]. The well-aligned ceramic NWs can boost the ion transport efficiency in batteries, and the assembly of carbon nanotubes (NTs) shows the strongest tensile strength of over 80 GPa [9]. Notably, the efficient synthetic strategies for uniform NWs with controlled morphology, components, and crystallinity are still crucial for exploring new phenomena.

The demand for improved dimensional control, purity, and reproducibility has motivated further study of how and why metals grow into 1D nanostructures. Researchers have undertaken exhaustive studies on how reaction temperature, reactant concentration, and the presence of different types of capping agents or halides affect the morphology of a metal nanostructure. This chapter focused on the Au/Ag/Cu NWs that can be obtained by controlling the assembly of atoms in the solution phase often referred to solution-phase or colloidal chemistry method. Overall, large-scale NWs were usually obtained with commonalities of high reactivity, uniform morphology, and high AR, as well as their transformation to a tremendous number of functional 1D NWs. Furthermore, the unique structure–property relationships of various 1D metal NWs enabling a multitude of applications will be highlighted.

5.2 One-Dimensional Metal Nanowires

1D metal nanostructures refer to those structures made of metals with two of the three dimensions between 1 and 100 nm. Different from their zero-dimensional (0D) and two-dimensional (2D) counterparts (i.e. nanoparticles and nanosheets, respectively), 1D nanostructures offer a unique platform for tailoring and utilizing the electronic, plasmonic, magnetic, electrical, mechanical, and thermal properties of metals to suit or enable various types of applications. For example, by simply adjusting the AR of Au NRs, one can tune their localized SPR_L peaks from visible to near-infrared wavelengths, enabling biomedical applications related to sensing, optical imaging, and cancer treatment [10]. For Ag NWs, they can be mixed with a polymer and cast as thin films with high optical transparency, excellent electrical conductivity, and good mechanical flexibility to allow for the fabrication of touchscreens and flexible solar cells. For both NRs and NWs comprised of Fe, Co, or Ni, their magnetic anisotropy is expected to bring new capabilities for applications related to actuation and information storage.

5.2.1 Ag Nanowires

Ag is a soft, white, lustrous transition metal that exhibits the highest electrical conductivity, thermal conductivity, and reflectivity among all the metals. The history of Ag dates back to 600 BC when Ag acted as the common currency or precious jewelry in many ancient countries. Ag is also a rather unreactive metal that occupies group 11 of the periodic table and exhibits similar physical and chemical properties to its vertical neighbors, Cu and Au.

5.2.1.1 Synthesis Approaches

Recently, remarkable progress has been made in research focused on the synthesis of Ag NWs, leading to well-controlled morphologies and ARs. Currently, Ag NWs are successfully synthesized by a variety of chemical approaches, such as template synthesis, ionic-liquid-assisted synthesis, microwave-assisted aerobic synthesis, and polyol reduction process. On the whole, Ag NWs show diameters ranging from 10 to more than 100 nm and an AR as large as 2000. Among all the synthesis methods, Ag NWs synthesized via the ionic-liquid-assisted strategy were mainly composed of small Ag NPs, which also exhibited a rough surface and inhomogeneous diameter. The microwave-assisted aerobic method could provide uniform heating condition and facilitate the formation of uniform Ag NWs with an average diameter of 45 nm and length in the 4–12 μm range. For the growth of Ag NWs with tunable diameter size, template-directed synthesis might be a promising strategy utilizing cylindrical channels to guide the growth of Ag crystals toward Ag NWs. As demonstrated by many research groups, templates, including 1D nanostructures, biological macromolecules, polymers, and porous materials, have been successfully adopted to synthesize Ag NWs. However, most of these template syntheses are complicated processes, and the Ag NWs will be destroyed with the removal of templates. Ag NWs synthesized via polymer templates could overcome the problem and ensured a very high AR (500–1000), but the cost was high and it was difficult to scale up.

The polyol synthesis has been developed as one of the most popular and versatile methods for the production of Ag NWs. The polyol reduction process usually contains the reduction of an Ag precursor in a homogeneous liquid medium with the assistance of a surfactant, such as polyvinyl pyrrolidone (PVP) (Figure 5.2a), which strongly bonds to the side planar of Ag NWs to facilitate the anisotropic growth and prevent the NWs from aggregating [11]. Early in 2002, Y. N. Xia group made significant progress in the large-scale synthesis of Ag NWs with a diameter ranging 30–40 nm and lengths up to 50 μm through a solution-phase process, in which ethylene glycol (EG) acts as both the solvent and the reducing agent [12]. In a typical synthesis process, the $AgNO_3$ was first reduced to Ag NPs with a bimodal size distribution due to the heterogeneous and homogeneous process, and the obtained NPs showed five twin planes with the assistance of PVP. Then, the Ostwald ripening process directed the particles growing into NWs, which were terminated by (111) facets, and the side surfaces were bounded by (200) facets [11].

By adjusting the molecular weight and molar ratio of the surfactant PVP, Ag NWs with uniform morphology and high AR could be obtained. Ag NWs with thin diameters and high aspect ratios up to 2000 can be synthesized by mixing halide ions with the Ag precursor. However, the Ag NPs were inevitably produced simultaneously and could only be removed through an extra selective precipitation process. Based on the employment of benzoin-derived radicals, the Yang group

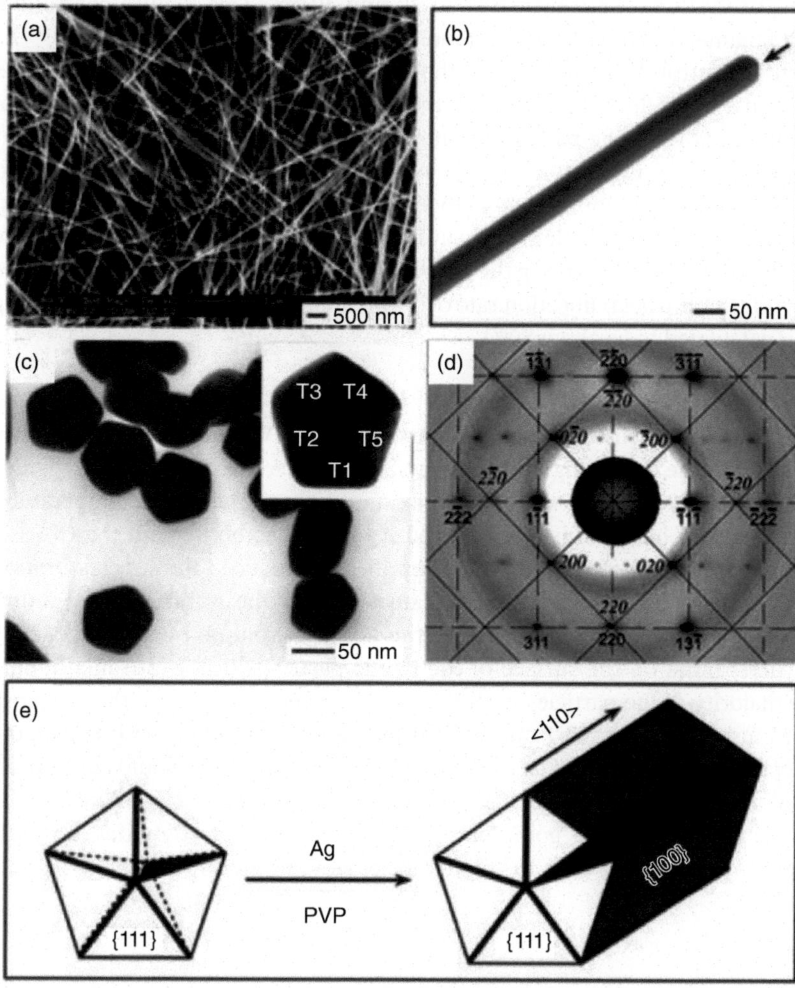

Figure 5.2 Solution-phase synthesis of Ag NWs. (a) SEM image of a random assembly of Ag nanowires purified via centrifuge. (b) TEM image of cross-section of the Ag nanowires. (c) TEM image of an individual Ag nanowire, clearly showing the straightness of the twin plane (indicated by the arrow). (d) SAED pattern taken from the Ag NW in (c) by aligning the electron beam perpendicular to one of the five side surfaces. (e) Schematic illustration of growth mechanism of Ag NWs with pentagonal cross-sections from decahedrons. Source: Sun et al. [11]/Reproduced with permission from @2003, American Chemical Society.

decreased the average diameters of Ag NWs to 13 nm and increased the ARs up to 3000 [13]. Through a modified facile protocol at atmospheric pressure, the Xia group successfully synthesized Ag NWs with tunable diameters from 20 to 40 nm in high-morphology purity [14] Ag NPs with quasi-spherical shape, and twin boundaries surface were first obtained within five minutes. When the reaction time was prolonged to 35 minutes, Ag NWs with diameters below 20 nm and lengths up to 20 μm were formed. With the reaction time extended beyond 35 minutes, the diameters of the NWs were increased to 30 and 40 nm. This mature controlling strategy of diameter and AR provides more potential for further application.

To further control the uniformity of the Ag NWs, our group report a facile and versatile route to produce Ag NWs with the tunable diameter with the assistant of Cl^- and Br^- ions [15]. The reduction kinetics was slowed down when the $AgNO_3$ was titrated with a syringe pump, promoting the uniform growth of the Ag NWs. In a typical synthesis, the capping agent (PVP) and the halide ions (Cl^- and Br^- ions) were mixed in EG under gentle stirring at 160 °C for 15 minutes. The Ag NWs were obtained by continuously titrating the Ag precursor ($AgNO_3$ in EG) into the mixture with a syringe pump at an injection rate of 0.2 ml min^{-1}. The yield of the original Ag NWs was calculated to be about 80%.

5.2.1.2 Growth Mechanism

Figure 5.3 illustrates the growth mechanism of Ag NWs via a modified polyol method [15]. At the beginning of the process, the Ag precursor was reduced by EG, which acts as a solvent and a source of the reducing agent under high temperature. In the presence of halide ions, the released Ag^+ could combine with the Cl^- and Br^- ions to form AgCl and AgBr, which significantly reduced the supersaturation of Ag^+ with time prior to reaching a plateau. When the number of Ag atoms reached supersaturation with increasing the addition volume of $AgNO_3$, the Ag nanoparticles burst on the surface of the halide particles. It is worth mentioning that the majority of the particles were five-fold twinned seeds due to the existence of halide ions. Because the PVP and halide ions were preferentially absorbed on the (200) facets rather than (111) facets, the newly formed Ag atoms would continually deposit along the [110] direction, while the lateral growth rate was inhibited.

With the titration process going on, the Ag atoms were slowly reduced by EG and kept at a low-level owing to the relatively slow injection rate of the $AgNO_3$

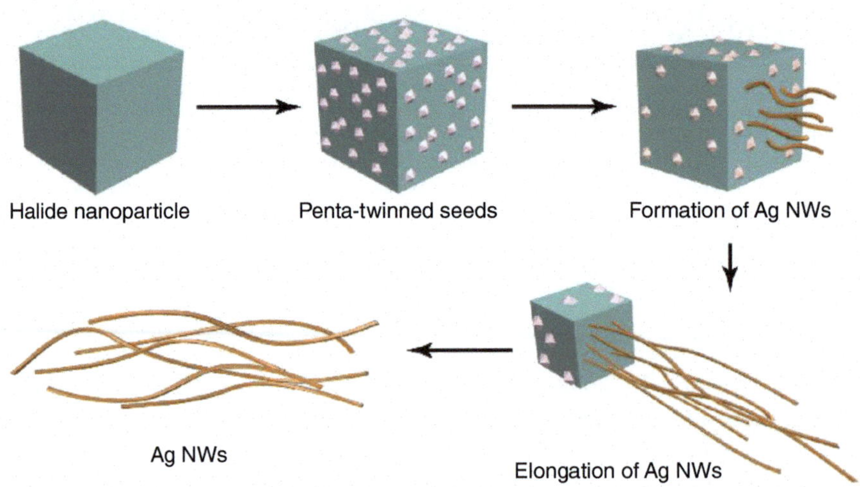

Figure 5.3 Schematic diagram illustrating the fabrication of the Ag NWs. Source: Reproduced from Zhu et al. [15]/with permission of American Chemical Society.

solution. In addition, the regulating capacity of halide particles further balances the concentration of Ag^+ ions, leading to the uniform growth of the Ag NWs. As a result, the growth of the Ag seeds was elongated into the ultralong Ag NWs with penta-twinned and pentagonal cross-sections. At the same time, AgCl and AgBr would dissolve or be consumed accompanying the growth of Ag NWs, keeping the concentration of the Ag^+ ions in the solution steady and tuning down the reduction kinetics. Therefore, Cl^- and Br^- ions play an important role in controlling the diameter of the Ag NWs, which is in agreement with the previous research.

The presence of Cl^- or Br^- ions plays a critical role in the formation of Ag NWs among the synthesis process. To verify the growth mechanism mentioned above, we further investigated the effect of the molar ratio of NaCl/NaBr on the synthesis of Ag NWs by conducting a set of experiments with different NaCl/NaBr ratios while keeping the total amount consistent. It is worth noting that the presence of NaBr caused the remarkable increase in the number of halide nanoparticles that could serve as heterogeneous nucleants, further promoting the increased number of Ag seeds, and thus the decreased diameter of the produced Ag NWs. In addition, the enhanced etching ability of Br^- ions with the increased concentration results in the etching of lots of small penta-twinned seeds, making huge amount of Ag nanoparticles produced at high Br^- ions concentration [16]. In summary, the optimum ratio of NaCl/(NaCl+NaBr) was 67%, which produced uniform and thin Ag NWs with relatively high yield. In addition, the thinner Ag NWs could be produced by increasing the total concentration of the halides while keeping the ratio of NaCl/NaBr constant. As shown in Figure 5.4, the diameter of the Ag NWs decreased with increasing concentration of the halide ions. For example, the diameter

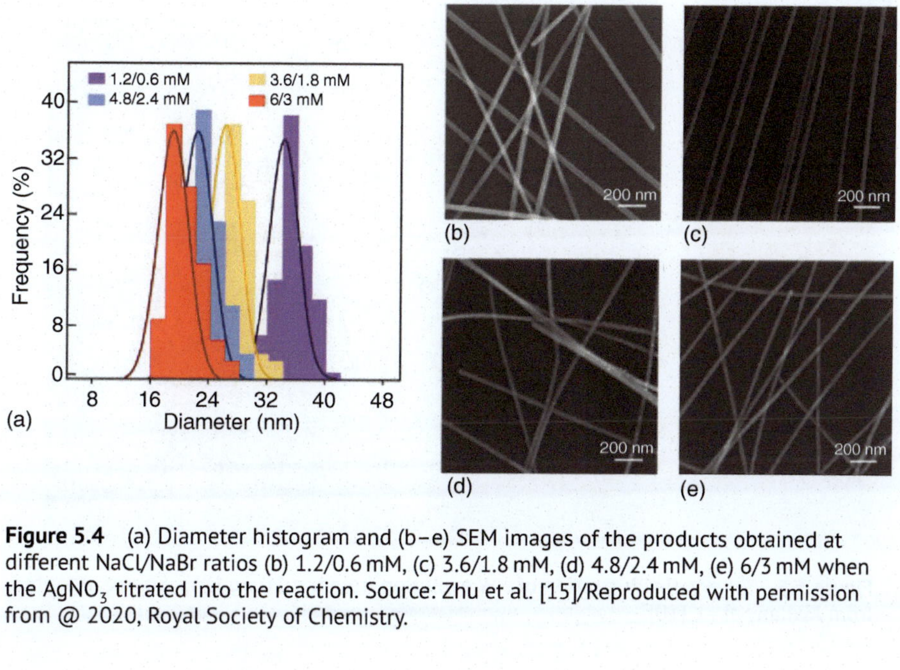

Figure 5.4 (a) Diameter histogram and (b–e) SEM images of the products obtained at different NaCl/NaBr ratios (b) 1.2/0.6 mM, (c) 3.6/1.8 mM, (d) 4.8/2.4 mM, (e) 6/3 mM when the $AgNO_3$ titrated into the reaction. Source: Zhu et al. [15]/Reproduced with permission from @ 2020, Royal Society of Chemistry.

decreased from 35 to 19 nm when the concentration of NaCl/NaBr increased from 1.2/0.6 to 6/3 mM, leading to an AR of up to 1500 for the Ag NWs.

Apart from the halide ions, the reaction temperature also plays an important role in the synthesis process. The product was mainly Ag nanoparticles with large size at low temperature. In contrast, Ag NWs were the main products at higher temperature, with the diameter increased from 35 to 42 nm along with the temperature. On the one hand, the reduction capacity of the EG was depended on the reaction temperature. The reduction capability of EG at lower temperature was too weak to outbreak the formation of penta-twinned seeds, decreasing the yield of Ag NWs dramatically. On the other hand, the density of chemisorbed Br^- ions on the Ag (200) facets was inversely proportional to temperature. At high temperature, the dynamic adsorption and desorption of Br^- ions on Ag (200) facets make the lateral surface exposed to Ag species in the solution, causing the lateral growth of Ag NWs.

5.2.1.3 Overgrowth

The capping agent used for controlling the morphology and sizes of Ag NWs was usually PVP, which also stabilized Ag NWs with high dispersion in a hydrophilic solvent. Thus, the good dispersion, the uniform morphology, and large-scale synthesis of Ag NWs make them a suitable template to guide the growth of other 1D nanostructures for further expanding the functionalization. The morphology and the components can then be delivered to next-generation nanomaterials with more complex systems. The schematic illustration of Ag NWs as the first generation (1G) NWs and the derived 2G–4G nanostructures is shown in Figure 5.5 [10]. Numerous types of materials can be decorated on the surface of Ag NWs, such as metals, semiconductors, magnets, and inert materials. The obtained Ag-NW-based hybrid

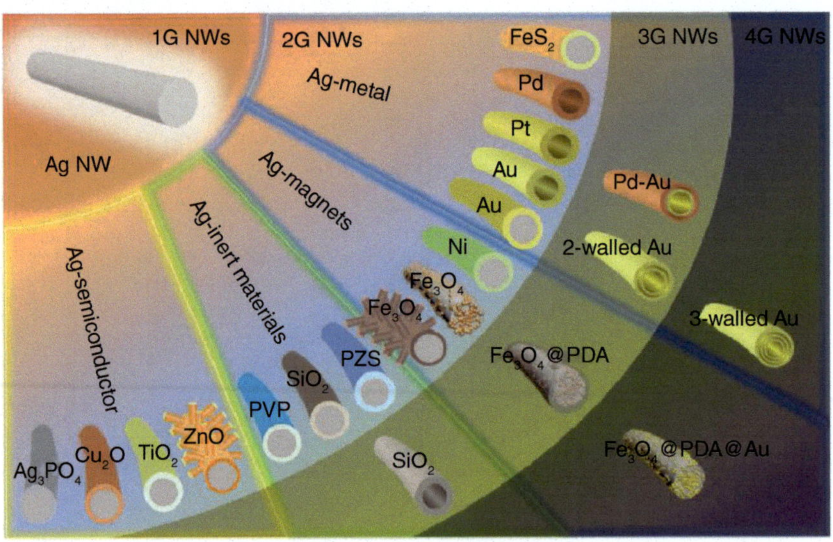

Figure 5.5 Schematic illustration of NW evolution from the 1D Ag NWs. Source: Adapted from Alkilany et al. [10].

nanomaterials combine the features of each component, bringing novel chemical and physical properties, and finally broadening their applications in many fields.

Due to the uniform size distribution and excellent dispersibility of Ag NWs, a series of core–shell nanostructures were fabricated via a standard physical templates process. For instance, Xia and coworkers reported the deposition of ferromagnetic sheaths of polycrystalline Ni onto the surface of Ag NWs via a solution-based method [17]. Compared with the pristine Ag NWs, the Ag/Ni NWs displayed a rough surface. These hybrid NWs exhibited a magnetic hysteresis property and could be aligned via an extra magnetic field. Moreover, when they were stored in ultrapure water or ethanol for six months, the Ni shell was easily oxidized into NiO nanoplates. The Zhan group reported a coprecipitation method to synthesize Ag-Fe_3O_4 core–shell NWs by using Ag NWs as the physical template [18]. Based on the Fe_3O_4-Ag hybrid NWs, a series of three-layered NWs nanocables with well-defined core–shell structure could be prepared via physical template process, such as Fe_3O_4-Ag@polyphosphazene, Fe_3O_4-Ag@SiO_2, and Fe_3O_4-Ag@polydopamine (PDA).

Then, 4G nanostructures could be prepared via two main processes, that is the etching or chemical converting of Ag. For example, by etching the Ag core in the Fe_3O_4-Ag@SiO_2 NWs, 4G Fe_3O_4@SiO_2 NTs were obtained. In addition, a series of 2G Ag-semiconductor NW hybrids were also fabricated by using Ag NWs as a physical template, such as the ZnO, TiO_2, $AgPO_4$, and Cu_2O. Among the semiconductors, ZnO is a wide-bandgap (E_g = 3.37 eV) semiconductor widely used in photodetectors, gas sensors, and photocatalysts. Since the small crystalline mismatch (2.68%) between the (001) plane of ZnO and the (111) plane of Ag, ZnO tends to nucleate and grow on the surface of Ag NWs. For example, the Lu group reported a thermal decomposition process for the fabrication of Ag/ZnO fluffy worm-like heterostructures (ZnO NRs on Ag NWs) [19].

Apart from acting as a physical template, the Ag NWs can also serve as a sacrificial chemical template for the synthesis of 1D nanostructures, such as noble metal alloy NWs or NTs, metal-sulfur NWs, and silver halide NWs. Notably, noble metal-Ag hybrid nanostructures have received increasing interest due to their wide applications in many areas, including catalysis and flexible transparent conductive films (FTCFs). These hybrid nanostructures are mainly synthesized via galvanic replacement reactions, which inherit the morphology, structure, and properties from the Ag NWs. It is worth noting that the standard reduction potential of Ag is lower than that of most of the noble metals (Au, Pt, and Pd). As a result, Ag NWs can transform into a series of 1D noble nanostructures via the galvanic displacement reaction.

As for the galvanic displacement with Au^{3+} ions, Ag NWs can react with the precursors of Au ($HAuCl_4$) to fabricate Ag–Au hybrid nanostructures at room temperature [20, 21]. In a typical procedure, the reaction started at the active sites on the (200) surface. First, Ag atoms in the Ag NW would continue to dissolve, while new surfaces with even higher surface energies were generated. Meanwhile, due to the good matching of crystalline structures and lattice constants between Ag and Au (the same fcc lattice with only 0.186% difference), the generated Au tended to be epitaxially deposited on the surface of the Ag NWs. By controlling the experimental conditions, the galvanic reaction could be suppressed to form Ag–Au core–shell NWs with a thin Au layer of 0–10 nm. The coating of Au exhibited excellent resistance to oxidation compared to that of Ag NWs. Without suppression, a fast galvanic reaction would lead to the formation of pinhole Ag–Au alloyed NTs. By adjusting the galvanic reaction to normal speed, Ag NWs would completely transform into Au NTs with smooth and uniform sheaths, which inherit a pentagonal cross-section shape from Ag NWs. Moreover, a conformal thin sheath of Ag could be deposited onto the obtained Au NTs via a simple electroless plating process [22]. After subjecting the Au@Ag NTs to a repeated galvanic replacement reaction, 3-layered nanostructures such as double-walled Au NTs and double-walled Au/Ag@Pd/Ag alloy NTs would be obtained. Moreover, by repeating this facile coating and etching method, 4-layered triple-walled Au/Ag alloy NTs could be easily designed.

Similarly, the Ag-Pt bimetallic hollow NTs were prepared through the galvanic replacement between Ag NWs and aqueous K_2PtCl_4 or H_2PtCl_6 solution. A tunable wall thickness of Ag-Pt NWs from 9 to 19 nm was obtained by simply increasing the concentration of the K_2PtCl_4 solution. When adding a small amount of aqueous ammonia into the Ag-Pt NWs solution, the silver proceeded to react with ammonia and resulted in hollow Pt NTs. The galvanic replacement reaction between Ag NWs and $Pt(CH_3COO)_2$ or $Pd(NO_3)_2$ was also performed and resulted in Pt NTs or Pd NTs with a rough morphology, respectively. Interestingly, when the length of Ag NWs was increased to 20 μm, only some segments of the NWs were converted into hollow structures due to the hydrolysis of Pd^{2+} ions before the reaction was completed. In addition, by introducing a reductant (such as ascorbic acid), Ag NWs were controllably modified with a small amount of Pd (5 at%), leading to excellent chemical stability against oxidation.

5.2.2 Cu Nanowires

Copper (Cu), residing in group 11 (the same as Au and Ag), exhibits a high malleability, ductility, electrical conductivity, and thermal conductivity. When compared with Ag, Cu is 1000 times more abundant with only 6% less conductivity and is 80 times less expensive. Based on these advantages, Cu nanostructures, especially the Cu NWs, have gained wide attention for the applications of electronics, optoelectronics, solar cells, and catalysts.

5.2.2.1 Synthesis Approaches

Recently, Cu NWs have been successfully synthesized via a variety of methods including chemical vapor deposition, template-assisted method, electrospinning, and solution-phase synthesis. Among these strategies, solution-phase synthesis has been the dominant one due to its advantages in product homogeneity and scalability, as well as the option for facile functionalization. Typically, the reaction solution for solution-phase synthesis is composed of a copper precursor, a reducing agent, and a capping agent.

Water-based Synthesis. In order to obtain high AR Cu NWs, it is important to control the synthesis parameters of NWs to increase the nucleation and crystal growth rate (Figure 5.6). The synthesis of Cu NWs requires a copper precursor

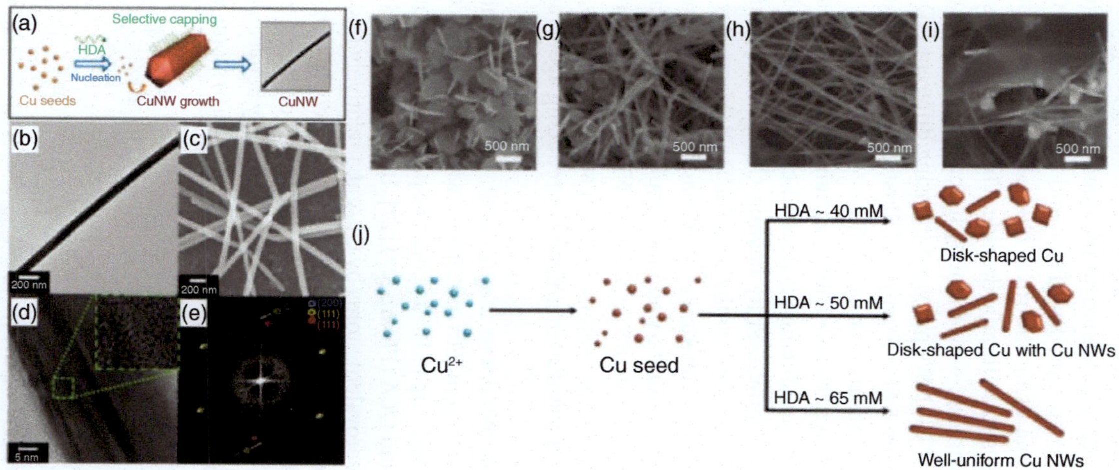

Figure 5.6 (a) Schematic image of the formation mechanism of Cu nanowires when using the HDA as capping ligand, (b) TEM, (c) SEM, (d) HRTEM, and (e) FFT spectra of Cu nanowires. Source: Reprinted with permission from Na et al. [23]; Copyright Elsevier (2018). The SEM images of prepared Cu nanowires by using different HDA concentrations: (f) 40 mM, (g) 50 mM, (h) 65 mM, and (i) 75 mM. (j) Schematic image of the Cu nanowire growth mechanism depending on the concentration of HDA. Source: Chee et al. [24]/Reproduced with permission from @2020, Springer Nature.

source, such as $Cu(NO_3)_2$ or $CuCl_2$, a reducing agent, such as hydrazine, ascorbic acid, glucose, or dextrose, and a capping ligand to control 1D growth, such as ethylenediamine (EDA), hexadecylamine (HDA), sodium dodecyl benzenesulfonate (SDBS), PVP, hexadecyltrimethylammonium bromide (CTAB), or octadecylamine (ODA) [23–25]. The concentrations of reducing agents and capping ligands play an important role in the formation of Cu NWs, and the heating temperature and reaction time also have a significant influence on the diameter and length of Cu NWs. Figure 5.6a–c shows the formation process, TEM, and SEM images of Cu NWs with the extension of heating time using the royal blue precursor solution containing $Cu(NO_3)_2$, EDA, hydrazine, and sodium hydroxide. The SEM images show the synthesized Cu NWs of uniform size with the length and diameter of 10 and 90 nm, respectively. SEM images of Cu NWs prepared by using different HDA concentrations were presented in Figure 5.6f–i. Figure 5.6j illustrates the morphologies of Cu product, depending on the concentration of HDA.

Organic Solvent-based Synthesis. Under certain reaction conditions, alkylamines can also serve as both reducing agents and capping agents. For example, penta-twinned Cu NWs were produced by heating CuCl in oleylamine (OLA) at 170 °C [26]. In the initial stage of the synthesis, a Cu(I)-OLA complex was reduced to generate Cu atoms and then decahedral seeds through homogeneous nucleation. Due to the high energy of twin boundary and the capping of Cu(200) facets by OLA, the decahedral seeds then grew into penta-twinned NWs with an average diameter of 63 nm and lengths around 30 μm (Figure 5.7). In such a synthesis, OLA played multiple roles as a solvent, a coordination ligand, a reducing agent, and a capping agent.

The introduction of other metal ions may sometimes help the generation of Cu NWs through galvanic replacement. For example, it was reported that Cu NWs with a diameter of 15 nm and lengths up to 10 μm could be produced by reducing $CuCl_2$ in the presence of $Ni(acac)_2$ and OLA (Figure 5.7) [27]. It was hypothesized that Ni(II) ions were first reduced to zerovalent atoms by OLA at a relatively low temperature, at which Cu(II) cannot be reduced to Cu(0) due to the stable complex between Cu(II) and ligands, such as OLA and Cl^-. The Ni atoms with high chemical activity would then rapidly reduce Cu(II) to Cu(0) atoms due to the synergy between the reduction by OLA and the galvanic replacement reaction, while the Ni(0) atoms were oxidized

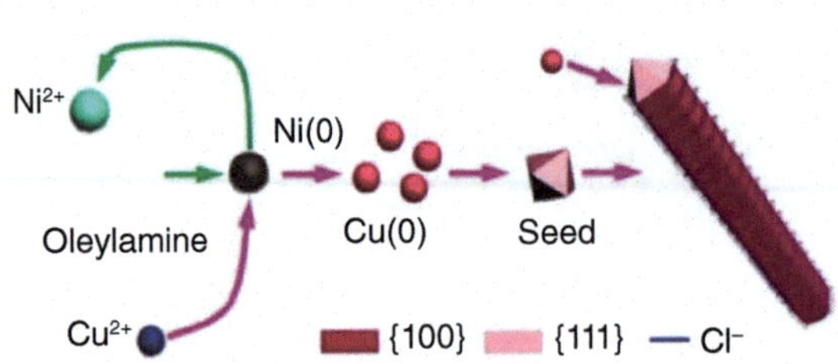

Figure 5.7 Schematic of the formation mechanism of Cu nanowires. The Ni^{2+} and Cu^{2+} ions play critical roles in promoting the growth of high aspect ratio Cu nanowires. Source: Reproduced from Guo et al. [27]/with permission of Springer Nature.

back to Ni(II). Eventually, nuclei made of Cu(0) atoms were formed, followed by their growth into decahedral seeds and penta-twinned NWs. Other additives such as $RuCl_3$ and $FeCl_2$ were shown to be able to protect the twinned seeds from oxidative etching, ensuring the evolution of NRs and then NWs from the decahedral seeds.

Organic radicals have also been explored as reducing agents for the formation of uniform Cu NWs with diameters ranging from 18 to 36 nm and lengths up to 20 μm [28]. It was demonstrated that, upon heating, benzoin decomposed into two radical segments, benzoyl radical, and α-hydroxybenzylradical, and the resulting radicals can act as efficient electron donors to reduce Cu(II) species to Cu(I) and then to Cu atoms. With the aid of OLA, Cu atoms nucleated to produce penta-twinned NWs with the side surface being covered by (200) facets. The reducing power of benzoin can be readily tuned by symmetrically adding different functional groups to the aromatic rings, with the introduction of electron-donating and electron-withdrawing groups to increase and decrease the reducing power, respectively.

Solvothermal and hydrothermal strategies are both effective ways for the synthesis of nanostructured copper. The solvothermal synthesis of Cu NWs is not readily oxidized due to the dense ligands attached to the NW surface, which have a great influence on the transmittance and conductivity of conductors. The conductive copper NW network can be obtained by rapidly volatilizing organic solvent, while it could be challenging to prepare large quantities of Cu NWs due to its relatively higher reaction temperature and energy-expensive processes. The conductive copper ink prepared by hydrothermal synthesis can be versatile, cost-effective, and environmentally friendly for printing, while the nanostructured copper materials can be dispersed and prepared for the aqueous inks.

5.2.2.2 Growth Mechanism

The schematic in Figure 5.8 exhibits the solution-phase synthesis of Cu NWs using an organic capping agent [29, 30]. The precursors were first reduced to decahedral-shaped Cu seed due to the lowest surface energy of the (111) plane. The capping agent directed the growth into NWs, probably by selectively adsorbing on the (200) surface and physically blocking the addition of Cu to that surface. Benefiting from the selective reduction of ionic Cu species onto the end of the NWs, the Cu NWs consisting of a pentagonal pyramid with five (111) planes would finally grow to be tens of micrometers long.

In a typical water-based synthesis, the Cu(II) precursor based on $CuCl_2$ was reduced by glucose in the presence of HDA, which was hypothesized to act as a facet-selective capping agent for the (200) facets on the side surface of Cu NWs. The capping agent not only directed the growth of nanostructures but also protected the surface from being oxidized in air. When reducing the amount of HDA by half, Cu nanocubes were generated due to the selective removal of multiply twinned, decahedral seeds via oxidative etching. Interestingly, using a similar protocol, tadpole-like NWs could be obtained by increasing the amount of glucose. Increasing the concentration of the reducing agent led to an accelerated reduction rate for the Cu(II) precursor, and thus the formation of decahedral seeds with a relatively

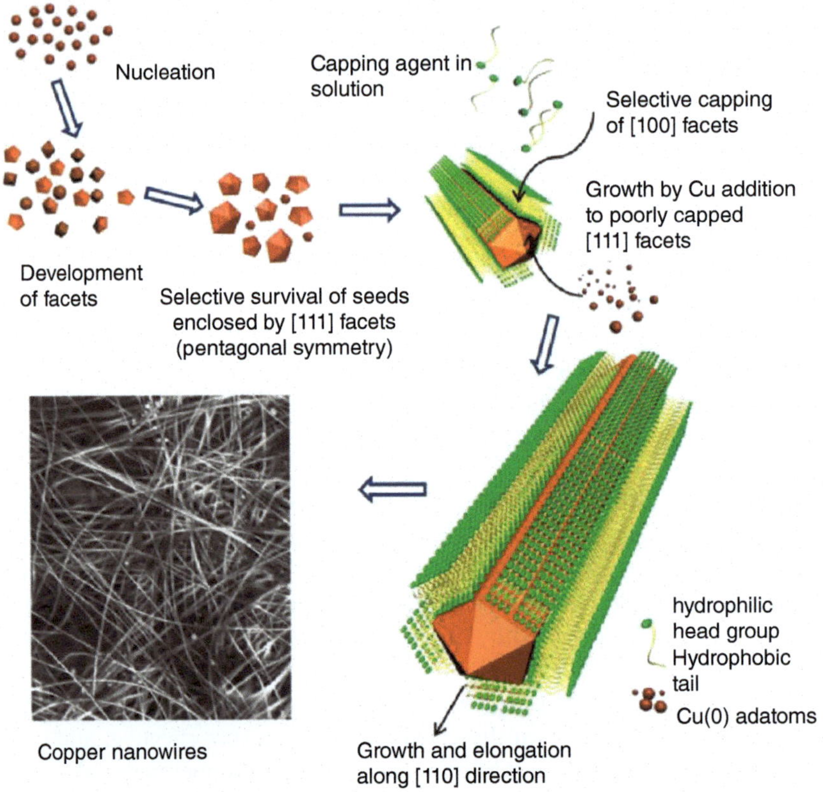

Figure 5.8 Schematic illustration of NW evolution from the Cu NWs and typical SEM image of the obtained Cu NWs. Source: Bhanushali et al. [29]/Reproduced with permission from JOHN WILEY & SONS, INC. Note: the facets should be (111) and (200), but not [111] and [100].

large size. As the synthesis proceeded, glucose was consumed and the reduction rate would be decreased. The anisotropic growth on the decahedral seeds finally resulted in the formation of pentagonal bipyramids, which further evolved into tadpole-like NWs.

More recent work has demonstrated that Cu NWs only formed in the presence of HDA when $CuCl_2$ rather than $Cu(NO_3)_2$ was used as a precursor to Cu [31]. This observation suggested that HDA by itself did not act as a facet-selective capping agent. Electrochemical measurements confirmed this hypothesis by showing that, in a solution of HDA and $Cu(NO_3)_2$, there is no difference in the charge transfer resistance between (111) and (200) of Cu crystal for the reduction of Cu(II). However, if Cl^- is added into the reaction solution, the charge transfer resistance was 14.7 times lower on the (111) than on (200) facet. The concentration of Cl^- that resulted in the lowest charge transfer resistance on (111) relative to (200) facet matched with the production of Cu NWs with the highest AR. DFT calculations further showed that coverage of 0.33 nm monolayer of Cl^- could selectively displace HDA from Cu

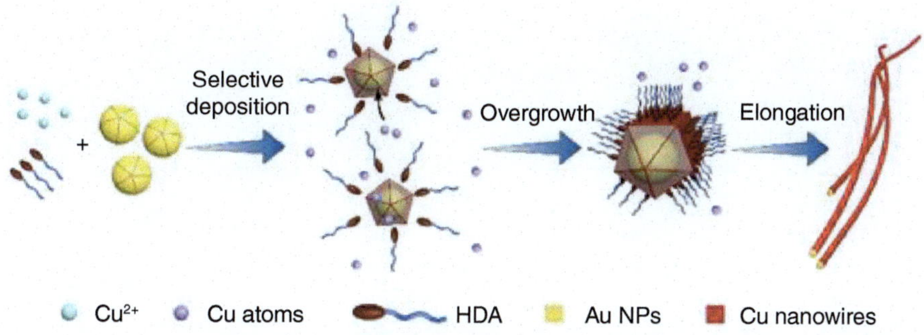

Figure 5.9 Schematic illustration of NW evolution from the Cu NWs. Source: Reproduced from Xu et al. [32]/with permission of Royal Society of Chemistry.

(111) surface, while allowing HDA to remain strongly bound to Cu (200) surface, corroborating the synthetic and electrochemical results.

Based on the growth mechanism proposed above, the ultrathin Cu NWs with a diameter of 7–10 nm could be obtained by introducing Au nanoparticles into the general growth system, as shown in Figure 5.9 [32]. The active Au atoms on the surface catalyze Cu(II) precursor into Cu(0) species with glucose as a reducing agent. The initially formed Cu(0) species land on the penta-twinned Au nanoparticles and act as the nucleation sites. Subsequent deposition of Cu(0) species preferentially takes place at these sites due to the absence of lattice mismatch. Moreover, the preferential bonding of HDA toward loosely packed (200) facets makes the (111) facets attractive for the deposition of Cu(0) species. This longitudinal deposition leads to the continuous increase of the AR of the Cu NWs.

5.2.2.3 Overgrowth

Similar to those of Ag NWs, the high chemical reactivity and scale-up advantages of Cu NWs make them suitable as one-generation (1G) materials as a physical or chemical template for next-generation nanostructures (Figure 5.10). As one typical 1G material, Cu NWs possess the characteristics of high reactivity, uniform morphology, and easily scalable synthesis. Hence, Cu NWs are intrinsically unstable, and the surface of Cu NWs is rapidly oxidized, even in a few hours, under common ambient conditions, leading to dramatic decreases in conductivity and limiting their practical applications. Tremendous efforts have been proposed to solve this challenge including the coating of inert materials or a metal shell onto the surface of Cu NWs by using Cu NWs as a physical template.

Similarly, various Cu@metal (Cu@M) NWs synthesized from Cu NWs can not only reserve the properties of Cu but also avoid being oxidized when exposed to air. A facile one-pot synthesis method developed by the Cai group successfully coated a series of metal onto Cu NWs, including Zn, Ni, Ti, and Ag metals [33]. Typically, after the growth of Cu NWs, metal acetylacetonate was injected into the reaction system for the shell coating. Then metal ions were rapidly reduced to metal atoms and subsequently nucleated on the (200) plane of the Cu NWs, which resulted in a

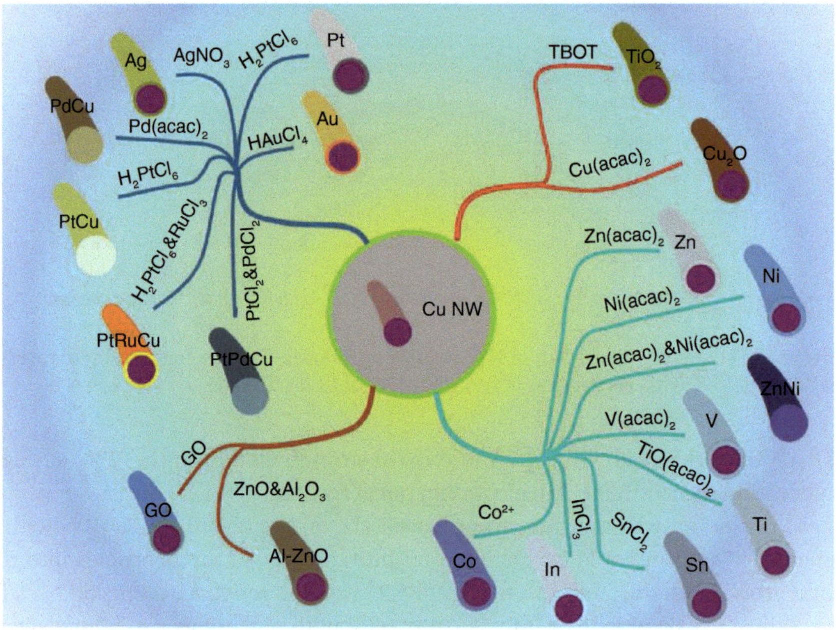

Figure 5.10 Schematic illustration of NW evolution from the 1G Cu NWs. Source: Adapted from Alkilany et al. [10].

uniform metal shell. When mixing two different precursors such as Ni(acac)$_2$ and Zn(acac)$_2$, Cu@NiZn NWs were obtained.

Furthermore, the Cu-semiconductor core–shell nanostructures can also be designed by using Cu NWs as a physical template. Similar to that of Ag-semiconductor nanostructures, the incorporation of Cu NWs facilitated a highly efficient separation and transfer of photoinduced charges, which improved the performance in photocatalytic degradation. As reported by the Yan group, Cu-TiO_2 core–shell NWs with a controlled shell thickness of 15–20 nm were synthesized via a facile hydrothermal method [34]. The localized SPR and plasmonic waveguide effect of the Cu-TiO_2 core–shell NWs significantly enhanced the light absorptions of dry molecules and the separation of carriers. When applied to dye-sensitized solar cells, the electrical network (optical path) formed by the Cu NW-core accelerated electronic transmission and light propagation, which eventually promoted a power conversion efficiency of 9.44% higher than that of the undoped one. Similar to the effect of the TiO_2 shell, the narrow-bandgap semiconductor Cu_2O NPs (bandgap of 2.17 eV) were assembled onto the surface of Cu NW via a facile reduction process [35]. Due to the strong capacity for oxygen activation of Cu NWs and the bandgap-adjusting effect of Cu_2O NPs, the obtained Cu-Cu_2O core–shell NWs showed remarkable enhancement in visible light photocatalytic activity toward the degradation of organic pollutants, including rhodamine B (RhB) and methyl orange [35].

Since the standard reduction potentials of noble metals (Au, Pt, Pd, Ag, and Ru) are higher than that of Cu, a galvanic replacement reaction will possibly take place when Cu NWs were mixing with the precursors of these metals. Meanwhile, many factors including temperature changes and the related ion concentrations can affect the actual value of the reduction potential, which eventually affects the occurrence of a galvanic replacement reaction. A series of Ag-capped Cu NWs with different morphologies were synthesized by controlling the reaction parameters [36]. For example, with a low concentration of Ag^+ ions (Cu/Ag molar ratio of 5 : 1), Ag NPs with an average diameter of 8.4 nm were deposited onto Cu NWs through the *in situ* substitutional reactions. When increasing the Cu/Ag molar ratio to 1 : 1, the reaction speed increased, and the Ag NPs grew into longer NRs. The Zhang group adopted the method by directly pouring the Ag-ammonia reagent into the Cu NWs dispersion solution instead of dropwise adding [36]. Clearly, the galvanic reaction quickly took place between Ag^+ ions and Cu NWs, which resulted in Ag nanosheets deposited onto Cu NWs. Moreover, the galvanic replacement of Cu by Ag^+ ions could be suppressed by adding a high concentration of the reducing agent (such as ascorbic acid). The resultant Cu-Ag NWs showed an increased shell thickness of 14 nm with less etching of the Cu core. When the amount of reducing agent was decreased and the reaction time prolonged, the galvanic replacement between Ag^+ ions and Cu eventually took place and resulted in Cu–Ag alloy NTs. The TCFs based on Cu-Ag NWs exhibited not only a comparable performance to that of pure Ag or Cu NWs but also excellent resistance to oxidation in dry air at 160 °C or under humid conditions (85% relative humidity) at 85 °C for 24 hours. By replacing $AgNO_3$ with $HAuCl_4$ or K_2PtCl_6 in water, this versatile synthetic strategy could be extended to the formation of Cu-Au and Cu-Pt core-shell NWs or NTs, respectively.

5.2.3 Ultrathin Au Nanowires

Au nanostructures have been widely studied due to the chemical inertness, low resistivity, and unique surface plasmon properties. The novel photoelectric properties of Au nanostructures depend on the size and morphology. When the size is larger than the critical size (2 nm), Au nanostructures exhibit localized surface plasmon resonance (LSPR) under light irradiation. The number of modes and wavelength of LSPR spectrum bands depend on the morphology of nanostructures, surface modification, and the surrounding environment. When the size of Au nanostructures is reduced below the critical size, the quantized electronic structure plays a role in their optical and photophysical properties. The diameter of Au ultrathin nanowires (Au UNWs), as a new 1D nanostructure, is less than the critical size of quantization (2 nm) and the length much larger than the critical value. The typical AR of Au UNWs is 100–1000. Compared with other 1D nanostructures and metal NWs, Au UNWs are expected to be used in future strain sensors, photovoltaic and light-emitting devices, and catalysts due to the unique morphology and special electronic, optical, and mechanical properties.

5.2.3.1 Structure of Au UNWs

The work on the synthesis of Au UNWs with a diameter of <2 nm and a length of the micron order were first reported in 2007. Ravishankar prepared Au UNWs by aging a mixture of AuCl-oleylamine complexes and ascorbic acid (Figure 5.11a) and proposed that the oriented attachment mechanism from the small Au nanospheres can be formed into 1D ultrathin structures (Figure 5.11e) [37]. Yang synthesized Au UNWs by simply mixing $HAuCl_4$ and OA (Figure 5.11b) and regarded that slow *in situ* reduction of 1D mesostructures can be formed from $HAuCl_4$ and OA, which led to the formation of Au UNWs in solution (Figure 5.11f) [38].

More currently, the Kan group reported that Au UNWs can be obtained by reducing polymeric strands of oleylamine-AuCl complexes formed via oleylamine soft template. The diameter of the Au UNWs can be adjusted by adding oleic acid and changing the volume ratio of oleic acid and oleylamine, Au NWs with different diameters (<10 nm) can be obtained. In conclusion, Au UNWs can be synthesized in organic solvents using oleylamine and other reagents through oriented attachment, slow in-situ reduction, aurophilic reaction, reverse-micelle, or soft template mechanism. In these synthesis methods, most of the Au UNWs are single crystalline with a diameter lower than 2 nm, with the lattice spacing of 0.23 nm growing along [111] direction. This is because of the low coordination number of atoms in (200) plane, the oleylamine bonding site obtained from (200) plane is significantly higher than that obtained from (111) plane. Therefore, in the synthesis process of Au UNWs, the oleylamine ligand preferentially combines with the (200) plane, resulting in the NWs growing longitudinally along the [111] direction. In addition, Au UNWs also contain polycrystalline domains composed of twinned boundary sliding (TBS)-attached face-centered cubic and hexagonal

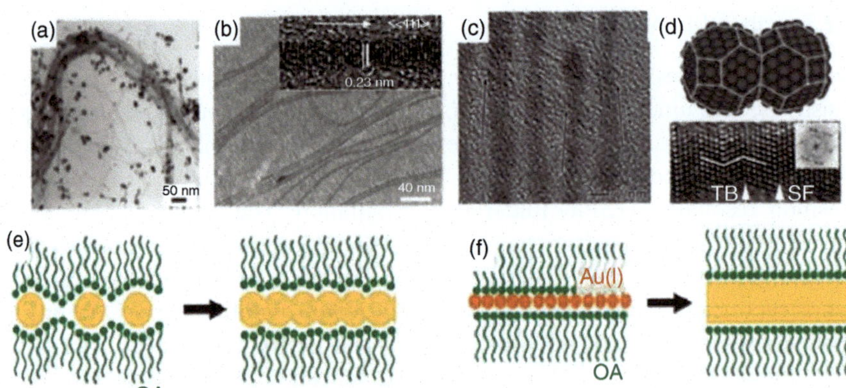

Figure 5.11 (a, b) TEM images of Au UNWs and the synthesis scheme of the oriented attachment. Source: Halder et al. [37]/Reproduced with permission from JOHN WILEY & SONS, INC. (c, d) TEM images of Au UNWs with a lattice spacing of 0.23 nm and the synthesis scheme of soft template mechanism. Source: Huo et al. [38]/Reproduced with permission from @2008, American Chemical Society. (e) HRTEM image showing [111] growth direction for most of the NWs. Source: Adapted from Halder et al. [37]. (f) The polycrystalline structure with twinned planes. Source: Adapted from Huo et al. [38].

close-packed (hcp) nanoparticles, as shown in Figure 5.11c,d. The two interactive phase structures change with the diameter of the NWs.

Au UNWs have become a research hotspot in nanoscience due to the large surface area, unique quantum size effect, and excellent electronic, optical and mechanical properties. However, it is difficult to synthesize high-quality Au UNWs, especially the Au UNWs will break under the irradiation of an electron beam. For to further improve the synthesis quality and explore the growth mechanism of Au UNWs, many experiments have been carried out by changing the organic solvent, adding the heterogeneous metal nanoparticles, and controlling the reaction temperature. For example, Xia prepared Au UNWs by reducing a 1D chain of [OA-Au(I)Cl] complexes by 10 nm Ag nanospheres. The usage of Ag nanospheres as a reducing agent significantly enhanced the yield of Au UNWs. The nucleation and growth of Au(0) produced by slow reduction were mediated by the 1D polymer strands produced by aurophilic interaction to generate Au UNWs. Liz-Marzán and Giersig produced Au UNWs by heating [(oleylamine)Au(I)Cl] complexes and proposed that NW growth was mediated by the initial formation of small nanoparticles, followed by surfactant-directed reduction of the existing nuclei [39].

Understanding of the atomic structure and stability of NWs is critical for their applications in nanotechnology, especially when the diameter of NWs reduces to ultrathin scale (1–2 nm). Yang [40] synthesized Au UNWs using a silane-mediated approach and reported a detailed atomic structure study of the Au UNWs contained large amounts of generalized stacking fault defects. By using aberration-corrected high-resolution transmission electron microscopy (AC-HRTEM), these defects evolve upon sustained electron exposure, and simultaneously the NWs undergo necking and breaking resulting from strain interaction in the breakdown process. In addition, the ligand-like morphology can be observed at the surface of the NWs. The coalescence dynamic of ultrathin Au NWs is demonstrated by *in situ* observations. As the NWs break, the ligands are separated from the metal surface under the irradiation of electron beam; the nanofragments were scattered, fused, and then emerged into together.

5.2.3.2 Critical Size Effect

Nanostructures exhibit unique electronic, optical, and catalytic properties due to size effect, surface effect, and macroscopic quantum tunneling effect [41, 42]. When the size of the nanoparticle is close to or smaller than the optical wave wavelength, de Broglie wavelength, the coherence length of the superconducting state, or the transmission depth, the periodic boundary conditions of the crystal will be destroyed, and the atomic density near the surface layer of the amorphous nanoparticle will be reduced, resulting in a smaller size effect of the photoelectric characteristics. When the size of the nanoparticle enters the nanoscale, these structural characteristics make nanomaterials produce many abnormal physical and chemical properties different from macroscopic objects and individual molecules and atoms within a certain size range. Especially when the scale is reduced to a few layers of atoms, ultrathin NWs, nanojunctions, or nanoclusters will be formed. The original structure and physical properties of condensed matter will no longer be maintained but show many amazing characteristics. For 1D ultrathin metal NW (atomic scale metal NW), especially when the dimension of Au nanostructure is reduced below a critical size (~2 nm), its quantized electronic structure comes into play in its optical and photophysical properties, as well as practical applications in a variety of devices.

Due to the peculiar structural behavior of Au characteristics, ultrathin Au NWs/NRs show a broad band in the NIR spectrum whose peak position was red-shifted with the AR of ultrafine NRs/NWs [37]. Au nanowires have potential applications in nano-electronic devices due to the good conductivity and unique surface plasmon resonance properties. Especially ultrathin NWs not only retain the small size effect of zero-dimensional nanoparticles but also behave the characteristics of single-electron transport and high catalytic activity, which is expected to connect macroscopic catalysts and nanostructures serving as a bridge [38, 43]. For example, Wang [41] successfully synthesized ultrathin Au NWs with a diameter less than 10 nm and studied the electrical conductivity of the circuit constructed by Au NWs. The nonlinear relationship between current and voltage indicated that Au NWs with diameters below 10 nm have good conductivity. Based on *in situ* aberration correction for HRTEM, EDS, and quantitative strain analysis, Wang [43] reported size-dependent grain-boundary structures improving both stabilities of electrical conductivity and mechanical properties in sub-10 nm-sized Au crystals, which provides a new idea for the design of microelectronic interconnect and nanodevices.

5.3 Conductive Films

Materials that are transparent and conductive are essential components for photovoltaics, OLEDs, smart windows, displays, transparent heaters, and transparent antennas. Indium tin oxide (ITO), an n-type semiconductor with a band gap of 4 eV, is the most widely used transparent conductor due to its high transmittance at low sheet resistances (>90% at 10 Ω sq^{-1}). Indium tin oxide films are usually produced with a sputtering process, in which high energy Argon ions bombard and vaporize an ITO target [44]. The sputtering process requires a vacuum, the film coating rates are relatively low (~1 cm s^{-1}), and most of the vapor is deposited on the walls of the chambers instead of on the film. There has also been concern over the cost and availability of indium since it has a similar crustal abundance as Ag, but for a 100 nm-thick ITO film, indium accounts for only 2% of the film cost. This means most of the cost of ITO film is due to the slow and inefficient sputtering process. The brittleness of ITO also means it cannot be used in applications that require flexibility or stretching capability.

Many new transparent, conducting materials have been developed to address these issues with ITO, including conducting polymers, carbon NTs films, graphene, and metal NWs [45, 46]. Among these materials, metal NW is the only material that can both exceed the performance of ITO and be coated from solution at rates >100 times faster than ITO sputtering, thereby eliminating the need for the costly sputtering process. Figure 5.12a,b show a picture and an SEM image of transparent

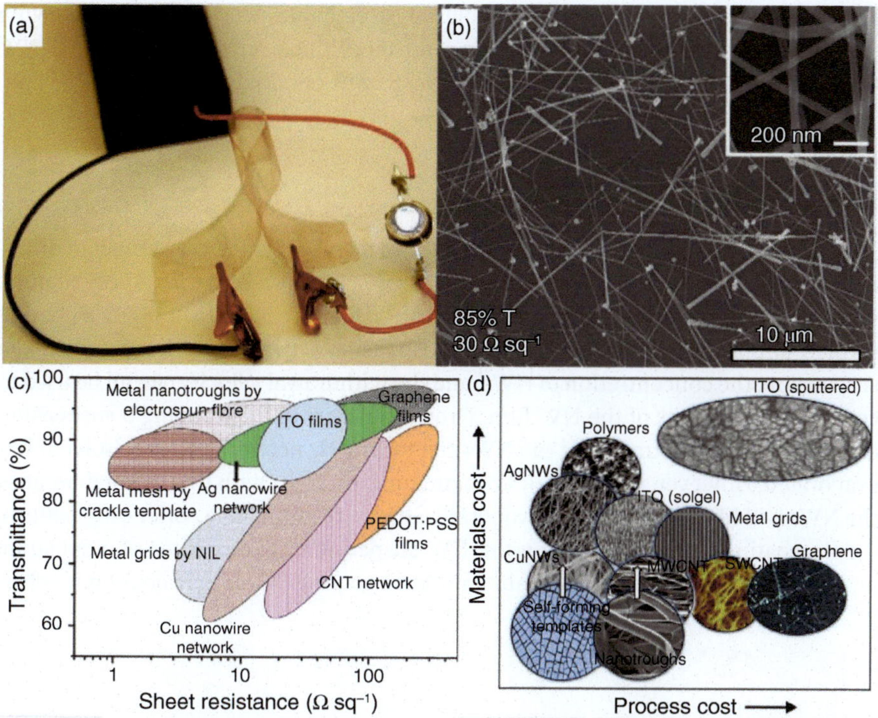

Figure 5.12 (a) A bent Cu NWs film completing an electrical circuit with a battery pack and a LED. (b). SEM image of uniformly dispersed Cu NWs networks from a film with 85% transmittance and 30 Ω sq^{-1} sheet resistance. Source: Rathmell et al. [46]/Reproduced with permission from @2011, JOHN WILEY & SONS, INC. (c) Schematic illustration for the clear trade off between transmittance and sheet resistance of different transparent conducting electrodes. (d) Comparison plot of materials cost versus process cost for different transparent conducting electrodes. Source: Kulkarni et al. [45]/Reproduced with permission from ELSEVIER.

conductors made of Cu and Cu NWs, respectively [46]. The transparent conducting film of Cu NWs on polyethylene terephthalate (PET) is particularly significant because PET is the most widely used substrate for fabrication of flexible, transparent conductive films. The bent film can easily carry enough current to power the LED. The optoelectronic properties of the various transparent conductors in Figure 5.12c is the schematic illustration for the clear tradeoff between transmittance and sheet resistance of different transparent conducting electrodes compared with ITO. Metal networks fabricated through self-forming crackles and electrospun nanotroughs, ultralong Ag nanowires and graphene thin films usually show a minimal decrease in transmittance while sheet resistances were reduced drastically. The cost comparison shown in Figure 5.12d highlights that the processing cost is significantly lower for NWs than for ITO. Their potential to be more cost-effective than ITO while showing comparable performance has motivated research into how to replace ITO with 2D networks of metal NWs in numerous applications. Researchers have also demonstrated that NW networks can be much more flexible and stretchable than

ITO, enabling the development of flexible and stretchable electronics. The following sections discuss the optical, electronic, and mechanical properties of metal NW networks, the methods for fabricating transparent conductors using metal NWs, and applications of the networks.

5.3.1 Preparation Approach

Transparent conducting films can be made from metal NWs using a number of different coating methods, including dip coating, drop casting, spin coating, spray coating, vacuum filtration, and coating with a Mayer rod (Figure 5.13) [47, 48]. Dip coating involves simply dipping a substrate into a suspension of the NWs. The viscosity of the suspension, the concentration of NWs, and the withdrawal velocity can all be used to control the thickness of the NW film. Drop casting involves dropping a suspension of the NWs on a substrate, and the NWs can form a 2D network after the solvent has evaporated. However, it is difficult to obtain uniform films with this method because the NWs will likely move due to evaporative flows, resulting in a coffee-ring pattern.

Spin coating involves dropping the NW suspension at the center of a substrate and spinning the substrate to obtain a uniform coating [47]. Spray coating involves

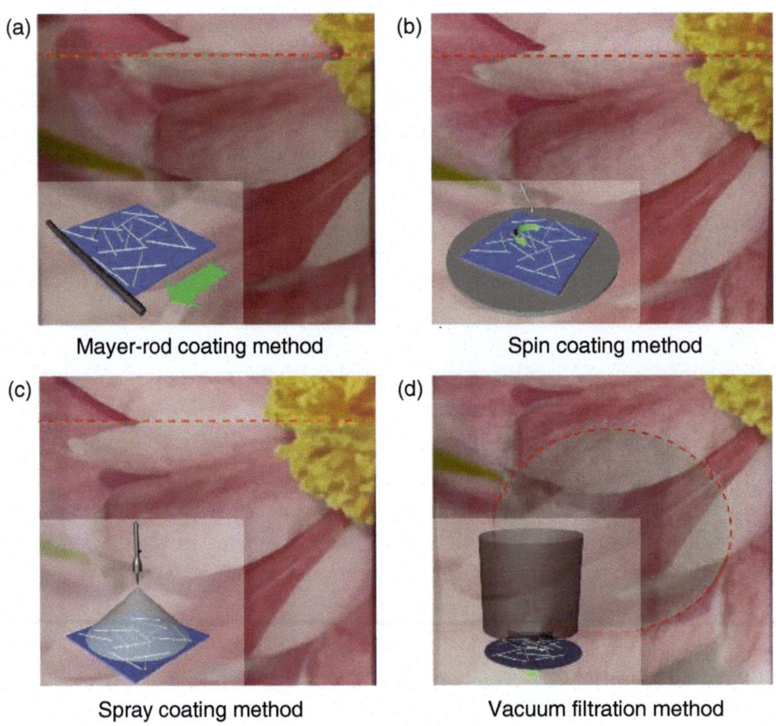

Figure 5.13 Photographs of Ag NW-based films prepared through (a) Mayer-rod coating, (b) spin coating, (c) spray coating, and (d) vacuum filtration. The inset are the corresponding diagrams of these four methods. Source: Zhang et al. [47]/Reproduced with permission from @2022, MDPI/CC BY 4.0.

spraying a NW suspension with an air gun or electrostatic spray. The number of NWs in the coating can be adjusted by the number of spraying and the concentration of the feed solution. NW films can also be formed through vacuum filtration followed by transferring the NWs from the filter to various substrates [48]. The filtration method can easily pattern the NWs by filtering the solution with a patterned mask, and the number of NWs in the film can be known and precisely controlled by changing the concentration of NWs in the feed solution. Mayer rod coating involves spreading a NW suspension across a surface with a wire-wound metering rod (known as a Mayer rod). The thickness of the wet film can be adjusted by using rods wrapped with wires of different sizes. Similar methods, such as doctor blade and slot die coating, are compatible with large-scale, roll-to-roll coating processes. For Ag NW-based FTCFs, these methods are suitable because of the high solubility of Ag NWs in various solvents. However, the Cu NWs-based FTCFs can only be fabricated through the vacuum filtration method, limited by the unstable Cu NWs in solvents.

Zhu et al. provided an approach to obtain Ag NWs-based FTCFs with excellent stability and EMI performance [49]. Figure 5.14a shows a schematic illustration for the fabrication of PET/Ag NW/PMMA FTCFs. First, the PET substrate was treated with a plasma cleaner to make the substrate electrically charged and hydrophilic, increasing the contact force between the Ag NWs and PET. In order to obtain a uniform film, the solvent of the purified Ag NW solution needs to be adjusted to maintain the evaporation rate of the solvent during the subsequent drying process. The mixture of water and isopropanol with the isopropanol content of 80% was found to be suitable as the solvent. Afterward, 0.6 ml Ag NW solution was dipped onto the PET substrate along the Mayer rod direction, followed by spread to cover the substrates by Mayer rod at uniform rate. After the evaporation of solvent in the oven, a uniform Ag NW percolation network formed on the PET substrate. However, the contact between the Ag NWs is too weak to obtain a satisfactory electrical conductivity due to the presence of PVP.

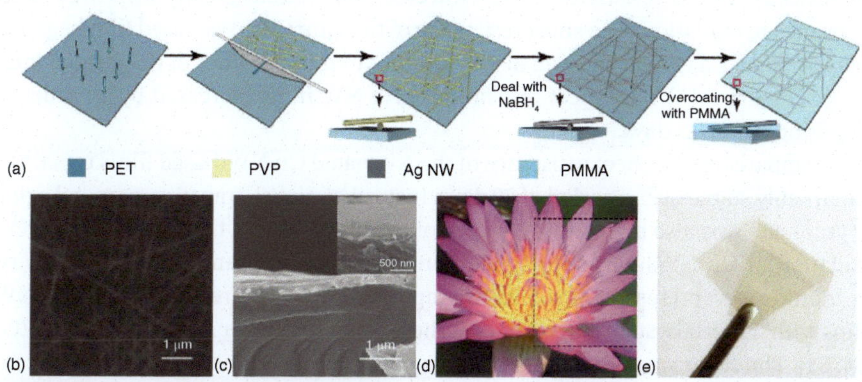

Figure 5.14 (a) Schematic diagram illustrating the fabrication of PET/Ag NW/PMMA FTCFs. (b, c) The surface and cross-sectional SEM images of the PET/Ag NW/PMMA FTCFs, respectively. (d, e) Digital photograph of the PET/Ag NW/PMMA FTCFs. The sample lies in the black dotted frame. Source: Zhu et al. [49]/Reproduced with permission from @2020, Royal Society of Chemistry.

To improve the connection of NWs junctions, the as-prepared Ag NWs film was then dipped into $NaBH_4$ solution to remove interfacial PVP ligands and produce uncovered NW contacts. Finally, 15 wt% PMMA was spin-coated on the Ag NW percolation network to prevent Ag NWs from oxidation and enlarge the contact area of Ag NWs. From the surface and cross-sectional SEM image of the composite samples, the Ag NWs were embedded in the PMMA layer without damaging the initial conductive Ag NW network, as illustrated in Figure 5.14b,c. In addition, the preferential permeation of PMMA into the pores of the Ag NW network would flatten the surface of the Ag NW mesh and improve the adhesion strength of the Ag NW network with the PET substrate. The digital photographs in Figure 5.14d,e showed the high transparency and excellent mechanical flexibility of the PET/Ag NW/PMMA FTCFs.

5.3.2 Stability of Cu Nanowire Film

For practical applications, NW-based transparent electrodes must also be chemically stable in electronic devices. The Ag surface is gradually converted to Ag_2S in the air due to reactions with ambient hydrogen sulfide and carbonyl sulfide [50, 51]. Metallic Cu nanostructures will also be corroded in air through reactions with oxygen and sulfides [52, 53]. Therefore, both Ag and Cu NWs must be protected against chemical corrosion when they were used for transparent conducting films. Methods to improve the stability of Ag and Cu NWs include coating with graphene, coating with shells of corrosion-resistant metals or transparent metal oxides, and coating with poly(methyl methacrylate) or dodecanethiol.

To prevent oxidation of the Cu NWs, Rathmell et al. reported the solution phase synthesis of Ni-coated Cu NWs via reducing Ni^{2+} ions on the surface of Cu NWs with hydrazine (N_2H_4) [52]. Coating Cu NWs with Ni resulted in small losses of transmittance and conductivity due to the enlarged diameter of the NWs and the intrinsically higher resistivity of Ni. However, the coating greatly improved their resistance to oxidation. The sheet resistance of the bare Cu NWs-based film began to increase after 1 day in a dry oven at 85 °C and soared by an order of magnitude after 5 days (Figure 5.15). The sheet resistance of Ag NW films increased by an order of magnitude after 13 days.

In comparison, the sheet resistance of the Ni-coated Cu NW-based films remained remarkably stable over a period of 30 days. Instead of nickel, transparent metal oxide or graphene was also applied to the shell of the Cu NWs and the shells can prevent oxidation of Cu NWs without degradation of the transmittance and conductivity. In addition, several researchers have attempted to cover the pre-synthesized Cu NW films with materials, such as reduced graphene oxide (RGO) or Al-doped ZnO (AZO) [54, 55]. The coverage of RGO and AZO can not only protect the Cu NWs from oxidation but also provide 2D pathways for charge transfer between non-percolated NWs, resulting in high stability and high conductivity.

On the other hand, thermal annealing is the most commonly used in improving the conductivity of the film. Temperatures between 80 and 180 °C are sufficient to fuse Ag NWs together, and the annealing can be performed in the air because of

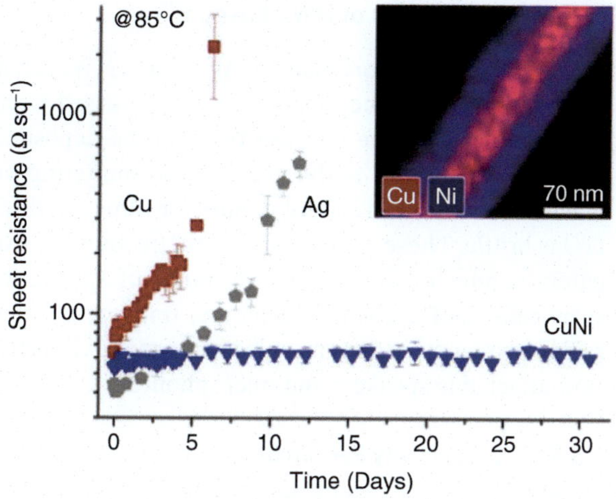

Figure 5.15 Plot of the sheet resistance versus time for films of Ag NWs, Cu NWs, and Ni-coated Cu NWs stored at 85 °C. The inset shows energy-dispersive X-ray spectroscopy images of Ni-coated Cu NWs. Source: Rathmell et al. [52]/Reproduced with permission from @ 2012, American Chemical Society.

the chemical stability of Ag. Cu NWs must be annealed in a reducing atmosphere, such as H_2, to remove the surface oxidation [46]. Annealing could not be used with plastic substrates that deform at the annealing temperature, and thin NWs are easily melted and fragmented into particles during the thermal annealing process. Therefore, the appropriate annealing conditions should be carefully designed to minimize thermal damage to the NWs and the substrate. Another method to improve NWs contacts is mechanical pressing. For example, the mechanical pressing of Ag NWs films can reduce the sheet resistance from 10^4 to $10\,\Omega\,sq^{-1}$ when the pressing pressure is 25 MPa. In the case of Cu NWs, the surface oxide must be removed prior to the mechanical pressing, which can be accomplished by washing the Cu NWs with aqueous H_2Asc. Ag NWs films can also be annealed by exciting a plasmon at the NW junctions with a tungsten halogen lamp at a high-power density. The intense electric field at the junctions results in localized heating and annealing of the NW junctions without damaging the NWs or the substrate.

As a cold-welding method, a plasmonic nanowelding process using a polarized laser was employed for the fabrication of a Cu NWs network-based transparent conductor. It was confirmed that the oxidation of Cu NWs was dramatically suppressed even under ambient conditions due to the localized heating at the NWs junction with rapid laser scanning (<100 μs). Recently, it was also found that by using a focused laser beam, photothermic reduction of Cu_xO NWs to Cu NWs could be induced rapidly at room temperature under ambient conditions without using any vacuum or inert gas environment. Interestingly, this laser photoreduction process is reversible; Cu_xO NWs can turn back into Cu NWs, repetitively. It indicates that oxidized Cu_xO NWs can be recycled as conductive electrodes. As another solution, a chemical treatment using acetic acid or lactic acid has been also developed for reducing the contact resistance between Cu NWs. By simply adding acids into the solution containing NWs or dipping a NW-coated sample in acids at room temperature, the passivating oxide/hydroxide and organic moieties on the surface of Cu NWs can be effectively removed, allowing direct contact between the NWs.

5.3.3 Applications of NWs-based Film

Transparent conductors made of Ag and Cu NWs can now match the optoelectronic properties of ITO. The fabrication of NW-based transparent conductor films can be more cost-effective because they can be deposited from a suspension at high coating rates relative to the vapor-phase sputtering process for ITO. Furthermore, metal NW-based transparent conductor films are flexible and stretchable, whereas ITO is brittle. These advantages are motivating the study of metal NW-based transparent conductor films in a wide range of applications, including touch screens, solar cells, OLEDs, supercapacitors, transparent memory, air filters, transparent radio frequency antennas, electromagnetic interference (EMI) shielding, and skin-attachable speakers and microphones.

5.3.3.1 Conductivity Performance

Optical transmittance and sheet resistance are two most important factors in estimating the optoelectrical properties of transparent conductive films. Figure 5.16a

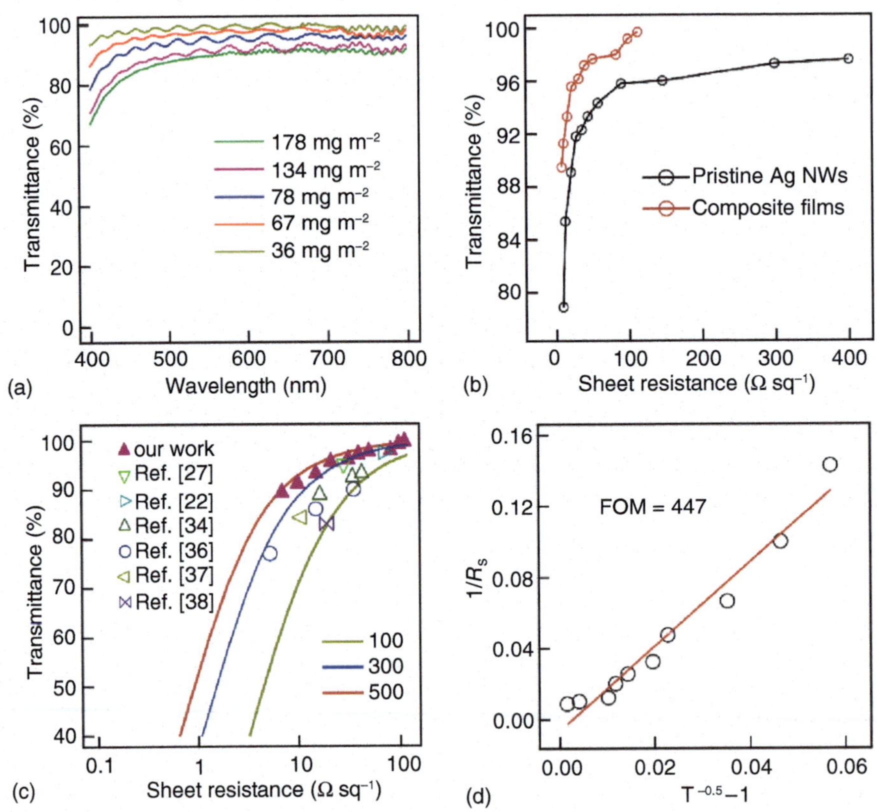

Figure 5.16 (a) Optical transmittance spectra of the PET/Ag NW/PMMA FTCFs with various Ag NW area densities. (b) Plots of optical transmittance ($\lambda = 550$ nm) versus sheet resistance for pristine Ag NW films and the PET/Ag NW/PMMA FTCFs. (c) A plot of the optical transmittance ($\lambda = 550$ nm) versus sheet resistance for PET/Ag NW/PMMA FTCFs. The solid line indicates the fit to Eq. (5.1) for $\sigma_{DC}/\sigma_{op} = 100$ (dark yellow), 300 (blue), or 500 (red). The previously reported values are also shown for comparison. (d) A plot of $1/R_s$ as a function of $(T^{-0.5} - 1)$ for the PET/Ag NW/PMMA FTCFs. Source: Reproduced from Zhu et al. [49]/with permission of Royal Society of Chemistry.

shows the optical transmittance of the obtained PET/Ag NW/PMMA FTCFs produced by our group with different area densities of Ag NWs. The transmittance of the FTCFs is decreased with the increase of Ag NW area density. One explanation for this phenomenon is that Ag NWs with high area density can give rise to a high density of Ag NW mesh. The decreased "open area" thus reduced the transmittance of incident light. Besides, more free electrons were excited from bound electrons by absorbing light, which contributes to the increase of the absorption of the thin film and thus the decrease in transmittance. The optical transmittance of the PET/Ag NW/PMMA FTCFs remains constant over the entire visible region. As shown in Figure 5.16b, the as-prepared PET/Ag NW/PMMA FTCFs have a great improvement in optoelectrical performance in comparison with pristine Ag NW films. The transmittance of the PET/Ag NW/PMMA FTCFs can reach 99.7% at a sheet resistance of $112\,\Omega\,\text{sq}^{-1}$ and still maintain a transmittance of 89.5% at the sheet resistance of $7\,\Omega\,\text{sq}^{-1}$, which is much better than those of commercial ITO ($50\,\Omega\,\text{sq}^{-1}$, 95%) [56].

The optoelectrical performance of the PET/Ag NW/PMMA FTCFs can be calculated by the FOM that is expressed by the following equation:

$$T(\lambda) = \left(1 + \frac{Z_0}{2R_s}\frac{\sigma_{op}(\lambda)}{\sigma_{DC}}\right)^{-2} \tag{5.1}$$

where λ is the wavelength, $T(\lambda)$ is the transmittance at λ nm, Z_0 is the wave impedance of free space (377 Ω), $\sigma_{op}(\lambda)$ is the optical conductivity at λ nm, and σ_{DC} is the direct current conductivity of the film. The FOM is expressed as the ratio of the direct current conductivity to the optical conductivity ($\sigma_{DC}/\sigma_{op}(\lambda)$). Higher FOM indicates a better optoelectrical performance of the films. The optical transmittance versus sheet resistance for PET/Ag NW/PMMA FTCFs are presented in Figure 5.16c, as well as the results for the previously reported transparent conductive films [57, 58]. All measured values are within the range of FOM = 300–500, which are at the same tier as the best performance reported in the literature and much higher than the requirements of practical applications. To further evaluate the optoelectrical performance of the PET/Ag NW/PMMA FTCFs, we calculated the FOM according to the variant form of Eq. (5.1) as follows:

$$\frac{1}{R_s} = \text{FOM} \cdot \frac{T(\lambda)^{-0.5} - 1}{188.5} \tag{5.2}$$

Figure 5.17d plots $1/R_s$ as a function of $(T^{-0.5} - 1)$ for the varied Ag NWs-based films, and the FOM value can be calculated from the slope of the linear fitting. The PET/Ag NW/PMMA FTCFs achieve a FOM value up to 447, which is better than those for the reported carbon NT, graphene, polymer, Cu NW, and Ag NW-based transparent conductive films [59]. Such a high FOM is attributed to both the treatment of $NaBH_4$ and the spin-coating of the PMMA. The excellent optoelectrical performance assures the PET/Ag NW/PMMA FTCFs with great potential as high-performance transparent EMI shielding materials.

Based on the high conductivity of the NWs-based conductors, touch screen panels consisting of two parallel metal NW-based transparent conductors have been reported by several groups. The two transparent conductors were separated by a spacer or optically clear adhesive, while the driving and sensing lines were fabricated by metal evaporation or screen printing. In addition to touch screens that can detect the location of touch, a force-sensitive touch screen was recently reported [60].

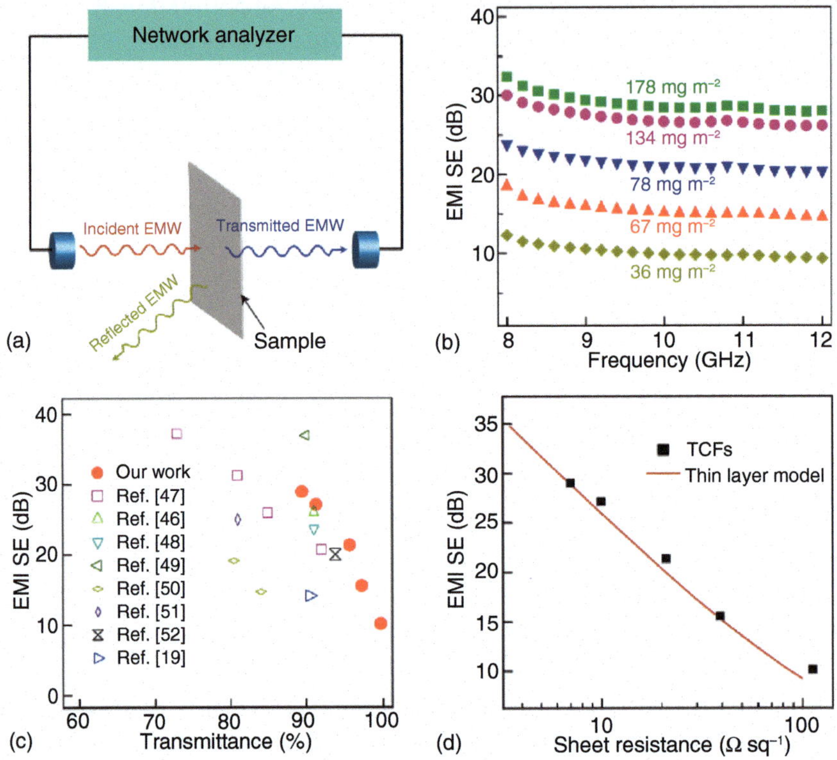

Figure 5.17 (a) Schematic illustrating the synthetic process of Au NBP@AgPt hollow nanostructures. (b–e) TEM images of the Au NBP@AgPt nanostructure samples produced at 30, 50, 70, and 90 °C, respectively. (f) Extinction spectra of the Au NBP sample and the Au NBP@AgPt nanostructure samples produced at different temperatures. (g) Variation of the longitudinal dipolar plasmon wavelength as a function of the varied temperature. The point at 0 °C represents the Au NBP sample. Source: Reproduced from Zhu et al. [49]/with permission of Royal Society of Chemistry.

A mechanochromic spiropyran-PDMS composite film was added to a touch screen made of Ag NWs. A touching force between 5 and 25 N was measured by analyzing the color change with a spectroradiometer. This enabled the touch screen to measure the distribution of writing forces used to draw a single letter, potentially enabling screens that can verify the authenticity of personal signatures using such force distributions.

Another promising application of transparent conductors based upon metal NWs is for OLEDs [61–63]. Replacing ITO with a metal NW-based transparent conductor enables the fabrication of an OLED that is flexible and stretchable and can be potentially manufactured with roll-to-roll fabrication. The primary technical hurdle for using metal NWs as the transparent electrode in an OLED is the higher surface roughness of a metal NW-based transparent conductor than ITO, which can lead to the formation of electrical shorts and a decrease in the efficiency of the OLED. Weak adhesion between metal NWs and the substrate can also reduce the mechanical stability of the devices. Similar solutions to those discussed for solar cells were employed to address these problems, including embedding metal NWs in a conductive polymer and coating of an additional buffer layer. In one example, Ag NWs were coated on a glass substrate, followed by overcoating and polymerization of a vinyl monomer. The composite electrode could then be peeled off from the substrate to reveal the NWs embedded in a smooth polymer surface. The NW composite electrodes exhibited moderate improvements in their maximum current efficiency and a 20–50% higher luminance relative to OLEDs fabricated using ITO. OLEDs produced with the composite Ag NW electrode largely retained their performance after 100 bending cycles.

Metal NW-based transparent conductors also enabled the fabrication of a flexible/transparent supercapacitor. However, bare Cu and Ag NWs are not stable because they are easily oxidized electrochemically at the voltages commonly used for a supercapacitor. In one example, Ag NWs were protected with an Au sheath with a thickness between nm. The Au sheath prevented the oxidation of Ag NWs during 500 cycles of charging/discharging. Subsequent work reported that a coating of polypyrrole on the Ag@Au core-sheath NW electrode improved the specific capacitance and created a transparent/bendable/stretchable pseudo-capacitor. WO_3 particles were also combined with an Ag NW transparent conductor to create a photoresponsive electrochromic supercapacitor. In this case, Ag NWs were deposited on cellulose nanofibers, followed by coating with RGO and WO_3. The performance of the electrochromic supercapacitor only dropped by 12.6% after 20 000 cycles and could retain its performance after bending. Such an electrochromic supercapacitor can power changes in coloration in mobile and wearable applications.

5.3.3.2 Electromagnetic Shielding

With the rapid development of electronic technology and the increasing utilization of electronic devices in our life, electromagnetic waves (EMWs) emitted by electronic equipment may affect the precision of sophisticated electronic equipment and even cause damage to the human health. Therefore, it is essential to apply EMI shielding materials in some electronic devices as detectors or sensors for space exploration, military, medical facilities, and civilian utilities [64]. As a conventional EMI shielding material, ITO has been used in many commercial EMI shielding applications for a long time, while it is limited by the abundance of indium and the brittleness of products. To meet the requirements of future applied electronic equipment, including transparency, stability, and flexibility, tremendous efforts have been devoted to the fabrication of EMI shielding materials, such as carbon-based materials, conductive polymer composites, and metal NWs. Typically, metal NWs, especially Ag NWs, show better performance in optoelectronic properties and flexibility in comparison with other EMI shielding materials. For example, the EMI shielding effectiveness (SE) and optical transmittance of most carbon-based materials were lower than the expected (15 dB @ 80%) [65]. Besides, the fabrication of the highly homogeneous carbon NTs or graphene is limited to small scale. Therefore, Ag NWs with high purity and large AR have attracted much attention.

Yun et al. reported a hierarchical graphene/Ag-mesh film for optically transparent EMI shielding, which achieved an EMI SE of 14.12 dB with a transmittance of 90.5% [66]. Hu et al. introduced a poly(ethersulfone) layer onto Ag NW films to prevent the shedding and corrosion of Ag NWs, with the resulting poly(ethersulfone)/Ag NW/polyethylene terephthalate (PET) sandwich-structured film reaching 23 dB at a transmittance of 81% [67]. Kim et al. prepared a polymer-coated/RGO/Ag NW EMI shielding film via a liquid-to-vapor pressure-assisted wet sintering method, which improved the NW–NW junctions, and the resulting EMI shielding film showed an EMI SE of 24 dB at a transmittance of 85% [68]. Although the performance of these EMI shielding films can be clearly improved by the posttreatment, it is still difficult to achieve SE of 20 dB and transmittance of 95% simultaneously for these composite films. Recently, our group achieved an EMI SE of 21 dB at a transmittance of 94% by using a polymer to deal with the Ag NW-based EMI shielding film [69]. However, the glass substrate limited the further application in flexible electronic devices. Besides, the stability of the Ag NW-based films in harsh environments with high temperature and humidity is unknown.

Kan's group provides an approach to obtain Ag NWs-based FTCFs with excellent stability and EMI performance [49]. The PET/Ag NW/PMMA FTCFs with excellent optoelectrical properties and great stability provide great potential as high-performance flexible transparent EMI shielding materials. The experimental setup for the EMI shielding measurement of the prepared PET/Ag NW/PMMA FTCFs is shown in Figure 5.17a. The network analyzer is connected to the X-band waveguide by the coaxial cables, and the sample is inserted between the waveguides. The total EMI SE, SE_T, is always used for evaluating shielding performance and can be defined as follows:

$$SE_T = 10\log(P_i/P_t) \tag{5.3}$$

where P_i is the incident power and P_t is the transmission power of the EMW. As shown in Figure 5.17b, the average EMI SE of the PET/Ag NW/PMMA FTCFs in the frequency range from 8 to 12 GHz was measured to be 10.2 dB at the lowest area density of 36 mg m^{-2} (optical transmittance at 99.7%). The EMI SE of the PET/Ag NW/PMMA FTCFs is obviously improved with the increase of the Ag NW area density. As the area density of Ag NW increased to 67, 78, 134, and 178 mg m^{-2}, the PET/Ag NW/PMMA FTCFs yield EMI SEs of 15.5, 21.3, 27.1, and 29 dB, with optical transmittances of 97.2%, 95.6%, 91.3%, and 89.5%, respectively. It is known that the EMI SE required for practical applications, such as mobile phones and laptop computers, is approximately 20 dB. The EMI SE value combined with high optical transmittance of the PET/Ag NW/PMMA FTCFs is sufficient to meet the requirement of the commercial transparent EMI shielding applications (20 dB @ 90%) [70].

The transparent Ag NW layer is the core material for effective EMI shielding. The EMI SE exhibits a strong dependence on the electrical performance of the Ag NW-based composite films. Similar to electrical conductivity, the electromagnetic shielding performance of a thin film can also be characterized as follows:

$$SE = 10\log\left(\frac{1}{4}\left[\begin{array}{c}\left(\frac{\sigma}{2\omega\varepsilon_0}\right)\left[\cosh\left(\frac{2d}{\delta}\right)-\cos\left(\frac{2d}{\delta}\right)\right]\\ +2\sqrt{\frac{\sigma}{2\omega\varepsilon_0}}\left[\sinh\left(\frac{2d}{\delta}\right)+\sin\left(\frac{2d}{\delta}\right)\right]\\ +2\left[\cosh\left(\frac{2d}{\delta}\right)+\cos\left(\frac{2d}{\delta}\right)\right]\end{array}\right]\right) \tag{5.4}$$

where σ is the conductivity, ω is the angular frequency, ε_0 is vacuum dielectric constant, μ $(= \mu_0\mu_r)$ is the absolute magnetic permeability, d is the thickness, and δ is the skin depth of the conductive materials. Because the thickness of Ag NW is much less than the skin depth, Eq. (5.4) can be reduced to:

$$SE = 20\log\left(1+\frac{Z_0}{2R_s}\right) \tag{5.5}$$

where Z_0 (377 Ω) is the wave impedance of free space. As plotted in Figure 5.18d, the theoretically calculated EMI SE of the PET/Ag NW/PMMA FTCFs increases with the decreased R_s and fits well with the experimental results.

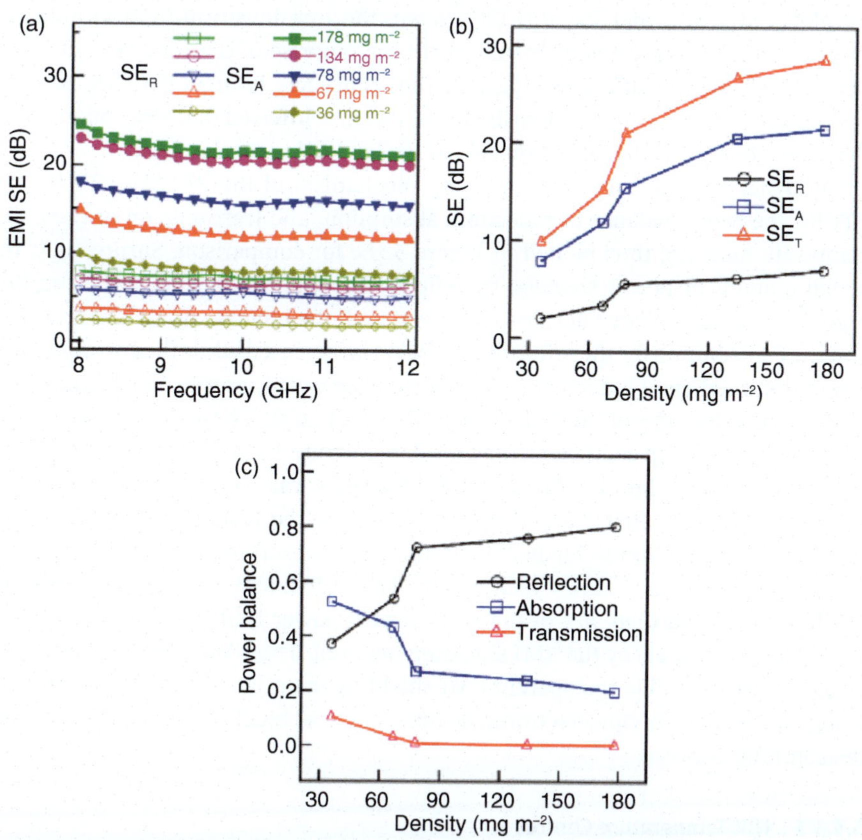

Figure 5.18 (a) SE$_R$ and SE$_A$ of the PET/Ag NW/PMMA FTCFs at different Ag NW area densities. (b) Three different EMI SEs of the PET/Ag NW/PMMA FTCFs at frequency of 10 GHz as a function of Ag NW area density. (c) Power balance of the PET/Ag NW/PMMA FTCFs at frequency of 10 GHz as a function of Ag NW area density. Source: Reproduced from Zhu et al. [49]/with permission of Royal Society of Chemistry.

In the EMI shielding theory, the SE_T is comprised of reflection (SE_R), absorption (SE_A), and multiple reflection (SE_M) components. The contribution of multiple reflections can be ignored due to the much thinner active layer than skin depth. Therefore, the SE_T can be expressed as:

$$SE_T = SE_R + SE_A = 10\log\left(\frac{1}{T}\right) \tag{5.6}$$

SE_R and SE_A can be calculated from the following equations:

$$SE_R = -10\ \log\left(\frac{T}{1-R}\right) \tag{5.7}$$

$$SE_A = -10\ \log(1-R) \tag{5.8}$$

Figure 5.18a shows the SE_R and SE_A of the PET/Ag NW/PMMA FTCFs with different Ag NW area densities. Both of SE_R and SE_A increased as the increase of the Ag NW area densities. This situation can be attributed to the increase of the free electrons in the composite films that can interact with the incident EMWs. Moreover, the plot of SE_R, SE_A, and SE_T at 10 GHz versus the area density of Ag NWs indicates that SE_A and SE_T have a similar trend with the increase of the Ag NW area densities, with SE_R a little difference (Figure 5.18b). Since SE_A is larger than SE_R over the frequency range for all five composite films, it can be inferred that absorption rather than reflection plays the dominant role in EMI shielding mechanism.

To further ascertain the EMI shielding mechanism of the PET/Ag NW/PMMA FTCFs, the power balance of reflection, absorption, and transmission of these five composite films are thus plotted in Figure 5.18c for comparison. Surprisingly, the actual amount of power shielded by reflection is higher than that by absorption except at low Ag NW area density. Moreover, with the increase of Ag NW area density, the contribution for the reflection of EMWs is increased while the corresponding absorption coefficient is decreased, which could be mainly ascribed to the improved electrical conductivity of the PET/Ag NW/PMMA FTCFs. These results on different EMI SEs and power coefficients clearly demonstrate the electromagnetic shielding mechanism of the PET/Ag NW/PMMA FTCFs, that is, a large portion of the incident EMWs are first blocked by reflection and the residual EMWs penetrate into the composite film to further be attenuated by absorption. For example, when the SE_R and SE_A are 10 dB, which means 90% of incident EM power is reflected (10 dB of SE_R) and then 90% of the nonreflected power is absorbed (10 dB of SE_A, 9% of incident power) by the EMI shielding material. The NWs-based FTCFs would be a promising flexible transparent EMI shielding material with great potential for using in a variety of optoelectronic devices, such as touch screens, displays, and telecommunications.

5.3.3.3 NIR Temperature Control

Heat management of electronic devices is essential for preventing their failure. A heat sink with a large surface area and high thermal conductivity is typically placed in contact with an electronic heat source, such as a central processing unit, to facilitate heat dissipation [71]. Air can be trapped between the rough surfaces of a heat source and heat sink, thereby reducing heat transfer, so a thermal interface material is usually applied to improve the thermal conductivity of this interface [72]. Common thermal interface materials include thermal grease, solder, and particle-laden adhesive tapes. Metal NWs elastomer composites have also been evaluated for use as thermal interface materials. The ability of metal NWs in percolating networks at low volume fractions means they can be added into an elastomer at volume fractions less than 2% and achieve the desired level of thermal conductivity. In contrast, spherical particles must be used at a volume fraction between 10% and 50%. Such a high loading causes the elastomer composite to become brittle and crack under pressure rather than conform to the interface between the heat sink and source.

A variety of effective medium models have been developed to account for the effect of a filler's AR, orientation, and Kapitza resistance (interfacial thermal resistance) on the thermal conductivity of composites. Similar to the case of electrical conductivity, the thermal conductivity of a NWs composite is predicted to increase with the AR, although the increase levels out at an AR of c. 1000 for a volume fraction of 1.8%. Experimental results qualitatively confirmed these predictions. As one comparison, a dimethicone composite containing 10 vol% of Cu NWs with an AR of ~22 exhibited a thermal conductivity 1.6 times greater than that of a composite containing Cu nanoparticle at the same volume fraction. Another study compared the thermal conductivity of composites containing short (<20 μm in length) and long (few tens to hundreds of micrometers) Cu NWs with a diameter of ca. 80 nm. The short NWs were produced by sonicating the long NWs. The addition of short NWs into polyacrylate increased the thermal conductivity by 350% at a volume fraction of 1.5%, whereas the long NWs increased the thermal conductivity by 1350% at a volume fraction of 0.9%.

Metal NWs networks can also be used to generate heat through the passage of electrical current, that is, Joule heating [73]. A transparent heater fabricated with metal NWs can be used for defogging windows or for a thermochromic display [74]. Since a metal NW-based composite can also be flexible and stretchable, it can be used for the fabrication of wearable heaters for personal thermal management or thermal therapy. Residential heating consumes 42% of global energy, much of which is wasted to maintain the temperature of empty space. An alternative approach would be to create a wearable thermal management device that could maintain the comfort level of building occupants at lower indoor temperatures. It was demonstrated that cloth impregnated with Ag NWs could be used for personal thermal management by passively blocking IR radiation from the body, as well as by actively heating the body through Joule heating [75]. Carbon NTs could be used for active heating but not IR reflection because of their low thermal emissivity. Figure 5.19 (left) shows IR images of gloves made of carbon NTs cloth and Ag NW cloth, respectively. The Ag NW-based glove appears to be several degrees colder in the image because of its low emissivity. The addition of Ag NWs to the cloth improved its thermal insulation by 21% due to the reduction of radiative loss. One sample of Ag NWs-based cloth with a size of 1 in. × 1 in. could also be heated from room temperature to 55 °C by applying only 1.2 V, whereas a carbon NTs-based cloth required a bias of 12 V to reach the

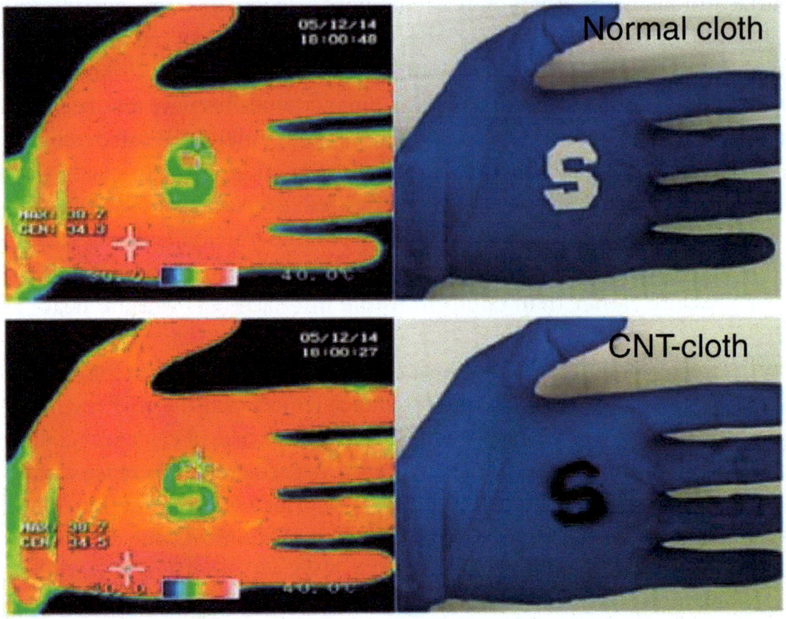

Figure 5.19 Thermal IR images (left) and photographs (right) of gloves fabricated using carbon nanotubes-based cloth (top) and Ag NW-based cloth (bottom), respectively. Source: Hsu et al. [75]/Reproduced with permission from American Chemical Society.

same temperature. Thus, the higher conductivity of Ag NWs could enable them to be used as more efficient heaters that use smaller, low-voltage batteries. Wearable devices made of Ag NWs are permeable to air and moisture, ensuring they are as comfortable as normal cloth.

Thermal therapy is another potential application for stretchable, conductive NW composites. Thermal therapy is often prescribed to alleviate the symptoms of painful or swollen joints, but current heat wraps are heavy, bulky, and inconvenient and do not maintain a desired temperature set point. Electrically programmable heaters on polyimide substrates cannot conform to the human body. To address this problem, researchers suspended Ag NWs in a copolymer SBS-based elastomer and molded the elastomer to create a stretchable, conductive serpentine mesh. This heater was connected to a battery and microcontroller to serve as a programmable heater for the wrist (Figure 5.19 (right)) [73]. An alternative approach to the construction of a stretchable heater is to start by creating stretchable, conductive yarn with Cu NWs. A helical yarn composed of polyester microfibers was coated with Cu NWs and then silicone by dip coating. A wearable heater was created by weaving the conductive threads through the fabric, and its temperature could be controlled with a phone via Bluetooth. The fabrics created using this conductive yarn were washable, breathable, and highly deformable.

5.4 Conclusion and Remarks

Over the last two decades, 1D metal nanostructures have a staggering amount of progress in synthesis, mechanistic understanding, processing, and application. Ag, Cu and Au NWs have garnered great interest as promising next-generation conducting electrodes due to their high conductivity, mechanical flexibility, and cost-effectiveness. As a result, ongoing efforts to develop a facile approach that allows the production of high-quality NWs in large quantities and their utilization in various electronic applications are being made. Numerous methods, such as wet chemical approaches, are considered to be facile, scalable strategies to produce metal NWs. For each technique, the growth mechanisms, properties of the obtained NWs, and key factors to determine the yield or quality of the products were essential to the applications of NWs. Concurrently, continuing efforts are required to overcome the inherent surface oxidation of NWs and facilitate controllability of the NWs, large-scale production on the kilogram scale, and cost-effective coating and patterning techniques before they can be widely used on the industrial scale. Such investigations may advance the realization of future electronics that can be transparent, flexible, and even stretchable.

Ag, the king among plasmonic materials, features low inelastic absorption in the visible-infrared (vis-IR) spectral region compared to other metals. In contrast, Cu is commonly regarded as too lossy for actual applications. Cu can substitute Ag in many applications due to the low cost and specific electrical resistance. However, in the nanometer structures, compared to those of the precious metals, Cu micro/nanostructures are attractive both from economic and especially catalysts in industrial points of view, although nanometer-sized Cu is easily oxidized during synthesis and subsequent storage.

Cu, due to the low cost and specific electrical resistance, can substitute Ag in many applications. Compared to those of the precious metals, Cu micro/nanostructures are attractive both from economic and especially catalysts from industrial points of view. But nanometer-sized Cu is easily oxidized during synthesis and subsequent storage. Nevertheless, Cu is emerging as an attractive, cheap material platform for high-quality plasmonics in elongated nanostructures. More recently, Mkhitaryan characterized Cu plasmons facilitated by EELS with nanometer spatial resolution [76]. The measurements show that elongated Cu nanostructures can sustain spectrally sharp vis-IR plasmons of similar quality factors. More precisely, they used high-resolution EELS to map and spectrally characterize plasmons in both Cu and Ag NWs. Cu NWs display plasmons at lower energies. More strikingly, Cu plasmons show a more retarded behavior than Ag plasmons; and EELS measurements are in excellent agreement with electromagnetic simulations.

Ag NWs have been researched extensively for use in flexible, transparent conducting thin films, which find application in touch screen panels, flexible displays, and solar cells. Significant research is also underway to facilitate their use in flexible and transparent heaters, artificial skin, smart windows, and implantable health-monitoring applications. Cu is slightly less conductive than Ag, but it is more earth-abundant and less expensive. Thus, Cu is seen as an alternative to Ag in many of the applications mentioned above.

References

1 Duan, X.F., Huang, Y., Agarwal, R., and Lieber, C.M. (2003). Single-nanowire electrically driven lasers. *Nature* 421: 241–245.

2 Wang, Z.L. and Song, J.H. (2006). Piezoelectric nanogenerators based on zinc oxide. *Science* 312: 242–246.

3 He, Z., Yang, Y., Liang, H.-W. et al. (2019). Nanowire genome: a magic toolbox for 1D nanostructures. *Adv. Mater.* 31: 1902807.

4 Yang, P.D., Yan, R.X., and Fardy, M. (2010). Semiconductor nanowire: what's next? *Nano Lett.* 10: 1529–1536.

5 Wang, J.-L., Hassan, M., Liu, J.-W., and Yu, S.-H. (2018). Nanowire assemblies for flexible electronic devices: recent advances and perspectives. *Adv. Mater.* 30: 1803430.

6 Wei, H., Pan, D., Zhang, S. et al. (2018). Plasmon waveguiding in nanowires. *Chem. Rev.* 118: 2882–2926.

7 Tian, B. and Lieber, C.M. (2019). Nanowired bioelectric interfaces. *Chem. Rev.* 119: 9136–9152.

8 Yao, J., Yan, H., Das, S. et al. (2014). Nanowire nanocomputer as a finite-state machine. *Proc. Natl. Acad. Sci. U. S. A.* 111: 2431–2435.

9 Bai, Y., Zhang, R., Ye, X. et al. (2018). Carbon nanotube bundles with tensile strength over 80 GPa. *Nat. Nanotechnol.* 13: 589–595.

10 Alkilany, A.M., Lohse, S.E., and Murphy, C.J. (2013). The gold standard: gold nanoparticle libraries to understand the nano-bio interface. *Acc. Chem. Res.* 46: 650–661.

11 Sun, Y., Mayers, B., Herricks, T., and Xia, Y. (2003). Polyol synthesis of uniform silver nanowires: A plausible growth mechanism and the supporting evidence. *Nano Lett.* 3: 955–960.

12 Sun, Y. and Xia, Y. (2002). Large-scale synthesis of uniform silver nanowires through a Soft, self-seeding, polyol process. *Adv. Mater.* 14: 833–837.

13 Niu, Z., Cui, F., Kuttner, E. et al. (2018). Synthesis of silver nanowires with reduced diameters using benzoin-derived radicals to make transparent conductors with high transparency and low haze. *Nano Lett.* 18: 5329–5334.

14 da Silva, R.R., Yang, M., Choi, S.-I. et al. (2016). Facile synthesis of sub-20 nm silver nanowires through a bromide-mediated polyol method. *ACS Nano* 10: 7892–7900.

15 Zhu, X.Z., Guo, A.Q., Xu, J., and Kan, C.X. (2020). The synthesis of silver nanowires with tunable diameters using halide ions for flexible transparent conductive films. *CrstEngComm* 22: 8421–8429.

16 Wu, F., Wang, W.H., Xu, Z.F., and Li, F.L. (2015). Bromide (Br)-based synthesis of Ag nanocubes with high-yield. *Sci. Rep.* 5: 10772.

17 McKiernan, M., Zeng, J., Ferdous, S. et al. (2010). Facile synthesis of bimetallic Ag/Ni core/sheath nanowires and their magnetic and electrical properties. *Small* 6: 1927–1934.

18 Ma, J., Wang, K., and Zhan, M. (2003). Growth mechanism and electrical and magnetic properties of Ag-Fe_3O_4 core-shell nanowires. *ACS Appl. Mater. Interfaces* 7: 16027–16039.

19 Wang, S., Yu, Y., Zuo, Y. et al. (2012). Synthesis and photocatalysis of hierarchical heteroassemblies of ZnO branched nanorod arrays on Ag core nanowires. *Nanoscale* 4: 5895–5901.

20 Kim, T., Canlier, A., Cho, C. et al. (2014). Highly transparent Au-coated Ag nanowire transparent electrode with reduction in haze. *ACS Appl. Mater. Interfaces* 6: 13527–13534.

21 Sun, Y. and Xia, Y. (2004). Mechanistic study on the replacement reaction between silver nanostructures and chloroauric acid in aqueous medium. *J. Am. Chem. Soc.* 126: 3892–3901.

22 Sun, Y. and Xia, Y. (2004). Multiple-walled nanotubes made of metals. *Adv. Mater.* 16: 264–268.

23 Na, W., Lee, J., Jun, J. et al. (2019). Highly sensitive copper nanowire conductive electrode for nonenzymatic glucose detection. *J. Ind. Eng. Chem.* 69: 358–363.

24 Chee, S.S., Kim, H., Son, M., and Ham, M.H. (2020). Aspect ratio control of copper nanowire via solution process and its flexible transparent conductive electrode applications. *Electron. Mater. Lett.* 16: 404–410.

25 Zhang, D., Wang, R., Wen, M. et al. (2012). Synthesis of ultralong copper nanowires for high-performance transparent electrodes. *J. Am. Chem. Soc.* 134: 14283–14286.

26 Yang, H.J., He, S.Y., and Tuan, H.Y. (2014). Self-seeded growth of five-fold twinned copper nanowires: mechanistic study, characterization, and SERS applications. *Langmuir* 30: 602–610.

27 Guo, H., Lin, N., Chen, Y. et al. (2013). Copper nanowires as fully transparent conductive electrodes. *Sci. Rep.* 3: 1–8.

28 Cui, F., Dou, L., Yang, Q. et al. (2017). Benzoin radicals as reducing agent for synthesizing ultrathin copper nanowires. *J. Am. Chem. Soc.* 139: 3027–3032.

29 Bhanushali, S., Ghosh, P., Ganesh, A., and Cheng, W. (2015). 1D Copper nanostructures: progress, challenges and opportunities. *Small* 11: 1232–1252.

30 Fu, Q.Q., Li, Y.D., Li, H.-H. et al. (2019). In-situ seed-mediated high-yield synthesis of copper nanowires on large scale. *Langmuir* 35: 4364–4369.

31 Kim, M.J., Alvarez, S., Chen, Z. et al. (2018). Single-crystal electrochemistry reveals why metal nanowires growth in solution. *J. Am. Chem. Soc.* 140: 14740–14746.

32 Xu, J., Zhu, X.Z., Xia, S.H. et al. (2022). Au-catalyzed ultrathin copper nanowires. *J. Mater. Chem. A* 10: 25431–25436.

33 Wang, H., Wu, C., Huang, Y. et al. (2016). One-pot synthesis of superfine core-shell Cu@metal nanowires for highly tenacious transparent LED dimmer. *ACS Appl. Mater. Interfaces* 8: 28709–28717.

34 Zhang, Y., Zhou, N., Zhang, K., and Yan, F. (2017). Plasmonic copper nanowire@TiO_2 nanostructures for improving the performance of dye-sensitized solar cells. *J. Power Sources* 342: 292–300.

35 Chen, H., Tu, T., Wen, M., and Wu, Q. (2015). Assembly synthesis of Cu_2O-on-Cu nanowires with visible-light-enhanced photocatalytic activity. *Dalton Trans.* 44: 15645–15652.

36 Li, X.K., Gittleson, F., Carmo, M. et al. (2012). Scalable fabrication of multifunctional freestanding carbon nanotube/polymer composite thin films for energy conversion. *ACS Nano* 6: 1347–1356.

37 Halder, A. and Ravishankar, N. (2007). Ultrafine single-crystalline gold nanowire arrays by oriented attachment. *Adv. Mater.* 19: 1854–1856.

38 Huo, Z.Y., Tsung, C.K., Huang, W.Y. et al. (2008). Sub-two nanometer single crystal Au nanowires. *Nano Lett.* 8: 2041–2044.

39 Pazos-Perez, N., Baranov, D., Irsen, S. et al. (2008). Synthesis of flexible, ultrathin gold nanowires in organic media. *Langmuir* 24: 9855–9860.

40 Yu, Y., Cui, F., Sun, J.W., and Yang, P.D. (2016). Atomic structure of ultrathin gold nanowires. *Nano Lett.* 16: 3078–3084.

41 Wang, C., Hu, Y.J., Lieber, C.M., and Sun, S.H. (2008). Ultrathin Au Nanowires and Their Transport Properties. *J. Am. Chem. Soc.* 130: 8902–8903.

42 Zhang, D., Du, Y., Yang, C. et al. (2021). Tuning plasmonic nanostructures in graphene-based nano-sandwiches using ultraviolet/ozone functionalization. *J. Mater. Sci.* 56: 1359–1372.

43 Wang, C.Y., Du, K., Song, K.P. et al. (2018). Size-dependent grain-boundary structure with improved conductive and mechanical stabilities in sub-10-nm gold crystals. *Phys. Rev. Lett.* 120: 186102.

44 Ye, S., Rathmell, A.R., Chen, Z. et al. (2014). Metal nanowire networks: the next generation of transparent conductors. *Adv. Mater.* 26: 6670–6687.

45 Kulkarni, G.U., Kiruthika, S., Gupta, R., and Rao, K.D.M. (2015). Towards low cost materials and methods for transparent electrodes. *Curr. Opin. Chem. Eng.* 8: 60–68.

46 Rathmell, A.R. and Wiley, B.J. (2011). The synthesis and coating of long, thin copper nanowires to make flexible, transparent conducting films on plastic substrates. *Adv. Mater.* 23: 4798–4803.

47 Zhang, J.Z., Zhu, X.Z., Xu, J. et al. (2022). Comparative study on preparation methods for transparent conductive films based on silver nanowires. *Molecules* 27: 8907.

48 Kim, B.S., Shin, K.Y., Pyo, J.B. et al. (2016). Reversibly stretchable, optically transparent radio- frequency antennas based on wavy Ag nanowire networks. *ACS Appl. Mater. Interfaces* 8: 2582–2590.

49 Zhu, X.Z., Guo, A.Q., Yan, Z.Y. et al. (2021). PET/Ag NW/PMMA transparent electromagnetic interference shielding films with high stability and flexibility. *Nanoscale* 13: 8067–8076.

50 Deng, B., Hsu, P.C., Chen, G. et al. (2015). Roll-to-roll encapsulation of metal nanowires between graphene and plastic substrate for high-performance flexible transparent electrodes. *Nano Lett.* 15: 4206–4213.

51 Graedel, T.E., Franey, J.P., Gualtieri, G.J. et al. (1985). On the mechanism of silver and copper sulfidation by atmospheric H_2S and OCS. *Corros. Sci.* 25: 1163–1180.

52 Rathmell, A.R., Nguyen, M., Chi, M., and Wiley, B.J. (2012). Synthesis of oxidation-resistant cupronickel nanowires for transparent conducting nanowire networks. *Nano Lett.* 12: 3193–3199.

53 Song, J., Li, J., Xu, J., and Zeng, H. (2014). Superstable transparent conductive Cu@Cu_4Ni nanowire elastomer composites against oxidation, bending, stretching, and twisting for flexible and stretchable optoelectronics. *Nano Lett.* 14: 6298–6305.

54 Kholmanov, I.N., Domingues, S.H., Chou, H. et al. (2013). Reduced graphene oxide/copper nanowire hybrid films as high-performance transparent electrodes. *ACS Nano* 7: 1811–1816.

55 Won, Y., Kim, A., Lee, D. et al. (2014). A highly stretchable, helical copper nanowire conductor exhibiting a stretchability of 700%. *NPG Asia Mater.* 6: e105.

56 Coskun, S., Ates, E.S., and Unalan, H.E. (2013). Optimization of silver nanowire networks for polymer light emitting diode electrodes. *Nanotechnology* 24: 125202.

57 Ran, Y., He, W., Wang, K. et al. (2014). A one-step route to Ag nanowires with a diameter below 40 nm and an aspect ratio above 1000. *Chem. Commun.* 50: 14877–14880.

58 Huo, D., Kim, M.J., Lyu, Z. et al. (2019). One-dimensional metal nanostructures: from colloidal syntheses to applications. *Chem. Rev.* 119: 8972–9073.

59 Sim, H., Kim, C., Bok, S. et al. (2018). Five-minute synthesis of silver nanowires and their roll-to-roll processing for large-area organic light emitting diodes. *Nanoscale* 10: 12087–12092.

60 Cho, S., Kang, S., Pandya, A. et al. (2017). Large-area cross-aligned silver nanowire electrodes for flexible, transparent, and force-fensitive mechanochromic touch screens. *ACS Nano* 11: 4346–4357.

61 Kang, S., Cho, S., Shanker, R. et al. (2018). Transparent and conductive nanomembranes with orthogonal silver nanowire arrays for skin-attachable loud speakers and microphones. *Sci. Adv.* 4: eaas8772.

62 Zilberberg, K., Gasse, F., Pagui, R. et al. (2014). Highly robust indium-free transparent conductive electrodes based on composites of silver nanowires and conductive metal oxides. *Adv. Funct. Mater.* 24: 1671–1678.

63 Lu, H., Sun, J., Zhang, H. et al. (2016). Room-temperature solution-processed and metal oxide-free nano-composite for the flexible transparent bottom electrode of perovskite solar cells. *Nanoscale* 8: 5946–5953.

64 Chen, Y., Zhang, H.B., Yang, Y. et al. (2016). High-performance epoxy nanocomposites reinforced with three-dimensional carbon nanotube sponge for electromagnetic interference shielding. *Adv. Funct. Mater.* 26: 447–455.

65 Hong, S., Kim, K., Kim, T. et al. (2012). Electromagnetic interference shielding effectiveness of monolayer graphene. *Nanotechnology* 23: 455704.

66 Han, Y., Liu, Y., Han, L. et al. (2017). High-performance hierarchical graphene/metal-mesh film for optically transparent electromagnetic interference shielding. *Carbon* 115: 34–42.

67 Hu, M., Gao, J., Dong, Y. et al. (2012). Flexible transparent PES/silver nanowires/PET sandwich-structured film for high-efficiency electromagnetic interference shielding. *Langmuir* 28: 7101–7106.

68 Kim, D.G., Choi, J.H., Choi, D.-K., and Kim, S.W. (2018). Highly bendable and durable transparent electromagnetic interference shielding film prepared by wet sintering of silver nanowires. *ACS Appl. Mater. Interfaces* 10: 29730–29740.

69 Zhu, X., Xu, J., Qin, F. et al. (2020). Highly efficient and stable transparent electromagnetic interference shielding films based on silver nanowires. *Nanoscale* 12: 14589–14597.

70 Ma, L., Lu, Z., Tan, J. et al. (2017). Transparent conducting graphene hybrid films to improve electromagnetic interference (EMI) shielding performance of graphene. *ACS Appl. Mater. Interfaces* 9: 34221–34229.

71 Hansson, J., Nilsson, T.M.J., Ye, L., and Liu, J. (2018). Novel nanostructured thermal interface materials: a review. *Int. Mater. Rev.* 63: 22–45.

72 Bhanushali, S., Ghosh, P.C., Simon, G.P., and Cheng, W. (2017). Copper nanowire-filled soft eastomer composites for applications as thermal interface materials. *Adv. Mater. Interfaces* 4: 1700387.

73 Choi, S., Park, J., Hyun, W. et al. (2015). Stretchable heater using ligand-exchanged silver nanowire nanocomposite for wearable articular thermotherapy. *ACS Nano* 9: 6626–6633.

74 He, X., Liu, A., Hu, X. et al. (2016). Temperature-controlled transparent-film heater based on silver nanowire-PMMA composite film. *Nanotechnology* 27: 475709.

75 Hsu, P.C., Liu, X., Liu, C. et al. (2015). Personal thermal management by metallic nanowire-coated textile. *Nano Lett.* 15: 365–371.

76 Mkhitaryan, V., March, K., Tseng, E.N. et al. (2021). Can copper nanostructures sustain high-quality plasmons? *Nano Lett.* 21: 2444–2452.

6

Normal and Novel Nanoplates for Understanding Growth Mechanism

Stacking Fault is Important in Understanding the Formation of Two-dimensional fcc Metal Nanomaterials

6.1 Introduction

To date, the production of Au and Ag-derived nanocrystals with controlled morphology has been widely produced. Their morphology influences not only the intrinsic optical, chemical, and catalytic properties of metal nanoparticles, but also their relevance for electronic, plasmonic, and sensing applications. Although 1D nanorods and nanowires have generally garnered the most versatile plasmonic nanoparticles [1–4], plate-shaped nanoparticles are appealing for numerous reasons relating to their unique morphology and properties [5–9]. Solution-phase colloidal nanoplates in this chapter are pseudo-two dimensional (2D). They are distinguished from 2D graphene and thin films by deposition methods. For colloidal nanoplates, understanding the growth mechanisms is an essential task in nanoscience. In a solution-phase approach, the growth process of metal nanocrystals can be divided into three distinct stages: (i) ions reduction and nucleation; (ii) evolution of aggregation and atomic addition into seeds; and (iii) oriented growth of seeds into nanocrystals [10]. In these cases, the current understanding of nanocrystal growth is far from being able to probe into atomistic details for the evolution pathways, even though we have various characterization methods, such as in-situ techniques in TEM and cell-TEM, HRTEM, STEM, UV–Vis–NIR spectrometer and SNOM. Usually, the final morphology of the metal nanocrystals is far from the thermodynamic theory in such a way that the internal structure of the corresponding seed is primarily determined by the requirement of the minimized surface energy, but rather the orientated growth of nanocrystals is determined by the kinetics such as the binding affinity of the capping agent [11, 12]. Here, searching for the

Plasmonic Metal Nanostructures: Preparation, Characterization, and Applications,
First Edition. Caixia Kan.

nucleation/growth mechanism that leads to the formation of metal nanocrystals with a specific shape has been the focus of current research interests.

In that regard, the properties of metal nanomaterials are closely related to their size, shape, and structure. Among the differently shaped metal nanocrystals, anisotropic Au and Ag nanoplates with flat surfaces and high aspect ratios are particularly appealing because of their decreased geometrical symmetry, which induces distinctive optical properties associated with the collective oscillations of conduction electrons [13]. From a geometric perspective, nanoplates synthesized to date have been flat, triangular, or hexagonal plates with large aspect ratios. It was known that nanoplates and nanowires are intrinsically higher in energy than the thermodynamically favored shapes, and their formation becomes favorable only in a slow reduction process. In the colloidal chemistry method with the presence of surfactant, when the reduction becomes slow enough, kinetic control will take over in both nucleation and growth of nanocrystals into anisotropic nanostructures with high surface energy.

Typically, nanoplate structures contain flat surfaces with edge lengths ranging from tens of nanometers to several micrometers and thicknesses in the nanometer scale. Their sharp corners and straight edges offer pronounced electric field enhancements. At a certain excitation frequency, this oscillation will be in resonance with the incident light, resulting in a strong oscillation of the surface electrons, known as localized surface plasmon resonance (LSPR) [14]. For example, the thin triangular Au nanoplates show superior LSPR with distinct in-plane and out-plane resonances. These structures are especially interesting because they have plasmonic features in the Vis–NIR regions and potentials in SERS, photo-thermal, and superior catalytic performance [15]. Therefore, metal nanomaterials with well-defined structures and tailorable sizes are crucial for exploring their potential applications. The past decade has seen a series of significant advances in the synthesis of Au and Ag nanoplates. From these studies, several possible growth mechanisms for nanoplates have been proposed.

The vast majority of literature examples of nanoplate particles possess highly symmetric crystal structures of face-centered cubic (*fcc*), which require a breaking of the unit cell symmetry to generate a 2D nanostructure. Consequently, the mechanistic processes that generate these particles remain the subject of ongoing research. This chapter will focus the discussion on the case of the *fcc* metals of Au, Ag, and Cu, which represent the benchmark for shape-controlled colloidal growth. In the works of literature, much attention was paid to the fabrication, growth mechanism, and optical properties of the Au and Ag nanoplates with normal shapes of triangular, hexagonal, and truncated triangular. Cu nanoplates and microplates were introduced briefly. For the *fcc* metal nanoplates, in addition to the normal triangular and hexagonal (or snipped triangular), novel star-like Au nanoplates were obtained through the wet-chemical method by introducing temperature variation. The results represent a significant step toward gaining a better understanding of the growth mechanism of nanoplates.

6.2 General Considerations for fcc Nanoplates

Nanoplates are among the most anisotropic structures available, which can be shown mathematically by defining a parameter called "anisotropicity" (α), as shown in Figure 6.1a. In a comparison of rod- and plate-shaped objects with a circular cross-section for a given aspect ratio, it was shown that the 2D structure maximizes the anisotropicity, providing the greatest degree of property control and anisotropic properties [5].

For *fcc* metals, atoms are arranged in a close-packed ABCABC stacking along the [111] direction. Three low-index surfaces, (111), (200), and (220), present decreasing coordination numbers with decreasing surface energy increases, i.e. $\gamma_{(111)} < \gamma_{(200)} < \gamma_{(220)}$ [17]. For the nanoplate with high symmetry, it requires symmetry breaking to generate a 2D nanocrystal. The most common manner in which asymmetry is created for cubic crystal systems is the appearance of 2D

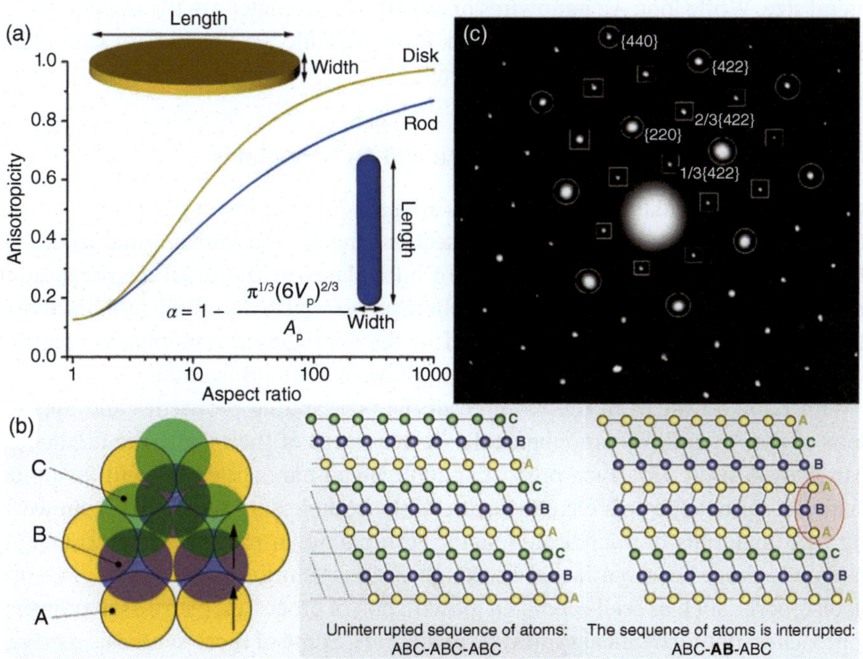

Figure 6.1 (a) Calculation of anisotropicity value (α was defined in the inset, where V_p and A_p are the volume and surface area of the particle, respectively) for a plate (golden solid line) and a rod (blue solid line) both with circular cross-section. Source: Reproduced from Scarabelli et al. [5]/with permission of American Chemical Society. (b) Atoms stacking order for *fcc* crystal structure was repeated ABC with uninterrupted (ABC–ABC…) and interrupted (ABC–AB–ABC…) atoms sequence [16]. Source: Reproduced with permission from the website (https://msestudent.com). (c) Diffraction pattern of an *fcc* nanoplate, showing the forbidden that in 1/3(422) reflections, indicating the formation of stacking faults. Source: Reproduced with permission from Kan et al. [8]; © 2010, John Wiley & Sons, Inc.

twin-plane defects. Such defects appear when the ordinary stacking of planes along a particular crystal axis (i.e. ABCABCABC) is disrupted in a way that creates a plane with mirror symmetry (e.g. ABCBA) (see Figure 6.1b) [16].

For highly symmetric systems, there is often a single family of planes for which twin formation is accompanied by a negligible energetic penalty. This is the case for the stacking of (111) planes along the [111] direction in an *fcc* crystal. The distinct defects for plate-like *fcc* nanoparticle morphologies of extended (111) facets are multiple parallel twins (sometimes referred to as "stacking faults"). Nanoplates of these two shapes are frequently confused because both can appear on the same two-dimensional (2D) plates. And these stacking faults can be distinguished by the selected area electron diffraction (SAED) patterns with forbidden (422)/3 and 2/3 (422) reflections, as illustrated by Figure 6.1c.

For *fcc* metal nanostructures, there are some interesting phenomena, particularly for Au and Ag with nearly the same crystal constants. With similar methods, Au tends to grow into nanorods or large-sized plates with several tens of micrometers in lateral size. While long Ag nanowires or small Ag nanoplates are the main product, which is also commonly observed for Cu. Monometallic, small-sized Au nanoplates and Ag nanorods have rarely been reported.

6.2.1 Synthetic Methodologies for Au and Ag Nanoplates

Although the colloidal metal nanoplates are pseudo-two dimensional, they are also defined as 2D nanostructures. In this section, the key parameters that affect the thermodynamics and kinetics growth of nanoplates in the different preparation methods, and the proposed growth mechanism were also described and discussed. Besides, the optical properties involved in the synthesized nanoplates were also provided a description toward the manipulation of their applications.

With rapid advances in the research of the outstanding properties and applications of the Au and Ag nanoplates, the development of their synthetic methods is extremely desired. However, preparing noble metal nanostructures with plate-like shapes has proved to be a challenge, due to the complexity that derives from working with materials of microscale length. Fortunately, in past decades, Au and Ag nanoplates have been synthesized through various solution-phase methods in previous reports, such as seed-mediated growth, polyol processes, hydro-/solvothermal approaches, photochemical synthesis, and others. Some of these synthesis methods are solution-based colloidal syntheses since they exhibit facile control over the morphological and crystallographic parameters of the obtained nanostructures. Many efforts have been made in the preparation of highly uniform metal nanoplates with tunable lateral sizes by suitably controlling the experimental parameters such as concentration of the reducing agents, metal precursor, stabilizers, and reaction conditions. Now we review the most developed synthetic methods and the corresponding experimental parameters for the final morphology of the synthesized Au and Ag nanoplates.

6.2.1.1 Seed-Mediated Growth

Even though different methods have been used for obtaining anisotropic nanostructures, seed-mediated growth is a widely used method to promote the anisotropic growth of nanostructures. Different morphologies are produced by seed-mediated growth, such as nanorods, nanowires, nanostars, and so on. Compared with the other synthetic methods, the seed-mediated growth method involves two or more steps. The first step is the synthesis of seed nanoparticles by simple reduction of the metal precursor. And then the seed nanoparticles are served as templates and centers for further nucleation or epitaxial growth for producing metal nanostructures. In this way, seeds nucleation and growth of the anisotropic crystals are temporally and physically separated, which permits very precise control over the morphology of the final product. Seed-mediated growth was initiated by Wiesner and Wokaun for the synthesis of anisometric Au nanoparticles in 1989 [18]. In this study, they modified the famous work of Faraday by reducing chloroauric acid ($HAuCl_4$) solutions with H_2O_2 in the presence of small colloidal nuclei. Later, Murphy's research group flourished the seed-mediated growth method to become a widely used high-yield methodology to synthesize anisotropic Au NRs at the beginning of the twenty-first century [19].

In addition to the mass production of the Au NRs through the seed-mediated method, the Au nanoplates were also obtained by modulating the growth solution of surfactant, capping agent, and mild reducing agent. In the second process of seed-mediated growth, Au^{3+} ions will be reduced and Au atoms deposited on the surface of the seed nanoparticles. External agents, such as various molecules or ions, can guide the growth direction of the nanoparticles and result in the formation of Au nanoplates. In the growth solution, CTAB, and AA were commonly used as capping agents and reducing agents, respectively. Here, the bromide ions (Br^-) from the CTAB adsorption appear to suppress the crystal growth along the [111] direction, resulting in rich Au(111) surface on the Au nanoplates [20, 21]. Research shows that the type of surfactant molecules and the concentration in the growth solutions play important roles in the size and shape determination of the nanostructures. For example, when the Br^- ions were replaced with Cl^- ions, the morphology notably changed from triangular nanoplates to novel nanorices [20].

Typical seed-mediated growth of Au nanoplates by Millstone et al. is an example of the key factors of surfactant concentration [21, 22]. In their approach, the small Au nanoparticles with a diameter of 5 nm were used as seeds for further growth in an aqueous solution with CTAB as the capping agent and AA as the reducing agent. Au nanoplates were obtained when saturated CTAB solution was only applied as a surfactant, which is a decisive parameter of the concentration dependence on the shape of the Au nanoplates. Otherwise, the edge length of the Au nanoplates, which determines their optical properties, can be controlled while preserving the crystal properties and original shape of the seeds. For this reason, the edge length of the Au nanoplates has been varied from 100 to 300 nm without changing their crystallinity or thickness by using plate-shaped nanoparticles as seeds and reinitiating

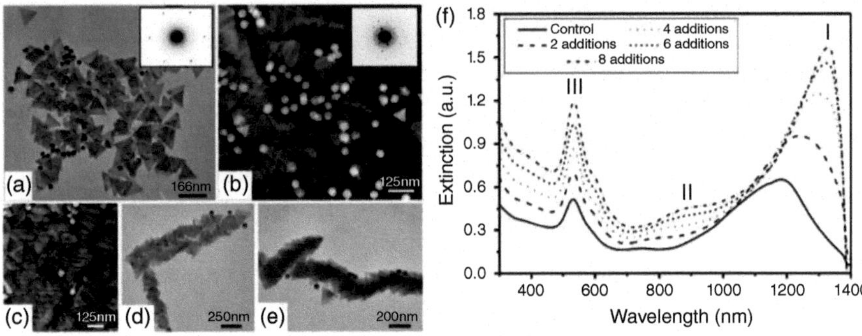

Figure 6.2 (a–e) TEM images of Au nanoplates with increasing edge lengths. (a) TEM image of Au nanoplates made without additional growth (control sample). The inset shows a SAED pattern of the nanoplates. (b) TEM image of nanoplates after two growth additions (×2). The inset shows the SAED pattern of ×2 nanoplates. (c) TEM image of ×4 nanoplates, (d) ×6 nanoplates, and (e) ×8 nanoplates. (f) UV–Vis–NIR spectra of nanoplates with different edge lengths. Source: Reproduced with permission from Millstone et al. [22]; © 2006, John Wiley & Sons, Inc.

the growth of the seeds in the presence of a reducing agent in a seed-mediated growth method [21]. TEM images in Figure 6.2a–e are of Au nanoplates with increasing edge lengths, and Figure 6.2f is the corresponding optical spectra of the Au nanoplates, showing two distinct SPR bands. Obviously, many nanoparticles can be observed in the product, which accounts for the fixed SPR band around 550 nm.

Besides Au nanoplates, Ag nanoplates with tunable edge lengths have also been successfully synthesized via a seed-mediated growth method in the presence of Ag seed particles and appropriate concentrations of CTAB, trisodium citrate, or poly(vinylpyrrolidone) (PVP) as capping agents at room temperature [23–25]. This synthetic procedure offers the advantage of generating Ag nanoplates in a controlled way, with the potential to tune the shape in the presence of suitable seeds and capping agents. As shown in Figure 6.3a–f, TEM images of the obtained Ag nanostructures are exclusively nanoplates without the discernible presence of particular morphologies. By adjusting the added seed amount, triangular Ag nanoplates with different edge lengths are readily synthesized. The trisodium citrate used as a capping agent is effective in selectively adsorbing on the (111) plane of Ag seeds and guiding the growth dimensions of the nanocrystals to high-purity plate-like Ag nanostructures. Figure 6.3g is one digital photo of a sol of the nanoplates and the corresponding TEM image of the triangular Ag nanoplates. The UV–Vis–NIR extinction spectra (see Figure 6.3h) of the Ag nanoplates with different edge lengths show that the main SPR band readily shifts to longer wavelengths as the edge length increases.

The study of seed-mediated growth method showed that the Au and Ag nanoplates have controllable edge length and thickness, and improved uniformity and high yield can be achieved by finely tuning the experimental parameters, such as the type and the concentration of capping agents, the volume of the seed solution, and the time interval between multiple steps of the synthesis process. In brief, it is

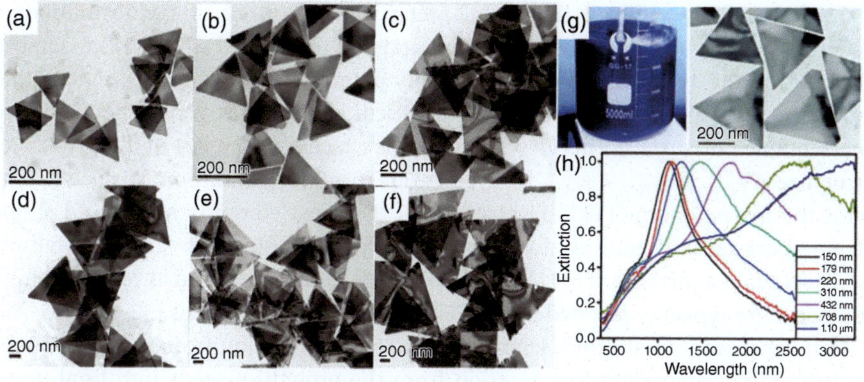

Figure 6.3 (a–f) TEM images of the triangular Ag nanoplates with different edge lengths (150–1500 nm) were obtained by tuning the volume of the seed solution. (g–h) A large-scale product of the triangular Ag nanoplates and UV–Vis–NIR extinction spectra of the triangular Ag nanoplates with different edge lengths. Source: Reproduced with permission from Liu et al. [24]; © 2014, Royal Society of Chemistry.

noteworthy that the seed-mediated growth process has become probably the most commonly used synthetic approach and has provided a versatile tool and robust route for generating Au and Ag nanoplates with widely tunable sizes.

6.2.1.2 Polyol Process

The polyol process is a convenient, versatile, and low-cost route for the synthesis of colloidal metal and alloy particles with various shapes and sizes. This method was originally developed in 1989 by F. Fievet et al. and G. Viau et al. [26, 27]. The present form of the polyol process, in which liquid polyol or diol acts both as a solvent for the precursors and as a mild reducing reagent, was modified and developed effectively in controlling the growth of metal nanostructures through the introduction of polymers, foreign ions, and seeds, as well as careful regulation of the reaction temperature [28, 29]. In the polyol process, polyol refers to alcohols containing multiple hydroxyl groups, such as ethylene glycol (EG), polyethylene glycol, propylene glycol (PEG), glycerols, and so on. The ability of polyols to dissolve many inorganic metallic salts, such as $HAuCl_4$ and $AgNO_3$, used as metallic salt precursors, makes them versatile and ideal for the synthesis of colloidal nanoparticles. Furthermore, the polyols with relatively high boiling points and highly temperature-dependent reducing power make them suitable solvents for the reduction of metallic salt precursors and controllable nucleation and growth processes through careful regulation of reaction temperature. For example, EG will get converted into glycolaldehyde upon heating in the presence of air, which acts as a dominant reductant for reducing the Au^{3+} and Ag^+ ions into atoms and thereby inducing the nucleation and growth of anisotropic nanostructures in the solution phase.

Up to now, it has been established that the formation of noble-metal nanoparticles with different shapes in a polyol process is mainly determined by the thermodynamically controlled initial seeds and kinetically controlled anisotropic growth. In the initial stage of a polyol process, the EG with elevated temperatures

(~150 °C) quickly reduced the metal precursors, resulting in the formation of different types of seeds. Both the experimental and theoretical studies have demonstrated that the shape of a seed is primarily determined by the requirement that surface energy must be minimized. Theoretical calculations generally agree that the sequence of surface energies for low-index facets of the Au and Ag with *fcc* structure is $\gamma_{(111)} < \gamma_{(200)} < \gamma_{(220)}$ [8, 17, 30, 31]. To achieve minimum surface energy with the lowest-energy (111) facet covered, the most common single and singly twinned-crystalline, multiply twinned, and plate with stacking faults-crystalline, which contain a mixture of (111) and (200) facets, can be obtained at relatively high temperatures typically reached in the EG solution-phase synthesis [32–34].

In this process, the thermodynamically controlled reaction from nuclei to seeds, and the kinetically controlled reaction from the growth of seeds into final metal nanostructures with well-defined shapes would be crucial stages in the polyol process. To thermodynamically control the synthesis of metal nanostructures, one should stabilize the product and allow sufficient time so that the reaction can approach equilibrium. Therefore, facet-selective capping agents selectively interacting with a specific crystallographic facet via chemical adsorption can promote the oriented growth of a particular shape. In a typical example, PVP is commonly used as the capping agent or stabilizer in polyol synthesis due to the presence of oxygen and nitrogen atoms in the pyrrolidone unit, which plays a critical role to prevent aggregation, adsorb onto the specific crystallographic facets of the colloidal nanoparticles, and induce anisotropic growth of Au and Ag nanostructures. In a typical polyol process, the thickness of the PVP coating and the location of PVP chains on the surface of a seed could be easily modified by adjusting the molar ratio between PVP and metal precursors. This modification altered the resistance of each facet to growth and led to the formation of Au and Ag nanostructures with distinct shapes. Take the Au nanoplates as an example, preferential adsorption of polar groups of PVP on (111) planes of nuclei leads to preferential growth within (111) planes into 2D structures. At the same time, with the additional capping agents added and easily temperature adjusted, shape-controlled metal nanostructures have been successfully synthesized by the polyol process.

6.2.1.3 Hydrothermal and Solvo-Thermal Approach

Hydrothermal and solvo-thermal approaches have been demonstrated for the synthesis of various kinds of metal nanomaterials attribute to their facile operation and simple process principles. In this method, water and various solvents (non-aqueous), serving as reductants and catalysts, were sealed in an autoclave at high pressure and elevated temperature. Especially, polar or non-polar solvents are used to widen the range of applicability of this approach. At the same time, the additives, such as PVP and CTAB, are employed to modify the initial properties of the solvents for the synthesis of metal nanomaterials. Up to date, hydro/solvo-thermal approaches have been successfully employed for several different metal nanoparticles.

For example, single-crystalline Au and Ag nanoplates were obtained by the hydrothermal method [36, 37]. Qin et al. obtained triangular Ag nanoplates with straight and curved edges from $AgNO_3$ with the presence of PVP in an aqueous

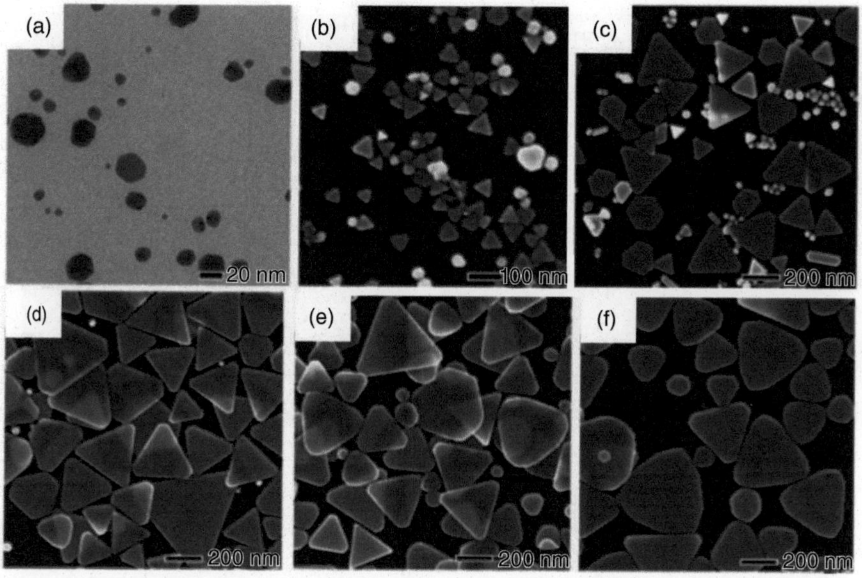

Figure 6.4 Au and Ag nanoplates were obtained by the hydrothermal method. (a–f) TEM and SEM images of Ag nanoparticles obtained from six standard hydrothermal syntheses stopped at different reaction times: (a) 1.5, (b) 2.5, (c) 3.5, (d) 4, (e) 5, and (f) 6 hours. (g) The corresponding UV–Vis spectra of Ag nanoparticles in (a–f). Source: Reproduced with permission from Yang et al. [35]; © 2013, John Wiley & Sons.

solution at 160 °C at different times [35], as shown in Figure 6.4a–f. Another study showed that well-defined Au nanoplates are grafted onto carbon nanotube sheets via a facile hydrothermal reduction process, during which bromine ions are employed as the surfactant [37]. Both the Au and Ag nanoplates show excellent SERS performance due to their large surface area, roughness, and sharp edges and corners.

The hydro/solvothermal approach has been verified to be a facile method for the production of metal nanoparticles, due to its low cost, modest equipment, environmental friendliness, and ease of controlling the yield and quality. Therefore, it can be used in the mass fabrication of Au and Ag nanostructures as well as other metal nanostructures, such as Rh, Cu, Pt, Pd, and so on [38–42].

6.2.1.4 Photochemical Synthesis

Photochemical synthesis is an effective and fascinating method for the mass synthesis of numerous Au, Ag, and bimetallic Au@Ag nanostructures [43–45]. Moreover, this method distinguishes itself from the discussed above syntheses methods by the fact that the shape transformation and growth of the nanostructures (especially to Ag nanoplates) are driven by incident light. Thus, photochemical synthesis is also defined as a plasmon-mediated method in some literature. Photochemical synthesis was initially discovered and reported by Mirkin et al. [46] in 2001 that the spherical Ag nanoparticles (~8 nm in diameter) can be transformed into triangular or hexagonal Ag nanoprisms/nanoplates in the presence of room light and low-concentration trisodium citrate as a reductant and capping agent, as shown in Figure 6.5a–c.

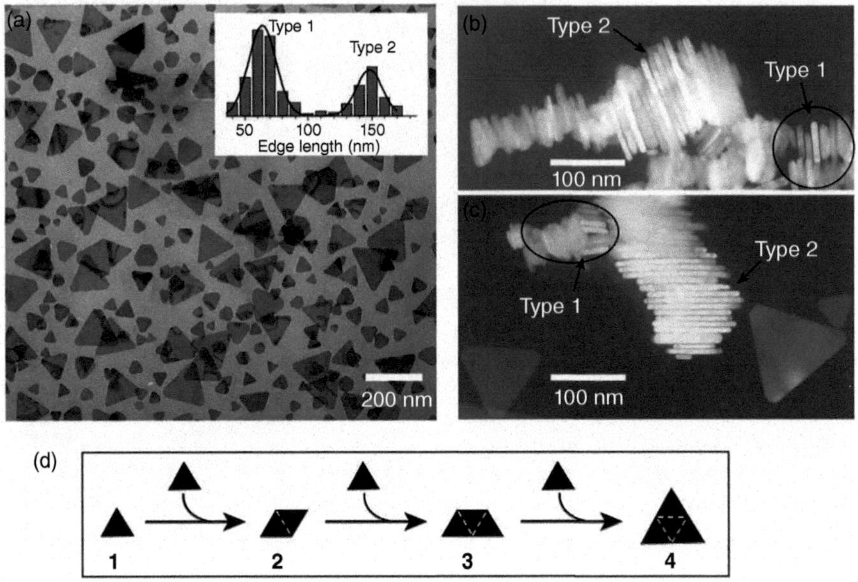

Figure 6.5 The bimodal growth and schematic illustration of the photochemical synthesis of Ag nanoplates. (a) TEM image of a sample of Ag nanoplates formed using single-beam excitation and histograms used to characterize the size distribution as bimodal (inset in a). (b and c) TEM images of nanoplate stacks show that nanoplates have nearly identical thicknesses. (d) Schematic diagram of the proposed light-induced fusion growth of Ag nanoplates. Source: Reproduced with permission from Jin et al. [47]; © 2003, Springer Nature.

In their following study, they demonstrated that the relatively monodisperse Ag nanoplates were obtained with a dual-beam illumination of the initial nanoparticles, which were driven by surface plasmon excitations in the form of a colloidal suspension. At the same time, the size of the Ag nanoplates with desired edge lengths from 30 to 120 nm can be manipulated by changing and tuning the primary excitation wavelength from 450 to 750 nm of the light source [47]. As the schematic diagram shown in Figure 6.5d, the observed bimodal growth process occurs through a proposed light-induced edge-selective particle fusion mechanism. Generally, the four type-1 nanoplates can fuse in a step-wise manner to form a type-2 nanoplate.

Further study pointed out that, when citrate was replaced by other chemically related molecules, such as citric acid, tricarballylate, citramalate, or aconitate in the spherical seed solution, the formation of triangular nanoplates seemed to be inhibited. The reason behind the effect of citrate on the formation of Ag nanoplates is not completely understood, but it might be consistent with the role of citrate in promoting the formation of twinned particles in wet-chemical syntheses of Ag nanoplates. In addition, although most studies have concluded that the conversion of spheres into nanoplates happens by a process of oxidation and re-deposition of Ag, a recent liquid cell TEM work demonstrated that adjacent spherical particles can coalesce with each other under electromagnetic field enhancement and form large nanoplates.

Based on Mirkin's initial work, the photochemical synthesis method was applied to control noble metal nanoparticle growth by using light irradiation. The wavelength of incident light can be used to exquisitely control the edge length of the resulting nanoplates. Moreover, the presence of trisodium citrate, the pH of the growth solution, the regulation of structural defects in the seed nanoparticles, precursor concentration, exposure time, and, importantly, light source and wavelength can also affect the light-driven synthesis of noble metal nanoparticles. This facile and size-controlled method allows for the preparation of Au and Ag nanostructures that have desired optical properties through fine-tuning their SPR spanning the entire visible and near-infrared (NIR) regions. Therefore, this synthetic strategy makes the obtained noble metal nanoparticles promising and attractive candidates for a variety of potential applications.

6.2.2 Growth Mechanisms for Au and Ag Nanoplates

It is well known that the growth of anisotropic metallic nanocrystals is possible through a slow reduction process, which provides a better chance to overcome the thermodynamically favored reaction toward the formation of truncated nanocubes or multiple twinned particles due to their intrinsically higher energy compared with other isotropic shapes. However, sophisticated controls on the kinetics of nucleation and growth are required.

As discussed above for the synthetic methodologies for Au and Ag nanoplates, the use of different synthetic strategies that improve nanoplate monodispersity and high yield was extensively studied. Generally, the normal triangular and hexagonal metal nanoplates are mostly made of close-packed metal atoms connected by metallic bonds. The synergistic effects of both thermodynamically controlled and kinetically controlled play a critical role in determining anisotropic nanoplates. The crystal growth is commonly controlled by the interplay of various thermodynamic and kinetic parameters to construct nanocrystals with a desired morphology. Different growth mechanisms for the formation of metal nanoplates have not been proposed yet. In this section, we will discuss and analyze the proposed growth mechanisms that generate the normal triangular and hexagonal metal nanoplates and highlight the factors that lead to the nanoplates growth in the different methodologies.

6.2.2.1 Au Nanoplates

The exact growth process of the Au nanoplates has yet to be established, although some groups have presented the proposed growth mechanism for anisotropic nanostructures synthesized by solution-phase methods. The standard and general explanation for the formation of nanoplates usually involves two stages: (i) the preformed "seeds" that are thermodynamically controlled, and (ii) the appropriate capping reagents that are used to kinetically control the growth rate of various facets of the preformed seeds.

In the polyol process for the synthesis of Au nanoplates, EG was often used as both a solvent and a reducing agent for the reduction of Au^{3+} ions when the reaction temperature was higher than 100 °C. Au atoms are obtained by reducing Au^{3+} ions with aldehyde during the reaction. When the concentration of Au atoms has reached supersaturation, it is reasonable to speculate that Au nuclei or seeds formed in the initial stage are kinetic structures with the lowest energy (111) surfaces. Under thermal equilibrium, twinned and multiplying (usually fivefold) twinned particles are widespread habits both in synthetic and natural materials. However, the defect energy that exists along the twinned facets will limit the growth of the twinned structure. This indicates that twinned seeds need to be confined to relatively small sizes. If these seeds expand in size rapidly, the low surface energy of (111) facets can no longer remedy the excessive energy, resulting in their transformation into single crystals. Thus, when the reduction of metal ions to atoms is fast, single-crystal seeds will prevail over their multiply twinned counterparts. On the other hand, in the hot system, the formed Au nuclei are exposed to oxygen, and the twinned seeds will be oxidized and dissolved in solution, possibly due to the critical oxidation size, the oxidizability of the reaction environment, or the instability of ultra fine particles [48, 49].

With the use of a capping agent in the synthesizing system, the growth rates of crystals are often controlled (kinetically rather than thermodynamically) by the sticking probability on a given face. The competition (and/or cooperation) between the surface energy difference and the coordination effect will confer the tendency to form single-crystalline nuclei bounded by a certain facet, which in turn not only determines the crystallographic growth of the particles but also drastically changes the kinetics of crystal growth. At a favorite PVP/Au molar ratio, the polar groups, such as pyrrole rings of PVP and hydroxyl groups of ethylene glycol molecules, adsorb preferentially on the sites of the (111) planes of Au nuclei and interact with the (111) planes, which also obviously slows (or prevents) the growth on the (111) plane and promotes highly anisotropic crystal growth, such as along [110] orientation, extending into triangular or hexagonal Au nanoplates with basal (111) planes [50]. The proposed growth mechanism and schematic illustration for the formation of Au nanoplates are shown in Figure 6.6.

It can be seen that the nuclei grow uniformly along six [110] directions if the growth environment is uniform, resulting in the formation of a hexagonal plate surrounded by three pairs of (220) lateral faces [8, 50]. However, in the open experiment system, the crystal growth would inevitably deviate from uniform growth along six [110] directions. Therefore, truncated triangular and triangular plates will be formed if the nuclei are grown along different [110] directions (or to some extent). Corresponding to the schematic drawing, Figure 6.6b presents some typical TEM images of the Au nanoplates observed. For Au crystals sampled at the initial growth stage, small Au nanostructures with triangular, and hexagonal shapes are observed, indicating the formation of different platy Au crystal seeds for further growth. Furthermore, less or excessive capping agent coverage on the surfaces of Au crystals would induce a simultaneous growth for different crystal faces to some extent (or even growth limitation) apart from the anisotropic growth within the (111) plane, leading to the formation of multi-shaped Au nanostructures.

Meanwhile, another accepted opinion suggested that at the second stage, from the seeds to large nanoplates, some small Au seeds with triangular shapes will be

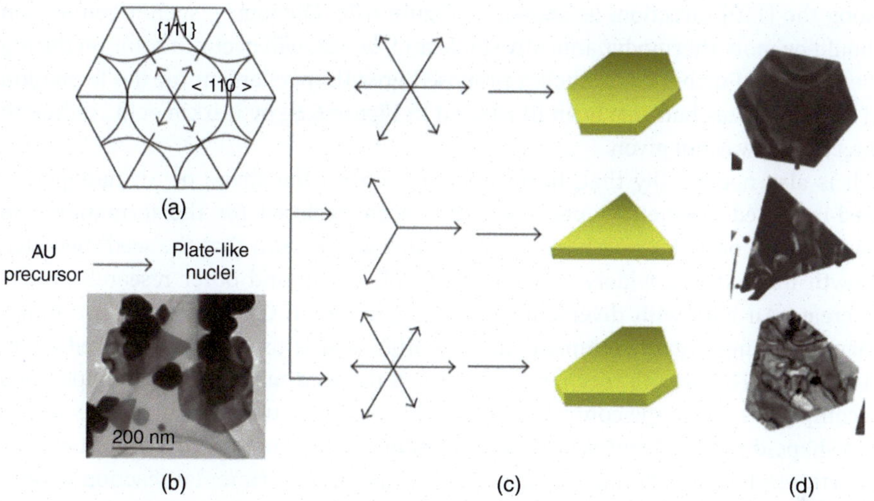

Figure 6.6 Proposed growth mechanisms and schematic illustration for the formation of Au nanoplates: (a) atoms arrangement of *fcc* (111) plane, (b) TEM image of Au crystal sampled at initial growth stage, (c) the formation of platy shapes along different [110] directions, and (d) the corresponding TEM images. Source: Figure modified with permission from Wang et al. [50]; © 2014, Editorial office of Acta Physico-Chimica Sinica.

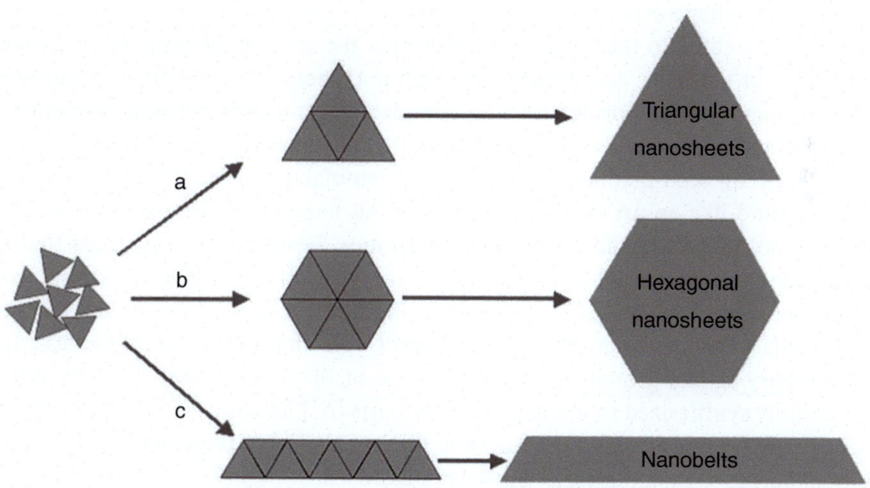

Figure 6.7 Schematic illustration for the formation of (a) triangular nanosheets/nanoplates; (b) hexagonal nanosheets/nanoplates; (c) nanobelts by the connection of small triangular nanosheets of the same size. Source: Reproduced from Li et al. [29]/with permission of John Wiley & Sons.

connected along the (220) lateral planes, as schematically illustrated in Figure 6.7. Finally, the very large, but thin, triangular or hexagonal nanoplates will be formed because of the relatively high surface energy of (220) planes. Also, a long and thin nanobelt was observed due to some small triangular nanoplates being connected along the [110] direction, as shown in Figure 6.7c. Obviously, such a connection should be more thermodynamically stable and, hence, will occur in solution during the reaction. Furthermore, such a proposed growth mechanism for the formation of Ag nanoplates had also been illustrated in Figure 6.5g by Mirkin et al., although direct proof was not given.

It is also noteworthy that, in an effort to control the shape of Au nanoplates, seed-mediated synthetic routes have also been explored for the formation and investigation of their growth mechanisms. It is well known that the seed-mediated growth method was widely applied by Murphy et al. and other research groups to create Au NRs with diverse aspect ratios, in which CTAB is used as a major shape-directing organic template, and ascorbic acid is used as a major reductant [19, 51–55]. The growth mechanism by which the seed-mediated growth approach accomplishes the shape control of anisotropic Au nanostructures by halide anions as a shape-directing agent is not yet well understood. Some controversies about the growth mechanism should be attributed to different experimental reaction parameters at the two growth stages. Over the past 10 years, many synthetic experiments and characterization analyses have been studied in detail by a variety of *in situ* and *ex situ* methods [56–60]. However, a detailed understanding of the chemical processes by which shape control is achieved remains elusive. Nonetheless, a variety of researchers claimed that changes in the halide and the specific adsorption of anions can completely change the morphology of the primary seeds and thus dominate the shape evolution of Au nanocrystals. For example, many researchers have recently come to the same conclusion that the specific adsorptions of halide ions (Cl^-, Br^-, I^-) on Au surfaces are well-documented in the field of surface chemistry and play important roles in complete control over the development of desired shapes in the seed-mediated growth of Au nanostructures [61–63].

However, in seed-mediated growth, the manufacture of Au nanostructures appears more like an art than a science, and Au nanoplates were synthesized in the presence of halide ions and their molar ratios. Research demonstrated that a minuscule amount of iodide ions was crucial for the formation of the Au nanoplates instead of nanorods [20, 64]. The growth mechanism of Au nanoplates by the seed-mediated method may be totally different from that in the polyol process. In the ordinary crystal growth with the presence of Br^- from CTAB, Au NRs were successfully synthesized in many previous reports [65]. In contrast, when a trace of iodide ions (I^-) was added to the reaction, the Au nanocrystals changed their shape from nanorods to nanoplates, as shown in Figure 6.8.

For the growth of Au nanoplates by the seed-mediated approach, the seed's crystalline nature and the applied surfactants are two key parameters for the final morphology. At the initial stage, the diversity of anisotropic nanostructures accessible from a single seed solution provides insight into the initial structure of the citrate-stabilized Au seeds. The formation of a stacking fault or twin plane in the lattice exposes a variety of nanocrystal facets, providing seed structures with distinct surface energies [66]. In solution, passivation of the Au surface by halide anions is expected to alter the growth kinetics of a particular crystal facet concerning other exposed facets through changes in the local redox potential. Thus, the facet-selective

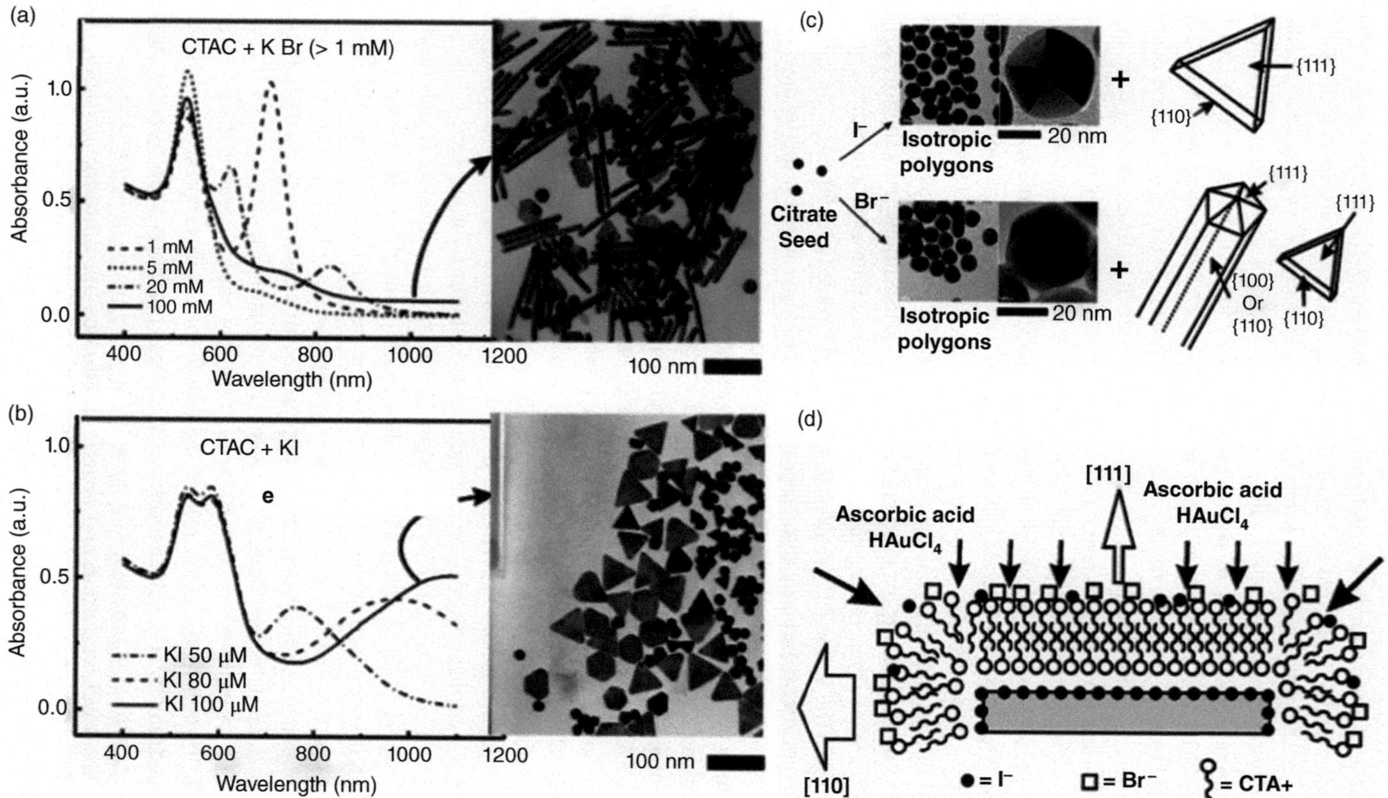

Figure 6.8 The general view of Au NRs and nanoplates is obtained in the absence and the presence of iodide ions, respectively. UV/Vis spectra and corresponding typical TEM images of diverse gold nanostructures formed under the CTAC template. (a) With higher concentrations of bromide ions (>1 mM). (b) In the presence of iodide ions (50–100 μM) instead of bromide ions. (c) Schematic illustration showing the shape-controlled syntheses of Au nanoplates and nanorods according to the presence of iodide ions under a CTAB template. (d) Schematic illustration describing the influence of the CTAB template, in conjunction with an iodide adlayer as a passive barrier for crystal growth during the formation of nanoplates. The white arrows indicate the rate of growth on specific facets. Source: Reproduced with permission from Ha et al. [20]. © 2007, American Chemical Society.

adsorption in the presence of halide anions leads to differential stabilization of Au nanocrystal surfaces, promoting the production of anisotropic nanostructures by restricting growth in specific crystallographic directions [67].

As schematically illustrated in Figure 6.8c and d, with the assistance of the CTAB template in the presence of I^- ions, faceted nanoparticles were simultaneously observed with triangular nanoplates. Besides, nanorods along with nanoplates were dominantly observed in the presence of Br^- ions. This distinction can be rationalized by the assumption that the absorbed Br^- ions cause both the (220) and (200) facets of Au to be more hindered than the (111) facets. Thus, the overall growth rates on the Br^- ions adlayer were much faster than those in the presence of iodide ions, as both nanorods and nanoplates were simultaneously found along with isotropic nanocrystals. This primary cause results from the adsorption tendency of the bromide ions, which is inferior to that of the iodide ions. Moreover, the twinned Au nanoparticles with a fivefold ridges domain in the presence of iodide ions were sharper than those obtained in the presence of bromide ions. This distinct characterization demonstrates that the anisotropic crystal growth was more controlled depending on the specific exposed crystal facet. It is also noteworthy that the pH or temperature elevations and the concentration of iodide ions during the growth stage can dominate the shape evolution of the Au nanoparticles [20, 60, 64]. The various control experiments indicate that the influence of iodide ions on the formation of nanoplates appears to occur within a narrow window of reaction parameters.

According to the above discussion, the proposed growth mechanism of the seed-mediated approach for the growth of Au nanoplates involves. Firstly, the addition of iodide ions results in the formation of a strong adlayer on Au facets, with a stable CTAB bilayer on this I^- ion adlayer preferentially formed due to its overwhelming adsorption tendency compared to other halide ions. This bilayer controls the supply of added chemicals to the Au (111) surface passively. On the other hand, at the edge faces of nanoplates (dominated by Au (220) and Au (200) faces), more spherical micelles should be formed as a result of geometric factors, which relatively expedites the addition of Au^+ on the sides of the planar-twinned seeds. Thus, the crystal growth rate in that direction seems to be more facilitated, which induces the high aspect ratio of triangular and hexagonal nanoplates. However, another research demonstrates that the I^- ions may not be essential for the formation of this plate-like nanocrystal, although they provide a substantial improvement in the yield of Au nanoplates because triangular nanoplates occur concomitantly with nanorods in the obtained samples. Thus, the reasonable explanation may be that these I^- ions do not promote the growth of nanoplates, but rather prevent the growth of nanorods. Therefore, the main role of the iodide ions is to suppress the growth of Au decahedrons by strongly adsorbing to the Au (111) facets, thereby allowing the epitaxial growth of planar-twinned seeds to the dominated Au nanoplates.

When the reaction system and growth conditions are uniform, Au nanoplates with normal triangle or hexagonal morphologies are the main product. However, fascinating nanostructures will be obtained when external interference or irradiation is introduced into the reaction system. For example, ultrathin, large-sized Au nanoplates with a single through a pore in the center or screw bending were

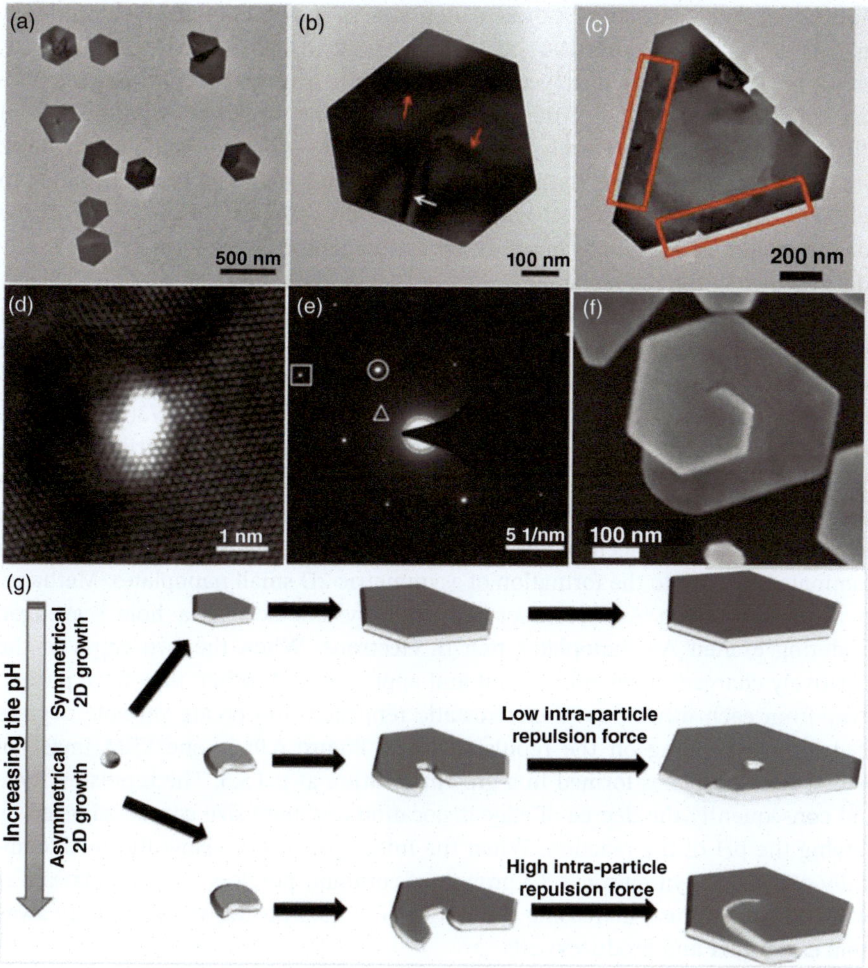

Figure 6.9 Characterization of Au nanoplate with a nanopore structure. (a) TEM image of the Au nanoplates with a nanopore. (b) Higher magnification TEM image of a single Au nanoplate with a nanopore. (c) TEM image of Au nanoplate with multiple nanopores near the edges. (d) HRTEM image of an Au nanopore. (e) The SAED pattern of an Au nanoplate with a nanopore. The spots indicated by the circle, box, and triangle correspond to the (220), (422), and 1/3(422) planes. Forbidden 1/3(422) Bragg reflections with a 2.50 Å lattice spacing are also observed as the sharp, weak spots in the center surrounded by the strong spots. (f) SEM images of coiling Au nanoplates. (g) Scheme of the formation mechanism of the Au nanoplate with or without a pore (or screw). Source: Reproduced with permission from Zhai et al. [68]; © 2017, John Wiley & Sons.

fabricated by Y. Zhai et al. as shown in Figure 6.9 [68]. They found crystal screw dislocation was created on the Au nanoplate during growth, and a void defect formed in the center part with the high crystal stress. The degree of screw dislocation can be manipulated simply by adjusting the pH value of the growth solution, and the nanopore size can be tuned from subnanometer to tens of nanometers. Structural characterization by HRTEM and SAED of the Au nanoplates confirmed the internal crystal structure of these nanoplates, as shown in Figure 6.9. Strikingly, an HRTEM image (Figure 6.9b) reveals a very small hole in the central region, with a unique boundary line (pointed by the white arrow) starting from the pore and radiating to the edge of the nanoplate. This line structure is different from other continuous and randomly oriented darker lines (pointed by the red arrows), which are commonly observed in thin Au nanoplates and have been attributed to bending. Figure 6.9c is one Au nanoplate with no nanopore in the center of the Au nanoplates. Interestingly, however, several nanopores were formed close to the edges (shown in the red rectangle).

Figure 6.9d shows the nanoplates (with a central pore) obtained here exhibit nearly perfect crystal lattices. This was further demonstrated by the SAED measurement (Figure 6.9e) taken from a single Au nanoplate. The spots indicated by the circle, box, and triangle correspond to the (220), (422), and 1/3(422) planes, respectively. These fractional forbidden diffraction patterns can be attributed to single crystalline Au with parallel stacking faults. It was found that the Au nanoplates were highly dependent on the pH value of the reaction solution. When the pH value and the precursor concentration are relatively high, kinetics-controlled growth dominates, leading to the formation of asymmetric 2D small nanoplates. Methanol in the photo-driven Au precursor reduction system acts as a hole scavenger, rendering excited Au nanoplates rich in electrons. When the two edges of the negatively charged plates roll around and approach each other, they have pushed away from each other due to electrostatic repulsion and create variable degrees of defects depending on the repulsion force. Figure 6.9f is one SEM image of coiling Au nanoplates formed in a growth solution at pH 5.5. The repulsion force and consequently the degree of edge dislocation and pore size can be adjusted by varying the pH of the reaction. When the initial pH was 3.5 slow thermodynamically controlled symmetrical 2D growth of common poreless Au nanoplates was obtained. In contrast, increasing the initial pH to 5.5 pushes the two edges away from each other and produces coils.

Investigations for the growth of Au nanoplates largely demonstrate that the reaction conditions or growth mechanisms by which they modulate crystal shapes are extremely complicated, and, have yet to be fully elucidated. Until now, scientists have paid much effort to understanding the complex physics that leads to the formation of nanocrystals with specific shapes. Therefore, further studies are still required to obtain more information about the rational design of synthesis strategies to promote anisotropic distributions of Au nanoplates.

6.2.2.2 Ag Nanoplates

Compared to most syntheses of plate-like Au nanostructures, which have been prepared by polyol processes and seed-mediated growth methods, it is particularly true that the photochemical synthesis has progressed from an effective method and fascinating way to highly controllable for the preparation of Ag nanoplates. This method introduced the possibility of using a different wavelength of light irradiation to exquisitely control the edge length of the resulting Au nanoplates. However, the accurate and underlying growth mechanisms of plate-like Ag nanostructures were not well understood until now, and many researchers engaged in studies aimed at elucidating the conceivable mechanisms. The exploration in this field steadily increased and arrived at conclusions about the understanding of growth mechanism in some publications by Mirkin and co-workers, Brus and co-workers, and so on, although they focused on different aspects of plate-like Ag nanostructures growth [69–71]. Essentially, the general growth process and the conversion of spherical Ag particles to Ag nanoplates were found to occur in three stages: (i) introduction, Ag nanoparticles smaller than the Ag seed particles are nucleated, (ii) growth, small nanoplates are formed and coexist with spherical nanoparticles, and (iii) termination, the reaction and conversion end, and the solution is composed entirely of Ag nanoplates without spherical nanoparticles. Importantly, this growth and conversion process requires many essential chemical and physical components acting together to yield the final Ag nanoplates, such as a source of Ag^+, the presence of citrate, plasmonic seeds, and visible light with a wavelength range between 350 and 700 nm.

Note that throughout this photochemical synthesis of the Ag nanoplates, it is clear that both citrate and plasmonic seeds are necessary for Ag^+ to be reduced to Ag^0 by visible light. This was demonstrated in a series of experiments performed that irradiating a solution of small Ag nanoparticles in the absence of citrate does not produce Ag nanoplates, and irradiating a solution of Ag^+ and citrate will not yield the reduction of Ag^+ to Ag^0 in the absence of plasmonic Ag seed particles [70, 72, 73]. Although citrate has been proven that it acts as both a particle stabilizing agent and a reducing agent due to its particular reduction potential and interaction with the Ag nanoparticle surface, the reason for citrate's importance in specific photochemical reactions is not completely understood. At the same time, the reduction of Ag^+ by citrate is extremely slow and unobservable at room temperature in the absence of Ag plasmonic seeds. As opposed to the reduction of Ag^+ ions to Ag atoms occurs when the seeds are present, implying that the Ag plasmonic seeds serve as photocatalysts to facilitate the reduction of Ag^+ ions and deposition of Ag atoms on the surface.

On one hand, a series of Ag redox cycles is primarily responsible for the conversion of Ag particles to nanoplates, where the oxidative dissolution of Ag particles to Ag^+ in solution, which is then reduced by citrate and redeposited onto the Ag particle surface when irradiated with visible light [70, 72, 74]. The process of oxidation and re-deposition of Ag is depicted as shown in Figure 6.10a–c. The roles of each factor in these experiments can be qualitatively defined by the conversion process from Ag nanoparticles to nanoplates, as originally proposed in Figure 6.10d. Thus, when irradiated with visible light in the presence of small Ag seed particles (about 5 nm in diameter), Ag^+ ions in solution are reduced by citrate and deposited onto the Ag particles. The oxidative dissolution, facilitated by capping agents such as PVP or bis(*p*-sulfonatophenyl)phenylphosphine dihydrate dipotassium (BSPP), is responsible for oxidizing the smaller Ag particles to provide a constant source of Ag^+ ions in solution. Finally, the large Ag nanoplates were obtained at the expense of the small Ag seed particles until they were entirely consumed. The details of these findings provide evidence against a spherical particle fusion mechanism.

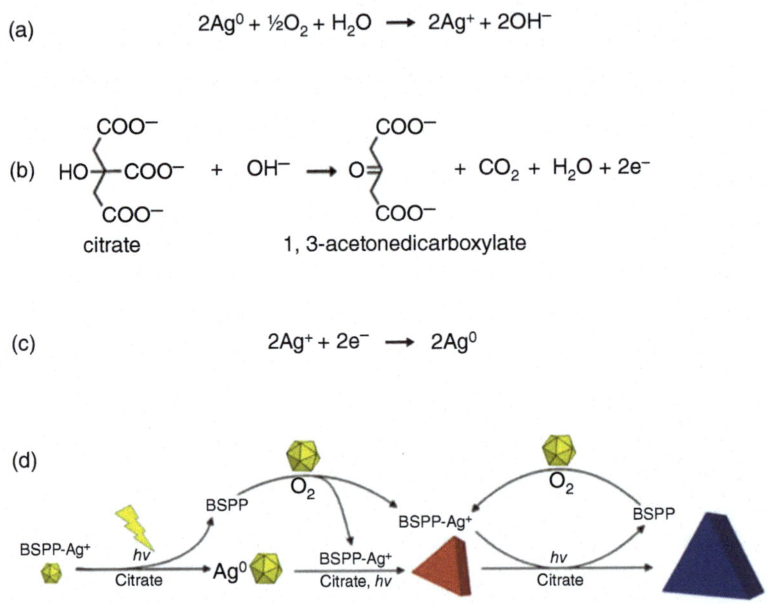

Figure 6.10 Primary chemical reactions involved in the photochemical synthesis of Ag nanostructures. (a) Dissolution of Ag nanoparticles by O_2 to Ag^+ and OH^-. (b) Oxidation of citrate to 1,3-acetonedicarboxylate, CO_2, and two electrons. (c) Reduction of Ag^+ to Ag^0. (d) Scheme illustrating the photochemical conversion of Ag spherical nanoparticles to Ag triangular nanoplates. Source: Reproduced from Millstone et al. [6]/with permission of John Wiley & Sons.

On the other hand, the excitation wavelength (λ_{ex}) of incident light in this photochemical reaction could be used to exquisitely tune the edge length of the resulting Ag nanoplates, with the nanoplates growing until their SPR is redshifted from λ_{ex}. Some reports showed that longer wavelengths of incident light can yield longer edge-length Ag nanoplates, although the full effects and comprehensive understanding of λ_{ex} are quite complicated and still not completely understood. Mirkin and co-workers found that the simultaneous use of two different excitation wavelengths can obtain the monodisperse Ag nanoplates close in size to one another and produce a monomodal SPR [47]. It was also found that by keeping a secondary λ_{ex} fixed at 340 ± 10 nm, the primary λ_{ex} could be tuned from 450 to 750 nm. The finally obtained Ag triangular nanoplates had an edge length that increased linearly with the λ_{ex} of the primary incident light (as shown in Figure 6.11). Thus, the proposed explanation for the ability to tune the edge length of Ag nanoplates may be that due to the SPR band will redshift from λ_{ex} which, coupled with a nanoplate, grows, causing its absorption to drop off and resulting in slowed growth. Meanwhile, the Ag nanoplates with short edge lengths will strongly absorb light and hence continue to grow rapidly, resulting in a size-focusing effect. Moreover, when the Ag nanoplates grow larger and their SPR band redshifts over the λ_{ex}, the oxidative etching may take place preferentially at the tips and will blueshift their SPR to the λ_{ex}. These processes, which enable size control through λ_{ex}, have been termed "self-limiting" growth by

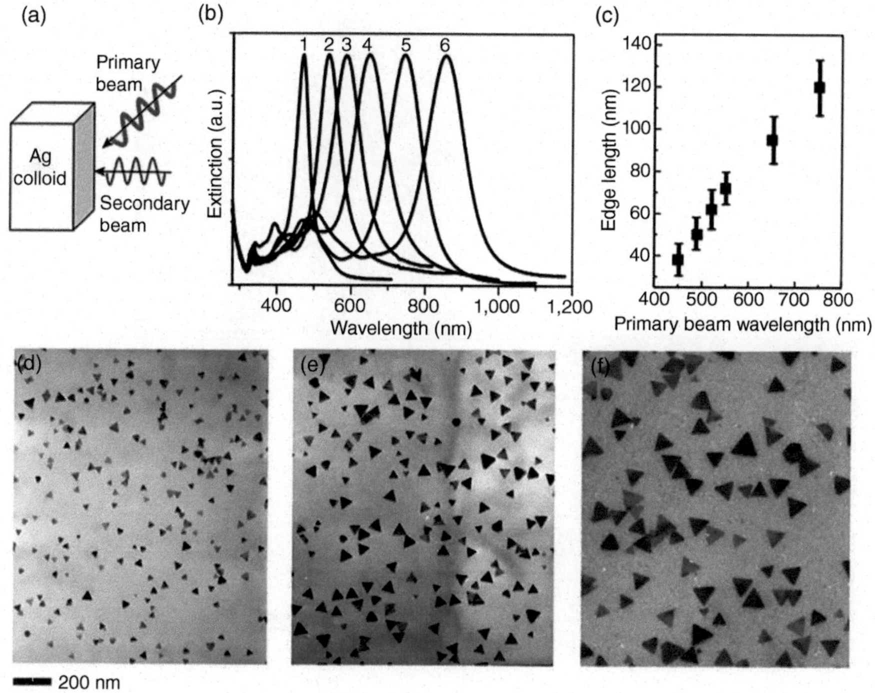

Figure 6.11 The size-controlled Ag nanoplates by the incident light wavelength in a dual beam excitation set-up. (a) Schematic diagram of dual-beam excitation. (b) The optical spectra (normalized) for six different-sized Ag nanoplates (1–6) were prepared by varying the primary λ_{ex} from 450 to 750 nm (central wavelength at 450, 490, 520, 550, 650, and 750 nm, respectively), coupled with a constant secondary excitation wavelength at 340 ± 10 nm. (c) Plot of the primary excitation wavelength versus the resulting prism edge length. (d and e) TEM images of Ag triangular nanoplates with average edge lengths of 38, 72, and 120 nm, respectively. The scale bar applies to images (d and e). Source: Reproduced with permission from Jin et al. [47]; © 2003, Springer Nature.

Xu and coworkers [75]. The results clearly show that the λ_{ex} has proven to be a key experimental parameter that can be used to effectively tune the edge length of the Ag nanoplates.

Based on the experimental parameters for the growth of Ag nanoplates discussed in the above section, the growth mechanisms were proposed in this photochemical synthesis process. For example, some researchers proposed that the capping agents preferentially bind to Ag (111) facets, thus restraining the Ag atom deposition and growth perpendicular to those facets, while guiding the lateral growth of the crystal [30, 73, 76]. Otherwise, early work reported that the irradiation wavelength drives the rapid formation of Ag nanoplates due to their strong absorbing light ability compared with that of other shapes, like Ag nanorods or nanospheres [74].

Consequently, considering the crystallographic and kinetic reactions for the conversion of Ag seeds to nanoplates in photochemical synthesis, it is generally acknowledged that the reaction kinetics and Ag seeds with planar defects drive the growth of these seeds into Au nanoplates. The characterizations have confirmed

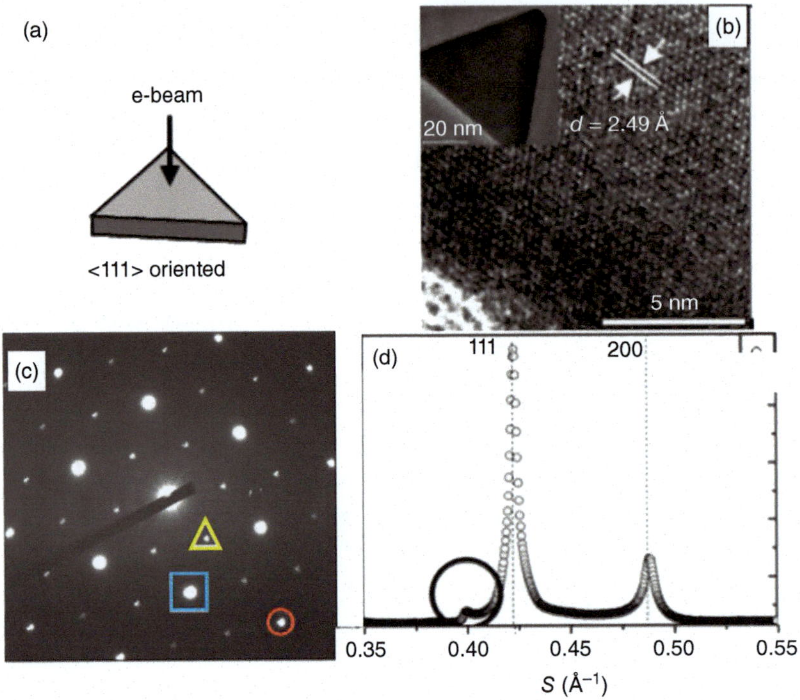

Figure 6.12 Structural characterization of the triangular Ag nanoplates. (a) Schematic illustration of the position relative to the electron beam. (b) HRTEM images of an Ag nanoplate (inset shows the low-magnification image). (c) SEAD pattern analysis of an Ag nanoplate with a hexagonal structure spot array indicates that the nanoplate is lying flat on the substrate with its (111) facet perpendicular to the electron beam. Marked reflections correspond to (220) (blue square), (422) (red circle), and 1/3(422) (yellow triangle). The intense spots are allowed (220) Bragg reflections with a 1.44 Å lattice spacing. Forbidden 1/3(422) Bragg reflections with a 2.50 Å lattice spacing are also observed as the sharp, weak spots in the center surrounded by the strong spots. (d) XRD pattern of Ag nanoplate has a reflection at 0.40 $Å^{-1}$ indicating that there may be multiple planar defects in the Ag nanostructure. Source: Reproduced with permission from Rocha and Zancchet [76]; © 2007, American Chemical Society.

that the nanoplates present an *fcc* lattice with the [111] direction perpendicular to the flat (111) faces bounded on the top and bottom. The lattice fringes observed in most [111]-oriented nanoplates showed a periodicity of 2.5 Å, as indicated by Figure 6.12a,b. The lattice fringes observed in the HRTEM image suggest the presence of structural defects. This periodicity also appears in the SEAD image, as relatively faint spots close to the direct beam. However, in addition to the typical (220) Bragg reflections, a weak kinematically forbidden 1/3(422) reflection that arises from the presence of stacking faults and twin planes was also observed. The result corroborated that the presence of twins and stacking faults in an *fcc* lattice generates *hcp* lamellas, which leads to the appearance of extra reflections, as illustrated by Figure 6.12c,d.

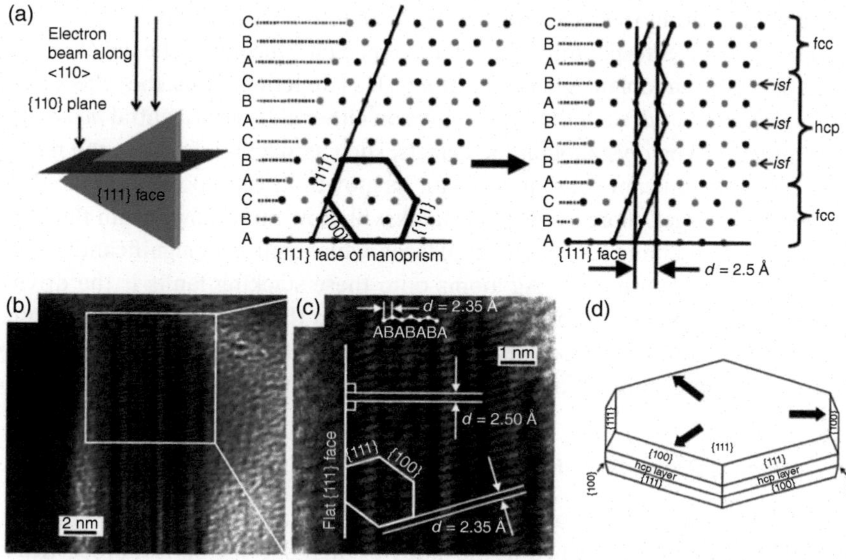

Figure 6.13 (a) Schematic illustrating how intrinsic stacking faults along [111]. The black dots represent atoms in the (110) plane while the gray dots represent atoms immediately below. (b) HRTEM image of a stack of vertically oriented silver nanoplate. (c) Analysis of the internal structure of nanoplate in (b). (d) Schematic of a hexagonal plate containing a defect-induced *hcp* layer sandwiched between two *fcc* layers. The black arrows indicated the proposed directions, which could potentially lead to the preferred growth of a triangular nanoplate. Source: Reproduced with permission from Aherne et al. [77]; © 2008, John Wiley & Sons, Inc.

To investigate the possible hcp arrangement of atoms, detailed HRTEM studies of vertically oriented Ag nanoplates were also observed [77]. In this orientation, two flat (111) planes and a (200) plane are aligned vertically to the electron beam, as illustrated schematically in Figure 6.13a. An *fcc* structure can be considered as a stacking of hexagonally close-packed (111) layers along the [111] direction with a pattern that follows an ABCABC……sequence. A stacking fault refers to the addition or the exclusion of one of the layers. For example, by introducing a series of intrinsic stacking faults (*isf*), it is easy to see the stacking arrangement of the atomic planes as ABABAB… sequence, giving rise to an *hcp* region.

Further evidence of the transformation of the crystal structure is demonstrated by HRTEM analysis of a nanoplate aligned along [111] direction, as shown in Figure 6.13b and c. The repeating stacking faults in the interior of the plates generated a local *hcp* ordering with a lamellar region about 1.5 nm thick, which was highlighted by the superposition of a zigzag pattern on the HRTEM image. These (111) planes also show discontinuities due to the numerous and repeated stacking faults between the (111) planes parallel to the face of this nanoplate. The flat (111) face is clearly indicated, and the lattice fringes can be seen propagating parallel to the (111)-labeled side with the correct spacing measured to be 2.35 Å for (111) planes and also for the alternate ABABAB…layers of the *hcp* region. There is an arrangement of atoms that is aligned perpendicular to the surface with a periodicity spacing of 2.50 Å, corresponding to the lattice spacing concerning the flat (111) face when a flat-lying nanoplate is observed along [111] direction. Thus, the schematic is shown in Figure 6.13d, such that the regions on either side of the central *hcp* region are asymmetric. The crystallographic defects, such as stacking faults in the nanostructure, have significant consequences for its growth. It is proposed that the three edges with the larger, more stable (111) faces will grow more slowly than the other three with the larger and less stable (200) faces, which will grow significantly faster. The preferential deposition of Ag atoms onto these stacking faults is the driving force for the highly anisotropic growth and leads to the formation of a triangular nanoplate. Therefore, it is important to note that these stacking faults in crystallinity can have enormous implications for the final morphologies of Ag nanostructures.

6.2.3 Optical Properties of Nanoplates

2D Au and Ag metal nanoparticles derived from the various synthetic strategies described above have flourished in the study of their specific optical properties as well as for the development of plasmon-based applications. Most of the attractive applications are based on the LSPR effect of nanoparticles. LSPR is a collective electronic oscillation of the surface conduction electrons of metal nanostructures excited by electromagnetic radiation, which strongly depends on their specific composition, size, and geometrical structure. In the last two decades, Au and Ag metal nanomaterials have been the most widely studied plasmonic materials due to their highly tunable LSPR wavelengths in the Vis–NIR spectral regions. In particular, Au and Ag nanoplates have received great attention because of their broad spectral absorption with an adjustable plasmon absorption in the NIR region. Their sharp corners and straight edges offer large electric field enhancements.

Taking Au nanoplates as an example, as shown in Figure 6.14a, the LSPR properties of Au nanoplates are significantly influenced by the thickness and the basal planes in triangular, hexagonal, or circular cross-sectional shapes. Figure 6.14a shows the triangular Au nanoplates, which present two plasmon resonance peaks at about 780 and 1230 nm. These two peaks are assigned to the out-of-plane quadrupolar mode and in-plane dipole resonance mode, respectively, which are characteristic of 2D metallic nanoparticles due to their plate-like morphology. With the geometrical structure of Au nanoplates changing, the sharp corners became round, which finally led to the formation of circular and hexagonal shapes with different thicknesses. The corresponding spectrum clearly revealed that the two LSPR peaks blue-shift to shorter wavelengths with increasing thickness and rounded shape transformation [7]. This discrepancy is ascribed to the inhomogeneous size distribution and agrees with the FDTD simulation results. Meanwhile, taking Ag nanoplates as an example, as shown in Figures 6.14b–d, experimental works also reported a blue shift of the LSPR peaks with shape transformation from a thin triangular Ag nanoplate with sharp corners to a thick round nanodisk [78]. Thus, the LSPR peaks of the Au and Ag nanoplates can be efficiently tuned over a wide range by changing their size and geometrical structure.

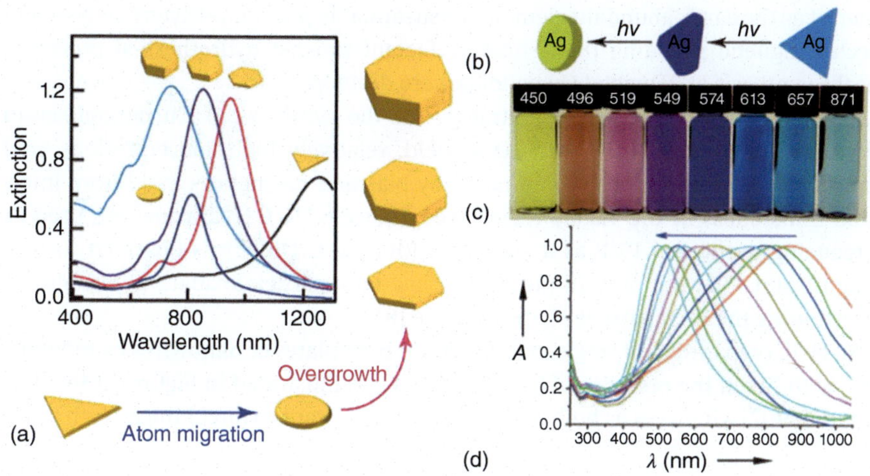

Figure 6.14 (a) Schematic illustrating the growing process through atom migration of Au nanoplates and the corresponding extinction spectra for the different Au nanoplate models. (b) Schematic illustration of the evolution process of Ag nanoplates with the sharp corners of triangular plates disappearing, and the thickness of the nanoplates increasing. Source: Adapted from Qin et al. [7]; © 2015, WILEY-VCH. (c) Digital photographs and (d) extinction spectra of the silver nanoplates. Source: Reproduced with permission from Zhang et al. [78]; © 2009, John Wiley & Sons, Inc.

6.2.4 Cu Nanoplates with Different Size

Nanoparticles of coinage metals (Au, Ag, and Cu) are known to display attractive optical properties, arising from localized surface plasmon resonances in the visible and NIR frequencies. However, Cu is less popular, mainly because the fabrication of chemically stable Cu colloids with intense plasmon resonance bands is far more complicated, first because they are prone to fast oxidation, but also because of the lower "free-electron character" of Cu. The free-electron behavior of Au and Ag colloids in the visible range is reflected in the fairly constant value of the imaginary part of their dielectric functions, which is responsible for the sharp and prominent extinction bands displayed by colloids of these metals. For Cu metal, the real and imaginary parts of the dielectric function vary markedly in the UV–Vis range, so that electronic interband transitions from the valence band to the Fermi level overlap the plasmon resonances up to 600 nm.

Cu has been considered the minor player of the coinage triad, due to its tendency to oxidize rapidly in the presence of oxygen and its lower plasmonic efficiency in the visible part of the electromagnetic spectrum. For these reasons, the colloidal chemistry of Cu is often focused on the use of Cu oxides and sulfides for the preparation of semiconductor nanoparticles [79]. Nevertheless, Cu nanocrystals can also be prepared with a fair control of size and shape. Although tuning of the optical response was claimed in these reports, the experimental absorption spectra do not display sharp and well-defined plasmon peaks, as expected from calculations. In recent years, Cu metallic nanoparticles have received renewed interest as the earth-abundant element for sustainable plasmon-driven catalysis and electromagnetic shielding [80]. Some surfactant-assisted hydrothermal processes for the synthesis of 2D metal Cu colloids were developed.

Liz-Marzán group presented a strategy for the synthesis of Cu nanoplates in dimethylformamide (DMF) with 100% yield, relatively high concentrations, and remarkable optical response (dominated by plasmon resonances away from interband transitions) in the visible region (see Figure 6.15a). Hydrazine was used as a reducing agent and PVP as a stabilizer. Moreover, DMF was observed to play a significant role, not only in preventing fast nanoparticle oxidation, but also in determining the final particle shape and size [81].

Xu et al. preparation of single-crystalline Cu nanoplates or nanowires by reducing Cu^+ with AA in the presence of CTAB or CTAC, as illustrated in Figure 6.15b. They found that CTAB serves as both Br source and surfactant to direct the growth of Cu

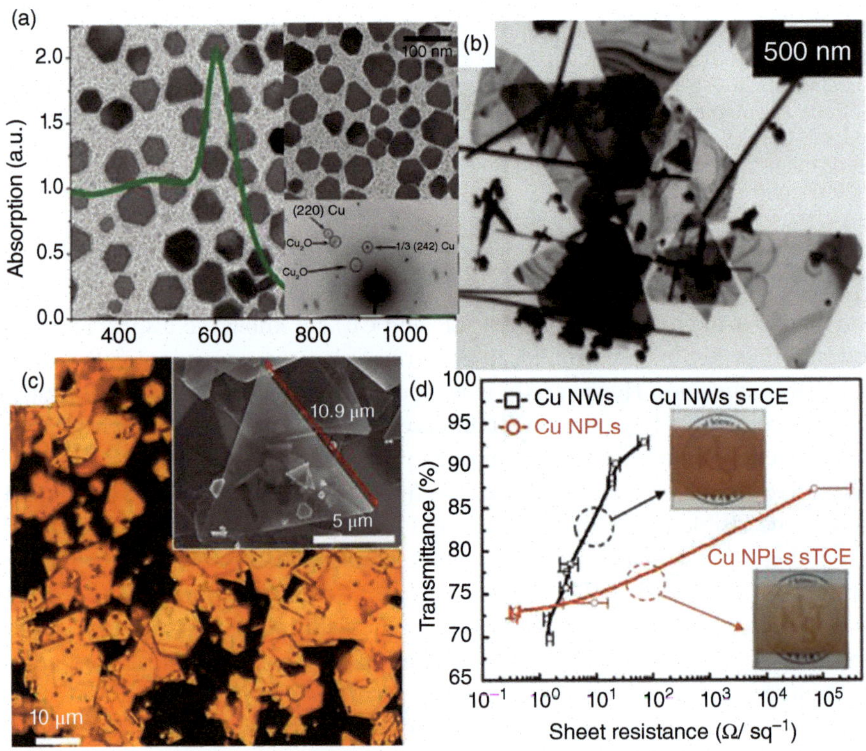

Figure 6.15 (a) TEM image and intense plasmon resonances of Cu nanoplates via reduction of Cu^{2+} ion with hydrazine in DMF at 60 °C using PVP as stabilizer, showing the forbidden diffraction patterns of the fcc nanoplates and Cu_2O). Source: Reproduced with permission from Pastoriza-Santos et al. [81]; © 2009, JOHN WILEY & SONS, INC. (b) TEM image of large sized Cu nanoplates or nanowires. Source: Reproduced with permission from Xu et al. [82]; © 2010, Elsevier. (c and d) Optical photograph image of Cu nanoplates by adding iodine (I) as an additive, and application as transparent electrode compared to Cu NWs. Source: Reproduced with permission from Lee et al. [83]; © 2018, John Wiley & Sons, Inc.

nanoplates. To further clarify the role of the Br^- anions in CTAB, they conducted the control experiments by substituting CTAB with CTAC under standard conditions. Cu nanowires were generated in the presence of CTAC, implying that CTA^+ alone is not sufficient for the shape control of nanoplates [82].

Then in 2018, Lee demonstrated a simple method for the dimensional control of Cu nanocrystals from 1D NWs to 2D nanoplates by adding iodine (I^-) as an additive. Further, 2D single-crystalline Cu nanoplates were prepared as a conductive filler for electronic ink applications [83]. The semi-transparent electrode proves its feasibility as a conductive filler, exhibiting very low sheet resistance compared to Cu NWs and a transmittance near 75%.

In 2022, Wu et al. proposed to establish a balance between oxidative oxygen gas and reductive glucose to control the formation of Cu nanodisks [84]. The reaction takes up to three days to complete, offering better control over size dispersity and shape yield (an average of 80 nm in edge length and 7 nm in thickness). The successful synthesis of Cu nanodisks relies on the balanced use of oxidative oxygen gas and reductive glucose. The obtained Cu nanodisks show promising catalytic activity for the electrochemical reduction of nitrite to ammonia property. DFT calculations demonstrate that a reconstructed Cu (111) surface is responsible for the enhanced activity of the nitrate reduction reaction. The surface triatomic Cu clusters are weak in the interaction of N—O and hence beneficial to the breakage of the N—O bond. The Gibbs free energy also indicates that nitrate reduction is preferable on the reconstructed Cu (111) surface.

Although these experimental approaches provide fundamental information on the crystal growth of Cu nanoplates and the possibility of using Cu nanoplates in practical applications, it remains a great challenge to achieve the well-defined, simple, eco-friendly, and dimension-controlled synthesis of Cu nanoplates, especially their large-size growth, up to tens of micrometers, for potential electronic ink applications.

6.3 Au Nanoplates with Novel and Well-Defined Shapes

For both fundamental aspects and potential technological applications, great efforts have been made to design synthetic methods for producing Au and Ag nanostructures of different morphologies. In most cases, the chemical strategies for the synthesis of noble metal nanostructures have yielded a great deal of literature featuring the anisotropic growth of crystals into defined morphologies, ranging from rods and plates to highly faceted particles or truncated versions of these shapes. Among a large number of synthetic methods available, the polyol process is a convenient, versatile, and low-cost route for the synthesis of metal nanoparticles. Metal nanostructures obtained by the polyol method are extremely rich in terms of morphology by changing any parameters, such as the introduction of polymers, foreign ions, and seeds, as well as the reaction temperature in the synthesis system.

With the development of colloidal chemistry, this method was improved or modified, through which the pre-formed crystal nuclei can be confined to grow into well-defined anisotropic nanostructures, depending on the experimental conditions and the nature of the pre-formed seeds. Until now, Au nanoplates formed in the general polyol process in an open atmosphere were regular hexagonal or triangular shapes with a multiple-twin-induced growth mechanism. However, once the conditions are sufficient for the anisotropic growth of nanostructures, the products may adopt a new shape. For instance, our research found that Au plates of novel shapes were usually mixed with the regular products if the hot polyol synthesis was disturbed by an obvious temperature fluctuation and other variables. Then, with the introduction of this temperature fluctuation to the synthesizing systems, Au plates with well-defined novel shapes were mass synthesized. Herein, in this section, new and well-defined Au nanoplates were obtained by introducing temperature variation in the early stage of crystal growth. Besides, a growth mechanism, and theoretical analysis that explain the observations are presented and discussed.

6.3.1 Growth Conditions of the Novel Au Nanoplates

Au nanoplates in triangular and hexagonal shapes with an average edge length of tens of micrometers and several nanometers in thickness were successfully and widely prepared in recent decades [29, 37, 85, 86]. In general polyol process at a high temperature of 150 °C (no obvious temperature fluctuation was introduced to the system), Au nanoplates with regular triangular, truncated triangular, and hexagonal shapes were obtained in EG with an application of binary surfactant PVP and CTAB.

Moreover, the representative FE-SEM images of the products obtained at different growth conditions are shown in Figure 6.16. The flat feature of nanoplates can be induced by their low and uniform contrast. In the general polyol process at high temperatures, Au nanoplates formed in an open atmosphere are in triangular and hexagonal shapes with an average edge size of >10 μm, including some truncated triangular plates as shown in Figure 6.16a. It is surprising that in the hot (150 °C) synthesizing system, which is disturbed by an obvious temperature fluctuation in the early crystal growth stage, Au nanoplates with novel shapes of hexagonal shields and other polygons (nonagons in most cases) are the main products, as shown in Figure 6.16b. Then, a series of experiments were performed at temperatures lower than 50 °C with different types of PVP surfactants (PVP-K30, PVP-K60, and PVP-K90) and obvious temperature variation in the early stage of crystal growth, as reported in our previous studies [86]. Many Au nanoplates with novel shapes, such as six-pointed stars, together with some plates of trigonal-star shape (Figure 6.16c), trifurcated star-like (Figure 6.16d), and sniped shield-like Au nanoplates (Figure 6.16e), were mass synthesized and found in the sample. Most of the star-like Au nanoplates were nearly uniform in size and shape. The results obtained with different PVPs revealed that the reduction rates of Au^{3+} ions and the growth rates of star-like Au nanoplates in the gels are ordered as PVP-K30 > PVP-K60 > PVP-K90, with the size order of the final nanoplates being PVP-K30 < PVP-K60 < PVP-K90. Therefore, these experiments reveal that the introduction of an obvious temperature variation in the early stage of crystal growth is crucial for the formation of Au nanoplates with these novel shapes.

Figure 6.16 FE-SEM images of Au nanoplates obtained from different polyol processes. (a) Truncated triangular, and hexagonal shape nanoplates. (b) Hexagonal shields and other polygonal nanoplates. (c) Six-pointed stars and trigonal-star shape nanoplates. (d) Trifurcated starlike nanoplates. (e) Sniped shield-like nanoplates. Source: Reproduced with permission from Kan et al. [8]; © 2010, John Wiley & Sons, Inc.

Moreover, for the exact outline of these newly discovered Au nanoplates, further characterization analyses were obtained in the TEM images, as shown in Figure 6.17. The exact angles of these new Au nanoplates can be obtained. In Figure 6.17a, the hexagonal shield-like Au microplates involve alternate angles of 90° and 150°. The Au nanoplates in Figure 6.17b are nonagons with 9 × 140° angles. Additionally, other interesting Au nanoplates were found in the sample, as shown in Figure 6.17c and d and inset in (d) for the plates that are not smooth and contain many big and small steps on the lateral surfaces. Star-like Au nanoplates synthesized with PVP-K30 have sharp corners with 6 × 90° angles (Figures 6.17e), while most of the star-like Au nanoplates synthesized with PVP-K90 are truncated (Figure 6.17f). If the six right angles of the star-like Au nanoplates are seriously truncated, the polygonal nanoplates were obtained with PVP-K60, as shown in Figure 6.17g and h. In addition to the above Au nanoplates, the polygonal plates of well-defined trigonal-star and trifurcated star-like shapes with concave or convex lateral surfaces but not regular triangular nanoplates were also obtained, as shown in Figure 6.17i–k.

Based on the above experimental results, the growth conditions for the well-defined Au nanoplates were expounded. In the past two decades, different opinions about the reduction of metal ions were proposed in the polyol process (usually with a temperature higher than 100 °C). The first one is that the EG, or agent derived from the EG is the reducing agent. At high temperatures, the molecule of EG may be decomposed to the reducing agent of glycollic aldehyde ($HOCH_2CHO$) with two aldehyde groups (–CHO) that have strong reductive properties [87]. Thus, the opinion that EG played the role of reducing agents in the hot system was proposed

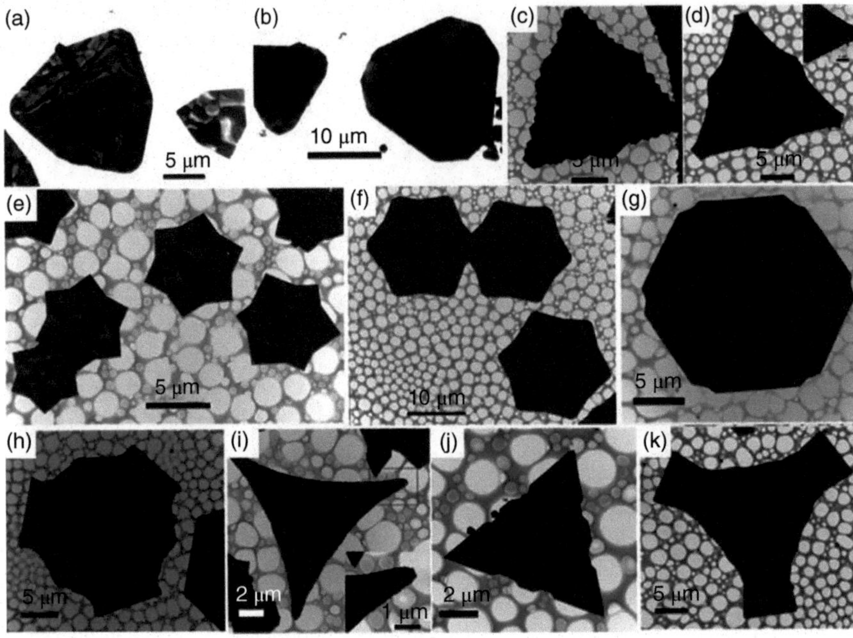

Figure 6.17 TEM images of the obtained Au nanoplates show the new and well-defined shapes. (a) Hexagonal shield-like and (b) nonagon Au nanoplates synthesized at high temperature with temperature variation. (c and d) Novel plates with many steps in lateral facets. (e–g) Star-like Au nanoplates formed at low temperatures with temperature variation using PVP-K30 (e), PVP-K90 (d), and PVP-K60 (g and h). (i–k) Well-defined trigonal star (i and g) and trifurcated star-like (k) shape nanoplates. Insets in (d) and (i) indicate many steps on the lateral facets. Source: Reproduced with permission from Kan et al. [8]; © 2010, John Wiley & Sons, Inc.

with priority. Recently, with the application of PVP in the polyol process, some groups proposed that PVP has a reducing ability on the metal ions [88]. In the previous study, it was found that Au and Ag nanostructures can be obtained at room temperature with the presence of PVP in the aqueous metal precursors (no special reducing agent or treatment was applied, and the EG molecules in the gels are stable at low temperatures). Therefore, the opinion that EG is the reducing agent is not reasonable for the case when the gels are kept at room temperature. Herein, it is believed that PVP polymer, usually used as a stabilizer, plays an important role in the reduction of metal ions, especially at room temperature.

In the study of the optical properties of the regular Au nanoplates synthesized by the polyol process at high temperatures, as shown in Figure 6.18a, it is found that with the reduction of Au^{3+} ions to Au atoms, the wide absorption extending to the NIR region increases without the obvious surface resonance of Au nanospheres in the region of 530–580 nm (even if it will be overlapped by the NIR absorption), which indicates the formation of Au nanoplates in the initial stage, as observed for the sample obtained 7 minutes after adding PVP. The broad absorption band centered at ~1050 nm is the mean effect of the absorption for both in-plane and

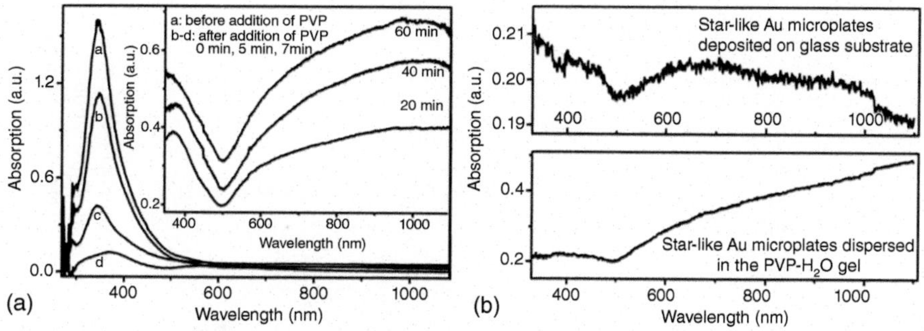

Figure 6.18 UV/Vis–NIR absorption spectra of (a) samples during the preparation of Au nanoplates at 150 °C and (b) star-like shapes formed at low temperatures with temperature variation. Source: Adapted from Kan et al. [8].

out-of-plane polarizations assigned to the dipole and quadrupole plasmon resonances of nanoplates in solution. It is noted that quadrupole plasmon resonance of Au nanostructures at ~820 nm is rarely observed because it is usually overlapped by the broad in-plane dipole resonance. At the same time, the optical absorption spectra of star-like Au nanoplates deposited on a glass slide and dispersed in H_2O-PVP gel are shown in Figure 6.18b. The yellow color of $HAuCl_4$ faded gradually with time, indicating that the slow reduction of Au^{3+} ions to Au^0 atoms proceed at low temperatures in the system. It is suggested that because the $HAuCl_4$ solution has strong acidity, only a very few free Au^{3+} ions were ionized from the stable complex $AuCl_4^-$ ions at room temperature. As a result, the complete reduction of Au^{3+} ions in $HAuCl_4$ takes a long time. The experimental results show that the color change (yellow to colorless) and the disappearance of 320 nm peak but no SPR of Au nanoparticles in the optical spectra suggest the reduction process of Au^{3+} ions involves two clear stepwise reduction stages of $Au^{3+} \rightarrow Au^+ \rightarrow$ Au atoms.

As evidenced by the above experimental results, some opinions can be reached on the reduction of metal ions. (i) The EG molecule has no reduction effect on metal ions at room temperature because the hydroxy groups in the EG molecule have no reduction property for metal ions. (ii) PVP, as a water-soluble polymer, plays a reduction effect on metal ions at low temperatures (<100 °C).

6.3.2 Growth Mechanism of the Novel and Well-Defined Au Nanoplates

Figure 6.19 presents the further characterization of the new-type Au nanoplates. In Figure 6.19a, the XRD patterns exhibit the regular hexagonal Au nanoplates (Figure 6.16a) and Au nanoplates with novel shapes. The diffraction peaks of the two types of nanoplates at 38.2° for the (111) lattice plane of the Au crystal display a similar profile and show an overwhelming intensity. These results demonstrate that the obtained Au nanoplates are single-crystalline with (111) planes as the two basal facets. The SAED patterns (inset of Figure 6.19c) recorded for the [111] incidence from the star-like Au nanoplates are the same for different parts of each plate (Figure 6.19b), thus indicating the single-crystalline nature of the

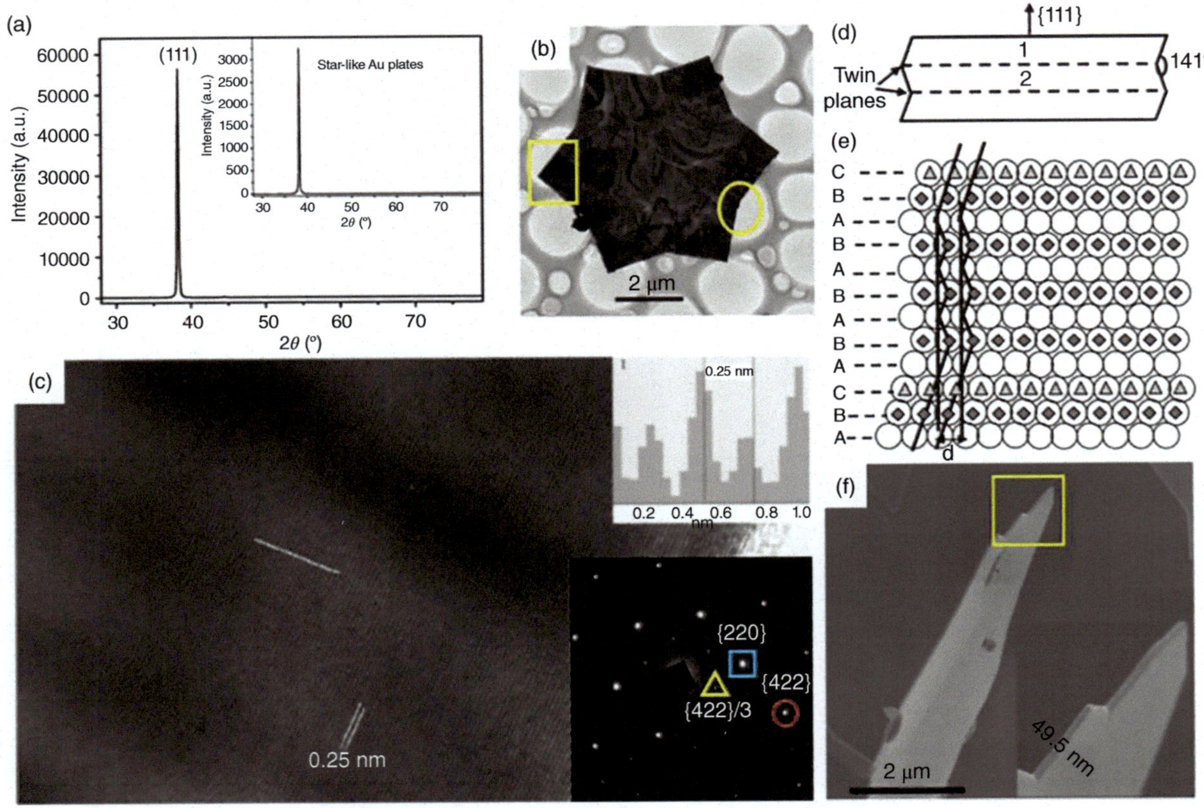

Figure 6.19 (a) XRD patterns of the Au nanoplates with hexagonal and star-like (inset) shapes indicate that the obtained Au nanoplates of different shapes are single-crystalline with two (111) basal planes. (b) A single star-like Au nanoplate with two distinct places, concave with an obtuse angle (yellow circle) and convex with a right angle (yellow rectangle). (c) HRTEM image and lattice spacing of one star-like Au nanoplate. Insets in the figure are lattice spacing histogram and SAED pattern analysis of the Au nanoplate with a hexagonal structure spot array. Marked reflections correspond to (220) (blue square), (422) (red circle), and 1/3(422) (yellow triangle). (d) Scheme illustration of cross-sectional coherent (111) twin planes and 141° reentrant. (e) Representation of the plate with multiple twin defects viewed along a [110] direction, which shows the fringe contrast (0.25 nm for Au). (f) SEM image of Au nanoplate with clear lateral surfaces, showing that the side facets are perpendicular to the basal planes. Inset corresponds to the yellow boxed part. Source: Reproduced with permission from Kan et al. [8]; © 2010, John Wiley & Sons, Inc.

obtained Au nanoplates. Besides the regular diffractions of (220) and (422) type, the SAED is dominated by the kinematically forbidden fractional (1/3)(422) and (2/3)(422) spots, which are commonly observed in the [111]-oriented flat-lying nanoplates. Similar forbidden reflections have been observed previously on thin films or platelets of *fcc* crystal structure with (111) surfaces and rather small thicknesses in the perpendicular direction [89]. Meanwhile, HRTEM images show the high crystallinity of the star-like Au nanoplates. The measured lattice spacing (0.25 nm) from the histogram (inset of Figure 6.19c), deviates distinctly from the (111) lattice spacing (0.235 nm) of the Au crystal. Instead, it agrees fairly well with $3 \times (422)$ superlattice spacing (0.249 nm) of the Au crystal and is consistent with the appearance of (1/3)(422) diffractions.

Here, the proposed formation mechanism of these large single-crystalline Au nanoplates is presented. The crystal structure, size, and morphology of the grown particles depend on the system and reaction parameters. In general chemistry, the nucleation step is a very fast process that takes place when the reagents are mixed. The subsequent crystal growth can be more readily examined. A certain shape occurs either to minimize the surface energy of the particles or due to the kinetics of growth. Under thermal equilibrium, the *fcc* metals are favorites for the formation of a twinned nanostructure with surfaces bounded by the lowest-energy (111) facets. However, if kinetics dominates, crystal growth is highly anisotropic.

In the case of platy metal nano/microstructures typically with a large flat (111) face, the presence of lattice imperfection in the (111) planes, that is, a twin-accelerated growth mechanism, was usually proposed to explain the growth of nanoplates. The defects in nuclei would act as active sites for adatom attachment to create preferred crystals. Assuming the crystal is bounded by (111) faces, multiple twins will provide a reentrant structure along the six faces of the crystal, which are intersected by the twins with 141° troughs, as schematically illustrated in Figure 6.19d. These troughs provide easy nucleation sites and act as preferential growth areas, which attain their maximum thickness in the early growth process, with subsequent growth occurring primarily in the lateral direction. From the schematic illustration in Figure 6.19e, the fringes (0.25 nm) can be calculated. Thus, it is clear that the plates, which contain at least two parallel (111) twin planes, can give rise to the formally forbidden reflection and the corresponding lattice fringes extending across the entire plates. Employing HRTEM measurements and simulations, the presence of multiple twins parallel to the (111) plane formed early in the growth process was confirmed and used to explain the forbidden reflection. With the crystal growth into structures of micrometer size, most of the troughs disappeared, and only the lateral side facet was observed to be perpendicular to the basal planes of the Au nanoplates, as shown in Figure 6.19f.

Therefore, based on the obtained results and the preferential adsorption of PVP on specific crystallographic planes, the competition and/or cooperation between the surface energy difference and the coordination effect will confer the tendency to form crystalline nuclei with parallel (111) twins, which in turn will not only determine the crystallographic growth but also dramatically change the growth kinetics of the crystal. Under favorable conditions, the preferential adsorption of polar groups of PVP on (111) planes of nuclei leads to preferential growth within (111) planes into 2D structures. Thus, the shape is not characterized by minimizing the surface energy, but rather by the growth rate of each face as determined by the kinetics.

As for the growth of nano/microplates, two mechanisms were proposed: (i) Most of the small nanoplates or nuclei, such as the perpendicular relationship between the side facets and the basal planes, will continue to grow within the (111) planes along the [110] direction, forming large nanoplates. As the previous Figure 6.6 schematically illustrated, when the growth environment is relatively uniform, nuclei will grow along (111)-close-packed [110] directions leading to the formation of a hexagonal, triangular, and truncated triangular nanoplate. (ii) Some small nanoplates will be connected along the (220) lateral planes, leading to the formation of large triangular or hexagonal nanoplates and nanobelts, as the previous Figure 6.7 schematically illustrated.

If some obvious variables, such as obvious temperature disturbance in different experimental stages, especially in the early crystal growth (or nuclei formation) stage, are introduced to the growth system, the crystal growth becomes more complex. Similarly, from the exact angles of each Au nanoplate and the growth of the (111) plane, the formation of these newly discovered Au nanoplates can be understood from the crystal growth along other directions in addition to the [110] directions, such as high-index [321] and [431] directions schematically illustrated in Figure 6.20.

Figure 6.20a presents the detailed atom arrangement of the (111) plane and different growth directions. Corresponding to the Au nanoplates labeled A–D in Figure 6.20b, Figure 6.20c–f illustrates the formation of these nanoplates. When the (111) crystal plane grows along three [110] directions (such as $[0\bar{1}1]$, $[\bar{1}10]$, and $[10\bar{1}]$) and three [211] directions (such as $[2\bar{1}\bar{1}]$, $[\bar{1}2\bar{1}]$, and $[\bar{1}\bar{1}2]$), shield-like Au nanoplates (plate A) will be formed with $3 \times 90°$ angles and $3 \times 150°$ angles, as shown in Figure 6.20c. When the (111) crystal plane grows along three [110] directions (such as $[\bar{1}01]$, $[01\bar{1}]$, and $1\bar{1}0$) and six high-index [321] directions (such as $[\bar{3}21]$, $[\bar{2}3\bar{1}]$, $[21\bar{3}]$, $[3\bar{1}\bar{2}]$, $[1\bar{3}2]$, and $[\bar{1}\bar{2}3]$) nonagonal Au nanoplates (plate B) with $9 \times 140°$ angles will be obtained in Figure 6.20d. Similarly, the formation of plate C and plate D can also be understood from Figure 6.20e and f, and high-index crystal planes and some typical angles are also provided. For example, a sniped shield-like plate (plate C with angles of 90°, 164°, and 166°) can be understood when the (111) plane grows along more high-index [431] directions in addition to [110] and [211] directions. Dodecagonal plate (plate D) with angles of 139°, and 161° can be illustrated when the (111) plane grows along [110] and [321] directions. All the angles in the schematic illustration are in accord with the measurements from the TEM images.

For the star-like Au nanoplates, the lateral facets are six (220) planes and six (422) planes. It can be seen that when the (111) crystal plane grows simultaneously and uniformly along six [110] directions and six [211] directions, the nanoplate of a regular dodecagon with $12 \times 150°$ angles will be formed. However, this growth cannot occur in an open experimental system. Au nanoplates with shield-like and shield-to-star transformation shapes were commonly observed in the products, which is very important for understanding the growth of star-like microplates.

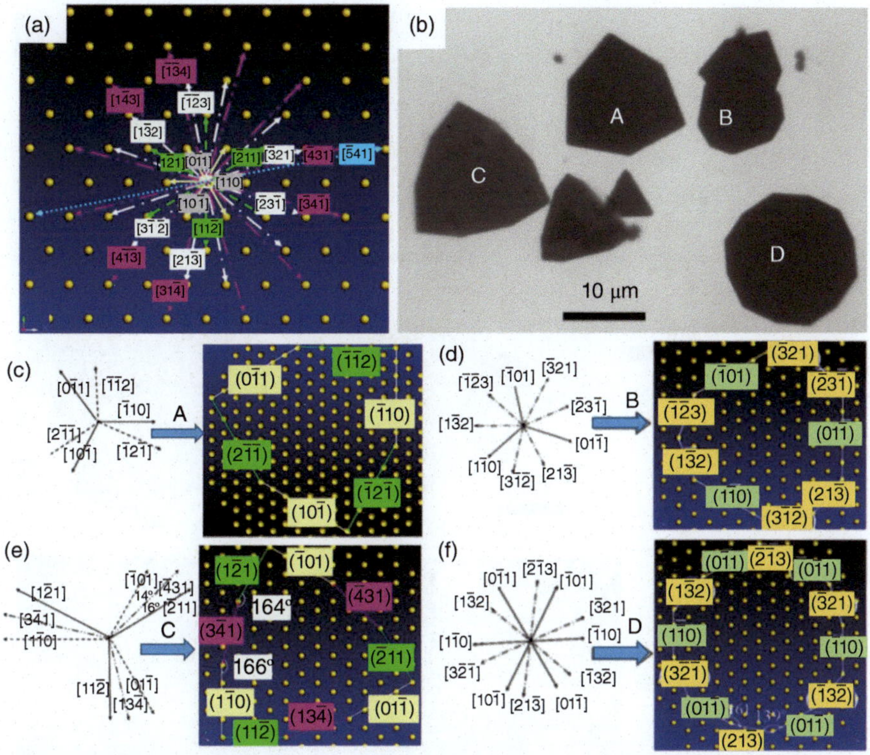

Figure 6.20 (a) Atoms arrangement and growth directions of (111) plane. (b) Optical microscopy image of some typical Au nanoplates, A–D is a hexagonal shield, nonagon, sniped shield-like nonagon, and dodecagon, respectively. (c–f) The schematic diagrams of the Au nanoplates with the corresponding shapes in (b). Source: Figure modified with permission from Kan et al. [8]; © 2010, JOHN WILEY & SONS, INC.. Wang et al. [50]; © 2014, Editorial office of Acta Physico-Chimica Sinica. Zhu et al. [86]; © 2011, Published by Elsevier.

When the (111) crystal plane grows along three [110] directions and three [211] directions, shield-like Au nanoplates will be formed, as illustrated in Figures 6.20c and 6.21a. Since the crystal can grow along the other three [110] directions and three [211] directions, with an increasing growth rate along other alternate [110] and [211] directions, plates with the shape of shield-to-star transformation will be formed (the outline of this morphology may be regarded as the composition of two shields of different sizes), as shown in Figure 6.21b1–b2. Therefore, Au nanoplates of a well-defined star-like shape (but not regular dodecagon) will be formed with $6 \times 90°$and $6 \times 210°$ angles when the growth rates of these two shield-like plates are similar, as illustrated in Figure 6.21d. In the case of the sniped shield and triple-branched Au nanoplates, as shown in Figure 6.21c1–c2, the formation process can also be understood from the growth of shield-like Au nanoplates. When the (111) plane grows along the other three pairs of alternate [110] and [211] directions (Figure 6.21e: outline for part of c1, and green outline for part of c2), sniped shields and triple-branched Au nanoplates will be formed.

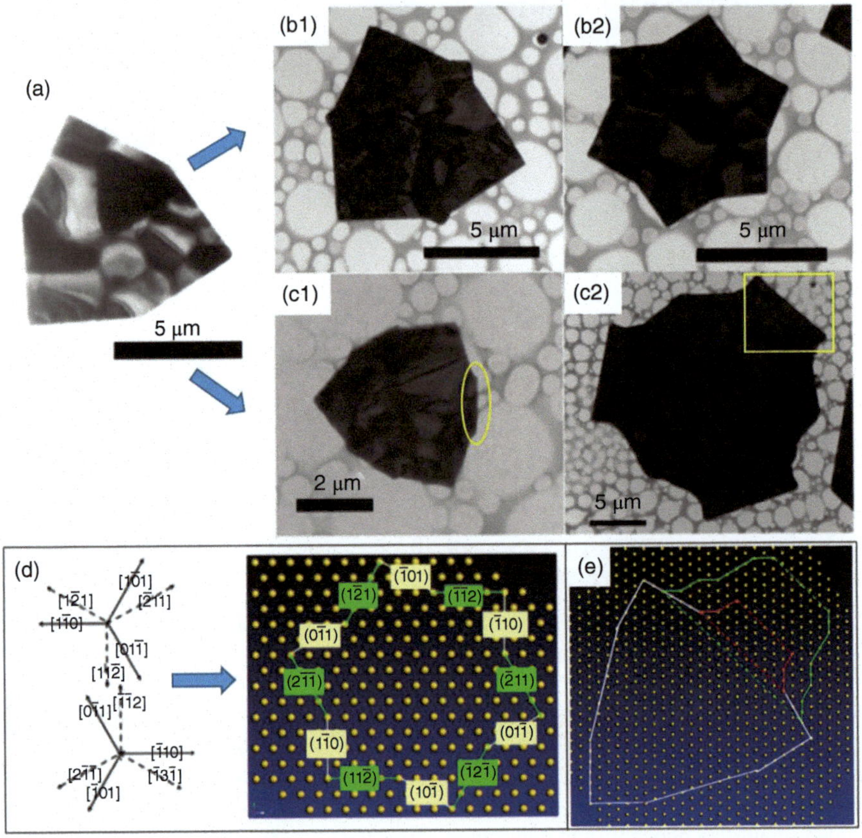

Figure 6.21 TEM images for two kinds of shape transformations from shield-like Au nanoplate. (a) Shield-like Au nanoplate. (b1–b2) Star-like Au nanoplate is formed with two sets of shield-like Au nanoplates. (c1–c2) The triple-branched Au nanoplate grew from the snipped shield-like Au nanoplate. (d and e) The schematic diagrams of the hexagonal star-like Au nanoplate and triple-branched Au nanoplate. Source: Figure modified with permission from Wang et al. [50]; © 2014, Editorial office of Acta Physico-Chimica Sinica. Zhu et al. [86]; © 2011, Published by Elsevier.

For further understanding of the (111) plane grown along high-index directions into these well-defined shapes, a theoretical calculation of the surface energy of several Au surfaces was presented in the framework of density-functional theory within the local density approximation as implemented in the Dmol3 package [90]. The calculated results indicate that (422) planes are more stable than (220) planes, and (321) planes are similar to (110) planes (Table 6.1). Therefore, it is reasonable that at low temperatures with obvious temperature fluctuation and suitable energy provided, Au nuclei will grow along [110] and [211] directions to form Au nanoplates of star-like shapes. At the same time, at high temperatures in the polyol process, slight temperature disturbance also provides the possibility for the Au crystal nuclei to grow along different directions with lateral facets of (422) and other higher-index planes in addition to (220) planes.

Table 6.1 Density-functional theory calculations of the surface energy of different planes of the Au crystal.

Crystal surface	(111)	(200)	(220)	(422)	(642)	(862)
Surface energy (eV·Å^{-2})	0.0587	0.0825	0.0991	0.0908	0.0995	0.1091
Stability order		(111) > (200) > (422) > (220) > (642) > (862)				

Source: Adapted from Kan et al. [8].

6.4 Summary of this Chapter

In this chapter, 2D metal nanoplates, especially Au and Ag nanoplates with two parallel extended facets, were described, including various bottom-up synthetic methods for normal nanoplates, the proposed growth mechanism, and their optical properties. Particularly important, single-crystalline Au nanoplates with normal and novel star-like shapes were synthesized at high temperatures and room temperatures with an obvious temperature fluctuation during the early stage of crystal growth. Further, with significant insight into the analysis of such crystallographic peculiarities and theoretical calculations on the surface energy, it has been possible to propose a growth mechanism such that the formation of these Au nanoplates can be understood from the growth of (111) plane along the [110], [211], and other higher-index directions, such as [321] and [431] directions. The exact angles and the morphological evolution of Au nanoplates with specific shapes are very important for understanding the growth process. This research portion is significant because understanding the growth mechanisms can serve as a set of design rules for the synthesis of novel metal nanostructures.

Although the research on the novel and well-defined Au nanoplates represents a significant step toward gaining a better understanding of the growth mechanism, there are still unresolved problems. On the other hand, the growth methods and mechanisms intensely depend on experimental parameters, such as concentration of the reducing agents, metal precursors, stabilizers, and reaction conditions. Research efforts are still required to fully understand the growth mechanisms, guide the further design of synthesis methods, and tight control over the geometry of metal nanoplates, thereby leading to a wide range of related applications and paving the way for the use of nanostructures in nanoscience and nanotechnology.

References

1 Zheng, J.P., Cheng, X.Z., Zhang, H. et al. (2021). Gold nanorods: the most versatile plasmonic nanoparticles. *Chem. Rev.* 121: 13342–13453.

2 Xu, H., Kan, C., Miao, C. et al. (2017). Synthesis of high-purity silver nanorods with tunable plasmonic properties and sensor behavior. *Photon. Res.* 5: 27–32.

3 Sun, Y., Yin, Y., Mayers, B.T. et al. (2002). Uniform silver nanowires synthesis by reducing $AgNO_3$ with ethylene glycol in the presence of seeds and poly(vinyl pyrrolidone). *Chem. Mater.* 14: 4736–4745.

4 He, H., Chen, R., Zhang, L. et al. (2020). Fabrication of single-crystalline gold nanowires on cellulose nanofibers. *J. Colloid Interface Sci.* 562: 333–341.

5 Scarabelli, L., Sun, M., Zhuo, X. et al. (2023). Plate-like colloidal metal nanoparticles. *Chem. Rev.* 123: 3493–3542.

6 Millstone, J.E., Hurst, S.J., Metraux, G.S. et al. (2009). Colloidal gold and silver triangular nanoprisms. *Small* 5: 646–664.

7 Qin, F., Zhao, T., Jiang, R. et al. (2015). Thickness control produces gold nanoplates with their plasmon in the visible and near-infrared regions. *Adv. Opt. Mater.* 4: 76–85.

8 Kan, C.X., Wang, C.S., Li, H.C. et al. (2010). Gold microplates with well-defined shapes. *Small* 6: 1768–1775.

9 Han, J., Zhang, J., Yang, M. et al. (2016). Glucose-functionalized Au nanoprisms for optoacoustic imaging and near-infrared photothermal therapy. *Nanoscale* 8: 492–499.

10 Xia, Y., Xiong, Y., Lim, B., and Skrabalak, S.E. (2009). Shape-controlled synthesis of metal nanocrystals: simple chemistry meets complex physics? *Angew. Chem. Int. Ed.* 48: 60–103.

11 Xia, Y., Xia, X., and Peng, H.-C. (2015). Shape-controlled synthesis of colloidal metal nanocrystals: thermodynamic versus kinetic products. *J. Am. Chem. Soc.* 137: 7947–7966.

12 Wang, Y., He, J., Liu, C. et al. (2015). Thermodynamics versus kinetics in nanosynthesis. *Angew. Chem. Int. Ed.* 54: 2022–2051.

13 Chen, L., Ji, F., Xu, Y. et al. (2014). High-yield seedless synthesis of triangular gold nanoplates through oxidative etching. *Nano Lett.* 14: 7201–7206.

14 Luo, X., Qiao, L., Xia, Z. et al. (2022). Shape- and size-dependent refractive index sensing and SERS performance of gold nanoplates. *Langmuir* 38: 6454–6463.

15 Chen, S., Xu, P., Li, Y. et al. (2016). Rapid seedless synthesis of gold nanoplates with microscaled edge length in a high yield and their application in SERS. *Nano Micro Lett.* 8: 328–335.

16 https://msestudent.com.

17 Chen, M., Wu, B.H., Yang, J., and Zheng, N.F. (2012). Small adsorbate-assisted shape control of Pd and Pt nanocrystals. *Adv. Mater.* 24: 862–879.

18 Wiesner, J. and Wokaun, A. (1989). Anisometric gold colloids. Preparation, characterization, and optical properties. *Chem. Phys. Lett.* 157: 569–575.

19 Jana, N.R., Gearheart, L., and Murphy, C.J. (2001). Seed-mediated growth approach for shape-controlled synthesis of spheroidal and rod-like gold nanoparticles using a surfactant template. *Adv. Mater.* 13: 1389–1393.

20 Ha, T.H., Koo, H.-J., and Chung, B.H. (2007). Shape-controlled syntheses of gold nanoprisms and nanorods influenced by specific adsorption of halide ions. *J. Phys. Chem. C* 111: 1123–1130.

21 Huang, Y., Ferhan, A.R., Gao, Y. et al. (2014). High-yield synthesis of triangular gold nanoplates with improved shape uniformity, tunable edge length and thickness. *Nanoscale* 6: 6496–6500.
22 Millstone, J.E., Métraux, G.S., and Mirkin, C.A. (2006). Controlling the edge length of gold nanoprisms via a seed-mediated approach. *Adv. Funct. Mater.* 16: 1209–1214.
23 Millstone, J.E., Park, S., Shuford, K.L. et al. (2005). Observation of a quadrupole plasmon mode for a colloidal solution of gold nanoprisms. *J. Am. Chem. Soc.* 127: 5312–5313.
24 Liu, X., Li, L., Yang, Y. et al. (2014). One-step growth of triangular silver nanoplates with predictable sizes on a large scale. *Nanoscale* 6: 4513–4516.
25 Chen, S. and Carroll, D.L. (2002). Synthesis and characterization of truncated triangular silver nanoplates. *Nano Lett.* 2: 1003–1007.
26 Fievet, F., Lagier, J.P., and Figlarz, M. (1989). Preparing monodisperse metal powders in micrometer and submicrometer sizes by the polyol process. *MRS Bull.* 14: 29–34.
27 Viau, G., Fiévet-Vincent, F., and Fiévet, F. (1996). Nucleation and growth of bimetallic CoNi and FeNi monodisperse particles prepared in polyols. *Solid State Ionics* 84: 259–270.
28 Wiley, B., Sun, Y., Mayers, B., and Xia, Y. (2005). Shape-controlled synthesis of metal snanostructures: the case of silver. *Chemistry* 11: 454–463.
29 Li, C.C., Cai, W.P., Cao, B.Q. et al. (2006). Mass synthesis of large, single-crystal Au nanosheets based on a polyol process. *Adv. Funct. Mater.* 16: 83–90.
30 Wiley, B., Sun, Y., Chen, J. et al. (2005). Shape-controlled synthesis of silver and gold nanostructures. *MRS Bull.* 30: 356–361.
31 Zhang, J.-M., Ma, F., and Xu, K.-W. (2004). Calculation of the surface energy of FCC metals with modified embedded-atom method. *Appl. Surf. Sci.* 229: 34–42.
32 Barnard, A.S., Young, N.P., Kirkland, A.I. et al. (2009). Nanogold: a quantitative phase map. *ACS Nano* 3: 1431–1436.
33 Lu, X., Rycenga, M., Skrabalak, S.E. et al. (2009). Chemical synthesis of novel plasmonic nanoparticles. *Annu. Rev. Phys. Chem.* 60: 167–192.
34 Xiong, Y. and Xia, Y. (2007). Shape-controlled synthesis of metal nanostructures: the case of palladium. *Adv. Mater.* 19: 3385–3391.
35 Yang, Y., Zhong, X.-L., Zhang, Q. et al. (2014). The role of etching in the formation of ag nanoplates with straight, curved and wavy edges and comparison of their SERS properties. *Small* 10: 1430–1437.
36 Zhang, Q., Yang, Y., Li, J. et al. (2013). Citrate-free synthesis of silver nanoplates and the mechanistic study. *ACS Appl. Mater. Interfaces* 5: 6333–6345.
37 Xin, W., Yang, J.-M., Li, C. et al. (2017). Novel strategy for one-pot synthesis of gold nanoplates on carbon nanotube sheet as an effective flexible SERS substrate. *ACS Appl. Mater. Interfaces* 9: 6246–6254.
38 Chang, C.-C., Wu, H.-L., Kuo, C.-H., and Huang, M.H. (2008). Hydrothermal synthesis of monodispersed octahedral gold nanocrystals with five different size ranges and their self-assembled structures. *Chem. Mater.* 20: 7570–7574.

39 Bari, B., Lee, J., Jang, T. et al. (2016). Simple hydrothermal synthesis of very-long and thin silver nanowires and their application in high quality transparent electrodes. *J. Mater. Chem. A* 4: 11365–11371.

40 Duan, H., Yan, N., Yu, R. et al. (2014). Ultrathin rhodium nanosheets. *Nat. Commun.* 5: 3093.

41 Adams, B.D., Wu, G., Nigro, S., and Chen, A. (2009). Facile synthesis of Pd–Cd nanostructures with high capacity for hydrogen storage. *J. Am. Chem. Soc.* 131: 6930–6931.

42 Saleem, F., Zhang, Z., Xu, B. et al. (2013). Ultrathin Pt-Cu nanosheets and nanocones. *J. Am. Chem. Soc.* 135: 18304–18307.

43 Kim, F., Song, J.H., and Yang, P. (2002). Photochemical synthesis of gold nanorods. *J. Am. Chem. Soc.* 124: 14316–14317.

44 Huang, W.-C. and Chen, Y.-C. (2008). Photochemical synthesis of polygonal gold nanoparticles. *J. Nanopart. Res.* 10: 697–702.

45 Osonga, F.J., Kariuki, V.M., Wambua, V.M. et al. (2019). Photochemical synthesis and catalytic applications of gold nanoplates fabricated using quercetin diphosphate macromolecules. *ACS Omega* 4: 6511–6520.

46 Jin, R., Cao, Y., Mirkin, C.A. et al. (2001). Photoinduced conversion of silver nanospheres to nanoprisms. *Science* 294: 1901–1903.

47 Jin, R., Charles Cao, Y., Hao, E. et al. (2003). Controlling anisotropic nanoparticle growth through plasmon excitation. *Nature* 425: 487–490.

48 Radha, B. and Kulkarni, G.U. (2011). A real time microscopy study of the growth of giant Au microplates. *Cryst. Growth Des.* 11: 320–327.

49 Xu, J., Kan, C.X., Lou, Y.K. et al. (2020). Structural transformation of Au seeds: the influence of temperature and surfactants. *J. Nanophotonics* 14.

50 Wang, C.S., Kan, C.X., Ni, Y., and Xu, H.Y. (2014). Facile preparation and growth mechanism of new-type gold nanoplates. *Acta Phys. Chim. Sin.* 30: 194–204.

51 Meena, S.K. and Sulpizi, M. (2016). From gold nanoseeds to nanorods: the microscopic origin of the anisotropic growth. *Angew. Chem. Int. Ed.* 55: 11960–11964.

52 Murphy, C.J., Sau, T.K., Gole, A.M. et al. (2005). Anisotropic metal nanoparticles: synthesis, assembly, and optical applications. *J. Phys. Chem. B* 109: 13857–13870.

53 Chang, H.H. and Murphy, C.J. (2018). Mini gold nanorods with tunable plasmonic peaks beyond 1000 nm. *Chem. Mater.* 30: 1427–1435.

54 Murphy, C.J., Thompson, L.B., Chernak, D.J. et al. (2011). Gold nanorod crystal growth: from seed-mediated synthesis to nanoscale sculpting. *Curr. Opin. Colloid Interface Sci.* 16: 128–134.

55 Orendorff, C.J. and Murphy, C.J. (2006). Quantitation of metal content in the silver-assisted growth of gold nanorods. *J. Phys. Chem. B* 110: 3990–3994.

56 Bullen, C., Zijlstra, P., Bakker, E. et al. (2011). Chemical kinetics of gold nanorod growth in aqueous CTAB solutions. *Cryst. Growth Des.* 11: 3375–3380.

57 Langille, M.R., Zhang, J., Personick, M.L. et al. (2012). Stepwise evolution of spherical seeds into 20-fold twinned icosahedra. *Science* 337: 954–957.

58 Lohse, S.E. and Murphy, C.J. (2013). The quest for shape control: a history of gold nanorod synthesis. *Chem. Mater.* 25: 1250–1261.

59 Personick, M.L., Langille, M.R., Zhang, J., and Mirkin, C.A. (2011). Shape control of gold nanoparticles by silver underpotential deposition. *Nano Lett.* 11: 3394–3398.

60 Langille, M.R., Personick, M.L., Zhang, J., and Mirkin, C.A. (2012). Defining rules for the shape evolution of gold nanoparticles. *J. Am. Chem. Soc.* 134: 14542–14554.

61 Magnussen, O.M. (2002). Ordered anion adlayers on metal electrode surfaces. *Chem. Rev.* 102: 679–726.

62 Millstone, J.E., Wei, W., Jones, M.R. et al. (2008). Iodide ions control seed-mediated growth of anisotropic gold nanoparticles. *Nano Lett.* 8: 2526–2529.

63 Ortiz-Castillo, J.E., Gallo-Villanueva, R.C., Madou, M.J., and Perez-Gonzalez, V.H. (2020). Anisotropic gold nanoparticles: a survey of recent synthetic methodologies. *Coord. Chem. Rev.* 425.

64 Lohse, S.E., Burrows, N.D., Scarabelli, L. et al. (2014). Anisotropic noble metal nanocrystal growth: the role of halides. *Chem. Mater.* 26: 34–43.

65 DuChene, J.S., Niu, W., Abendroth, J.M. et al. (2013). Halide anions as shape-directing agents for obtaining high-quality anisotropic gold nanostructures. *Chem. Mater.* 25: 1392–1399.

66 Sau, T.K. and Rogach, A.L. (2010). Nonspherical noble metal nanoparticles: colloid-chemical synthesis and morphology control. *Adv. Mater.* 22: 1781–1804.

67 Zhang, D., Diao, P., and Zhang, Q. (2009). Potential-induced shape evolution of gold nanoparticles prepared on ITO substrate. *J. Phys. Chem. C* 113: 15796–15800.

68 Zhai, Y., Zhang, F., Zhang, B., and Gao, X. (2017). Engineering single nanopores on gold nanoplates by tuning crystal screw dislocation. *Adv. Mater.* 29: 1703102.

69 Wu, X., Redmond, P.L., Liu, H. et al. (2008). Photovoltage mechanism for room light conversion of citrate stabilized silver nanocrystal seeds to large nanoprisms. *J. Am. Chem. Soc.* 130: 9500–9506.

70 Zhang, J., Langille, M.R., and Mirkin, C.A. (2010). Photomediated synthesis of silver triangular bipyramids and prisms: the effect of pH and BSPP. *J. Am. Chem. Soc.* 132: 12502–12510.

71 Tan, T., Zhang, S., Wang, J. et al. (2021). Resolving the stacking fault structure of silver nanoplates. *Nanoscale* 13: 195–205.

72 Xue, C., Métraux, G.S., Millstone, J.E., and Mirkin, C.A. (2008). Mechanistic study of photomediated triangular silver nanoprism growth. *J. Am. Chem. Soc.* 130: 8337–8344.

73 Sun, Y. and Xia, Y. (2003). Triangular nanoplates of silver: synthesis, characterization, and use as sacrificial templates for generating triangular nanorings of gold. *Adv. Mater.* 15: 695–699.

74 Maillard, M., Huang, P., and Brus, L. (2003). Silver nanodisk growth by surface plasmon enhanced photoreduction of adsorbed [Ag^+]. *Nano Lett.* 3: 1611–1615.

75 Zheng, X., Xu, W., Corredor, C. et al. (2007). Laser-induced growth of monodisperse silver nanoparticles with tunable surface plasmon resonance properties and a wavelength self-limiting effect. *J. Phys. Chem. C* 111: 14962–14967.

76 Rocha, T.C.R. and Zanchet, D. (2007). Structural defects and their role in the growth of ag triangular nanoplates. *J. Phys. Chem. C* 111: 6989–6993.

77 Aherne, D., Ledwith, D.M., Gara, M., and Kelly, J.M. (2008). Optical properties and growth aspects of silver nanoprisms produced by a highly reproducible and rapid synthesis at room temperature. *Adv. Funct. Mater.* 18: 2005–2016.

78 Zhang, Q., Ge, J., Pham, T. et al. (2009). Reconstruction of silver nanoplates by UV irradiation: tailored optical properties and enhanced stability. *Angew. Chem. Int. Ed.* 48: 3516–3519.

79 Xie, Y., Riedinger, A., Prato, M. et al. (2013). Copper sulfide nanocrystals with tunable composition by reduction of covellite nanocrystals with Cu^+ ions. *J. Am. Chem. Soc.* 135: 17630–17637.

80 Hao, Y., Kang, S.-Z., Liu, X. et al. (2017). An efficient Noble-metal-free photocatalyst for visible-light-driven H_2 evolution: Cu/Ni-codoped $Cd_{0.5}Zn_{0.5}S$ nanoplates. *ACS Sustainable Chem. Eng.* 5: 1165–1172.

81 Pastoriza-Santos, I., Sanchez-Iglesias, A., Rodriguez-Gonzalez, B., and Liz-Marzan, L.M. (2009). Aerobic synthesis of cu nanoplates with intense plasmon resonances. *Small* 5: 440–443.

82 Xu, S., Sun, X., Ye, H. et al. (2010). Selective synthesis of copper nanoplates and nanowires via a surfactant-assisted hydrothermal process. *Mater. Chem. Phys.* 120: 1–5.

83 Lee, J.W., Han, J., Lee, D.S. et al. (2018). 2D single-crystalline copper nanoplates as a conductive filler for electronic ink applications. *Small* 14.

84 Wu, K., Sun, C., Wang, Z. et al. (2022). Surface reconstruction on uniform Cu nanodisks boosted electrochemical nitrate reduction to ammonia. *ACS Mater. Lett.* 4: 650–656.

85 Liu, B., Xie, J., Lee, J.Y. et al. (2005). Optimization of high-yield biological synthesis of single-crystalline gold nanoplates. *J. Phys. Chem. B* 109: 15256–15263.

86 Zhu, J.J., Kan, C.X., Li, H.C. et al. (2011). Synthesis and growth mechanism of gold nanoplates with novel shapes. *J. Cryst. Growth* 321: 124–130.

87 Skrabalak, S.E., Wiley, B.J., Kim, M. et al. (2008). On the polyol synthesis of silver nanostructures: glycolaldehyde as a reducing agent. *Nano Lett.* 8: 2077–2081.

88 Pastoriza-Santos, I. and Liz-Marzán, L.M. (2009). N,N-dimethylformamide as a reaction medium for metal nanoparticle synthesis. *Adv. Funct. Mater.* 19: 679–688.

89 Courty, A., Henry, A.I., Goubet, N., and Pileni, M.P. (2007). Large triangular single crystals formed by mild annealing of self-organized silver nanocrystals. *Nat. Mater.* 6: 900–907.

90 Delley, B. (2000). From molecules to solids with the DMol3 approach. *J. Chem. Phys.* 113: 7756–7764.

7

Hollow and Open Nanostructures with Enhanced Activity

Solid noble metals are expensive and scarce,
while hollow nanocrystals present more possibilities.

7.1 Introduction

Global warming and climate change issues arising from fossil fuel usage pose formidable challenges to the preservation of the already fragile global eco-system and the maintenance of human quality of life. Heterogeneous catalysis built around noble-metal nanocrystals is central to the chemical, energy, and automobile industries. Over the past decades, nanocrystals made of Pd, Pt, Rh, Ru, Ag, Au, and combinations of those have received great interest owing to their favorable electronic structures for achieving high catalytic activity and/or selectivity toward a broad spectrum of reactions. These materials often exhibit their resistance to corrosion, sustaining long-term durability in a wide variety of catalytic processes. However, because of their extraordinary low abundances in the Earth's crust and the ever-increasing demands from various industries, there is an urgent need to utilize this extremely rare resource in a cost-effective and sustainable manner without compromising the catalytic performance. To this end, it is pivotal to maximize the mass-based catalytic activity of the noble-metal nanocrystals.

The most popular approach has been the preparation of noble metal based alloy nanocatalysts, which enable a degree of control over the subtle balance between the binding and desorption of the reaction intermediates via d-band engineering [1]. Nanoparticle facet control and the structural design concept of core-shell morphology are further tools to fine-tune the surface energy of nanocatalysts. Facet-controlled nanoparticles exhibit different catalytic activities from those of spherical nanoparticles due to the differences in facer-dependent surface energy states. Additionally, lattice mismatch between the different phases in a core-shell structure creates tensile strain or compressive strain on the surface atoms [2]. The nature of this strain is critical to the energy states of the surface atoms and thereby their catalytic properties.

Plasmonic Metal Nanostructures: Preparation, Characterization, and Applications,
First Edition. Caixia Kan.

On the other hand, the catalytic events occurring on the surface of nanoparticles are intrinsically heterogeneous. Therefore, increasing the number of active sites and the overall surface area of a nanoparticle would be of immense value in boosting the catalytic activity. From this logic, high dispersion of metal atoms on supporting material is crucial for fabricating the efficient heterogeneous electrocatalysts because all catalysis events occur on the surface of a catalyst particle. To date, the impregnation reduction method or functionalization of the carbon surface are regarded as effective methods to prepare the noble metal nanoparticles on carbon carriers [3, 4]. However, these preparation methodologies have been limited when we developed rationally designed complex nanostructures using various different strategies. Alternatively, several groups adopted the incipient-wetness procedure in order to prepare the heterogeneous catalysts with high dispersion. While the above-described approaches allow for enhanced electrocatalytic activity as well as reduced noble metal content, they are all based on solid nanoparticles, in which a significant fraction of the noble metal atoms are buried under the surface. The past few years have witnessed the development of hollow nanostructures as a novel nanocatalyst design modification (Figure 7.1) [5–7]. In general, the term "frames" refers to nanostructures comprised of only edges without facets, and such structures can allow external reactants to access the interior of the nanostructures. The term "cages" is used for nanostructures whose side facets have large cavities. These terms have been hardly differentiated, so we have re-defined the term "hollow" to include both "frames" and "cages."

Hollow nanostructures can provide a multitude of advantages for catalysis over solid nanoparticle catalysts. They can expose a very high fraction of surface sites, have strain-induced highly reactive surfaces, and allow for increased collision frequency by confining reactants within nanoscale spaces. Furthermore, hollow nanoparticles would require significantly lower amounts of expensive noble metals, representing a noble-metal economic design. Hollow nanostructures can have metal alloy compositions, enabling very high catalytic activities. Intense research efforts have been directed toward fortifying flimsy hollow nanostructures, and some

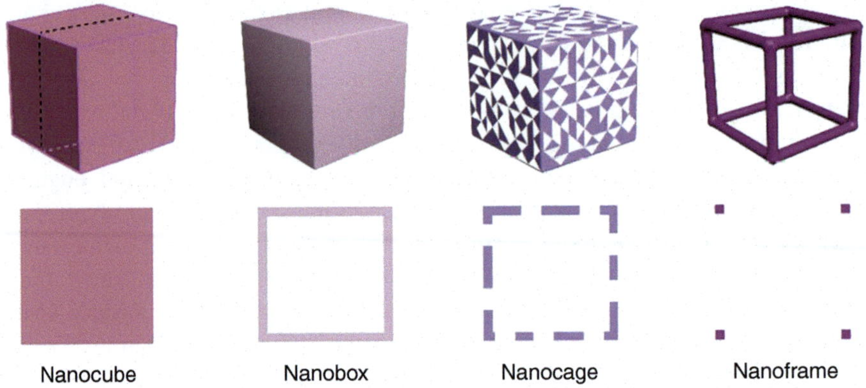

Figure 7.1 Model of a nanocube, nanobox, nanocage, and nanoframe in three dimensions (top) and cross-section (bottom) views.

promising outcomes in the development of highly active yet robust electrocatalysts have been reported.

The first synthesis of noble-metal nanoframes was reported in 2007 [8]. The authors fabricated Au-based cubic nanoframes by selectively removing the Ag atoms from Au–Ag alloy nanoboxes using an aqueous etchant based on $Fe(NO_3)_3$ or NH_4OH. Thanks to the incredible progress made in recent years, it is now feasible to design and rationally fabricate hollow nanostructures with tunable compositions and morphologies to achieve the desired properties for catalytic and plasmonic applications [9–11]. Notable examples include those mainly comprised of one of the noble metals, as well as their bi- and trimetallic combinations. Note that it is insufficient to only use "composition" to describe a bi- or trimetallic system, because their properties are also strongly dependent on the spatial distributions and atomic ordering of the constituent elements. With the recent development in nanochemistry, it is feasible to control the spatial distributions of constituent elements and, thus, the directionality of nanoscale phase segregation in bimetallic hollow nanostructures [12, 13]. This ability allows one to control the compositional heterogeneity on the surface of a nanoframe, promising enhanced catalytic performance. Today, nanoframes with many distinctive morphologies have been reported, including those in the architecture of a three-dimensional cube, cuboctahedron, octahedron, rhombic dodecahedron, decahedron, icosahedron, tripod, hexapod, and pentagonal rod, as well as a two-dimensional triangle, hexagon, and circle. Additionally, nanoframes such as octopods enclosed by high-index facets have been prepared in high yields. Remarkably, hollow nanostructures featuring complex and/or hybrid structures have also been realized [14–16]. These new variants have enabled a wide variety of applications, including those related to catalysis for energy conversion and environmental protection, as well as plasmonically enhanced spectroscopy and photothermal cancer therapy.

Herein, we aim to summarize recent advances in the synthetic concepts for nanoframes and nanocages, along with the synthetic toolkits available, followed by a description of the overall performance of the nanoframe and nanocage catalysts toward specific electrocatalytic reactions as compared to those of nanocatalysts of different structural types. Recent research efforts to enhance the structural robustness and catalytic activity of nanoframe and nanocage catalysts will then be described. Finally, a concluding summary and suggestions for further study directions will be presented.

7.2 Hollow Nanostructures

The synthesis of hollow nanostructures largely consists of three steps: the synthesis of a template, the formation of over-layers such as shell or frame structures on the template, and the removal of the inner template. The template, which can be either pre-synthesized or prepared *in-situ*, plays a crucial role in the synthesis of hollow nanostructures, which determines the final morphology of the hollow nanostructure and affects its atomic composition. While the synthetic methodologies used for hollow nanostructures vary according to the type of template material, the types of reactions used with a pre-synthesized template are different from those used on an *in-situ* formed template. Since the pre-synthesized templates are already thermodynamically stable. Heteroepitaxial growth, galvanic replacement, and nanoscale Kirkendall effect processes are feasible with pre-synthesized templates. However, *in situ* formed templates require delicate kinetic control of the decomposition of several precursors so that the template will be formed significantly ahead of the deposition of over-layers. Without such delicate kinetic control, only simple alloys containing all the components in the reaction vessel would be formed, and none of the many material design concepts, such as phase mixing/segregation and metastability-empowered etching to produce hollow structures, would be possible.

7.2.1 Synthetic Approaches

7.2.1.1 Selected Carving of Solid Nanostructures

Figure 7.2 illustrates a wet-chemical methods that leverage the intrinsic surface inhomogeneity to selectively carve atoms from solid nanocrystals for their transformation into nanoframes [17]. For a monometallic nanocrystal, surface inhomogeneity exists among atoms located at faces, edges, and vertices due to their differences in coordination numbers. It can also arise from crystallographic defects such as a vacancy, dislocation, and twin/grain boundary. In general, the site with the highest surface free energy is most vulnerable to oxidative etching.

In general, the surface inhomogeneity originating from elemental composition determines the selectivity of carving for the different sites of a nanocrystal. Specifically, when the faces of the rhombic dodecahedron are enriched by metal A with a lower reduction potential, while the edges and vertices are dominated by metal B with a higher reduction potential, the oxidative etchant will selectively attack and dissolve atoms from the faces, while the edges and vertices will be retained.

7.2.1.2 Selected Deposition of a Different Metal on a Template

Different metals can also selectively deposit onto the edges and vertices of the template nanostructures for the generation of a core-frame nanocrystal, followed by the removal of the core through oxidative etching.

Figure 7.3 illustrates the transformation of a cubic nanocrystal into a nanoframe when the metal A in the original nanocrystal is less reactive than the deposited

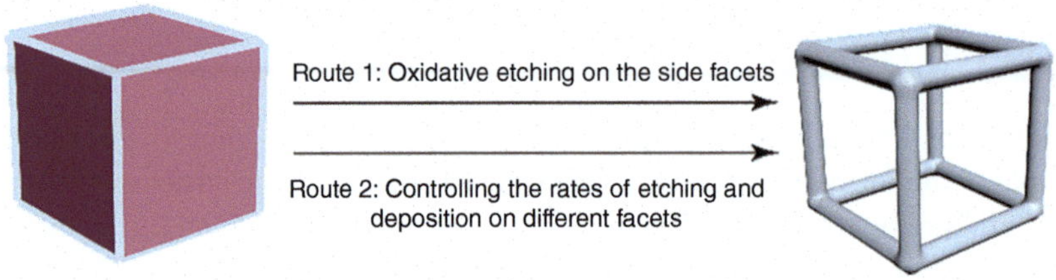

Figure 7.2 Schematic illustration of two protocols that rely on the carving of atoms from side faces for the transformation of monometallic octahedron.

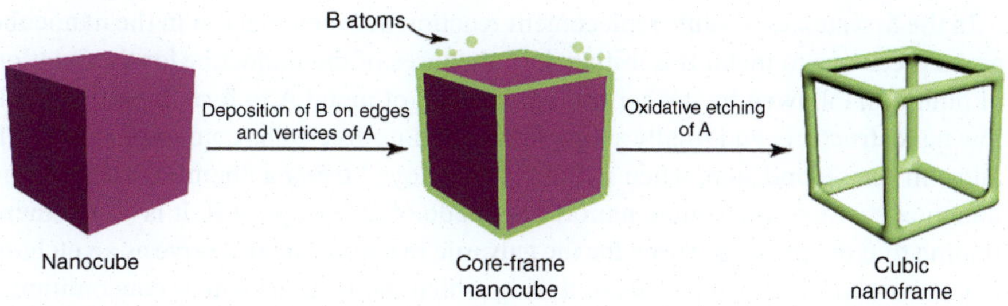

Figure 7.3 Schematic illustration showing the fabrication of cubic nanoframes made of metal B by selectively depositing it on the edges and vertices of a nanocube made of metal A, followed by the selective removal of metal A from the core via etching.

metal B [18]. For a cubic nanocrystal made of an fcc metal, the surface energies decrease in the order of edges, vertices, and faces when the faces are passivated by a capping agent. The B atoms derived from the reduction of a precursor should only nucleate and grow from the edges and vertices with higher surface energy. If the rate of atom deposition is faster than that of surface diffusion to faces, the deposited B atoms will be confined to their original sites for the generation of an A@B core-frame nanocube. Upon etching of metal A from the core while leaving the frames intact, a cubic nanoframe based on metal B will be obtained. When the amount of metal B being deposited is controlled, the ridges of the nanoframes can be made as thin as a few nanometers.

7.2.1.3 Galvanic Replacement

A solid alloy nanocrystal can also be transformed into a nanoframe by selectively dissolving one type of atom from the side faces through oxidative etching. In this case, the site selectivity is typically achieved by manipulating the elemental distributions among the faces, edges, and vertices when the alloy nanocrystals are synthesized. This requirement can be relaxed by switching to hollow alloy nanocrystals, in which the atoms situated on faces are typically more reactive than those at the edges and vertices due to the presence of a tensile strain on the face with an ultrathin thickness. Figure 7.4 illustrates this concept by showing the transformation of a solid nanocube into an alloy nanobox and then a nanoframe [19].

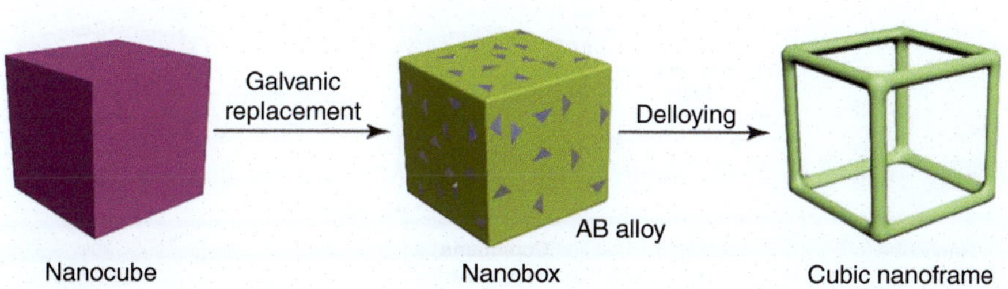

Figure 7.4 Schematic illustration showing the formation of an A–B alloy nanobox through a galvanic replacement reaction and then a B-based cubic nanoframe by dealloying metal A from the walls of the A–B nanobox with a chemical etchant.

In the first step, a galvanic replacement reaction between metal A in the nanocube and a precursor to metal B is initiated on the faces of the nanocube for the creation of pinholes, followed by the continuous carving of metal A to form a partially hollow nanostructure, and finally a nanobox with pore-free walls composed of an A–B alloy. In the second step, when metal A is dealloyed using a chemical etchant, the nanobox evolves into a cubic nanoframe made of almost pure B. It is worth mentioning that the precursor used for the galvanic reaction can also serve as an etchant to facilitate the dealloying process, during which the B atoms will be concomitantly produced and deposited onto the surface. However, the coupling between the dealloying of metal A and the deposition of B makes it difficult to precisely control the wall thickness and porosity of the resultant nanoframe. The quality of the hollow alloy nanocrystals fabricated using a galvanic replacement reaction also plays an essential role in determining the success of such a transformation process. In most cases, the final nanoframes are comprised of an alloy rather than a pure metal due to the difficulty of achieving complete dealloying.

7.2.1.4 Combining the Reduction and Galvanic Replacement Reactions

In order to enhance the controllability of the deposition and growth of B atoms, combining the reduction and galvanic replacement reactions together in a reaction system is the general synthesis approach. As shown in Figure 7.5, in the first step, a galvanic replacement reaction between metal A in the nanocube and a precursor to metal B is initiated on the faces of the nanocube for the creation of pinholes, followed by the continuous carving of metal A to form a partially hollow nanostructure, and finally a nanobox with pore-free walls composed of an A–B alloy [20]. It is worth mentioning that the released A atoms and newly produced B atoms will be concomitantly produced and deposited onto the surface. In the second step, when metal A is dealloyed using a chemical etchant, the nanobox evolves into a cubic nanoframe made of almost pure B. However, the coupling between the dealloying of metal A and the deposition of B makes it difficult to precisely control the wall thickness and porosity of the resultant nanoframe. The quality of the hollow alloy nanocrystals fabricated using a galvanic replacement reaction also plays an essential role in determining the success of such a transformation process. In most cases,

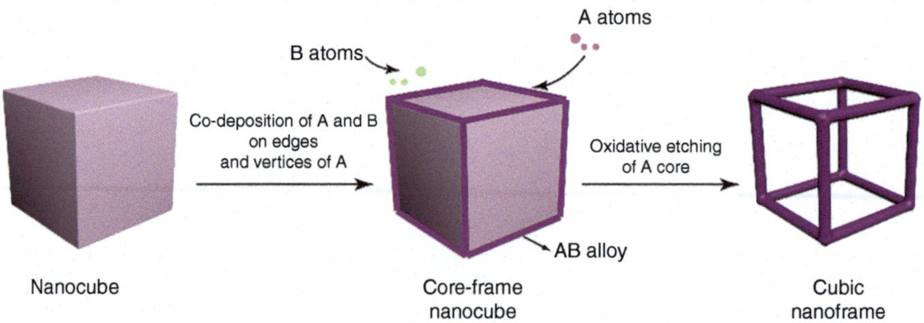

Figure 7.5 Schematic illustration showing the fabrication of nanoframes made of an A–B alloy by titrating B^{n+} ions into an aqueous suspension of A@A–B core-frame nanocubes to induce a galvanic replacement with metal A in the core.

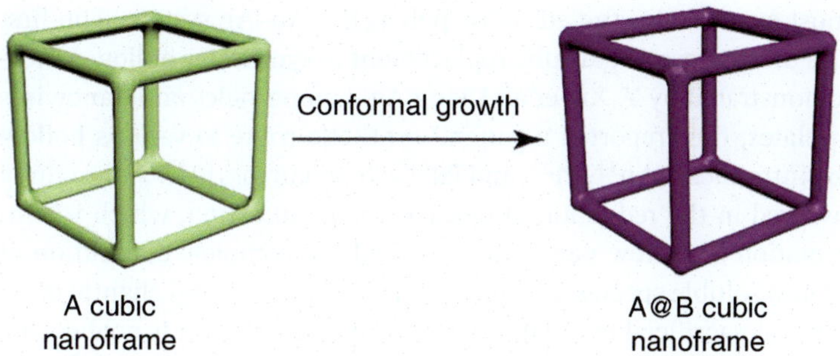

Figure 7.6 Schematic illustration showing the use of a preformed metal nanoframe as a seed for the deposition of another metal. Depending on the growth mode, the deposited metal can evolve into a complete shell on the ridges of the nanoframe.

the final nanoframes are comprised of an alloy rather than a pure metal due to the difficulty of achieving complete dealloying.

7.2.1.5 Hollow Nanostructure-Directed Growth

In principle, nanoframes can directly serve as a template for the deposition of a different metal to increase the diversity in terms of composition while improving their mechanical strength and even enhancing their catalytic and/or plasmonic properties. This methodology has been applied to the fabrication of bi- and multimetallic nanoframes with a variety of compositions. Figure 7.6 outlines the general concept, which is essentially the same as the conventional seed-mediated growth except for the use of hollow and porous nanocrystals as the seeds [5]. Depending on the growth mode, the deposited metal can either evolve into a complete shell on the ridges of the nanoframe or grow into islands or spikes randomly distributed on the skeleton of the nanoframe.

The final morphology is mainly determined by the ratio between the deposition rate ($R_{deposition}$) and the surface diffusion rate ($R_{diffusion}$) of the adatoms. In general, faster diffusion favors the formation of a complete shell, whereas faster deposition favors the island growth mode. During the synthesis, the precursor for the deposited metal is typically titrated into the suspension of nanoframes in order to eliminate homogeneous nucleation. Again, there should be no galvanic replacement reaction between the nanoframes and the precursor being added. Otherwise, the nanoframes could be easily destroyed due to their ultrathin ridges. A similar protocol should be extended to the deposition of other functional materials, such as semiconductors, onto the surface of metal nanoframes. In practice, this method has not been widely explored in the literature, primarily because of the limitations arising from the small quantities of nanoframes as well as their structural fragility.

7.2.2 Typical Hollow Nanostructures

7.2.2.1 Ag Nanostructures-Directed Hollow Nanostructures

Silver (Ag) has been widely employed as a template material for nanoframe formation as the morphology of Ag nanocrystals can be easily controlled and selectively removed. The most extensively studied redox pair is the Au–Ag system. The first example of the use of a nanoscale galvanic replacement to synthesize hollow nanostructures was demonstrated by Y. Xia et al. Using Ag nanoparticles and nanowires as sacrificial templates, they reported a general synthetic route to various hollow nanostructures by introducing Au^{3+}, Pt^{2+}, and Pd^{2+} salt precursors [21]. Two or three Ag atoms are involved in the reduction of one Pd, Pt, or Au cation, which is conducive to the formation of hollow nanostructures and the complete dissolution of the Ag templates into soluble species. In subsequent works, they used slightly truncated Ag nanocubes as sacrificial templates to form Au nanoboxes. It is noteworthy that the galvanic replacement occurred preferentially on the facets with high surface free energy. Because the surface free energy of metals with an fcc crystal structure generally decreases in the order of $\gamma_{(220)} > \gamma_{(200)} > \gamma_{(111)}$, the facets, on which the galvanic replacement reaction would occur, for particular metal nanocrystals can be predicted. However, it is also feasible to make the reaction occur only on the desired facets by capping the undesired facets.

In addition to facet selectivity, various phenomena such as alloying/dealloying and the Kirkendall effect can be coupled with galvanic replacement to add further diversity to the elemental composition and architecture of the hollow nanostructures. González et al. reported simultaneous or sequential galvanic replacement and Kirkendall effect using Ag nanocubes to synthesize complex hollow nanostructures of Au–Ag, Ag–Pd, and Au–Ag–Pd [22]. The initial Au deposition on the Ag nanocube via galvanic replacement reaction results in the formation of pinholes, which enable the dissolution of the inner portion of Ag nanocube. As the reaction continues, Au deposition on the interior surface of the cavity occurs, so that a thin Ag layer is enclosed between two Au shells. At this stage, both galvanic replacement and Kirkendall effect are in action simultaneously due to the faster diffusion rate of Ag^{+} ions, leading to the formation of a Kirkendall void between the two Au shells. They also demonstrated a sequential galvanic replacement and Kirkendall effect in which an Ag nanocube first undergoes a galvanic replacement reaction with Pd, and then the addition of Au promotes the formation of cavities, primarily driven by the Kirkendall effect. Kitaev and Coombs et al. demonstrated that selective Au deposition onto decahedral Ag nanoparticles under optimized reducing conditions and subsequent Ag etching offers a reliable pathway to well-defined gold nanoframes [24]. In the initial stage, when Au deposition occurs on the Ag templates, the Au–Ag alloying process is favored owing to the stronger bonding between Au and Ag as compared to Ag–Ag and Au–Au, and the low energy barrier for interdiffusion between Ag and Au. The overall reaction process can be kinetically controlled to stop at Au–Ag nanocages or to form Au nanoframes via further etching process (Figure 7.7a and b) [23, 25]. In the case of Pt, due to the Pt–Pt bonding being stronger than Pt–Ag bonding, the growth of Pt islands on the Ag template is more favored than alloying.

7.2.2.2 Pd Nanostructures-Directed Hollow Nanostructures

Palladium (Pd) has been widely used as a template material for nanoframe fabrication by a seed-mediated growth strategy. Ye et al. reported the synthesis of platinum (Pt) cubic nanocages using Pd nanocubes as the sacrificial template

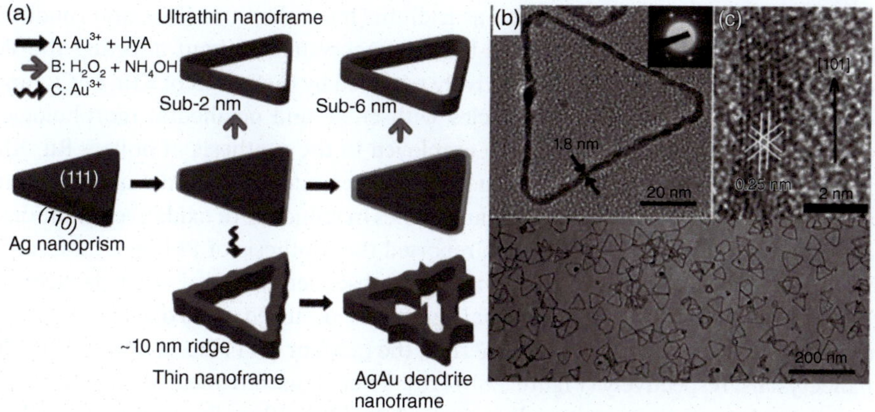

Figure 7.7 (a) Schematic illustration of the structure evolution based on the silver nanoprism template toward various nanoframe structures. (b) TEM and HRTEM images of ultrathin gold nanoframes. Source: (a) Reproduced with permission from González et al. [22]; © 2011 American Association for the Advancement of Science - AAAS. (b) Reproduced with permission from Shahjamali et al. [23]; © 2013, John Wiley & Sons, Inc.

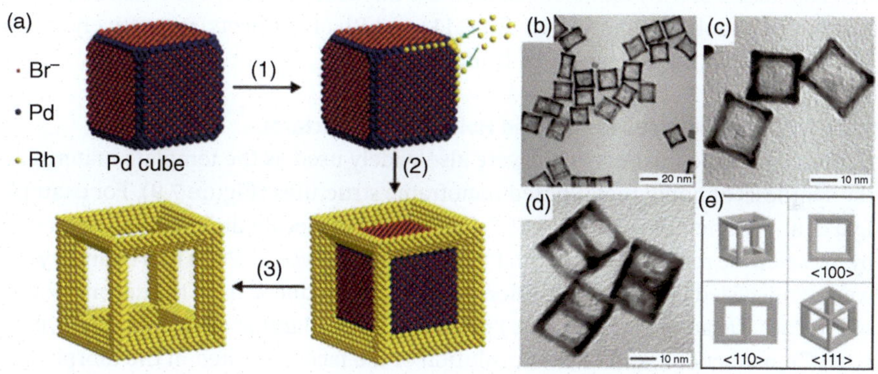

Figure 7.8 (a) The three major steps involved in the synthesis of Pd-Rh core–frame nanocubes with concave faces and Rh cubic nanoframes. (b) TEM images of Rh cubic nanoframes obtained by selectively etching away the Pd cores from the PdRh core–frame nanocubes. (c,d) TEM images of Rh cubic nanoframes projected along <100> and <110> zone axes, respectively. (e) A 3D model of the Rh cubic nanoframe and its projections along <100>, <110>, and <111> zone axes. Source: Reproduced with permission from Xie et al. [18]. © 2012, John Wiley & Sons, Inc.

(Figure 7.8) [26–28]. The Pt content obtained from the precursor Na_2PtCl_6 was selectively deposited on the edges and corners of Pd nanocubes, with Br^- ions from KBr acting as blocking agents for the six facets of Pd nanocubes. The Pd core in Pd@Pt core@frame nanocubes was then selectively removed by a mixture of H_2O_2, HCl, and Br^- ions, to form Pt cubic nanoframes. The final morphology of Pt nanoframes can be varied by altering the initial shape of the Pd template nanocrystals. Park et al. reported the synthesis of Pt octahedral nanoframes using Pd octahedral nanocrystals as the template.

Other noble metal elements, such as iridium (Ir), ruthenium (Ru), and rhodium (Rh), could be deposited on the Pd-based nanocrystals without affecting the Pd template. The Pd core could be selectively removed by the action of KBr, $FeCl_3$, and HCl. Facet-controlled Pd nanoparticles with cubic and octahedral morphologies of cubes and octahedrons have been employed in the synthesis of hollow Ru, Rh, Pt, and Ir nanostructures [25, 26]. Sneed et al. fabricated Rh on Pd nanocubes and synthesized PdRh nanoboxes or Rh nanoframes in reductive or oxidative conditions, respectively [26]. Similarly, Xie et al. reported the synthesis of Pt-based nanocages via Rh overgrowth on Pd nanocubes and the subsequent removal of Pd via etching. The growth of Rh on Pd nanocrystal templates occurred epitaxially so that the synthesized Rh-based nanocages inherited the (200) or (111) facets of the cubic Pd nanocrystals, respectively (Figure 7.8b) [18]. Ye et al. used truncated Pd octahedral nanocrystals to grow Ru epitaxially (Figure 7.8c) [28]. While Ru itself prefers a hcp structure, the epitaxially grown Ru phase exhibited a fcc structure with (200) and (111) facets due to the presence of the fcc Pd template, and the fcc Ru nanoframe remained intact after the removal of the Pd core via etching. Concave Pt was used for the heteroepitaxial growth of fcc Ru and Rh phases. The selective growth of Ru and Rh on the (111) vertices of the Pt nanocube template was induced by capping the (200) facets of the Pt nanocubes with CO molecules. The synthesized octahedral Ru and Rh nanoboxes showed fcc Ru (111) and Rh (111) facets, respectively, whose atom packing motif was inherited from Pt (111).

7.2.2.3 Cu Nanostructures-Directed Hollow Nanostructures

Copper (Cu)-based nanocrystals were also widely used as the template material for the synthesis of noble-metal-based nanoframe structures (Figure 7.9). For example, Lyu et al. reported the synthesis of PtCu rhombic dodecahedral (RDH) nanoframes by the selective removal of Cu from PtCu RDH nanocrystals. PtCu RDH nanocrystals were synthesized from the deposition of Pt atoms, produced from the decomposition of $Pt(acac)_2$ (acac: acetylacetonate) precursor in oleylamine (OAm), on presynthesized Cu nanocubes, and further evolution of the facets resulted in the morphology change from cubic toward RDH structures. The selective removal of the Cu-based core was achieved by HNO_3 treatment. Cu has also been widely applied as a template material for the solvothermal synthesis of nanoframes. Moreover, a Cu-based template can be formed *in situ* and subsequently removed to fabricate various nanocages and nanoframes via the one-pot process. Ding et al. reported the one-pot preparation of highly open PtCu RDH nanoframes by simultaneous decomposition and reduction of $Pt(acac)_2$ and $CuCl_2 \cdot 2H_2O$ precursors with glucose in OA and OAm at 180 °C [29]. In the initial step of the reaction, irregular Cu-based nanocrystals were formed, followed by the galvanic replacement of Pt content, forming RDH structures. Subsequently, the Cu content in the as-formed RDH nanocrystals was dissolved by the action of Cl^- ions present in the Cu precursor, leading to the formation of PtCu RDH nanoframe structures.

The utilization of Cu-based templates is extendable to other Pt-group elements, such as Ir, Rh, and Pd. Pei et al. reported the one-pot fabrication of IrCu nanoframes via *in-situ* galvanic replacement of Ir in the preformed Cu nanoparticles at 170 °C. Zhang et al. reported the synthesis of octahedral RhCu nanoframes.

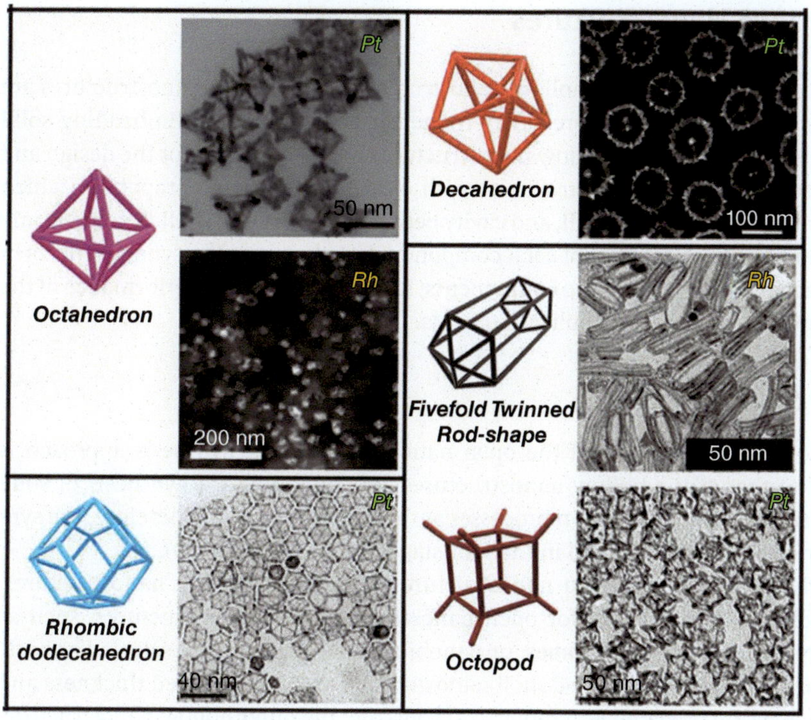

Figure 7.9 Summary of bimetallic nanoframe structures via solvothermal route including *in situ* formed Cu templates. Source: Reproduced with permission from Refs. [29, 30], © 2015 Royal Society of Chemistry; Reproduced with permission from Part et al. [31], © 2016 Wiley-VCH.

RhCu octahedral nanocrystals were formed by the coreduction of $RhCl_3{\cdot}3H_2O$ and $CuCl_2{\cdot}2H_2O$ precursors in OAm at 180 °C. RhCu octahedral nanoframes were obtained by the selective removal of Cu content in RhCu octahedral nanocrystals using diluted HCl at 80 °C [30]. Park et al. reported the one-pot fabrication of 3D RhCu nanoframes with truncated octahedral, cubic, and fivefold twinned morphologies. The simultaneous thermal decomposition and coreduction of $Rh(acac)_3$ and $Cu(acac)_2$ precursors with stearic acid, CO, and OAm at 250 °C resulted in the formation of RhCu nanoframes [31]. Irregular Cu-based nanoparticles were formed in the initial step of the synthesis, after which the growth and shape evolution toward morphology occurred, assisted by the growth of Rh content on Cu-based nanoparticles, as the decomposition of the Rh precursor is slower than that of the Cu precursor. The subsequent *in-situ* galvanic replacement of pre-existing Cu nanoparticles by Rh induced the *in-situ* dissolution of the Cu content, which resulted in the one-step fabrication of RhCu nanoframes. The morphologies of RhCu nanoframes could be altered by controlling the reaction kinetics of Rh and Cu precursors during synthesis. The addition of 1,2-hexadecanediol (1,2-HDD) to the reaction mixture, which facilitates the reduction of Rh and Cu precursors, resulted in the formation of RhCu hollow nanocube structures. Moreover, RhCu fivefold twinned nanoframes were obtained by aging the reaction mixture at 100 °C under Ar atmosphere.

7.3 Open Nanostructures

Compared with the various hollow nanostructures above, open nanostructures are more interesting for their core-cavity-frame (or cage) structure. Embedding solid structures in the center of hollow nanostructures points out a way for the design and preparation of hollow nanostructures with excellent plasmonic properties, which are combined by the core, shell, and cavity between the core and shell. It can not only maintain the characteristics of each component but also show the synergy effects in the nanogaps between any two components, further enhance the performance of the structure, and expand the application fields.

7.3.1 Synthetic Approach

Based on the morphology of the open nanostructures, the synthesis approach is similar to that of the hollow nanostructures, except adding the synthesis of solid nanostructures and overgrown processes at the beginning stage. Therefore, the synthesis procedure is concluded in the four stages below (Figure 7.10).

Firstly, choose an ideal Au nanostructure as the base of open nanostructures. Depending on the demand for open nanostructures, Au nanostructures, such as nanosphere, nanorod, nanocubes, or nanobipyramids could all be selected as basic structures. Secondly, the core@shell nanostructures with the desired thickness and morphology are prepared as templates to construct the outer hollow nanostructure. In particular, the distance between the inner core and outer hollow shell could be adjusted by controlling the thickness of shell. Thirdly, one strategy is to fabricate core@metal@shell sandwich nanostructures. The other strategy involves the construction of core@metal nanostructures, followed by the coating of a shell material via the galvanic-replacement reaction. Finally, the metal interlayer can be partially or completely removed by various etching methods. The open nanostructure is synthesized.

7.3.2 Au Nanosphere-Embedded Open Nanostructures

Nowadays, the synthesis of open nanostructures is more concentrated on metal/semiconductor or metal/metallic oxide nanostructures. While two metals

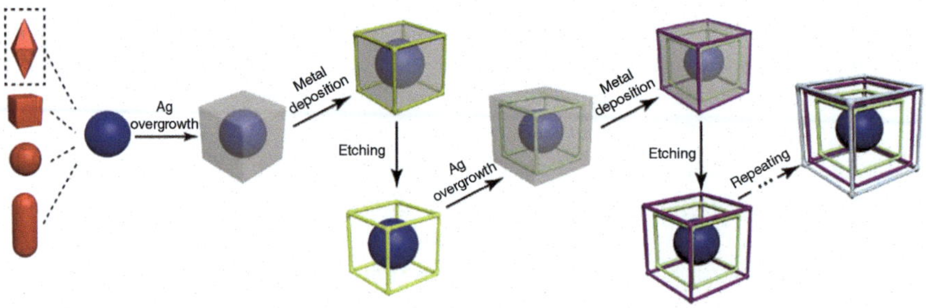

Figure 7.10 Schematic diagram of the synthesis of open nanostructures.

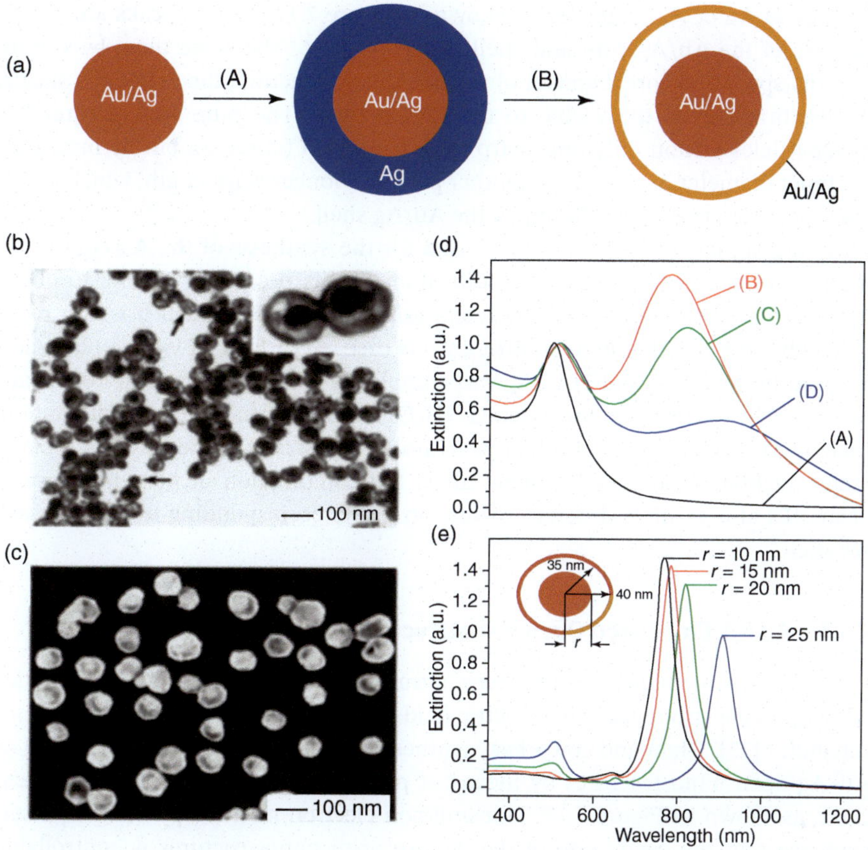

Figure 7.11 (a) Schematic synthesis procedure for Au/Ag alloy YS NPs, (b) TEM and (c) SEM images of YS NPs, (d) UV–Vis–NIR extinction spectra of YS NPs that were prepared by titrating the same amount of Ag-coated Au/Ag colloids with different volumes of 1 mM $HAuCl_4$ solutions of (A) 0, (B) 0.4, (C) 0.5, and (D) 0.6 mL (as labeled in d). All of the spectral curves were normalized against the intensities of peaks around 510 nm. (e) Extinction spectra calculated for a YS whose core and shell contained 75% of Au and 25% of Ag. The inner and outer diameters of the shell were fixed at 70 and 80 nm, respectively. The diameter of the core was varied from 20 to 50 nm. As a result, the spacing between the core and the inner surface of the shell varied from 25 to 10 nm, respectively. Source: Reproduced with permission from Sun et al. [32], © 2004 American Chemical Society.

as individuals or alloys are present in the form of the open structure, the optical properties of the resultant structure are also better than those of the core/shell structure. In the first reported study of alloy-type open NPs, both core and shell structures were made of the Au/Ag alloy via a galvanic replacement reaction between silver and chloroauric acid, as shown in Figure 7.11a [32]. The study showed the development of new optical features that varied across a broad range of the optical spectrum, extending to the visible and near infrared regions, compared to conventional solid or hollow NPs. Additionally, the obtained structure showed two distinct extinction peaks at ~510 nm and another at the NIR region, similar to those of Au or Ag nanorods, as shown in Figure 7.11d. These peaks are mainly because of the Au/Ag core and shell, respectively. At the same time, by varying the void space and core diameter of the open nanostructure, the extinction spectra can be tuned from the visible to the NIR regions. The pure Au/Ag core/shell nanoparticles exhibit only one sharp peak at 502 nm (curve a), but by increasing the inner diameter, it extends to another peak at a longer wavelength, which can be described from the Mie scattering of the Au/Ag shell.

Later, a similar principle was also used for the synthesis of the Au core inside the Pt/Ag alloy shells. A small modification of this process was done to obtain pure Au/Au open NPs, where the composition of the core/shell was controlled by varying the reaction temperature. In this case, AgCl formed via a replacement reaction between Ag and $HAuCl_4$ was completely soluble in water at 100 °C and can be easily removed by an aqueous NH_4OH wash. Further, many combinations such as Au/Au, Au/Pd, and Au/Pt were also synthesized via the galvanic replacement reaction. Apart from a metal alloy, the combination of pure metal/metal open NPs also exhibits different extinction peaks corresponding to the core and the shell.

7.3.3 Au NR-Embedded Open Nanostructures

Porous plasmonic nanostructures have attracted interest for their built-in cavities, which provide strong localized electric field enhancement and additional plasmon tunability [33]. Chen and coworkers synthesis a series of nanostructures based on Au–Ag open nanostructures by means of progressive galvanic replacement reactions, as shown in Figure 7.12. Transmission electron microscopy images present evidence that the cavity size of the Au–Ag open nanostructures is controllable. Moreover, the synthetic method enables us to design the longitudinal plasmon modes of the Au–Ag open nanostructures progressively varied from 660 to 970 nm with nearly the same aspect ratio. To explore the origin of the spectral shift, we analyze the size, the composition, and the void effects of the Au–Ag open nanostructures using computational electrodynamics simulations. This analysis indicates that the longitudinal plasmon modes are less sensitive to the frame composition of the open nanostructures, in contrast to their susceptibility to the size and the void effects.

In addition, the facile preparation of Au/(PdAg alloy) electrocatalysts with diverse architectures, including nanodumbbells, yolk/shell nanoframes, and yolk/shell nanoboxes, from Au NR core/Ag shell bimetallic nanostructures are also studied by Fang and coworkers [33]. The growth routines are determined by the adding sequence of the surfactants and the Pd precursor in the growth solution, as well as the thickness of the Ag shell in the starting Au/Ag nanorod seeds.

7.3.4 Au Nanobipyramid-Embedded Open Nanostructures

Au NBPs with two sharp end tips show much larger local electric field enhancements and narrower peak widths than Au NRs. Au NBPs are therefore more

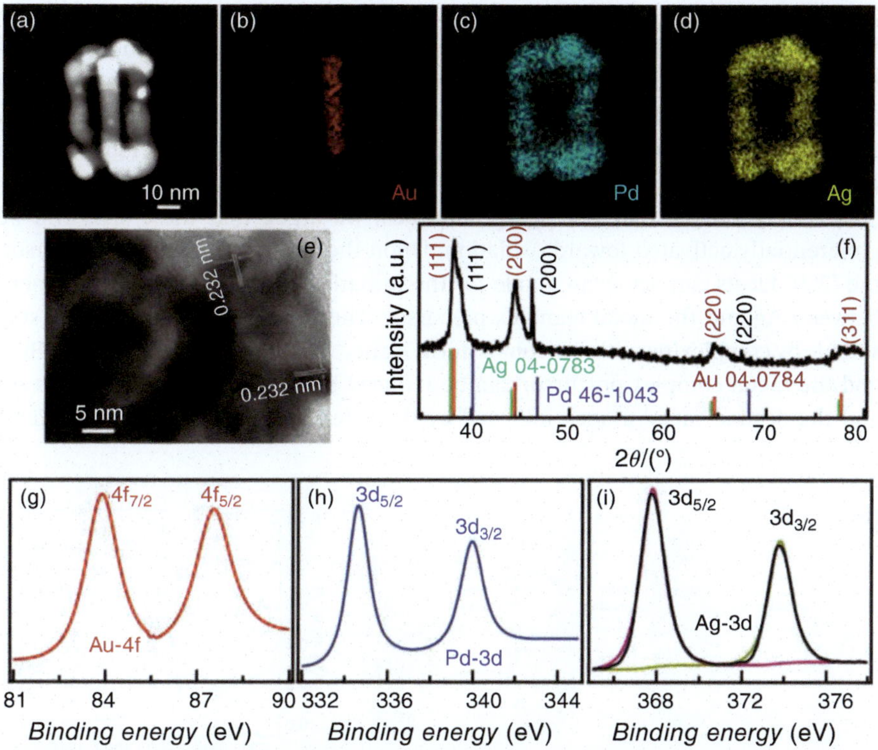

Figure 7.12 Structure and composition analysis of the Au/PdAg yolk/shell nanoframe nanocrystals. An individual nanoframe was imaged by (a) HAADFSTEM, (b–d) EDX elemental mapping, and (e) HRTEM. The bulk samples of the same nanostructure were further characterized by (f) XRD and (g–i) XPS spectroscopies. Source: Reproduced with permission from Fang et al. [33]; © 2018, John Wiley & Sons, Inc.

advantageous for photonic, biotechnological, and spectroscopic applications [34, 35]. The structure of Au NBP is built up from two pentagonal pyramids base-stacked together symmetrically, with a total of 10 triangular faces and fivefold rotational symmetry about the length axis. The unique geometry of Au NBPs endows them with desired plasmonic features in the field of nanoplasmonics. By tailoring the length-to-diameter aspect ratio, the longitudinal dipolar plasmon wavelength of Au NBPs exhibits high tunability toward the infrared region. The plasmon peak of a single Au NBP is also narrower than that of a single Au NR with a similar longitudinal dipolar plasmon wavelength because of the smaller plasmon dephasing time of the Au NBP. The high monodispersity and large local electric field enhancement of Au NBPs enable them to possess higher refractive index sensitivities and higher values of the figure of merit (the index sensitivity divided by the FWHM of the plasmon peak) for index change-based plasmon sensing compared with other Au nanoparticles. In addition, the anisotropic shape and high uniformity of Au NBPs allow for large-area ordered assemblies of Au NBPs to manifest their collective plasmon responses. Therefore, Au NBPs are the excellent basic structures for the open nanostructures with excellent plasmonic properties.

7.3.4.1 Au Nanobipyramid-Embedded Nanoframes

The Au NBPs samples were prepared through a stepwise combination of seed-mediated growth and subsequent depletion force-induced purification. The deposition of ultrathin metal nanoframes is realized through the synergetic action of galvanic replacement and co-reduction of metal ions (Figure 7.13) [36]. Because Br^- ions derived from CTAB can selectively bind to the side facets, that is, the (200) facets, on the (Au NBP)@Ag nanorods, we argue that these sites will be preferentially activated toward oxidation. Upon the addition of a metal precursor, the (200) facets can act as an anode for the initiation of the galvanic replacement between Ag and the metal complex precursor. The released Ag^+ ions should stay soluble by complexing with Br^- ions to form $AgBr_2^-$. At the same time, both $AgBr_2^-$ and the metal complex precursor can be reduced by ascorbic acid to generate Ag and the desired metal atoms, followed by their deposition on the (Au NBP)@Ag

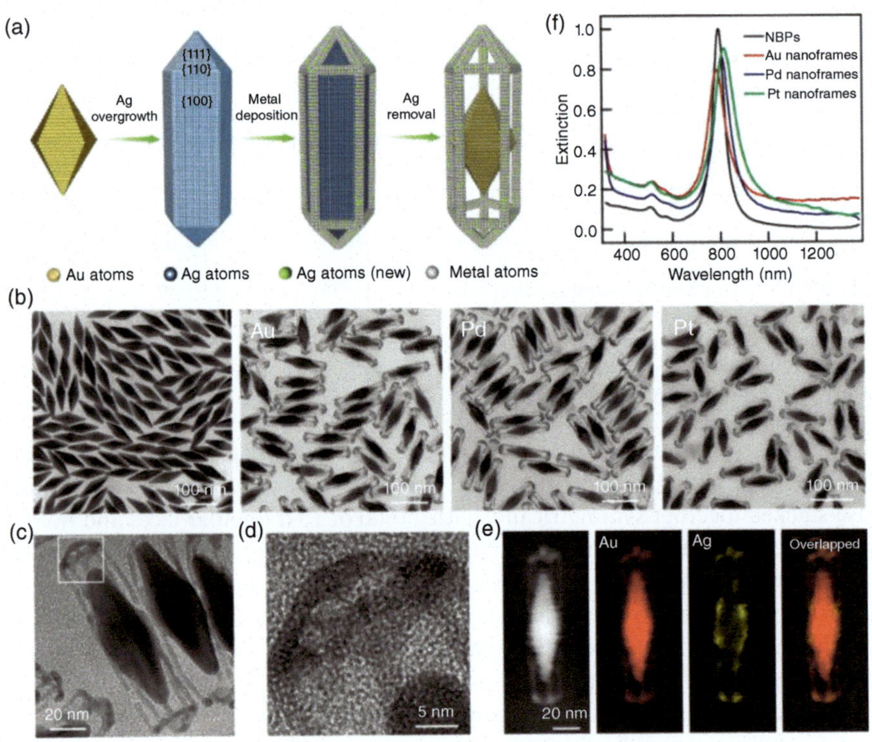

Figure 7.13 Au NBP-embedded ultrathin metal (Au, Pd, Pt) nanoframes. (a) Schematic illustrating the synthetic process. (b) TEM images of the Au NBP sample and the Au NBP-embedded metal (Au, Pd, Pt) nanoframe samples. (c) TEM image of several Au NBP-embedded Au nanoframes. (d) HRTEM image recorded in the region indicated with the white box in (c). (e) HAADF-STEM and elemental mapping images of a single Au NBP-embedded ultrathin Au nanoframe. The elemental mapping images have the same scale bars as the HAADF-STEM image. (f) Extinction spectra of the Au NBP sample and the Au NBP-embedded metal (Au, Pd, Pt) nanoframe samples. Source: Reproduced with permission from Zhu et al. [36]; © 2020 Royal Society of Chemistry.

nanorods [37]. Because the side facets are involved in the galvanic replacement reaction, the co-deposition of Ag and the desired metal atoms is automatically directed and confined to the (220) facets at the edges and corners because of their higher free energies. The Ag atoms are increasingly carved away from the side facets, leading to the formation of (Au NBP)@Ag@AgM (M = Au, Pd, Pt) nanoframes. Upon the etching of the pure Ag segments in the core with H_2O_2, the (Au NBP)@Ag@AgM nanoframes are transformed into (Au NBP)@AgM nanoframes.

TEM images of the Au NBP sample and the Au NBP-embedded metal nanoframes revealed the uniform morphologies and narrow size distributions of all samples (Figure 7.13b). The purified Au NBPs have an average length of 98 ± 3 nm and a waist width of 28 ± 1 nm. Each nanostructure contains one Au NBP at the center, with an ultrathin metal nanoframe surrounding the NBP core. The ultrathin metal nanoframes possess a three-dimensional geometric shape with a virtually pentagonal morphology. Such a deposition behavior is not affected by the type of metal precursor added. The difference in the morphology for the Pt nanoframes can be attributed to the much higher bonding energy of Pt–Pt (307 kJ mol^{-1}) than that of Pt–Ag (218 kJ mol^{-1}). In contrast, the bonding energies of Pd–Pd (100 kJ mol^{-1}) and Au–Au (226 kJ mol^{-1}) are smaller than those of Pd–Ag (137 kJ mol^{-1}) and Au–Ag (229 kJ mol^{-1}), respectively, favoring the formation of Pd–Ag and Au–Ag alloys rather than the separation into the different phases.

In order to find out the composition of the ultrathin metal nanoframes around the Au NBPs, we performed HAADF-STEM imaging and elemental mapping (Figure 7.13c–e). HRTEM images (Figure 7.13c and d) reveal that most of the Ag atoms are selectively removed, leaving behind the pentagonal nanoframes with a unique open structure. The ridges in the nanoframes are only 2–3 nm in thickness. Elemental mapping clearly shows the presence of Au and Ag atoms on the nanoframes, while the Au atoms located dominantly at the center maintain the bipyramidal shape. The metal components of the nanoframes are therefore Au-enriched Au–Ag alloy, as evidenced by the complete overlap of the elemental distributions of Au and Ag on the elemental maps. The residual Ag atoms at the gaps between the nanoframes and the middle of the Au NBPs can be ascribed to the weak oxidation ability of H_2O_2 in the etching process. The results for the Au NBP-embedded Pd and Pt nanoframes further reveal that the nanoframes are composed of metal (Au, Pd, Pt)–Ag alloys, which verifies the schematic shown in Figure 7.13a. For simplicity, the nanoframes are named as the Au, Pd, and Pt ones according to the salt precursor used in the synthesis. The deposition of the ultrathin metal nanoframes on the Au NBPs causes a slight shift of the longitudinal plasmon peak, which is accompanied by slight peak broadening and intensity reduction (Figure 7.13f). For the longitudinal plasmon peaks, the FWHMs are 62, 87, 80, and 107 nm for the Au NBPs, Au NBP-embedded ultrathin Au, Pd, and Pt nanoframes, respectively. The transverse plasmon peak, however, shows slight shifts. The peak broadening and intensity reduction arise mainly from the increase in size distribution upon the growth of the ultrathin metal nanoframes because the starting Au NBPs are highly uniform in shape and size.

7.3.4.2 Au Nanobipyramid-Embedded Nanocages

Varied the added volume of Pd precursor, the obtained Pd open nanostructures could change nanoframes into nanocages [38]. The reduced Pd atoms were deposited on the surface of the Au NBP@Ag nanorods, leading to the formation of Au NBP@Ag@Pd nanorods. After selective etching of the Ag component, the Au NBP@AgPd hollow nanostructures were successfully fabricated. As shown in Figure 7.14a, the longitudinal plasmon peak of Au NBP@AgPd nanostructures slightly red-shifted as the volume of Pd precursor increased, which was accompanied by peak broadening. In addition, the intensity of plasmon peaks of the Au NBP@AgPd nanostructures was similar to that of the original Au NBPs, indicating that the presence of the Pd nanostructures had rare influence on that of Au NBPs. The longitudinal plasmon peak position and FWHMs were plotted in Figure 7.14b as a function of the added H_2PdCl_4 volume. It is clear that the FWHMs display an almost identical variation trend to the plasmon peaks. Furthermore, the TEM images of the fabricated Au NBP@AgPd nanostructures with different volumes of H_2PdCl_4 are shown in Figure 7.14c–h. All samples show uniform size and shape distributions due to the high purity of the Au NBPs. When the volume was 1.2 mL, the reduced Pd atoms were too few to form robust Pd nanostructures (Figure 7.14c). Parts of the formed Pd particles were attached to the head of Au NBPs, causing the obvious broadening of the plasmon resonance. As the volume reached 1.8 mL, the deposition of Pd atoms on both the head and edges of Au NBP@Ag nanorods

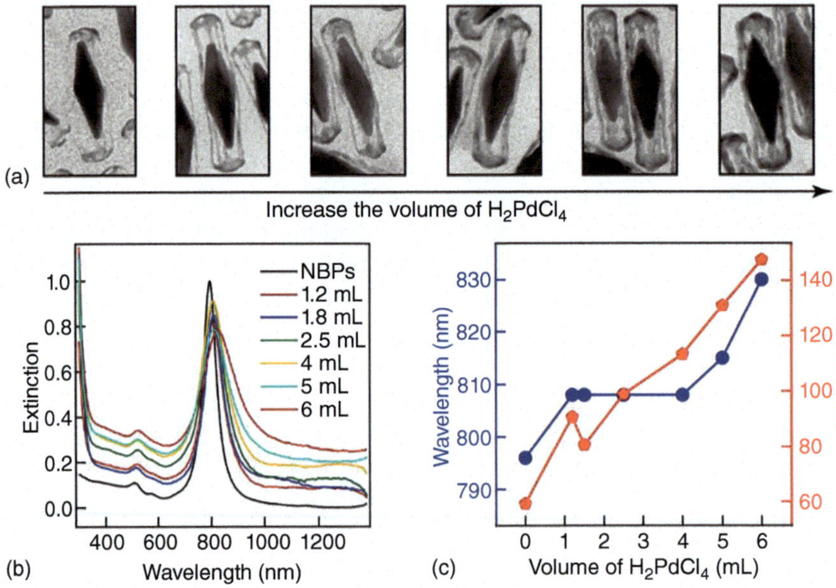

Figure 7.14 (a) TEM image of Au NBP@AgPd nanostructure samples. (b) Extinction spectra of the Au NBP@AgPd nanostructure samples produced with different amounts of H_2PdCl_4. (c) Variations of the longitudinal plasmon wavelength and FWHM as a function of the volume of added H_2PdCl_4. Source: Reproduced with permission from Xu et al. [38]; © 2023 Royal Society of Chemistry.

promoted the formation of an intact AgPd nanoframe after selectively etching the Ag component. The weak nanoframes became robust with the continuous deposition of Pd atoms (Figure 7.14d–f). Particularly, the Pd atoms began to appear at the side facets, leading to the formation of Au NBP@AgPd nanocages when the volume of H_2PdCl_4 increased to 5 mL (Figure 7.14g). The density of Pd atoms was obviously increasing as the volume of Pd precursor increased (Figure 7.14h). It is worth noting that the deposition of Pd atoms on Au NBP@Ag nanorods was incomplete with pores because of the high bonding energy of Pd–Pd, which provide access to etch Ag components and form hollow nanocages. The disappearance of the characteristic peak of Ag at 400 nm indicated that the Ag atoms were almost etched.

In comparison, the Pt nanocages are synthesized by adjusting the reaction temperature of the reaction system to promote the diffusion rate of Pt atoms [39]. Figure 7.15a outlines the proposed pathway for the selective deposition of Pt atoms on the Au NBP@Ag nanorods and, at different temperatures, the formation of open nanostructures after etching. The Br^- ions derived from CTAB selectively absorb on

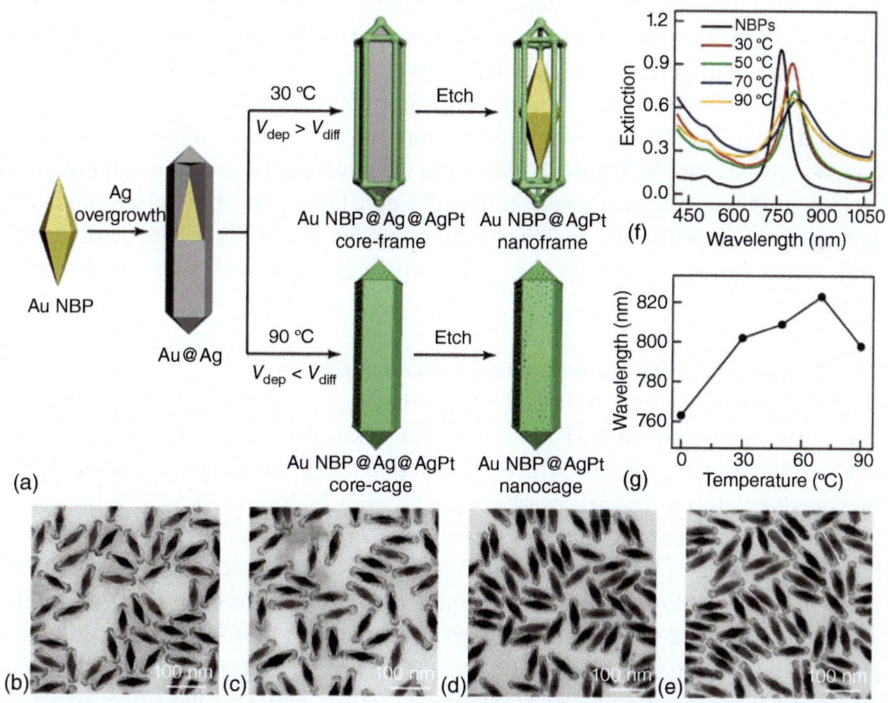

Figure 7.15 (a) Schematic illustrating the synthetic process of Au NBP@AgPt hollow nanostructures. (b–e) TEM images of the Au NBP@AgPt nanostructure samples produced at 30, 50, 70, and 90 °C, respectively. (f) Extinction spectra of the Au NBP sample and the Au NBP@AgPt nanostructure samples produced at different temperatures. (g) Variation of the longitudinal dipolar plasmon wavelength as a function of the varied temperature. The point at 0 °C represents the Au NBP sample. Source: Reproduced with permission from Xu et al. [39]; © 2020 Royal Society of Chemistry.

the (200) facets, leading to a decrease in surface energy and preferentially activating oxidation. The galvanic replacement reaction between Ag from (200) facets and $PtBr_6^{2-}$ could release Ag^+ ions to form $AgBr_2^-$. Meanwhile, both the $AgBr_2^-$ and $PtBr_6^{2-}$ were reduced to generate Ag and Pt atoms, followed by the selective deposition at the corners and edges of nanorods. At low temperatures, the reduced Ag and Pt atoms stay at the corners and edges with less diffusion, leading to the formation of (Au NBP@Ag)@AgPt core-frame nanostructures. With the increase in temperature, the Ag and Pt atoms diffuse to the side surfaces to form (Au NBP@Ag)@AgPt core-shell nanorods [27]. The Au NBP@AgPt nanoframes and nanocages will be finally formed upon the removal of Ag.

Due to the limitation of the etching ability of H_2O_2, another stronger dealloying agent, $Fe(NO_3)_3$, was used to remove pure Ag to generate Au NBP@AgPt hollow nanostructures. The corresponding TEM images of the Au NBP@AgPt hollow nanostructures are shown in Figure 7.15b–e. It is clearly shown that all the nanoparticles contain an Au NBP at the center, with the ultrathin AgPt nanostructure surrounding the Au NBP core. The AgPt atoms changed from nanoframes to nanocages. At 30 °C, the amount of Pt atoms derived from Pt precursor was selectively deposited at the corners and edges. As the temperature increases, the Pt atoms diffuse from the heads to the center. When the reaction temperature reaches 90 °C, the Pt atoms are homogeneously dispersed on each side surface to create uniform Au NBP@AgPt nanocages. The longitudinal SPRs of Au NBP sample and the Au NBP@AgPt nanostructure samples are shown in Figure 7.15f. With the increase in reaction temperature, the longitudinal plasmon peak constantly redshifts and then blueshifts. The deposition of an ultrathin AgPt nanoframe on the Au NBPs causes a slight red shift of the longitudinal plasmon peak, which is consistent with our previous findings.

7.3.4.3 Synthesis of Pd Nanorod Arrays on Au Nanoframes

At the beginning of the Au NBP-embedded Au nanoframes (Figure 7.16a), the deposition of Pd arrays on Au NBP-embedded Au NFs was realized with the assistance of CPC [40]. Particularly, the Au NFs with open morphology and abundant active sites were crucial for the anisotropic deposition of Pd atoms. The TEM images of Figure 7.16b–f reveal the detailed distribution of Pd atoms on Au NFs with the increased volume of Pd precursor. It is clearly shown that there are no obvious Pd nanoparticles on the Au NFs when the added H_2PdCl_4 is lower than 30 μL. As the volume is increased to 60 μL, significant Pd arrays with uniform distribution have occurred on the edges of Au NFs (Figure 7.16d). Longer and coarser Pd arrays were obtained with the volume further increased to 160 μL, while the density of Pd arrays showed no distinct change (Figure 7.16e and f). Taken together, the size of the Pd arrays could be controlled by the volume of the added Pd precursor. It seems that most of the reduced Pd atoms preferred to deposit on the outside Au NFs, while rare Pd atoms were seen on the surface of Au NBPs.

To further identify the deposition sites of Pd atoms, HRTEM imaging and elemental mapping were performed and shown in Figure 7.16g–i. The results clearly show the presence of Au, Ag, and Pd atoms on the outside Au NFs, while the map of

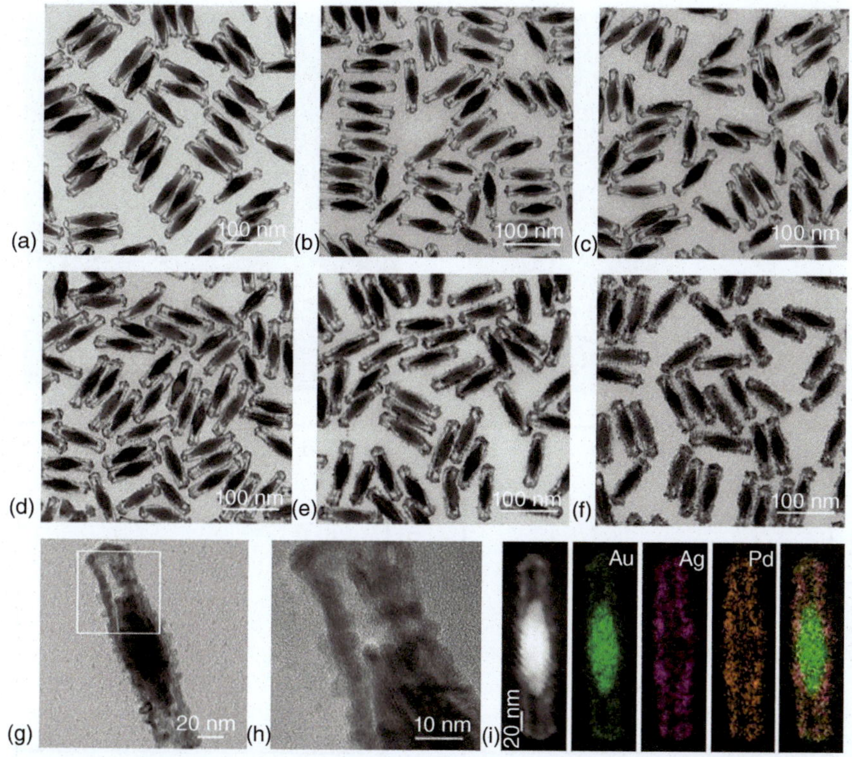

Figure 7.16 (a–f) TEM images of the Au NBP-embedded Au NFs and Au NF@Pd arrays synthesized with 10, 30, 60, 90, and 160 μL H_2PdCl_4 (1 mM), respectively. (g) TEM image of a single Au NF@Pd array. (h) HRTEM image recorded in the region indicated with the white box in (g). (i) HAADF-STEM and elemental mapping images of a single Au NF@Pd array. The rightmost image is the overlapped image of the Au, Ag, and Pd elemental maps. The elemental mapping images have the same size scale as the HAADF-STEM images. Source: Reproduced with permission from Yun et al. [40]; © 2022 Royal Society of Chemistry.

the majority of Au atoms maintains the bipyramidal shape. Pd atoms were found to appear on the surface of both Au NBPs and Au NFs, with most of them on the Au NFs in the array arrangement.

Additionally, the detailed structure and composition of the Au NF@Pd arrays were identified by HRTEM, SAED pattern, XRD, and XPS characterizations. It is clearly shown that the Pd atoms prefer to selectively deposit on the surface of Au NFs to form Pd arrays (Figure 7.17a). The lattice fringe of the Pd array is 0.23 nm, which corresponds to the lattice spacing of (111) plane of Pd. Meanwhile, the deposition of Pd arrays had little influence on the original structure of Au NFs. The SAED of Au NF@Pd arrays exhibited a series of concentric rings corresponding to the (111), (200), and (220) planes, respectively (Figure 7.17b). However, the content of Pd was too low to be identified through an XRD pattern. The XPS measurement was performed to investigate the surface composition and metal valence of Au NF@Pd arrays. The two symmetric peaks located at 83.1 and 86.8 eV in the high-resolution Au 4f spectrum are identified as the Au $4f_{7/2}$ and Au $4f_{5/2}$ signals and therefore demonstrate the metallic Au^0 state (Figure 7.17c. In addition,

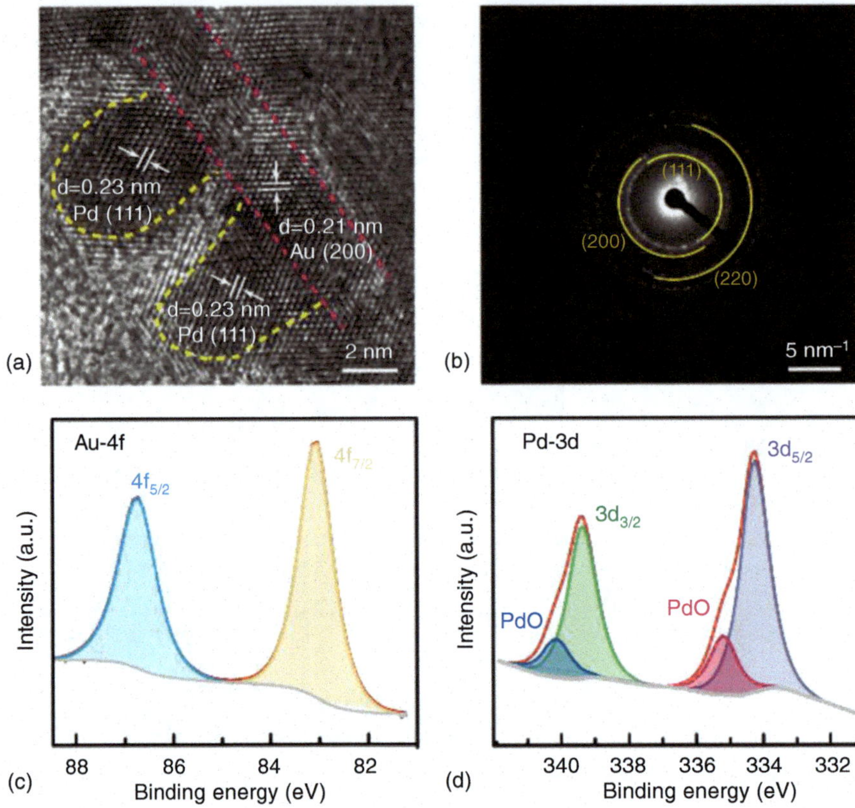

Figure 7.17 (a) HRTEM image of Pd arrays on Au NFs. (b) SAED image of the corresponding Au NF@Pd arrays. (c) Au 4f and (d) Pd 3d XPS spectra for the Au NF@Pd arrays. Source: Reproduced with permission from Yun et al. [40]; © 2022 Royal Society of Chemistry.

the peaks at 334.3 eV (Pd $3d_{5/2}$) and 339.4 eV (Pd $3d_{3/2}$) are attributed to the metallic state of Pd, whereas those at 335.2 and 340.1 eV are assigned to the Pd^{2+} species (Figure 7.17d. The percentage of Pd^0 is about 74.3%, implying the metallic state is predominant in the Au NF@Pd arrays.

7.3.5 Deposited Mechanism of Au Atoms on Ag Nanostructures

Based on the results above, galvanic replacement and reduced reaction are the general reaction types for the preparation of open nanostructures. Among the reaction systems, the deposition of Au atoms on Ag nanostructures is influenced by the surfactant, reaction temperature, reaction types, and the morphology of the template. Therefore, the deposited mechanism of Au atoms depends on the reaction conditions, such as facet selective absorption and pH of the reaction system.

7.3.5.1 Facet Selectivity

Up to date, the atomic deposition mechanisms of the single reaction system, including the spontaneous galvanic replacement reaction and the sole reduction reaction, are well understood and perspicuous. However, for the system involving galvanic replacement and reduction reactions, the detailed deposition mechanism of Au atoms on Ag nanostructures is complicated and unclear. Recently, Qin and coworkers systematically studied the growth mechanism of a number of Ag nanocube-embedded Au–Ag alloy nanoparticles, such as hollow and concave nanostructures, in terms of the surfactants in the hybrid system. Xue also demonstrated the transformation of Ag nanoprisms into Au–Ag alloy nanoprisms in the presence of hydroxylamine as a reducing agent. In these experiments, with the assistance of surfactant, the newly formed Au and Ag atoms were selectively deposited at the corners and edges of the Ag nanocubes or nanoprisms, which are the sites with high surface energy and high curvature simultaneously. The deposition behavior of these atoms was generally attributed to the synergetic action of the galvanic replacement and reduction reactions, one of the vital reasons for which is the overlap of the facets with high surface energy and high curvature on the Ag nanostructures. Actually, it is difficult to distinguish the relationship between the reduced metal atoms and the type of the reactions because there is no distinction between the newly formed metal atoms produced by the two reactions, making it challenging to clarify the growth mechanism of the final morphology in this system. The ambiguous deposition mechanism of the newly formed metal atoms is disadvantageous for guiding the construction of novel Au–Ag nanostructures using this system. Therefore, the investigation, proposal, and verification of the deposition mechanism of Au atoms on Ag nanostructures involving simultaneous galvanic replacement and reduction reactions are in strong demand.

The Ag nanorods were produced through the overgrowth of Ag on Au NBPs, which have a penta-twinned structure bounded by five (200) facets on the side surfaces, 10 (111) facets at the two ends, and the (220) facets at the edges [41]. In particular, in the presence of cetyltrimethylammonium bromide (CTAB), the (220) facets located at the edges have the highest surface energy, while the (111) facets located at the ends of the nanorods are high curvature sites. The design of the Ag template with high surface energy facets and high curvature sites located at different sites allows us to study the detailed atomic deposition mechanism by tracing the location of the newly formed Au atoms on the Ag nanorods.

7.3.5.2 pH of the System

Under the ambient conditions, the Au precursor was easily reduced through the spontaneous galvanic replacement reaction with the Ag atoms or the direct reduction reaction by H_2Asc. It is well known that the reducing ability of the H_2Asc was enhanced with an increase in pH values. The ratio between the galvanic replacement and reduction reactions could thus be adjusted by tailoring the pH value of the aqueous solution in the experiments. The reaction will transform from a galvanic replacement reaction dominant to a reduction reaction dominant when the pH values increase from 1.88 to 12.95 (Figure 7.18). At the pH value of 1.88, the products exhibited voids and concave facets in most of the nanorods, which could be attributed to the dominant galvanic replacement reaction between the Au precursor and the (200) facets of the Ag nanorods (Figure 7.18a). After the removal of the Ag template, it is obvious that the newly formed Au atoms are mainly deposited at the

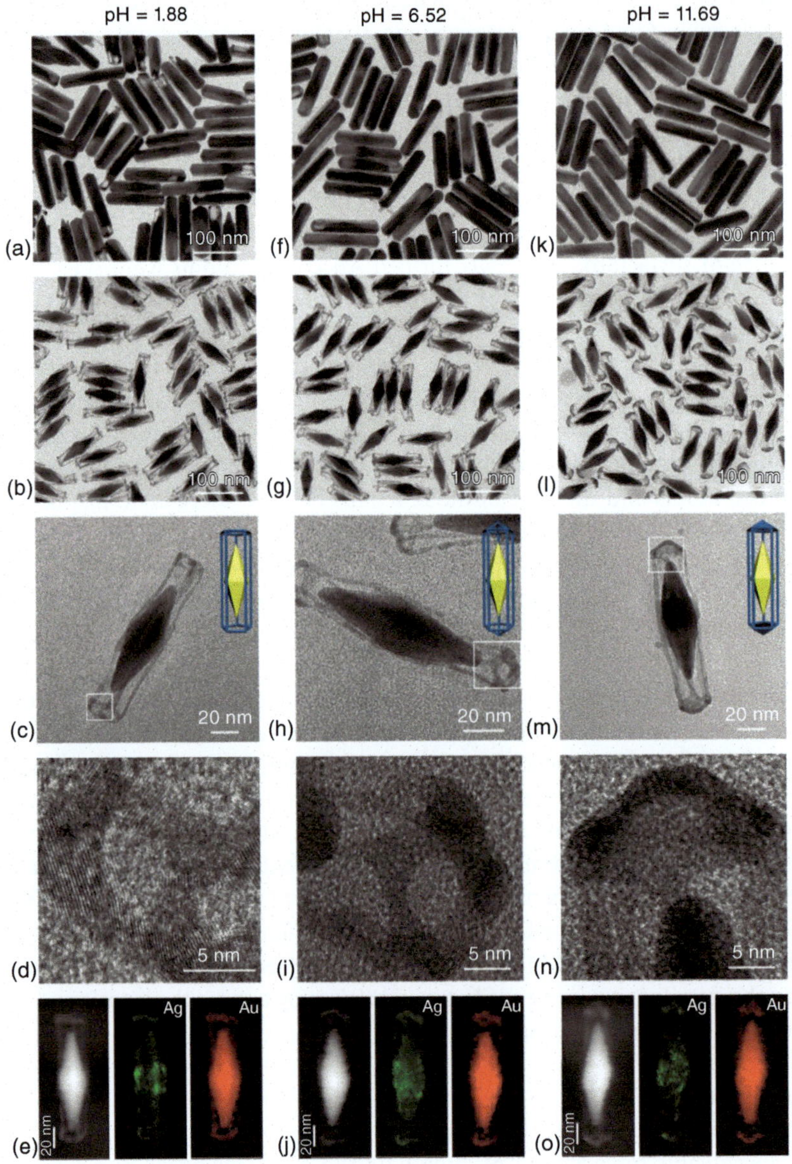

Figure 7.18 Shape evolution with the pH values. (a, f, k) TEM images of the Ag nanorod@AuAg nanoframe samples and (b, g, l) the corresponding Au NBP@AuAg nanoframe samples at pH of 1.88, 6.52, and 11.69, respectively. (c, h, m) TEM images of a single Au NBP@AuAg nanoframe sample at pH of 1.88, 6.52, and 11.69, respectively. The insets were the models of the corresponding nanoparticles. (d, i, n) HRTEM images recorded in the region indicated with the white box in (c, h, m), respectively. (e, j, o) High-angle annular dark-field scanning transmission electron microscopy (HAADF-STEM) and elemental mapping images of a single Au NBP@AuAg nanoframe sample at pH of 1.88, 6.52, and 11.69, respectively. The elemental mapping images have the same scale bars as the HAADF-STEM image. Source: Reproduced with permission from Xu et al. [41]; © 2021 Royal Society of Chemistry.

edges ((220) facets), that is, the highest energy facets of Ag nanorods (Figure 7.18b). Particularly, the head of the nanoframe was a hollow flat pentagon with no Au atoms deposited, indicating that the ends of Ag nanorods with high curvature have no attraction for the Au atoms formed from the galvanic replacement reaction, as presented by Figure 7.18c and d.

Further elemental mapping clearly shows that the Au atoms located dominantly at the center maintain the bipyramidal shape, while the metal components of the nanoframes are Au–Ag alloys (Figure 7.18e). The obtained nanoframes have no specific change at a pH value of 3.03. The control experiment with the galvanic replacement reaction only (in the absence of H_2Asc) at a pH value of 1.88 shows that the obtained nanostructures were the same as those synthesized in the presence of H_2Asc, further confirming that the reduction reaction is suppressed in the system under strong acidic conditions. These results suggest that when the galvanic replacement reaction dominates under acidic conditions, the newly formed Au atoms would deposit at high surface energy facets of the Ag nanorods.

The turning point occurred at the pH value of 6.52, where the reduction reaction started to take effect and compete with the galvanic replacement reaction, leading to the decrease of the voids and concave facets on the Ag nanorods after titration (Figure 7.18f). Part of the Au atoms appeared at the ends of the Ag nanorods, while most of the Au atoms were still located at the edges. These atoms support the framework at the ends to give the nanoframes a porous hat, as indicated by Figure 7.18g–j. The reduction reaction became dominant as the pH value further increased to 11.69, where the galvanic replacement reaction was too weak to give rise to the voids on the Ag nanorods (Figure 7.18k). The dense ends and deformation of nanoframes indicate that the deposition sites of the newly formed Au atoms are reversed, with a large proportion of Au atoms deposited at the ends of Ag nanorods (Figure 7.18l). Evidently, the deposition sites are related to the curvature rather than surface energy, with the Au atoms deposited at not only the (220) facets but also the (111) facets of the ends (Figure 7.18m and n). The metal components of the nanoframes are therefore an Au-enriched Au–Ag alloy due to the weakened galvanic replacement reaction (Figure 7.18o), which is consistent with our previous result. Taken together, in the system involving simultaneous galvanic replacement and reduction reactions, the newly formed Au atoms originated from the galvanic replacement reaction will deposit at high surface energy facets of the Ag nanorods, while the other originated from the reduction reaction at high curvature sites, forming the final Ag nanorod@AuAg nanoframes and the Au NBP@AuAg nanoframes.

On the basis of these experimental results, a detailed atomic deposition mechanism of Au atoms on the Ag nanostructures involving simultaneous galvanic replacement and reduction reactions were proposed (Figure 7.19): (i) The Au atoms will originate from two parts: one is from the galvanic replacement reaction, and the other is from the reduction reaction. (ii) The Au atoms originated from the galvanic replacement reaction will deposit at surface energy-related facets of the Ag nanostructures, while the other originated from the reduction reaction at high curvature sites. These two types of Au atoms are independent of each other. (iii) The morphology of the final products was determined by the ratio between

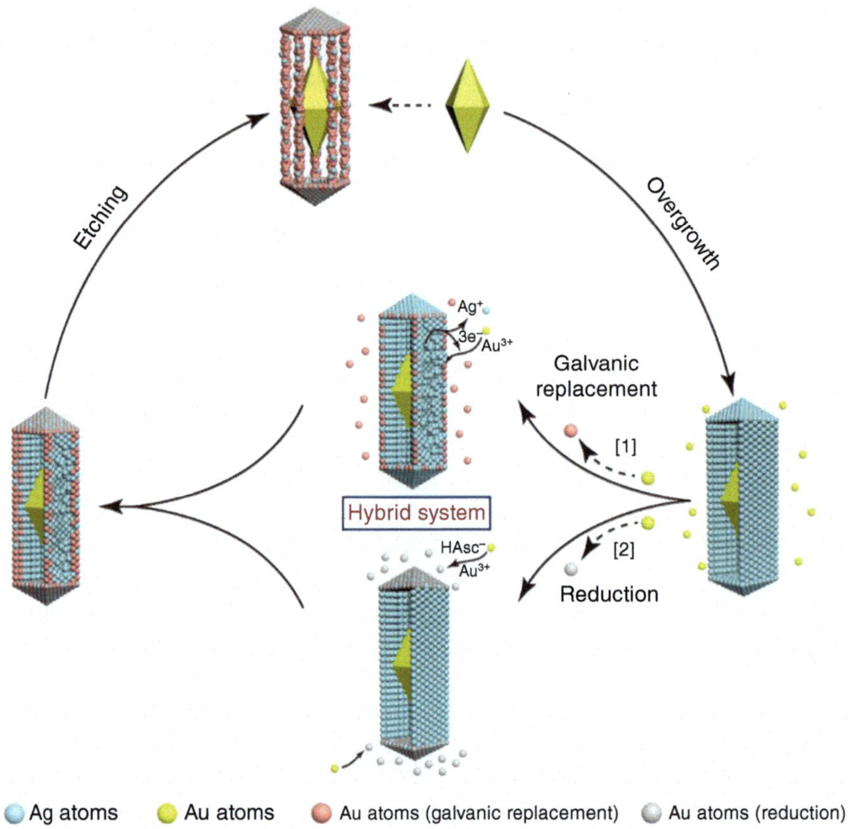

Figure 7.19 Schematic illustrating the proposed deposition mechanism of Au atoms on Au NBP@Ag nanorods involving simultaneous galvanic replacement and reduction reactions. Source: Reproduced with permission from Xu et al. [41], © 2021 Royal Society of Chemistry.

the two reactions. In the specific case of this model, for the galvanic replacement reaction in this system, it tends to occur between the Au precursor and the Ag atoms on the (200) facets of Ag nanorods due to the absorption of Br^- ions, while the resultant Au atoms prefer to deposit on the highest surface energy facets, that is, the (220) facets of Ag nanorods [37, 42]. In comparison, for the reduction reaction in this system, the Au atoms reduced by H_2Asc will selectively deposit on the high curvature sites, that is, the ends of Ag nanorods [6, 30]. Based on the cooperation of these two types of the reactions, the newly formed Au atoms could deposit on both the edges and ends of the Ag nanorods to form a hatted AuAg nanoframe, while the final morphology mainly depends on the ratio between galvanic replacement and reduction reactions in this system. The further etching of Ag atoms could make the deposition sites of Au atoms in this system clearer.

The successful applicability of this atomic deposition mechanism to Ag nanostructures under different reaction conditions and with different shapes suggests the generality of this mechanism. Our study not only clarified the principle of atomic deposition in this system but also provided a creative guideline for precise shape control of Au–Ag nanostructures. We believe this mechanism reported here can also be extended to the synthesis of composite nanostructures of other metals and the development of the atomic deposition mechanism in other systems.

7.4 Properties of Au NBP-Embedded Nanostructures

7.4.1 Optical Properties

7.4.1.1 Hollow Nanostructures

Metal nanoframes have unique optical properties due to their localized surface plasmon resonances (LSPR), involving the coherent oscillation of the free conduction band electrons in resonance with an incident electromagnetic field. In comparison to other morphologies, metal nanoframes have distinct LSPR properties. Cheng and coworkers have investigated the plasmon modes of Ag nanoboxes, nanocages, and nanoframes. The results show that the plasmon modes (dipole and quadrupole) of a single Ag nanocage are red-shifted relative to the nanobox as the surface porosity of the nanocage increases. In particular, the wavelengths at the peaks of these modes almost linearly depend on the surface porosity: the larger the surface porosity, the more redshifted these modes. For an Ag nanoframe whose hole length on the side faces is larger than half of the edge length, the dipole mode is within the near-infrared region.

Recall that the size, shape, and composition of metal nanostructures can all influence the position of their corresponding SPR peak. Thus, the galvanic replacement of Ag nanocubes with $HAuCl_4$ provides a facile way of tuning the SPR peak of the resultant nanostructure, as both the morphology and composition of the material are controllably altered. Figure 7.20a and b show the colloid and corresponding extinction spectra that were obtained by titrating 30 nm Ag nanocubes with different volumes of $HAuCl_4$ solution (0.2 mM) [42]. Significantly, the position of the SPR peak is continuously shifted from the visible to the near-infrared as more $HAuCl_4$ solution is added. Discrete dipole approximation (DDA) calculations indicate that Au nanocages have large scattering and absorption cross-sections. It is clear that for the 30 nm nanocage, light absorption dominates, whereas for the 50 nm nanocage, light scattering becomes more significant. Additional DDA calculations indicate that the SPR peak position is highly sensitive to wall thickness, while the magnitude of both the scattering and absorption cross-sections depends only slightly on wall thickness (Figure 7.20c and d). The magnitude of both the scattering and absorption cross-sections does, however, depend on the porosity of the nanocages, with both decreasing as the number of holes increases, the SPR peak position does not change with porosity. As these calculations indicate, in addition to being able to finely tune the SPR peak position, the magnitudes of both the absorption and scattering components can be tailored by simply controlling the size and porosity of the nanocages. This flexibility is invaluable when preparing such materials for different applications.

Xia and co-workers observed UV–Vis spectra of the intermediate products prepared through the galvanic replacement reaction between Ag nanocubes and

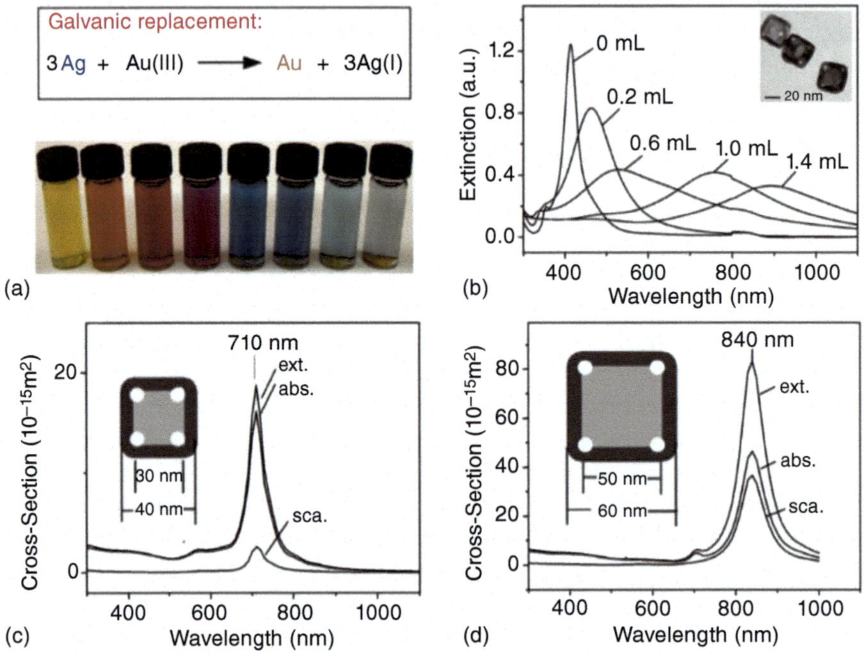

Figure 7.20 Some typical colloid photos of Ag and Au nanostructures, and the corresponding optical spectra. (a and b) Photograph and extinction spectra of Ag nanoplates after the addition of PSS at room temperature. The concentration of PSS in the systems was marked on the container. (c and d) Photograph and extinction spectra of Au NRs with increasing aspect ratios. Source: Adapted from Skrabalak et al. [42].

different amounts of $HAuCl_2$ (Figure 7.21a) [37]. To determine how these factors, specifically the size and shape, influence the LSPR peak, the scattering spectra of individual nanoframes were calculated in air using the DDA method. After reviewing many spectra, we found a correlation between the peak position and the ratio (R) between the outer edge length (l) and the ridge thickness (t). As the value of R increased, the peak position was red-shifted while the peak intensity slightly increased (Figure 7.21b). Figure 7.21c shows the DDA-calculated extinction, absorption, and scattering spectra for the nanoframe with $R = 3.37$, and geometry as shown in the inset.

To confirm the theoretical calculations, the scattering spectra of nanoframes plotted in Figure 7.21d show that the spectra were red-shifted as the R value increased, validating the trend found by DDA calculations. This result is expected as a decrease in ridge thickness or increase in edge length will increase the charge separation, thus reducing the restoring force for electron oscillation and resulting in red-shifting for the resonance peak. As previously reported for hollow nanospheres, thinner shells have stronger coupling between charges inside and outside of the shell, causing greater charge separation and a further red-shift in peak position. As for the calculated spectra, the peak intensity was higher for larger R values; this trend is also expected as larger particles scatter more light.

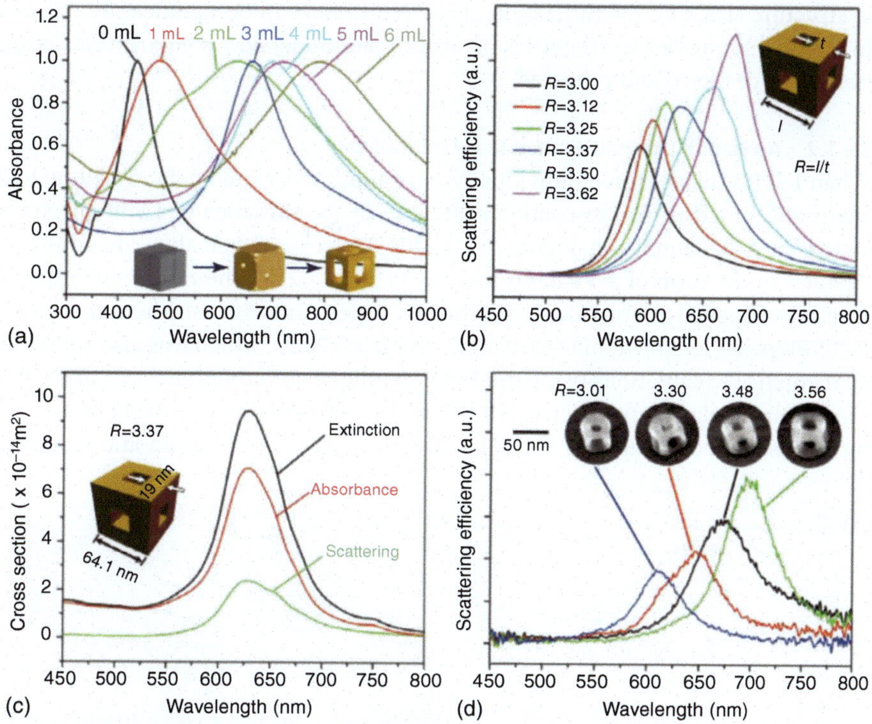

Figure 7.21 (a) UV–Vis spectra taken from aqueous suspensions of the structures, which were synthesized by titrating Ag nanocubes with 1–6 mL of 0.1 mmol l^{-1} $AuCl_2$. (b) DDA-calculated scattering spectra for nanoframes with an edge length of 57.0, 59.4, 61.8, 64.1, 66.5, and 68.9 nm, having a constant ridge thickness of 19 nm, together with a fixed composition of 89% Au and 11% Ag. The inset shows a drawing of the nanoframe used in these calculations, which has both sharp corners and edges. The red-shifts of the plasmon resonance peak increase with an increasing ratio (*R*) between the outer edge length (*l*) and the ridge thickness (*t*). (c) DDA-calculated extinction, scattering, and absorption spectra for a nanoframe with $R = 3.37$. (d) Scattering spectra of individual Au nanoframes and the SEM images of the corresponding nanoframes. For DDA calculations in (b) and (c), the nanoframes were filled and surrounded by air. Source: Reproduced with permission from Skrabalak et al. [42]; © 2008 Springer Nature.

In fact, at this stage, the application of hollow structures mainly focuses on the electrocatalysis, which is less involved than photocatalysis. Solar-driven photocatalysis technology also has bright prospects for solving energy and environmental challenges. Its performance is highly dependent on the design of photocatalysts. Reasonably designed photocatalysts can achieve efficient light collection, promote charge separation and transmission, and accelerate surface reactions. Due to their unique structural characteristics, open nanostructures have many advantages in photocatalyst design, including improving light scattering and collection, shortening the distance of charge migration and directional charge separation, and having abundant shell surface reaction sites. However, the improvement of surface plasmon properties of open nanostructures needs to be compensated by the increase in structure size, and the utilization of surface atoms will be significantly reduced. In order to improve the properties of surface plasmons, new composite structures need to be designed and prepared.

7.4.1.2 Au Nanobipyramid-Embedded Nanoframes

When LSPR was measured individually, Au nanospheres and Pd nanoshells exhibit sharp and broad peaks, respectively, because of the presence of LSPR spectra as shown in A–B samples in Figure 7.22a [43]. However, when they are combined together in the form of a YS nanostructure (Figure 7.22b), their plasmonic spectra hybridized and led to a broadening of the LSPR spectrum with a dip between Au and Pd LSPR peaks as shown in C–D samples in Figure 7.22a, which was also supported by theoretical calculations using the DDA technique with different sizes of YS NPs. This complex plasmonic behavior of YS can be well explained by the hybridization model, as shown in Figure 7.22c. The Pd nanoshells exhibit plasmonic properties

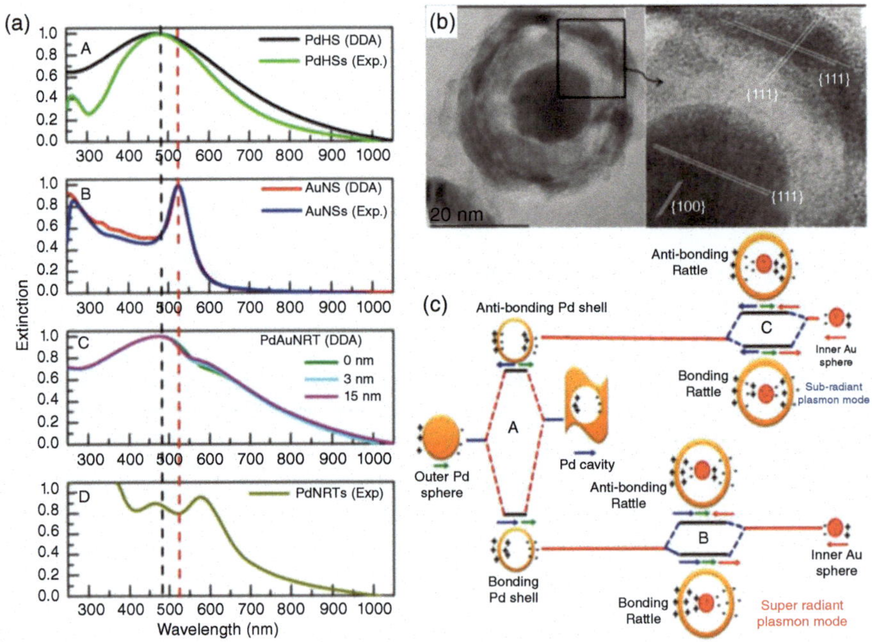

Figure 7.22 (a) Extinction spectrum of (up to bottom: A–D) 55.5 ± 6.3 nm Pd hollow NPs experimentally measured (green spectrum) and 77.5 nm Pd hollow NPs with 5 nm wall thickness calculated by DDA technique (black spectrum). (B) 29.8 nm Au NPs were experimentally measured (blue spectrum) and calculated theoretically (red spectrum). (C) Pd-Au YS consist of 77.5 nm Pd nanoshell and 29.8 nm Au NP when the inside separation distance between them is 0 (green spectrum), 3 (cyan spectrum), and 15 nm (magenta spectrum) calculated theoretically. (D) Measured for Pd-Au YS of 29.8 ± 2.4 nm inside Au NPs and 77.5 ± 7.2 nm Pd outer nanoshells. (b) High-resolution TEM image of PdAu YS NPs of 40.8 nm diameter and 3.2 nm wall thickness; the inner AuNS is 17.1 nm in diameter. Lattice-resolved image of PdAu YS. (c) Schematic diagram that describes the hybridization of the plasmon resonance. Source: Reproduced with permission from Mahmoud [43], © 2014 American Chemical Society.

due to their inner and outer surfaces, which are like a Pd nanosphere and nanocavity. When Au NPs are encapsulated inside the Pd shell, the plasmon mode of Au NPs hybridizes with the bonding and anti-bonding modes of the Pd nanoshell. The reason for broad LSPR with a dip can be understood on the basis of the hybridization model. Au/Pd YS NPs showed two bonding plasmon modes: one with lower energy and large values of dipole moment (a bright, super-radiant broad plasmon mode), and the other with high energy and a low value of dipole moment (dark, sub-radiant narrow plasmon mode). The interference between narrow dark and bright broad plasmon modes results in the formation of a dip in the LSPR spectrum of Au/Pd YS and a Fano resonance.

To further understand the nearly identical plasmon resonance before and after the deposition of the ultrathin metal nanoframes, finite-difference time-domain (FDTD) simulations were performed to ascertain the plasmonic properties of the Au NBP-embedded ultrathin Au nanoframes (Figure 7.23) [36]. The nanoframes were assumed to be made of pure Au for simplification, which has been shown to be reasonable in the simulation of the plasmonic properties of metal alloy nanostructures. The geometrical models employed in the simulations were set according to the shapes observed on the TEM images. The length and waist width of the penta-twinned Au NBP were set according to the TEM images. The ultrathin Au nanoframe has pentagonal symmetry, with the ridge being 2 nm × 2 nm × 119 nm.

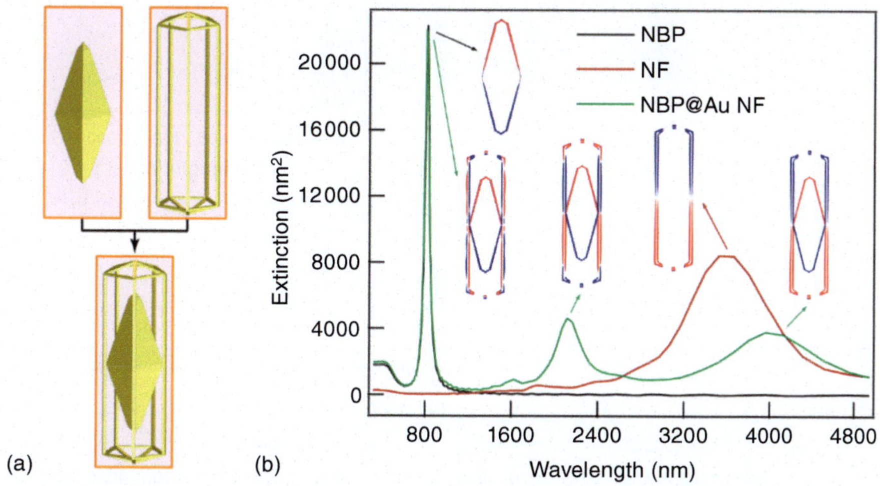

Figure 7.23 FDTD simulations. (a) Schematics of the simulation models. (b) Simulated extinction spectra of the Au NBP, Au nanoframe, and Au NBP-embedded Au nanoframe. The insets are the charge distribution contours calculated at the peak wavelengths under the longitudinal excitation. If the five side ridges of the nanoframe are labeled as 1–5 consecutively, the cross-section for showing the charge distribution contours is determined by the ridges 1 and 3, and it is perpendicular to the propagation direction of the excitation light. The red and blue colors represent positive and negative charges, respectively. Source: Reproduced from Zhu et al. [36]/with permission of Royal Society of Chemistry/Public Domain CC BY 4.0.

The nanoframe is in contact with the middle corners of the Au NBP for the Au NBP-embedded Au nanoframe.

The excitation light direction was perpendicular to the length axis, with the electric field polarized along the length axis. As a result, only the longitudinal plasmon modes were excited. The Au NBP has a longitudinal plasmon peak at 815 nm, which can be unambiguously attributed to the dipolar mode (Figure 7.23b). For the ultrathin Au nanoframe without the Au NBP, the charges on the inner and outer surfaces oscillate in phase, producing a dipolar bonding mode at 3523 nm according to the plasmon hybridization model. The antibonding mode cannot be effectively excited by far-field light owing to a nearly zero net dipole moment. Interestingly, when a gold NBP is embedded in the ultrathin Au nanoframe, both of the antibonding and bonding dipolar modes of the nanoframe can be excited. They couple with the dipolar mode of the Au NBP, resulting in two peaks at 808 and 4133 nm, respectively. The presence of the Au NBP at the center also induces the excitation of an octapolar mode of the nanoframe, which is usually a dark mode. A new peak at 2116 nm can therefore be assigned to the combination of the NBP dipolar and the nanoframe

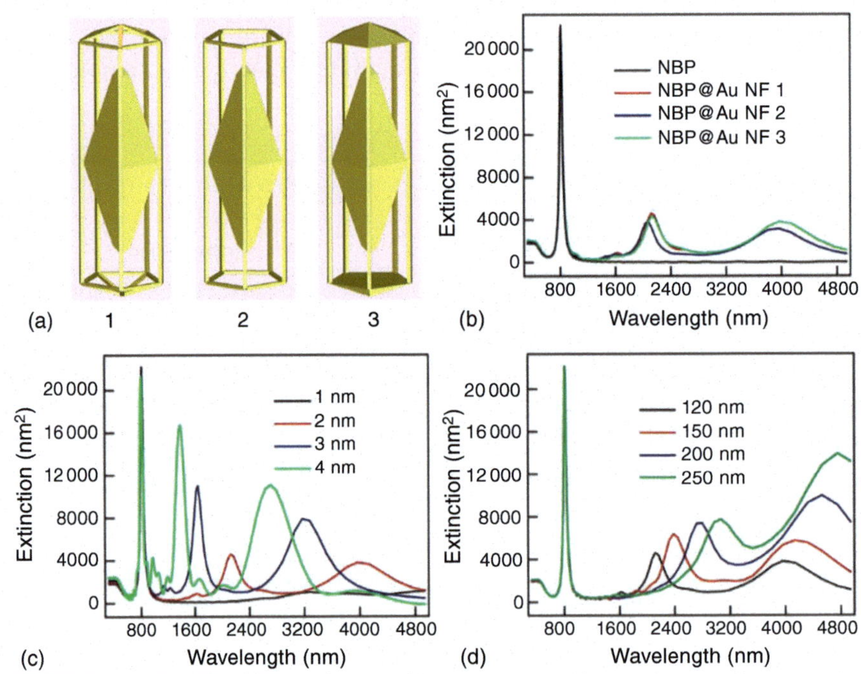

Figure 7.24 FDTD simulations. (a) Schematics of the models used in the simulations. (b) Simulated extinction spectra of the Au NBP-embedded Au nanoframes 1, 2, and 3. (c) Simulated extinction spectra of the Au NBP-embedded Au nanoframes with the ridges having different widths, as indicated in the inset. (d) Simulated extinction spectra of the Au NBP-embedded Au nanoframes with the side ridges having different lengths, as indicated in the inset. The excitation polarization direction is along the length axis of the Au nanoframe. Source: Reproduced from Zhu et al. [36]/with permission of Royal Society of Chemistry/Public Domain CC BY 4.0.

octapolar modes. Simulations were further conducted to clarify the nature of the plasmon resonances of the Au NBP-embedded Au nanoframe.

The structural changes at the ends of the nanoframe were found to have little effect on the plasmon resonances of the Au NBP-embedded Au nanoframe, indicating that the interaction mainly takes place between the side ridges of the nanoframe and the Au NBP (Figure 7.24) [36]. The two low-energy peaks of the ultrathin Au nanoframes in the mid-infrared region exhibit gradual redshifts as the width of the ridges is decreased or the length of the side ridges is increased, while the highest-energy plasmon peak remains almost unchanged at ~800 nm. The transmission measurements in the mid-infrared region show that the two low-energy peaks observed in the simulations cannot be detected at the ensemble level, which can be ascribed to the inherent low intensities of the two peaks and the inhomogeneous distributions in shape and size. On the other hand, the simulated results are generally in good agreement with the experimental ones for the longitudinal plasmon wavelength in the visible region.

7.5 Conclusion and Remarks

In this chapter, some recent work on the fabrication, properties, and applications of typical hollow nanocages and open nanoframes was delivered. In fact, in nanocluster science, many clusters were designed in the shape of cage structures. With the development of fabrication, more and more bimetallic nanocluster cages were achieved as novel nanocatalysts. In general, hollow nanostructures consist of only edges without facets, which allow external reactants to freely access the interior of the structures. Compared with the hollow nanostructures, open nanostructures with a core-cavity-frame (or cage) composite are more interesting. They are also defined as ship-in-bottles. Usually, open nanostructures have an excellent plasmonic nanostructure in the core part, such as the Au NBP described in this chapter. Therefore, they exhibit plasmon-enhanced performance and synergy effects in the nanogaps between any two components. They are widely used as catalytic materials toward reactions critical to fuel cells, water splitting, as well as dehydrogenation, hydrogenation, Suzuki coupling, and pollutant degradation. Despite the rapid progress on the open and hollow nanostructures over the past decade, the widespread use of this novel class of nanomaterials still faces a number of issues.

(a) Preparation with high-abundance and low cost metals. Most of the reported nanoframes are based on low-abundance and high-cost noble metals of Au, Pd, Pt, Rh, and Ru, although they are based on solid Cu and Ag nanocrystals. It will be beneficial to have the synthesis extended to metals that are more abundant and less expensive than other noble metals while still offering key plasmonic and catalytic properties. However, it is extremely difficult to synthesize nanoframes dominated by Cu and Ag metals, which are also promising in catalysis. It is urgent to develop a general strategy for the synthesis of nanoframes made of Ag, Cu, and many other earth-abundant metals with various morphologies and further explore their plasmonic and catalytic properties.

(b) Precise control over the nanostructures with a clean surface. In the fabrication process, the size, shape, and atom arrangement are far from ideal designing structures. It remains a daunting challenge to precisely control the atomic composition and arrangement on the outermost surface of a nanoframe. Moreover, in most cases, the solution-phase synthesis of noble-metal nanoframes involves the use of capping agents or stabilizers. Inevitably, the agents will compromise or dampen the performance of the hollow and open nanostructures in heterogeneous catalysis.
(c) Stability and recyclicity. Usually, nanoframes are far-from equilibrium structures and tend to reconfigure or aggregate, especially when they are subjected to high-temperature catalysis. Thus, it is important to enhance the structural stability (chemical or thermal) and catalytic durability of the nanostructures.
(d) Scaling up the production. Although a wide variety of metal nanoframes have been successfully synthesized in high quality, the protocols mainly rely on conventional batch reactors with small reaction volumes in labs. Uniform growth conditions, including temperature, distributions of chemicals, will be considered in the scaled-up fabrication.

Therefore, there are still a number of challenges that need to be addressed before elevating noble-metal hollow and open nanostructures to the next level application success. It is hoped that the concepts and case studies presented in this chapter can serve as a hot spot for researchers who are interested in developing new forms of noble-metal based nanostructures.

References

1 Greeley, J. and Mavrikakis, M. (2004). Alloy catalysts designed from first principles. *Nat. Mater.* 3: 810–815.
2 Chaudhuri, R.G. and Paria, S. (2012). Core/shell nanoparticles: classes, properties, synthesis mechanisms, characterization, and applications. *Chem. Rev.* 112 (4): 2373–2433.
3 Kumar, S.M.S., Hidyatai, N., Herrero, J.S. et al. (2011). Efficient tuning of the Pt nano-particle mono-dispersion on Vulcan XC-72R by selective pre-treatment and electrochemical evaluation of hydrogen oxidation and oxygen reduction reactions. *Int. J. Hydrog. Energy* 36 (9): 5453–5465.
4 Cho, J., Xu, L., Jo, C., and Ryoo, R. (2017). Highly monodisperse supported metal nanoparticles by basic ammonium functionalization of mesopore walls for industrially relevant catalysis. *Chem. Commun.* 53 (27): 3810–3813.
5 Yang, T.-H., Ahn, J., Shi, S. et al. (2021). Noble-metal nanoframes and their catalytic applications. *Chem. Rev.* 121: 796–833.
6 Wang, W., Dahl, M., and Yin, Y. (2013). Hollow nanocrystals through the nanoscale Kirkendall effect. *Chem. Mater.* 25 (8): 1179–1189.
7 Anderson, B.D. and Tracy, J.B. (2014). Nanoparticle conversion chemistry: Kirkendall effect, galvanic exchange, and anion exchange. *Nanoscale* 6 (21): 12195–12216.

8 Lu, X., Au, L., McLellan, J. et al. (2007). Fabrication of cubic nanocages and nanoframes by dealloying Au/Ag alloy nanoboxes with an aqueous etchant based on $Fe(NO_3)_3$ or NH_4OH. *Nano Lett.* 7 (6): 1764–1769.

9 Nosheen, F., Anwar, T., Siddique, A., and Hussain, N. (2019). Noble metal based alloy nanoframes: syntheses and applications in fuel cells. *Front. Chem.* 7: 456.

10 Fang, Z., Wang, Y., Liu, C. et al. (2015). Rational design of metal nanoframes for catalysis and pasmonics. *Small* 11 (22): 2593–2605.

11 Wang, X., Ruditskiy, A., and Xia, Y. (2016). Rational design and synthesis of noble-metal nanoframes for catalytic and photonic applications. *Natl. Sci. Rev.* 3 (4): 520–533.

12 Chen, C., Kang, Y., Huo, Z. et al. (2014). Highly crystalline multimetallic nanoframes with three-dimensional electrocatalytic surfaces. *Science* 343 (6177): 1339–1343.

13 Oh, A., Baik, H., Choi, D.S. et al. (2015). Skeletal octahedral nanoframe with Cartesian coordinates via geometrically precise nanoscale phase segregation in a Pt@Ni core–shell nanocrystal. *ACS Nano* 9 (3): 2856–2867.

14 Park, J., Sa, Y.J., Baik, H. et al. (2017). Iridium-based multimetallic nanoframe@nanoframe structure: an efficient and robust electrocatalyst toward oxygen evolution reaction. *ACS Nano* 11 (6): 5500–5509.

15 Yoo, S., Kim, J., Choi, S. et al. (2019). Two-dimensional nanoframes with dual rims. *Nat. Commun.* 10: 5789.

16 Han, C., Ge, L., Chen, C. et al. (2014). Site-selected synthesis of novel Ag@AgCl nanoframes with efficient visible light induced photocatalytic activity. *J. Mater. Chem. A* 2 (31): 12594–12600.

17 Wang, Z., Wang, H., Zhang, Z. et al. (2017). Synthesis of Pd nanoframes by excavating solid nanocrystals for enhanced catalytic properties. *ACS Nano* 11 (1): 163–170.

18 Xie, S., Lu, N., Xie, Z. et al. (2012). Synthesis of Pd–Rh core–frame concave nanocubes and their conversion to Rh cubic nanoframes by selective etching of the Pd cores. *Angew. Chem. Int. Ed.* 51 (41): 10266–10270.

19 Yang, Y., Liu, J.Y., Fu, Z.-W., and Qin, D. (2014). Galvanic replacement-free deposition of Au on Ag for core–shell nanocubes with enhanced chemical stability and SERS activity. *J. Am. Chem. Soc.* 136 (23): 8153–8156.

20 Ahn, J. and Qin, D. (2019). Fabrication of nanoscale CGE cubes by drilling orthogonal, intersected holes through all six side faces of Ag nanocubes. *Chem. Mater.* 31 (21): 9179–9187.

21 Sun, Y.G., Mayers, B.T., and Xia, Y.N. (2002). Template-engaged replacement reaction: a one-step approach to the large-scale synthesis of metal nanostructures with hollow interiors. *Nano Lett.* 2 (5): 481–485.

22 González, E., Arbiol, J., and Puntes, V.F. (2011). Carving at the nanoscale: sequential galvanic exchange and Kirkendall growth at room temperature. *Science* 334 (6061): 1377–1380.

23 Shahjamali, M.M., Bosman, M., Cao, S.W. et al. (2013). Surfactant-free sub-2 nm ultrathin triangular gold nanoframes. *Small* 9 (17): 2880–2886.

24 McEachran, M., Keogh, D., Pietrobon, B. et al. (2011). Ultrathin gold nanoframes through surfactant-free templating of faceted pentagonal silver nanoparticles. *J. Am. Chem. Soc.* 133 (21): 8066–8069.

25 Zhang, J.W., Winget, S.A., Wu, Y.R. et al. (2016). Ag@Au concave cuboctahedra: a unique probe for monitoring Au-catalyzed reduction and oxidation reactions by surface-enhanced Raman spectroscopy. *ACS Nano* 10 (2): 2607–2616.

26 Sneed, B.T., Brodsky, C.N., Kuo, C.H. et al. (2013). (2013) Nanoscale-phase-separated Pd-Rh boxes synthesized via metal migration: An archetype for studying lattice strain and composition effects in electrocatalysis. *J. Am. Chem. Soc.* 135 (39): 14691–14700.

27 Zhang, L., Roling, L.T., Wang, X. et al. (2015). Platinum-based nanocages with subnanometer-thick walls and well-defined, controllable facets. *Science* 349 (6246): 412–416.

28 Ye, H., Wang, Q., Catalano, M. et al. (2016). Ru nanoframes with an fcc structure and enhanced catalytic properties. *Nano Lett.* 16 (4): 2812–2817.

29 Ding, J., Zhu, X., Bu, L. et al. (2015). Highly open rhombic dodecahedral PtCu nanoframes. *Chem. Commun.* 51 (47): 9722–9725.

30 Zhang, Z.-P., Zhu, W., Yan, C.-H., and Zhang, Y.-W. (2015). Selective synthesis of rhodium-based nanoframe catalysts by chemical etching of 3d metals. *Chem. Commun.* 51: 3997–4000.

31 Park, J., Kim, J., Yang, Y. et al. (2016). RhCu 3D nanoframe as a highly active electrocatalyst for oxygen evolution reaction under alkaline condition. *Adv. Sci.* 3 (4): 1500252.

32 Sun, Y., Wiley, B., Li, Z.-Y., and Xia, Y. (2004). Synthesis and optical properties of nanorattles and multiple-walled nanoshells/nanotubes made of metal alloys. *J. Am. Chem. Soc.* 126 (30): 9399–9406.

33 Fang, C.H., Zhao, G.L., Zhang, Z.J. et al. (2018). Morphology engineering of Au/(PdAg alloy) nanostructures for enhanced electrocatalytic ethanol oxidation. *Part. Part. Syst. Charact.* 35 (10): 1800258.

34 Li, Q., Zhuo, X.L., Li, S. et al. (2015). Production of monodisperse gold nanobipyramids with number percentages approaching 100% and evaluation of their plasmonic properties. *Adv. Optical Mater.* 3 (6): 801–812.

35 Chow, T.H., Li, N.N., Bai, X.P. et al. (2019). Gold nanobipyramids: an emerging and versatile type of plasmonic nanoparticles. *Acc. Chem. Res.* 52 (8): 2136–2146.

36 Zhu, X., Xu, J., Zhang, H. et al. (2020). Gold nanobipyramid-embedded ultrathin metal nanoframes for in-situ monitoring catalytic reactions. *Chem. Sci.* 11 (12): 3198–3207.

37 Au, L., Chen, Y., Zhou, F. et al. (2008). Synthesis and optical properties of cubic gold nanoframes. *Nano Res.* 1: 441–449.

38 Xu, J., Xu, H.Y., Xu, L.H. et al. (2023). Plasmonic and catalytic Au NBP@AgPd nanoframes for highly efficient photocatalytic reactions. *Phys. Chem. Chem. Phys.* 25: 13189–13197. https://doi.org/10.1039/D3CP01153D.

39 Xu, J., Yun, Q.R., Wang, C.S. et al. (2020). Gold nanobipyramid-embedded silver-platinum hollow nanostructures for monitoring stepwise reduction and oxidation reactions. *Nanoscale* 12 (46): 23663–23672.

40 Yun, Q.R., Xu, J., Wei, T.C. et al. (2022). Synthesis of Pd nanorod arrays on Au nanoframes for excellent ethanol electrooxidation. *Nanoscale* 14 (3): 736–743.
41 Xu, J., Yun, Q.R., Zhang, H. et al. (2021). A novel deposition mechanism of Au on Ag nanostructures involving galvanic replacement and reduction reactions. *Chem. Commun.* 57 (67): 8332–8335.
42 Skrabalak, S.E., Chen, J.Y., Sun, Y.G. et al. (2008). Gold nanocages: synthesis, properties, and applications. *Acc. Chem. Res.* 41 (12): 1587–1595.
43 Mahmoud, M.A. (2014). Plasmon resonance hybridization of gold nanospheres and palladium nanoshells combined in a rattle structure. *J. Phys. Chem. Lett.* 5 (15): 2594–2600.

8

Metal–Semiconductor Nanocomposite

Plasmonics can modulate the energy flow, trap light and inject hot-electrons for semiconductors.

8.1 Introduction

Metal nanostructures with plasmonic properties, such as Au and Ag nanocrystals, have been attracted significant attention for their excellent ability to interact with photon. Compared with single component, the hybrid nanostructures have triggered great interest among researchers due to their superior properties and extend functionalities. For example, in Au–Pd bimetallic nanostructure, Au can produce the surface plasmon related effects and Pd acts as highly efficient catalytic sites. The hybrid nanostructure possesses simultaneously excellent catalytic activities and stabilities [1]. In hydrogen sulfide detection, Au–Ag bimetallic nanostructure exhibits great sensitivity and selective for sulfur ions. The plasmonic resonance wavelength and intensity of Au nanocrystal are good nanoprobes for chemical and biological sensing. Ag and sulfur ions react to form silver sulfide, which causes the variation of plasmon resonance wavelength and intensity by the changes of dielectric function from the conversion of Ag to Ag_2S [2]. Moreover, in the past period, large numbers of hybrid nanostructures have been designed and prepared, and one of the most influential is a series of metal–semiconductor composites. According to the geometrical configuration of nanohybrids, the composites can be divided into metal decorated semiconductors, core-shell structures, yolk-shell structures, Janus, etc. [3].

Nanomaterials of semiconductor have been studied for decades due to their unique physical and chemical properties. Semiconductor nanostructures with remarkable optoelectronic and physicochemical properties are foundation of optoelectronic devices and have wide application in areas of photocatalysis, solar cells, photodetectors, light-emitting diodes. However, single component semiconductors often suffer from weak spectral response, especially in visible region, because of wide band gap, such as TiO_2, ZnO, etc. Numerous efforts have been made to treat these issues, including coupling noble metal nanostructures with local surface

Plasmonic Metal Nanostructures: Preparation, Characterization, and Applications,
First Edition. Caixia Kan.

plasmon resonance (LSPR), to enhance the visible light response of semiconductor nanostructures. Metal–semiconductor composite nanostructure have attracted extensive attentions. At the same time, the hybrid nanostructures of metal coating with semiconductors can effectively protect the metal core and improve the stability of materials. For example, SiO_2 generally uses as coating material on account of its chemical stability, biological compatibility. In addition, as a gain medium for plasmonic property, SiO_2 can also increase plasmonic metal property [4]. These unique properties make metal–semiconductor composite nanostructures widely investigated in the fields of photocatalysis, solar cells, SERS, sensor, and so on.

In this chapter, we will focus on metal–semiconductor composite nanostructure, including metal decorated semiconductor (oxide semiconductor, perovskite semiconductor, perovskite quantum dots and 2D materials) and metal–semiconductor core-shell nanostructures. The developments of metal–semiconductor composite nanostructure are described. The properties and preparations process will be discussed, especially the nanostructure and morphology. The high performance of materials depends on the shape, size and morphology of nanostructures. Finally, we provide a conclusion for metal–semiconductor composite nanostructures.

8.2 Metal Decorated Semiconductor

To date, metals and semiconductors have been extensively investigated in various fields. Incorporating plasmonic metal and semiconductor is an effective pathway to modulate the optic properties and carrier behaviors of nanomaterials, which enhanced the application of photoelectric conversion, detection and energy harvesting. Noble metals contain Au, Ag, Cu, etc. and semiconductors contain transition metal oxide and perovskite, etc. are the most commonly used to prepare metal–semiconductor hybrid nanostructures. In this part, metal decorated semiconductor of transition metal oxide and perovskite quantum dots will be discussed.

8.2.1 Metal-Oxide Semiconductor

As one of the earliest explored nanomaterials, transition metal oxide semiconductors have been extensively studied. Many oxide semiconductors, like TiO_2, SiO_2, ZnO, etc., have been widely used in the fields of photocatalysis, SERS, dye sensitized solar cells, and so on. However, most of oxide semiconductors have small light spectral response range due to the wide band gap. Especially in the visible region, oxide semiconductors interact with light weakly. Fortunately, noble metal nanostructures with the LSPR can interact with light from visible to near-infrared region and can couple with semiconductor to enhance the visible light response. Moreover, metal–semiconductor interaction, such as Schottky junction, strong metal-support interactions, is helpful for regulating charge behaviors of nanostructure. Therefore, metal-oxide semiconductor nanostructures have attracted extensive attentions.

For example, TiO_2, as one of the earliest studied n-type semiconductors with outstanding chemical stability, nontoxicity and low cost, has been widely used in

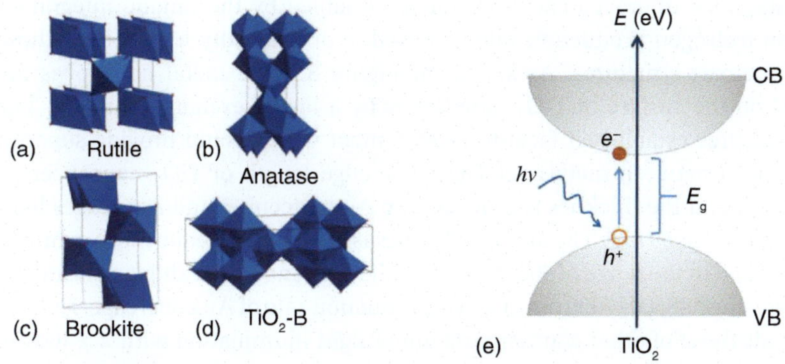

Figure 8.1 Schematic unit cell of TiO_2 polymorphs: (a) Rutile, (b) Anatase, (c) Brookite, and (d) TiO_2-B. (e) Schematic diagram of a typical excitation of electrons from the filled TiO_2 VB to the unoccupied CB via band-to-band excitation. Source: (a–d) Reproduced from Cai et al. [5]/with permission of John Wiley & Sons. (e) Reproduced from Guo et al. [6]/with permission of John Wiley & Sons.

environmental purification, hydrogen reaction evolution and photovoltaic device. TiO_2 polymorphs are mainly anatase, rutile, brookite and TiO_2(B) in nature and all four phases are composed of TiO_6 octahedron with different distortion of the octahedron units and share edges and corners in different manners, as shown in Figure 8.1a–d [5, 7]. Octahedron of anatase arranging in zigzag chains along (221) share four edges while octahedron of rutile share only two edges and connect in linear chains parallel to (001). Both corners and edges are connected in brookite. For TiO_2(B), layered structure is composed of both edges and corners shared TiO_6 octahedron. The different lattice structures lead to distinct mass densities and electronic band structures of TiO_2 in different phases. The bandgap energy of TiO_2 is 3.0–3.2 eV depending on the crystalline structure [8]. When TiO_2 absorbs photons with energy equal to or greater than its E_g, the electrons in valence band (VB) will be excited to conduction band (CB), leaving the holes in VB (Figure 8.1e) [6]. This electron–hole pair generation process can be formulated as follows:

$$\mathrm{TiO_2} + hv \rightarrow e^-(\mathrm{TiO_2}) + h^+(\mathrm{TiO_2})$$

However, the wide bandgap of TiO_2 can only absorb the UV region of the solar spectrum which limits conversion efficiency of light energy. In addition, high recombination of photogenerated charge carriers also affects the practical application of TiO_2. To solve these problems, plasmonic metal loading is a commonly adopted method which can effectively facilitate the proton reduction stemming from the formation of Schottky barriers and the localized surface plasmon resonance.

Jiang et al. reported an approach to integrate dual plasmonic nanostructures with anatase TiO_2 semiconductor nanosheets for photocatalytic hydrogen production in visible and near-infrared spectral regions [9]. The Au nanocubes and nanocages can harvest visible and near-infrared light, respectively, and generate and inject hot electrons into TiO_2. Meanwhile, Pd nanocubes that trap the energetic electrons from TiO_2 and efficiently participate in the hydrogen evolution reaction are employed as co-catalysts for improved catalytic activity. Enabled by the unique integration design, the hydrogen production rate achieved is dramatically higher than those of its counterpart structures. As shown in Figure 8.2, the metal nanocubes are assembled on the surface of TiO_2 nanosheets by a ligand-exchange method. The hybrid structures established facet-to-facet contact which could provide superior electric contact between metal and TiO_2. The edge length of TiO_2 nanosheets is above 200 nm and the thicknes is 5–7 nm. The Au nanocubes and nanocages have an edge length of ~50 nm. The size of Pd cubes is ~10 nm. Au nanocubes exhibit a plasmonic band in the range of 500–600 nm while Au nanocages exhibit a plasmonic band in the range of 600–1000 nm. The H_2 production rate of Au cube/cage-TiO_2-Pd cube exceeds those of other counterparts under light illumination with $\lambda > 400$ nm and $\lambda > 800$ nm. This unique quarternary hybrid structure with excellent photocatalytic performance further confirms the optimization of metal–semiconductor nanostructures.

As the simplest structure in the various metal-oxide semiconductor nanostructures, metal nanoparticle directly decorated on oxide semiconductors has been widely developed. The structure is commonly fabricated though the deposition or growth of metal nanoparticle on prepared semiconductors. This type of the nanostructures has great advantages for controlled the shape and size of metal and semiconductor nanocrystals precisely. In the past, Ag/ZnO hybrid nanostructures are often used for SERS substrate as the LSPR of Ag nanoparticles and self-cleaning of ZnO by photodegradation of organic molecules. Tang et al. uses the ion sputter method to simultaneously load small Ag nanoparticles and large Ag spheres onto the side surface and the top ends of large-scale vertically aligned cone-shaped ZnO nanorods, respectively (Figure 8.3) [10]. The 3D hybrid substrate manifests high SERS sensitivity to rhodamine (R6G) with concentration as low as 10^{-12} M and a detection limit as low as 10^{-11} M to polychlorinated biphenyl (PCB)-77. This excellent performance benefits from three kinds of inter-Ag-nanoparticles gaps in 3D geometry creating a huge number of SERS "hot spots", including (i) gaps between the small Ag-NPs growing on the same cone-shaped ZnO nanorods; (ii) gaps between the Ag nanoparticles growing on the two neighboring ZnO nanorods; (iii) gaps between the large Ag spheres on the top ends of two neighboring ZnO nanorods. Moreover, the supporting chemical enhancement effect of ZnO nanorods and the better enrichment effect ascribed to the large surface area of the substrate also help to achieve a lower detection limit. The 3D Ag/ZnO hybrid structures show great potentials in SERS with rapid trace detection and monitoring.

Moreover, the synthesis of Ag/ZnO hybrid nanostructures by ion sputter method could be used in the field of photocatalysis. The content of Ag can be controlled by the ion sputter time and repeat periods. Ag nanoparticles are uniformly loading on the surface of ZnO nanorodes. The resonance absorption peak of Ag is range from 400 to 800 nm, which is beneficial to the application of visible-light-driven photocatalytic water splitting [11]. In addition, the ultrasonication-assisted *in-situ* deposition strategy is a facile method to deposit Ag nanoparticle on TiO_2 nanotube to construct Ag/TiO_2 hybrid nanostructure. The efficient surface plasmon resonance photosensitizers under visible light irradiation and charge separated at the interfacial between

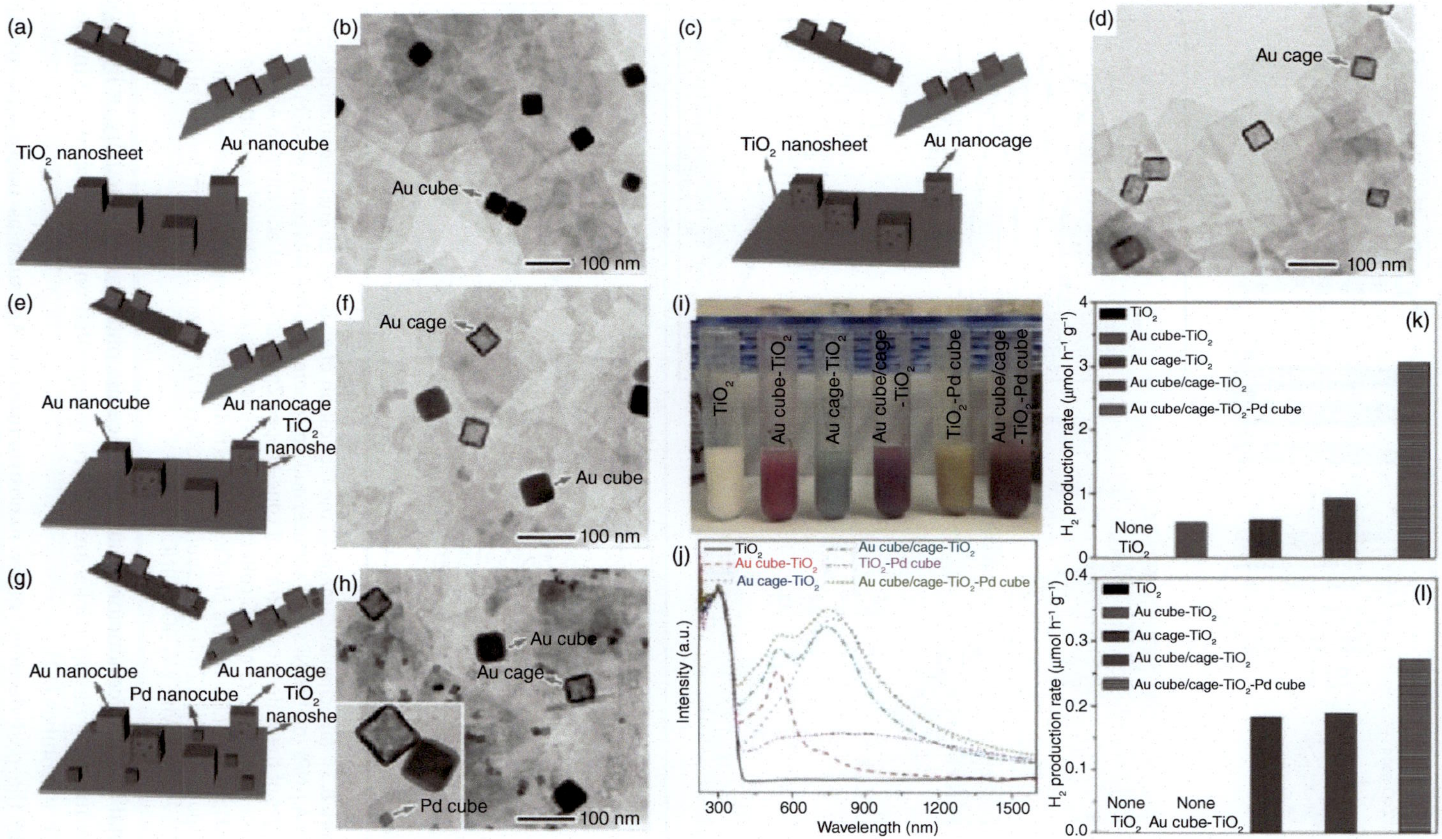

Figure 8.2 Schematic illustrating the hybrid structure and TEM images. (a, b) TiO_2-Au cube, (c, d) TiO_2-Au cage, (e, f) Au cube/cage-TiO_2, (g, h) Au cube/cage TiO_2-Pd. (i) Photograph of suspensions, and (j) UV–vis–NIR diffuse reflectance spectra of hybrid structures. Photocatalytic average rates of hydrogen production under light irradiation with (k) $\lambda > 400$ nm and (l) $\lambda > 780$ nm. The power densities of light are measured to be 100 and 18 mW cm^{-2} for $\lambda > 400$ nm and $\lambda > 780$ nm, respectively. Source: Reproduced with permission from Jiang et al. [9]. ©2016, JOHN WILEY & SONS, INC.

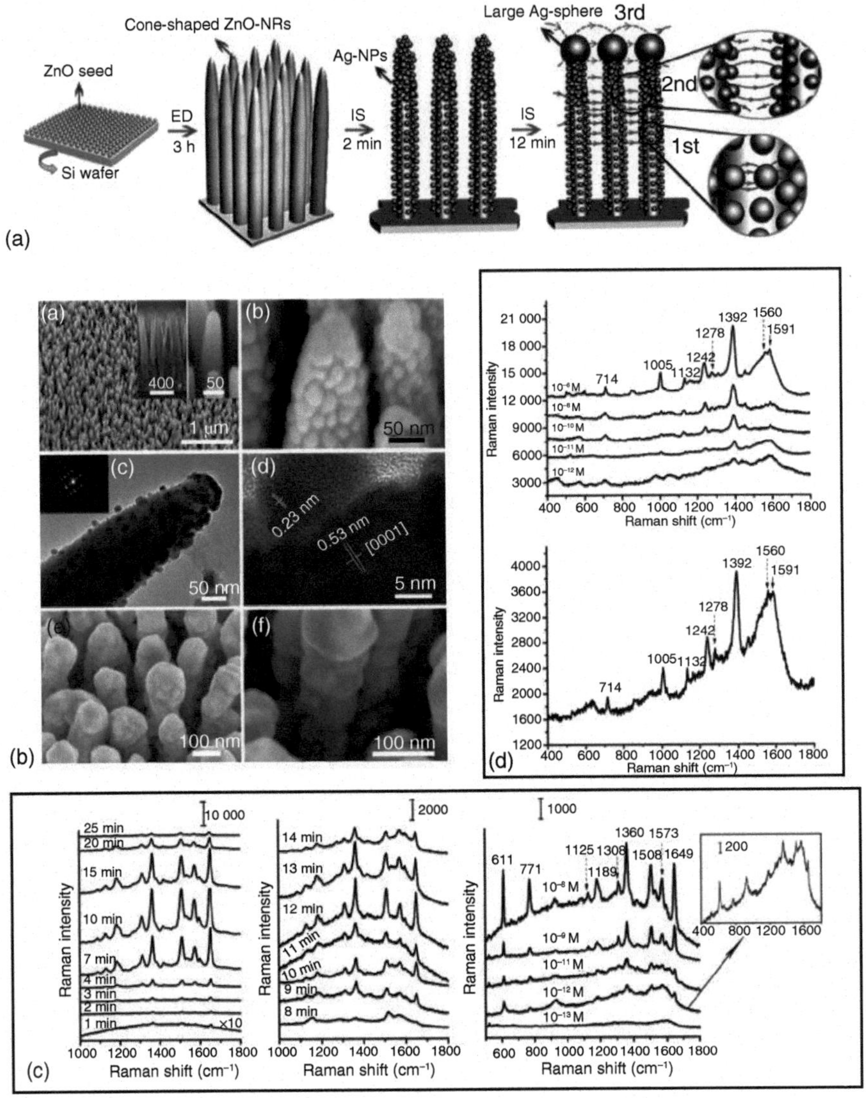

Figure 8.3 (a) Schematic image of the fabrication of an arrays of ZnO nanorods decorated with Ag-NPs (on the NRs side surface) and Ag spheres (on the NRs tops). (b) SEM and TEM images of the wafer-scale arrays of cone-shaped ZnO nanorodds and after an Ag-sputtering. (c) The SERS spectra of R6G collected on the substrates with different Ag-sputtering durations. (d) SERS spectra of PCB-77 collected on the as-prepared substrates with Ag-sputtering for 12 minutes. *Source:* Reproduced with permission from Tang et al. [10]. ©2012, JOHN WILEY & SONS, INC.

metal and semiconductor greatly enhance the photocatalytic activity of hydrogen evolution. Ag nanoparticles are uniformly deposited on both surface and inside of TiO_2 nanotubes [12].

In the past decades, other nanostructures of Ag, such as nanosheet, are also prepared. Ag nanosheet assembled hollow tubular array on ZnO substrate is obtained by electrodeposition method [13]. Au is another hot spot nanomaterial for construct metal–semiconductor hybrid nanostructure. Up to date, a large number of researches have been reported. Au with shape manipulating are decorated on TiO_2 nanowires by changing the synthesis process. Au nanoparticles decorated TiO_2 is obtained by immersed TiO_2 into Au precursor solution and then annealed. Au NRs decorated TiO_2 is obtained by immersed TiO_2 into Au NR solution and dried. Au NRs are prepared in advance by seed-growth method. A mixture of Au nanoparticles and Au NRs decorated TiO_2 can be formed by coupling the two approaches above [14]. Au nanostars decorated on SiO_2 nanostructures are also prepared and used for applications [15]. Moreover, noble metal decorated SiO_2, Fe_2O_3, CeO_2, SnO_2 nanostructures have also been explored [16–19].

8.2.2 Metal-Perovskite Semiconductor

Perovskites are materials described by the formula ABX_3, where X is an anion (I^-, Br^-, Cl^-), A is an organic or inorganic cation (methylammonium [MA, $CH_3NH_3^+$], Sr^{2+}, Pb^{2+}, Ce^{3+}) and B is a divalent cation (Pb^{2+}, Sn^{2+}) (A being larger than B). B and X form a BX_6 octahedral, where X lies in the corners around B, which is located at the center of the octahedral. The BX_6 octahedral forms a three-dimensional extended system by linking all-corners with each other. The A cation is located in a vacancy within the octahedral and maintains the charge neutrality of the entire network system, as shown in Figure 8.4a,b [20, 22, 23]. In the

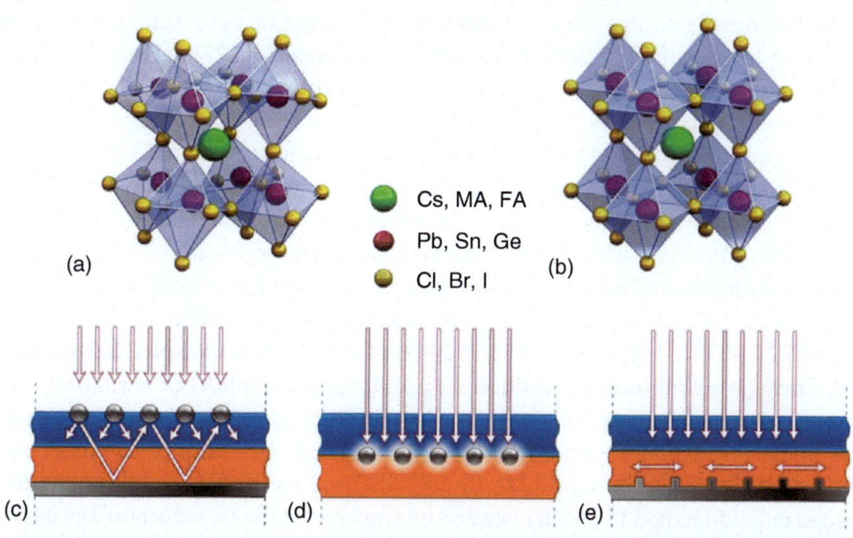

Figure 8.4 Crystal structures of orthorhombic (a) and cubic (b) metal halide perovskites. Note that orthorhombic perovskite has a distorted structure with tilted octahedra. (c) the scattering form metal nanoparticles, (d) the excitation of localized surface plasmons in metal nanoparticles, (e) the excitation of localized surface plasmons at metal/semiconductor for thin-film solar cells. Source: (a, b) Reproduced from Zhang et al. [20]/with permission of John Wiley & Sons. (c–e) Reproduced from Atwater and Polman [21]/with permission of Springer Nature.

recent years, perovskite semiconductors have emerged as important materials using in the field of solar cells, photodetectors and lasers due to their unique characteristics, such as defect tolerance, emission color purity, tunable band gap, and long carrier transport. However, in order to further improve efficiency of perovskite optic devices, coupling plasmonic metal is an effective way by boosting light absorption and carrier transport efficiency. For example, in perovskite solar cells, plasmonic structures provide at least three mechanisms of decreasing the physical thickness of the photovoltaic absorber layers while maintaining the optical thickness constant. First, metal nanoparticles can be used to scatter subwavelength and capture the incident solar light, subsequently fold it into a thin absorber layer (Figure 8.4c). Second, metal nanoparticles can be used as subwavelength antennas to increase the cross-section area of effective absorption by the plasmonic near-field coupled with semiconductor (Figure 8.4d). Third, a corrugated metallic film on the rear exterior of a thin photovoltaic absorber layer can convert solar light into surface plasmon polariton modes at the metal/semiconductor interface and guide modes into the semiconductor slab. Scilicet, light is converted to photocarriers in the semiconductor (Figure 8.4e) [21].

Moreover, the physical and chemical properties of perovskites are not only related to their composition, but also depended on their morphology and sizes. To date, various morphologies of perovskites have been designed and synthesized from zero dimension quantum dots to three dimension nanostructures [24]. Quantum dots will be discussed in the next section. In this part, we will focus on the nanorods, nanosheets and nanocube. For example, as shown in Figure 8.5, Guo et al. demonstrates a synthetic strategy toward cesium-based inorganic perovskite nanorods by exploiting composition-controlled crystal phase engineering [25]. It is accomplished for Cd-rich mixed-cation $CsPb_{1-x}Cd_xBr_3$ nanocrystals, where the initial 1D hexagonal perovskite phase drives the growth of the 1D nanorods. The average length and width of nanorods is 84 and 16 nm, corresponding to an aspect ratio of 5.25. The band gaps of the nanorods are tunable by varying the Cd-content, and the highly uniform $CsPb_{0.08}Cd_{0.92}Br_3$ nanorods exhibit a true blue-color emission centered at 460 nm, with a high quantum yield of 48%. Moreover, the Fermi levels in the films made of $CsPb_{1-x}Cd_xBr_3$ alloyed nanocrystals are also related with Cd content. Besides, Zhou et al. reported that high quality all inorganic cesium lead halide alloy perovskite nanorods with the length of 2–20 μm are synthesized by vapor-phase deposition [26]. Through solvothermal conditions, 2D all-inorganic $CsPbX_3$ (X = Cl, Br, I) perovskite nanocrystals are prepared by the assembly of corresponding 1D nanorods. The thickness of nanocrystals was about 3.5 nm and the width of squares could be controlled from 0.1 to 1 μm [27]. By solvothermal method, Li et al. successfully formed cuboid-shaped $MAPbBr_3$ perovskite single crystals. The morphology of perovskite is evolved under different reactive time [28].

Hybrid metal/semiconductor heterostructures are proved to be great candidates for a variety of applications. Therefore, researches have also focused on metal/perovskite hybrid nanostructures for the past years. Ag/perovskite hybrid nanostructures are widely investigated for optoelectronic devices. Huang et al. reported that Ag nanoparticles are deposited on the surface of perovskite semiconductor to construct hybrid nanostructures. $CsPbBr_3$ nanocrystals are prepared as substrate

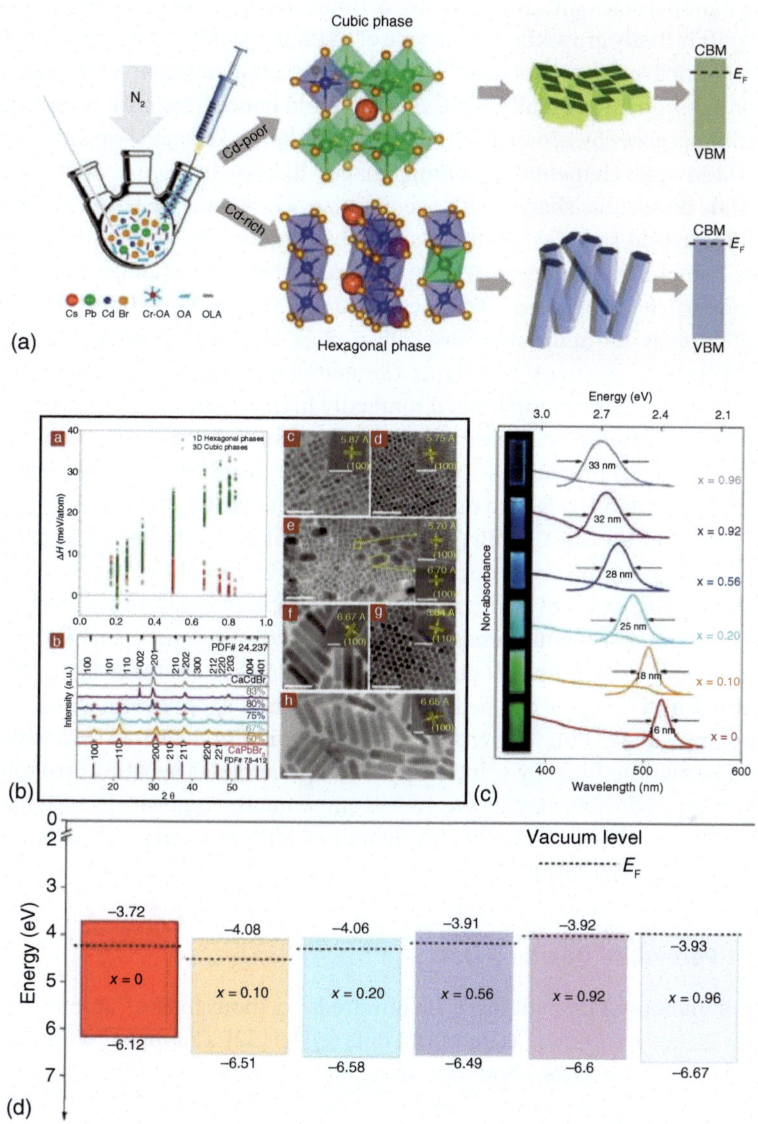

Figure 8.5 (a) Illustration of the synthetic strategy to modulate the crystallographic structure of alloyed CsPb$_{1-x}$CdxBr$_3$ perovskite. In the Cd-poor synthetic environment, the alloyed NCs maintain the 3D cubic crystallographic structure of CsPbBr$_3$, while adopting 1D hexagonal crystallographic structure under Cd-rich conditions. (b) Calculated formation enthalpies, ΔH (in meV per atom), of the alloyed CsPb$_{1-x}$Cd$_x$Br$_3$ ($0 < x < 1$) as a function of the Cd content (x), XRD patterns, TEM and HRTEM images of CsPbBr$_3$ and alloyed CsPb$_{1-x}$Cd$_x$Br$_3$ with x values of 0.10, 0.56, 0.92, and 0.96. Scale bars are 50 nm for the main frames and 5 nm for the insets, respectively. (c) Absorption and emission (excitation wavelength 365 nm) spectra of alloyed CsPb$_{1-x}$Cd$_x$Br$_3$ with different Cd content (x). (d) Energy band diagrams (with the Fermi levels EF indicated by dashed lines) of alloyed CsPb$_{1-x}$Cd$_x$Br$_3$ NCs with different Cd content (x). Source: Reproduced with permission from Guo et al. [25]. ©2020, JOHN WILEY & SONS, INC.

and then Ag nanoparticles are deposited by a three-step procedure as shown in Figure 8.6a. AgBr is firstly grown on the surface of $CsPbBr_3$ after Ag ions introduced, and small Ag nanoparticles are formed by the photodecomposition of AgBr under UV irradiation subsequently. Finally, Ag-$CsPbBr_3$ hybrid nanocrystals are obtain via Ostwald ripening process by prolonged UV irradiation to make small Ag nanoparticle grown. TEM and elemental mapping images display in Figure 8.6b [29]. The nanocrystals have cubic shapes with average sizes of ~13 nm for both $CsPbBr_3$ and Ag-$CsPbBr_3$ hybrid samples. Dark dots are observed on the surfaces of cubic $CsPbBr_3$ NCs, which are the deposited Ag nanoparticles. The average diameter of the Ag nanoparticles is ~5.5 nm. The local mode of SPR in Ag-$CsPbBr_3$ hybrid NCs is manifested as an additional absorption peak with a full width at half maximum (FWHM) of ~130 meV at 424 nm. Compared with conventional systems, Ag-$CsPbBr_3$ hybrid nanocrystals exhibited markedly high efficiency of hot electron transfer benefited from the increased metal/semiconductor coupling. Both the plasmon-induced hot electron and the resonant energy transfer processes can occur on a time scale of less than 100 fs with quantum efficiencies of $50 \pm 18\%$ and $15 \pm 5\%$, respectively. Based on the Ag-$CsPbBr_3$ system, Zhang et al. constructed the plasmonic perovskite light-emitting diodes. The morphology of $CsPbBr_3$ nanocrystals were formed with cubic shape of 10 nm and Ag nanorod structures were synthesized with the length of ~50 nm. Ag nanorods enhanced the electroluminescence (EL) of $CsPbBr_3$ perovskite light-emitting diode by the plasmonic near-field effect [30]. Wu et al. used Au NRs to enhance the near-infrared amplified spontaneous emission (ASE) performance of $MAPbI_3$ polycrystalline films. Au NRs doped PMMA layer was coated on perovskite films by solution-processing method. The ASE threshold of $MAPbI_3$ film was reduced from 26.5 to 16.9 $\mu J\,cm^{-2}$ and the intensity of ASE had increased by 13.9-fold due to the LSPR wavelength of Au NRs overlapped with the band-edge emission of $MAPbI_3$ [31].

8.2.3 Metal-Perovskite Quantum Dots

Semiconductor nanocrystals composed of hundreds to thousands of atoms with sizes from 2 to 20 nm are known as quantum dots (QDs) [32]. Compared with bulk materials, semiconductor QDs show the size-, shape-, and composition-tunable electronic and optical properties by the virtue of their quantum and dielectric confinement effects. In the past few years, semiconductor QDs are widely used in solar energy harvesting, display, detection, photovoltaics and medical imaging technologies. Recently, perovskite QDs such as $CsPbBr_3$ QDs have emerged as the most promising semiconductor QDs with superior performance. As the high quantum yields, narrow emission peaks, short exciton lifetimes and good gain medium, perovskite QDs have great potential for application in photovoltaic and optoelectronic devices. As shown in Figure 8.7, Li et al. reported the synthesis of $CsPbX_3$ (X = Cl, Br, I, or mixed Cl/Br and Br/I) QDs by direct ion reactions in ether solvents, and the synergetic effects of solvent polarity and reaction temperature on the nucleation and growth of QDs are demonstrated [33]. The use of solvent with a low polarity enables controlled growth of QDs, which facilitates the synthesis of

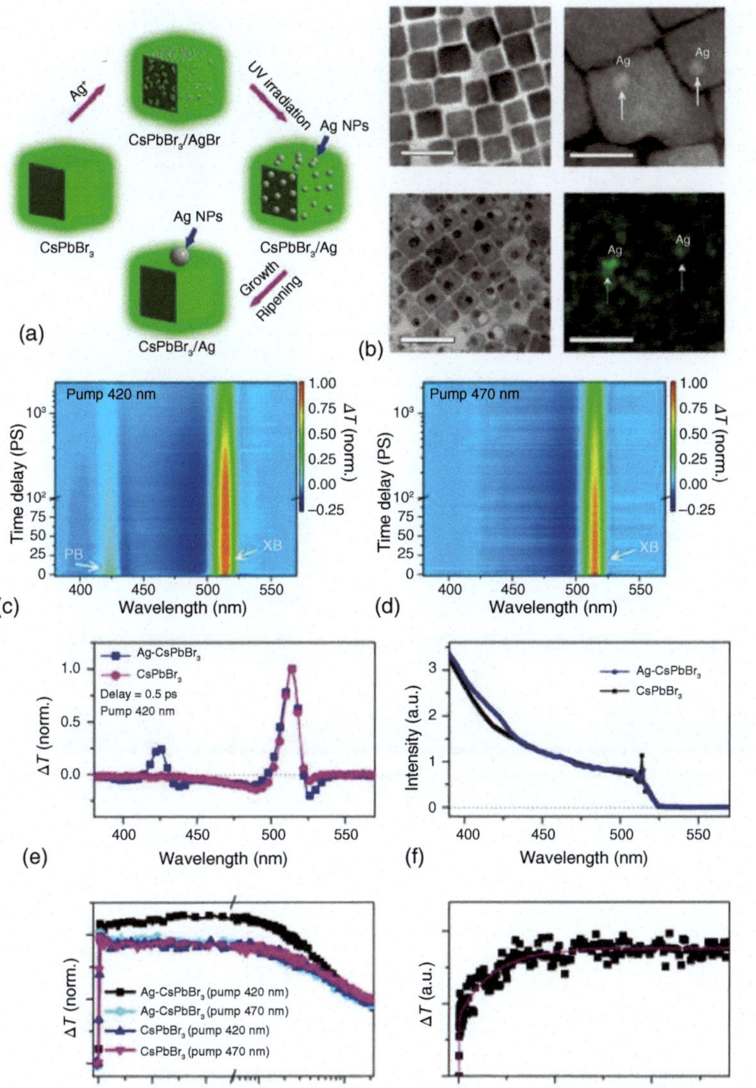

Figure 8.6 (a) Schematic of formation process of Ag-$CsPbBr_3$ hybrid NCs. (b) TEM images of Ag-$CsPbBr_3$ hybrid NCs (scale bar: 20 nm) and elemental mapping images for element Ag at the same spot (scale bar: 10 nm). Ultrafast transient absorption recorded from a solution sample of Ag-$CsPbBr_3$ NCs with pump wavelengths of (c) 420 nm and (d) 470 nm, respectively. (e) Ultrafast transient absorption spectra normalized to the signal amplitude at 514 nm recorded at a time delay of ~0.5 ps from $CsPbBr_3$ and Ag-$CsPbBr_3$ NCs, respectively. (f) PL excitation spectra recorded from $CsPbBr_3$ and Ag-$CsPbBr_3$ NCs. The spectra are normalized to the PL intensity under 470 nm excitation. (g) Kinetic curves probed at 514 nm recorded from $CsPbBr_3$ and Ag-$CsPbBr_3$ NCs pumped at the wavelengths of 420 and 470 nm, respectively. The curves are plotted normalized to the signal amplitude recorded at a delay of 2 ns. (h) Scaled kinetics of carrier transfer from Ag to $CsPbBr_3$ NCs in hybrid NCs obtained by subtracting the contribution of $CsPbBr_3$ from that of Ag-$CsPbBr_3$ NCs, whereas the signals for the two samples are normalized to the amplitude at a time delay of 2 ns. Source: Reproduced with permission from Huang et al. [29]. © 2019, Springer Nature/CC BY 4.0.

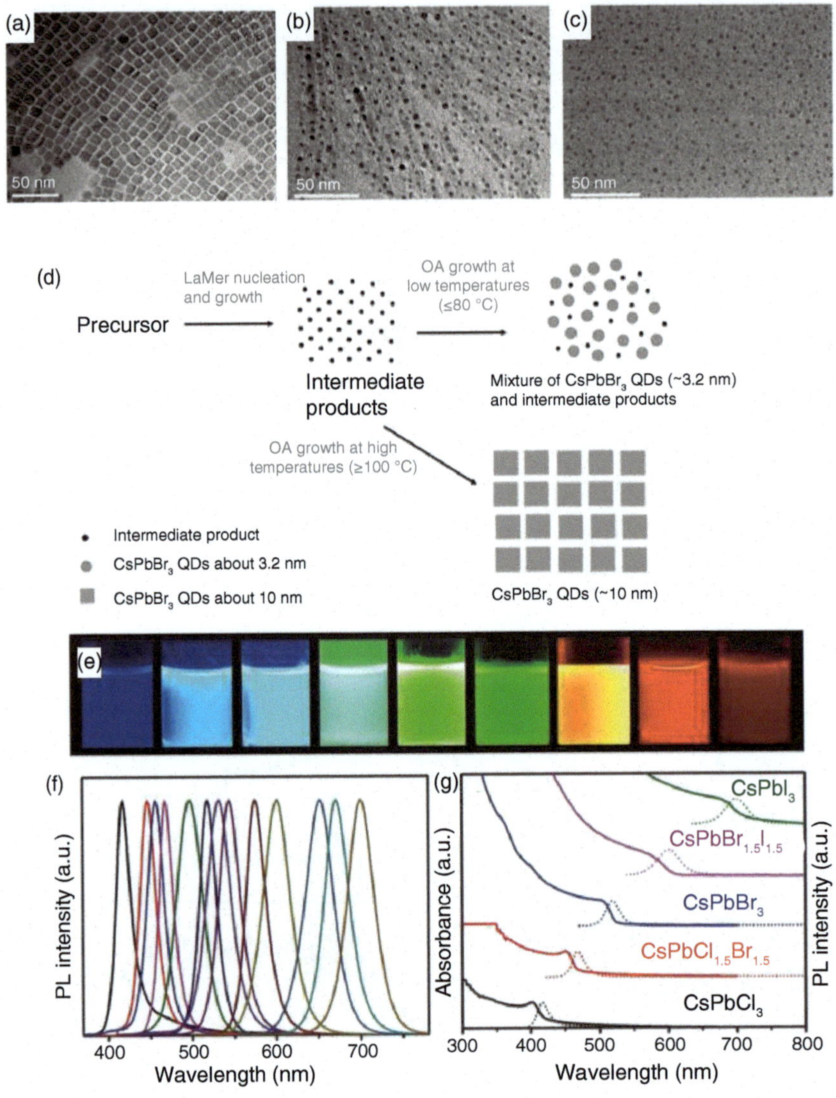

Figure 8.7 (a–c) The $CsPbBr_3$ QDs synthesized with different solvent. (d) The growth processes of $CsPbBr_3$ QDs with the low polarity at different temperatures. (e) Color-tunable $CsPbX_3$ (X = Cl, Br, I, and mixed Cl/Br and Br/I) QDs. Colloidal solutions in tolune under UV irradiation ($\lambda = 365$ nm). (f) Representative PL spectra of $CsPbX_3$ QDs. (g) Typical UV–vis absorption (solid) and PL spectra (dot) of $CsPbX_3$ QDs with different halogen element compositions. Source: Reproduced with permission from Li et al. [33]. @2016, JOHN WILEY & SONS, INC.

high-quality $CsPbX_3$ QDs with broadly tunable luminescence, narrow emission width, and high quantum yield. The size distribution of $CsPbX_3$ QDs ranges from 3.2 to 10.5 nm when synthesized in different solvents. In Figure 8.7d, the proposed growth procedures for $CsPbX_3$ QDs based on LaMer mechanism and OA mechanisms in the solvents with different polarity are presented. The $CsPbX_3$ QDs with wide color gamut are synthesized by adjusting the composition and size, of which the spectral range is from 410 to 700 nm.

Plasmon enhanced perovskite QDs optical properties has also attracted the attention of many researchers. In various fields, perovskite QDs coupling with noble metal with LSPR have emerged great application potential. Ye et al. reported that benefiting from the increased absorbance of ultraviolet or blue light by the Ag induced plasmonic near-field effect, the enhanced photoluminescence intensity of $CsPbBr_3$@Ag hybrid nanocrystals was achieved compared with that of pure $CsPbBr_3$ nanocrystals. The optimizing $CsPbCl_xBr_{3-x}$@Ag hybrid nanocrystals further promoted the advantageous effect of plasmonic resonance enhancement by surface trap states. The 2–3 nm Ag nanoparticles were coated on the surface of $CsPbCl_xBr_{3-x}$ with the cubic shape [34]. Moreover, Au NRs were also be used to combined with $CsPbBr_3$ QDs. Juan et al. prepared plasmonic Au NRs/$CsPbBr_3$ QDs film by spin-coating method. The photoluminescence intensity had twofold increased after combined with Au NRs [35]. In laser investigation, for example, Hsieh et al. constructed a plasmonic nanolaser thought the coupling of perovskite QDs and plasmonic metal. The threshold was 1.9 W cm^{-2} under 120 K [36].

Shi et al. reported a strategy of combining plasmonic metal with perovskite QDs to addresses simultaneously the emission efficiency and stability issues for perovskite light-emitting diode. The enhanced electroluminescence was related with the increased spontaneous emission rate and improved internal quantum efficiency induced by exciton-LSPR coupling (Figure 8.8) [37]. TEM, HRTEM and FFT image of the $CsPbBr_3$ QDs shows in Figure 8.8a–c, indicating the uniform size and single crystalline nature of the QDs. The size distribution of the $CsPbBr_3$ QDs can be tuned in a range of 3.6–9.0 nm, corresponding to an emission range from 483 to 514 nm. Absorption spectra of $CsPbBr_3$ QDs, Au NPs and PL spectra of $CsPbBr_3$ QDs were presented in Figure 8.8d. The insets show the photographs of perovskite $CsPbBr_3$ QDs and Au NPs solutions. An obvious exciton absorption peak occurs at 503.5 nm. The processing procedures for the preparation of coaxial core-shell NW heterojunction architecture and the Au/*p*-NiO/$CsPbBr_3$ QDs/MgZnO/Au NPs/*n*-ZnO/n^+-GaN heterostructure LED are shown in Figure 8.8e. A five main processing procedures are involved: growth of well-aligned ZnO NWs, Au NPs decoration, MgZnO layer deposition, spin-coating of inorganic $CsPbBr_3$ QDs, and sputtering of NiO layer. The cross-sectional SEM image shows the structure of the NiO/$CsPbBr_3$ QDs/MgZnO/Au NPs/ZnO coaxial NWs. Compared the EL spectra of the plasmonic LED with Au NPs decoration and the reference device without Au NPs, the EL performance of the LED device can be enhanced by the plasmonic Au NPs. The plasmonic perovskite LED reaches up a luminance of 10 206 cd m^{-2}, an external quantum efficiency of ≈4.626%, and a current efficiency of 8.736 cd A^{-1}. The exciton-LSP coupling contributes to the EL enhancement.

8.2.4 Metal-Two-Dimensional (2D) Materials

2D materials have emerged as promising rising star nanomaterials due to their unique properties that originate from their atomically thin thickness and 2D morphological features, such as a high surface area and specific physicochemical

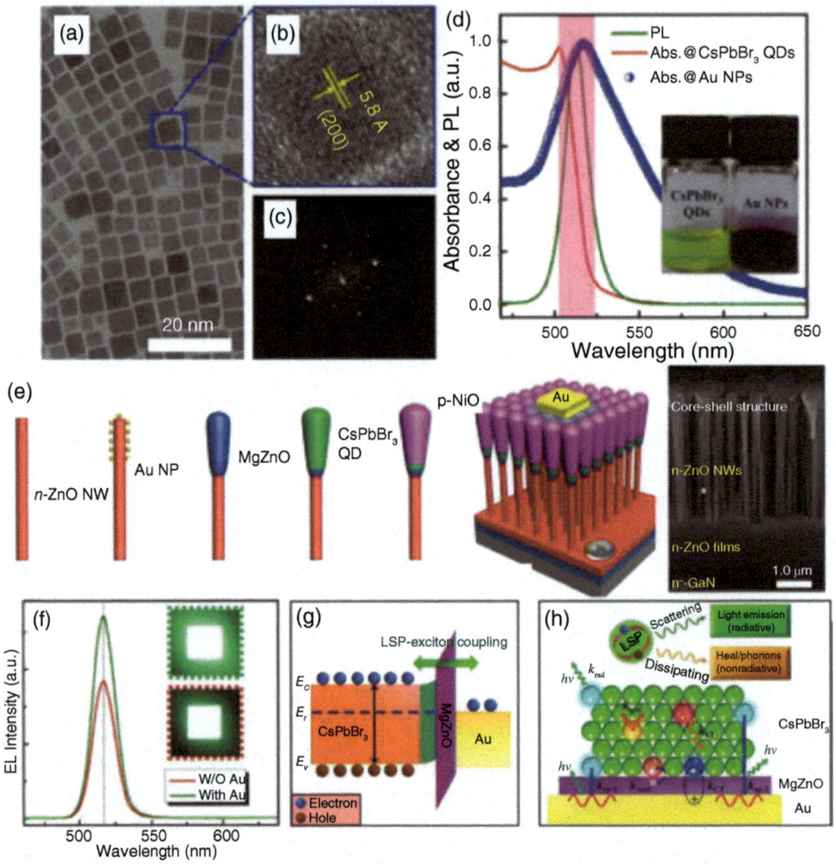

Figure 8.8 (a–c) TEM, high-resolution TEM and FFT image of the $CsPbBr_3$ QDs. (d) Absorption spectra of $CsPbBr_3$ QDs, Au NPs and PL spectra of $CsPbBr_3$ QDs. The insets show the photographs of perovskite $CsPbBr_3$ QDs and Au NPs solutions. (e) Processing procedures for the preparation of coaxial core–shell NW heterojunction architecture and the Au/*p*-NiO/$CsPbBr_3$ QDs/MgZnO/Au NPs/*n*-ZnO/n^+-GaN heterostructure LED and cross-sectional SEM image of the NiO/$CsPbBr_3$ QDs/MgZnO/Au NPs/ZnO coaxial NWs. (f) EL spectra of the plasmonic LED with Au NPs decoration and the reference device without Au NPs. The insets present their corresponding emission photographs of an emitting unit ($2 \times 2\ mm^2$). (g) Band alignment of the Au/MgZnO/$CsPbBr_3$ heterostructures. (h) Detailed relaxation pathways of excitons in $CsPbBr_3$ QDs on Au plasmonic resonators with a MgZnO spacer layer (bottom right). Source: Reproduced with permission from Shi et al. [37]. © 2018, John Wiley & Sons.

properties. In recent years, a large number of 2D materials have been extensively studied in photovoltaic devices, photocatalysis, batteries, supercapacitors, sensors, fuel cells, and SERS, including graphene, transition metal dichalcogenides, phosphorene, borophene, graphitic carbon nitride (g-C_3N_4), boron nitride (BN), and so on. Although 2D materials have superior optical and electrical properties, the ultra-thin structure still limits the absorption of light for some applications in photocatalysis, optical sensors and visual images. Plasmonic noble metal coupled 2D materials have once again attracted the attention of researchers as the different functionalities of two individual components and the synergies of hybrid nanostructures. These hybrid nanostructures also provide an attractive, multifunctional platform for various applications with enhanced performance [38].

Based on graphene and its derivatives, researchers have conducted to extensively study of metal-2D material hybrid nanostructures in fuel cells, electrochemical sensors, and SERS. For example, Kabeer Jasuja and Vikas Berry synthesized snowflake-shaped Au nanostructures on GO surface by the reduction of $HAuCl_4$ with hydroxyl-amine as the reduction agent [39]. These nanostructures achieve a twofold Raman enhancement and realize the control on the band gap and Schottky barrier. Zhang's group reported the synthesis of the fluorescent Au nanodots, ultrathin Au nanowires and Au square sheets on GO and reduced GO (rGO) [40–42]. The size of Au nanodots is ~2 nm, the diameter of Au nanowires is 1.6 nm and the edge length and thickness of Au nanosquare is 200–500 nm and ~ 2.4 nm, respectively. As shown in Figure 8.9a, a photochemical method is used to synthesize Au NR

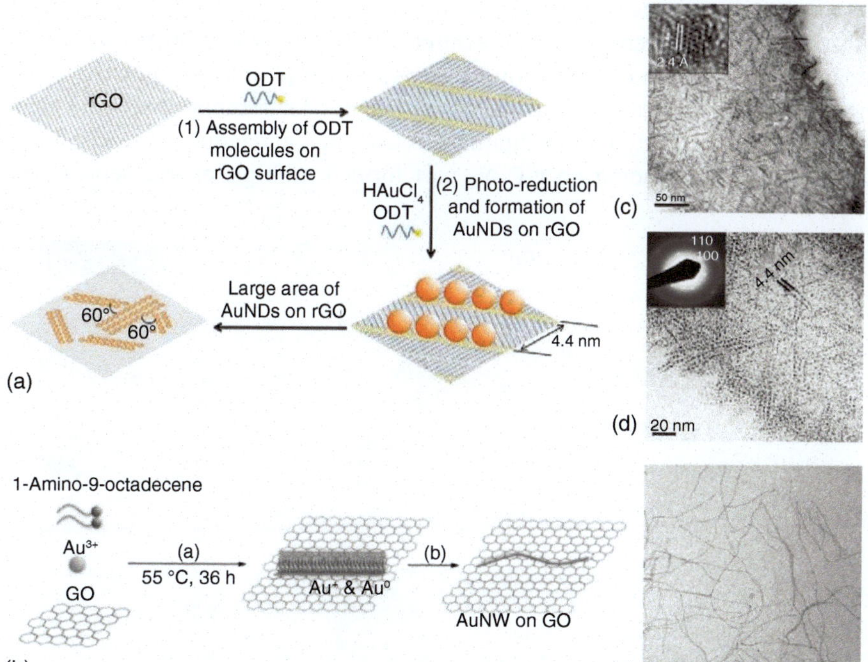

Figure 8.9 (a) Schematic illustration of the *in-situ* synthesis and assembly of AuNDs on rGO. (b) Schematic illustration of the formation process of AuNW on GO. (c, d) TEM and magnified TEM image of Au nanodots on reduced GO. (d) TEM image of Au nanowires grown on GO. Source: (c, d) Reproduced with permission from Huang et al. [40]. © 2010, JOHN WILEY & SONS, INC. (e) Reproduced with permission from Huang et al. [41]. © 2012, JOHN WILEY & SONS, INC.

chains on reduced GO. The 1-octadecanethiol (ODT) molecules first self-assemble into an ordered herringbone structure on the rGO surface. Under light exposure, Au^{3+} is progressively reduced to Au^+ and Au^0. Due to the strong Au–S covalent bonding, the nucleation of AuNDs preferentially occurs on the rGO surface, where the nuclei of the AuNDs bond to the thiol groups. The rGO sheet serves as a template to direct self-assembly of ODT molecules into ordered structures, which then are used as a secondary template for the *in-situ* synthesis and assembly of AuNDs. In Figure 8.9b, Au nanowires are synthesized on GO sheet templates. The Au seeds in the chain-like assembly possibly nucleate from the reduction of Au^+ in the chain-like structure of the [(1-amino-9-octadecene) AuCl] complex. After one more hour, short rod-like structures began to grow from the small Au seeds, and then arrange themselves into discontinuous chains. Continuous aging and reduction of the chain-like structure formed from [(1-amino-9-octadecene) AuCl] finally led to the formation of ultrathin AuNWs on the GO surface.

Moreover, 2D materials of graphene-like MoS_2 and $MoSe_2$ decorated with Au, Ag, Pt nanoparticles have been explored in the past few years [43]. BN coated Au nanoparticles for improving SERS performance has been investigated [44]. Kan's group constructed hybrid heterostructures based on Au nanoparticles, Au/g-C_3N_4/TiO_2 heterostructures exhibits excellent photocatalytic activity for pollutant removal and water splitting (Figure 8.10) [45]. The simultaneous and cooperative dual roles of Au as electron mediator for forming Z-scheme charge transfer and as cocatalyst for facilitating electron traps and reduction reaction when Au nanoparticles loading on g-C_3N_4 part formed multiple charge separation in

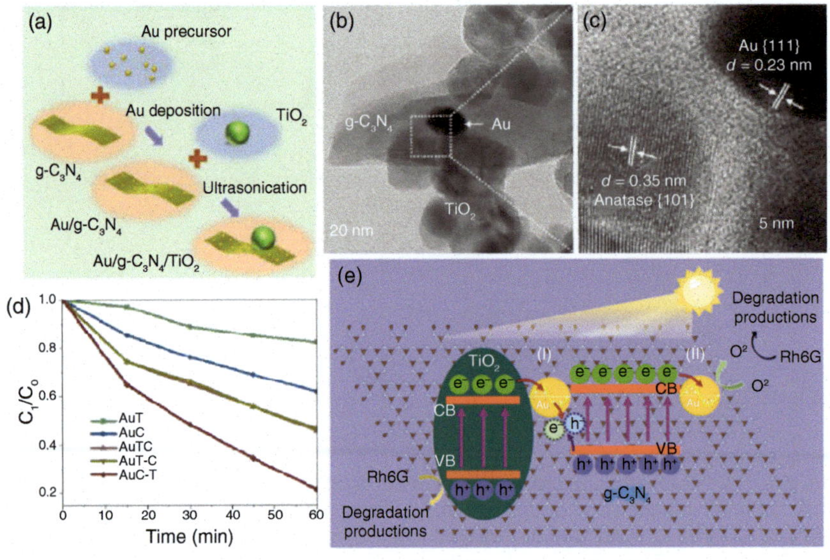

Figure 8.10 (a) Schematic representation of the synthetic procedures for the composites. (b, c) HR-TEM images and (d) Photocatalytic activity of the composites. (e) Schematic diagrams of multiple interfacial charge transfer mechanism. Source: Reproduced with permission from Wei et al. [45]. ©2021, ELSEVIER.

$g\text{-}C_3N_4/TiO_2$ heterostructure for enhancing photocatalytic efficiency. To date, the explorations of this field of metal-2D material nanostructures are still developing rapidly and attracting the attention of scientists.

8.3 Core–Shell Structure and Properties Modulation

8.3.1 Mesoporous SiO_2 Coating Metals for Molecular Delivery

Incorporating plasmonic metal and semiconductor can greatly improve activity of nanomaterials, but the hybrid nanostructures exposed to harsh environments are often difficult to maintain their activity for a long time due to the shedding and loss of noble metals on the semiconductor surface which will affect the recycling performance. Therefore, core–shell structure has been proposed to retort these issues, which can effectively reduce the loss of metals in the core. Moreover, the regulation of shell layers components and thicknesses can produce variety charge effects, which makes the core-shell structure show great advantages in the field of biomaterials, sensing, and photocatalysis. In this part, two typical sorts of metal/semiconductor core-shell structures, SiO_2 and TiO_2 coated noble metals core-shell structure, will be discussed.

Due to high chemical stability, high surface area and large pore volume, mesoporous SiO_2 has attracted much attention as shell materials. Over the past decade, core–shell noble metal@mesoporous SiO_2 has been widely been applied in various fields, such as drug, antibodies delivery, bioimaging and catalysis. Generally, to form mesoporous SiO_2 core-shell structure, a typical method is that the core structures contained nanomaterials (such as Au, Ag, etc.) are firstly coated with silica layer thought silica condensation in precursor solution, and then another template molecule is used to fabricate mesoporous SiO_2 coating layer. Moreover, for molecular delivery, such as drug delivery, controlled release of loaded molecular is a key step. Based on polymers contacted to SiO_2 surface via covalently or non-covalently, various mechanical controls of mesoporous SiO_2 has been explored, including (i) Changing in configurations of polymer chains assembled on particle surface; (ii) Forming and breaking of covalent bonds between molecules; (iii) Coupling and decoupling of bulky units; (iv) Assembling and de-assembling of polymer; (v) Moving of a cyclic device between two positions along a molecular pole next to pore mouths; (vi) Jigging transition of the azo groups linked to the pore walls induced by light irradiation [46].

So far, lots of mesoporous SiO_2 core-shell structures for molecular delivery has been reported. For example, Deng et al. constructed Au NR coated with mesoporous SiO_2 structures and enabled a precise control over the aspect ratio by oxidative etching and metal overgrowth. The plasmonic properties and the particle stability/affinity have been improved simultaneously [47]. The size parameters of initial AuNRs were 75.3 ± 9.4 nm in length (L), 17.6 ± 2.1 nm in diameter (D), and thus had an aspect ratio (AR) of 4.3 ± 0.7. Based on etched AuNRs@meso-SiO_2, different metals, like Ag, Pd, and Pt, could be epitaxially grown on the Au NRs surface, forming a multi-metal core–shell structure. Zheng's group reported a smart drug delivery system integrating both photothermal therapy and chemotherapy for killing cancer cells [48]. The delivery system is based on a mesoporous silica-coated Pd@Ag nanoplates composite. The Pd@Ag nanoplate core can effectively absorb and convert near infrared (NIR) light into heat. The mesoporous silica shell is provided as the host for loading anticancer drug, doxorubicin (DOX). Figure 8.11a shows the procedure for the synthesis of mesoporous silica-coated Pd@Ag nanoparticles. Before coating the Pd@Ag nanoplates with mesoporous SiO_2, a layer of dense silica was grown on the nanoplates using a modified Stöber method. To synthesize mesoporous

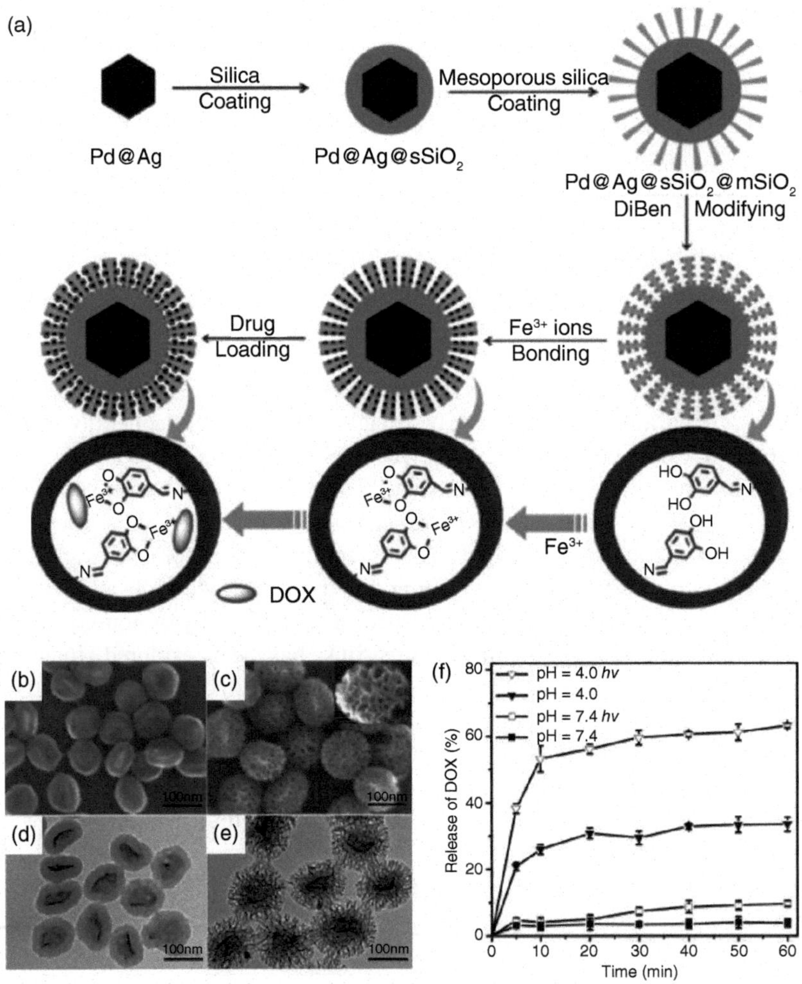

Figure 8.11 (a) Synthetic Process of Pd@Ag@sSiO$_2$@mSiO$_2$-DihBen/DOX Nanocomposites. (b), (c) SEM and (d), (e) TEM images of Pd@Ag@sSiO$_2$ and Pd@Ag@sSiO$_2$@mSiO$_2$ nanoparticles. (f) Release kinetics of DOX from Pd@Ag@sSiO$_2$@mSiO$_2$ in PBS buffer at pH values of 7.4 and 4.0 with or without laser irradiation at 808 nm. Source: Reproduced with permission from Fang et al. [48]. © 2012, JOHN WILEY & SONS, INC.

SiO_2-coated Pd@Ag nanoparticles, a mesoporous SiO_2 was then deposited on Pd@Ag@sSiO_2 particles. The average diameter of Pd@Ag@sSiO_2@mSiO_2 particles is 150 nm and the pores in the mesoporous SiO_2 shells are 10 nm. This hybrid structure allows the DOX loading as high as 49% in weight and DOX molecules are loaded in the mesopores shell through coordination bonds that are responsive to pH and heat. The release of DOX from the core-shell delivery vehicles into cancer cells can be therefore triggered by the pH drop caused by endocytosis and also NIR irradiation.

The plasmonic properties of the core-shell structure were investigated by experimental and simulated extinction spectra. These special structures showed a variety of potential applications, such as sensing and catalysis, and could regulate the plasmonic responses of composite nanostructure based on specific applications by adjusting parameters [47]. Chen et al. reported that mesoporous SiO_2 core-shell structure was a potential platform for simultaneous anticancer drug delivery and MRI cell imaging. Fe_2O_3 and Au nanomaterials acted as core composition and then SiO_2 mesoporous shell structure was formed. Moreover, cavity between core and shell was construct by selective etching method. This mesoporous core-shell structures showed both high loading capacity (20%) and efficiency (up to 100%) for doxorubicin simultaneously because of the formation of the cavity, enhanced surface area/pore volume and the electrostatic interaction between DOX molecules and mesoporous silica surface. Besides, the prepared structures as contrast agents of MRI was demonstrated both *in vitro* and *in vivo*, indicating the simultaneous imaging and therapeutic multifunctionalities of the core–shell composition [49].

8.3.2 Photocatalytic Property of TiO_2 Coating Metal

Noble metal nanostructure with LSPR and tunable resonance absorption in a wide range of incident light can act as excellent candidate photocatalysts, especially in vis–NIR light spectral region. For instance, Au NRs and Au nanobipyramids have strong plasmonic properties, of which the resonant absorption peak can be adjusted in extensive wavelength region by controlled the size and shape of nanostructure. At the same time, in order to further improve the photogenerated charge transfer efficiency and noble metal stability, coating metal oxides is an effective way.

TiO_2 as an outstanding photocatalyst with high photochemical stability and catalytic activity has received much attention in the past years. The hybrid nanostructure of mesoporous TiO_2 coating noble metal can significantly improve the photocatalytic activity and stability. Kan's group constructs a new AuNBP-Pt@TiO_2 nanostructure by coating mesoporous TiO_2 on Au nanobipyramids with Pt clusters decorated, shown in Figure 8.12 [50]. Sodium dodecyl sulfate (SDS) plays a key role in the synthesis of TiO_2 shell. The size of AuNBPs is 100 nm $\pm$ 2 nm in length and 34 nm in diameter. The thickness of the TiO_2 shell is ~30 nm. This unique nanostructure not only provides a superior reaction space for the photocatalytic reaction but also promotes the separation of carriers. Au nanobipyramids absorbed

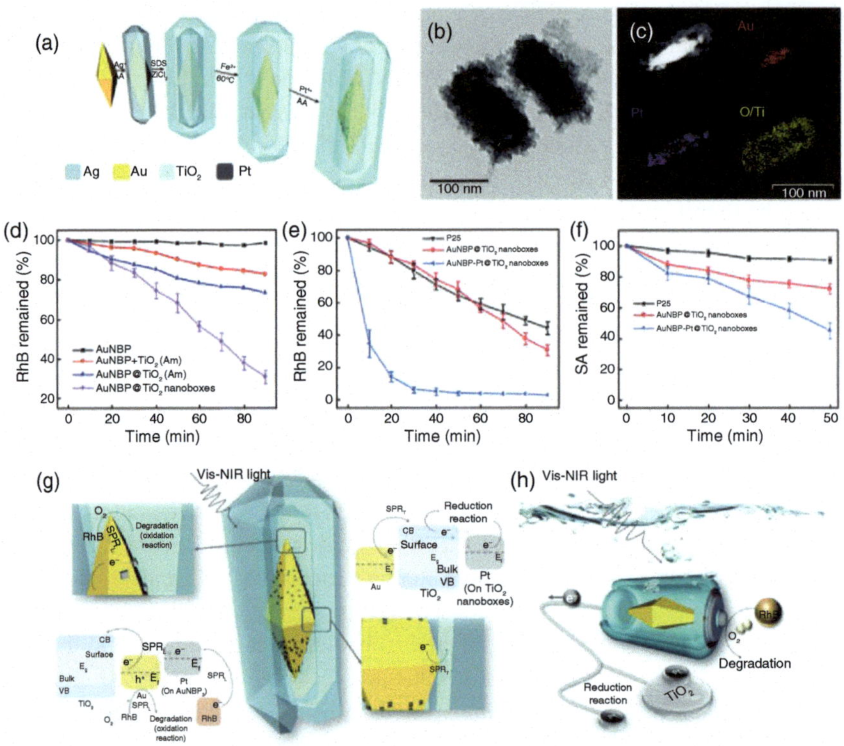

Figure 8.12 (a) The diagram of the synthesis process of the AuNBP-Pt@TiO_2 nanoboxes. (b) TEM and (c) HAADF-STEM images of AuNBP-Pt@TiO_2 nanoboxes. (d–f) Photodegradation performance of RhB and SA. (g) Photocatalysis schematic diagram and energy band diagram and (h) model analogy diagram of the platinum modified nanoboxes. Source: Reproduced with permission from Guo et al. [50]. ©2021, Royal Society of Chemistry.

light and excited electron–hole pairs firstly. Then electrons were injected into TiO_2 nanoboxes as the band alignment between TiO_2 and metals. This process achieved the rapid separation of carriers. The rhodamine-B degradation efficiency has been increased by 6.5 times.

In addition, Pt clusters increase the performance of photodegradation of organics. Recently, photocatalytic CO_2 reduction is one of the hottest studies to deal with the global energy and environmental problems. Noble metal core-shell structures act as candidate catalysts showing great potential. Pougin et al. reported that Au@TiO_2 core-shell structure with Au nanoparticle modulation for the photocatalytic CO_2 reduction. TEM images characterize Au@TiO_2 core-shell structure with Au deposition obviously, displayed in Figure 8.13. Photocatalytic measurement shows that methane was the main product formed on all samples and Au@TiO_2 core-shell structure with Au deposition is much more active [51]. The electronic interaction between the Au core, the TiO_2 shell, and the Au NPs on the outer surface contributed to the improved activation.

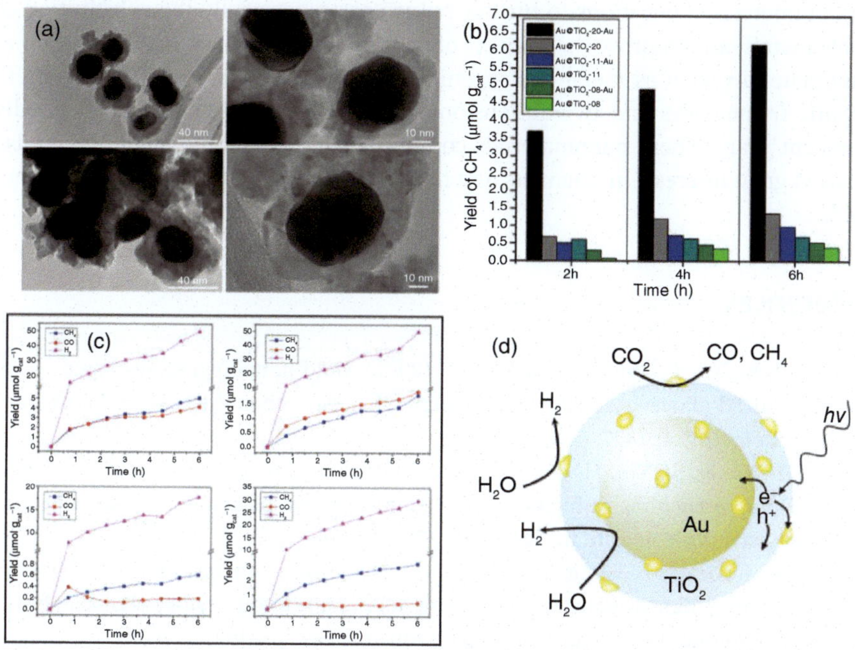

Figure 8.13 (a) TEM images of Au@TiO_2 and Au@TiO_2-Au core-shell structure. (b) Activity screening for methane formation. (c) Activity testing in CO_2 reduction. (d) Schematic representation of the mode of action of Au@TiO_2-Au core–shell structure. Source: Reproduced with permission from Pougin et al. [51]. ©2018, JOHN WILEY & SONS, INC.

8.4 Conclusion

In this chapter, we discussed the metal-semiconductor composite nanostructure which exhibit excellent photoelectric properties in energy utilization, photodetector, biomedical and other fields. Plasmonic noble metals acted as a source of electrons can interact with photons ranged from visible to infrared to generate electron–hole pairs. When the electrons injected into the CB of semiconductors, this process will realize the efficient separation of carriers by Schottky junctions and electric fields formed at the metal–semiconductor interface. Semiconductors coupling with noble metal can not only form metal–semiconductor interfaces, but also provide metal attachment sites, which is conducive to the stability of metal nanostructures in the actual reactive environment, especially the protection of the semiconductor shell to the metal core in the core-shell structure. The fast growth of these hybrid structures is benefit to the industrial application of the metal–semiconductor structures. However, some challenges should be considered and addressed in further researches: (i) It is necessary to further develop a fine and controllable synthesis of metal nanostructures, metal–semiconductor composite nanostructure. (ii) The mechanism of metal–semiconductor interaction should be further explored in different working environment based on controllable synthesis. (iii) It is crucial for commercial applications to further reduce costs, especially the substitution of noble metals, and improve the stability of hybrid nanostructures. In addition, other nanomaterials such as metal chalcogenides, metal–organic frameworks, covalent organic frameworks and two-dimensional materials have also developed rapidly in recent years. These nanomaterials combined with plasmonic metals have also aroused great interests in scientists and have been extensively explored in different fields.

References

1 Hanqi, B., Xu, J., Zhu, X., and Kan, C. (2022). Gold nanobipyramids doped with Au/Pd alloyed nanoclusters for high efficiency ethanol electrooxidation. *Nanoscale Adv.* 4: 1827.

2 Mi, H., Wang, S., Yin, H. et al. (2020). (Gold triangular nanoplate core)@(silver shell) nanostructures as highly sensitive and selective plasmonic nanoprobes for hydrogen sulfide detection. *Nanoscale* 12: 20250.

3 Liu, X., Iocozzia, J., Wang, Y. et al. (2017). Noble metal-metal oxide nanohybrids with tailored nanostructures for efficient solar energy conversion, photocatalysis and environmental remediation. *Energ. Environ. Sci.* 10: 402.

4 Liu, J., Kan, C., Cong, B. et al. (2014). Plasmonic property and stability of core-shell Au@SiO_2 nanostructures. *Plasmonics* 9: 1007.

5 Cai, J., Shen, J., Zhang, X. et al. (2019). Light-driven sustainable hydrogen production utilizing TiO_2 nanostructures: a review. *Small Methods* 3: 1800184.

6 Guo, Q., Zhou, C., Ma, Z., and Yang, X. (2019). Fundamentals of TiO_2 photocatalysis: concepts, mechanisms, and challenges. *Adv. Mater.* 31: 1901997.

7 Ma, Y., Wang, X., Jia, Y. et al. (2014). Titanium dioxide-based nanomaterials for photocatalytic fuel generations. *Chem. Rev.* 114: 9987.

8 Naldoni, A., Altomare, M., Zoppellaro, G. et al. (2019). Photocatalysis with reduced TiO_2: from black TiO_2 to cocatalyst-free hydrogen production. *ACS Catal.* 9: 345.

9 Jiang, W., Bai, S., Wang, L. et al. (2016). Integration of multiple plasmonic and Co-catalyst nanostructures on TiO_2 nanosheets for visible-near-infrared photocatalytic hydrogen evolution. *Small* 12: 1640.

10 Tang, H., Meng, G., Huang, Q. et al. (2012). Arrays of cone-shaped ZnO nanorods decorated with ag nanoparticles as 3D surface-enhanced Raman scattering substrates for rapid detection of trace polychlorinated biphenyls. *Adv. Funct. Mater.* 22: 218.

11 Huo, D., Chen, B., Li, M. et al. (2021). Template-assisted fabrication of Ag-nanoparticles@ZnO-nanorods array as recyclable 3D surface enhanced Raman scattering substrate for rapid detection of trace pesticides. *Nanotechnology* 32: 145302.

12 Ge, M.-Z., Cao, C.-Y., Li, S.-H. et al. (2016). In-situ plasmonic Ag nanoparticle anchored TiO_2 nanotube arrays as visible-light-driven photocatalysts for enhanced water splitting. *Nanoscale* 8: 5226.

13 Zhang, H., Wang, J., Li, G. et al. (2019). Fabrication of Ag-nanosheet-assembled hollow tubular array and their SERS effect. *J. Alloys Compd.* 772: 663.

14 Pu, Y.C., Wang, G., Chang, K.D. et al. (2013). Au nanostructure-decorated TiO_2 nanowires exhibiting photoactivity across entire UV-visible region for photoelectrochemical water splitting. *Nano Lett.* 13: 3817.

15 Li, X., Xing, L., Zheng, K. et al. (2017). Formation of gold nanostar-coated hollow mesoporous silica for tumor multimodality imaging and photothermal therapy. *ACS Appl. Mater. Interfaces* 9: 5817.

16 Mukherjee, S., Zhou, L., Goodman, A.M. et al. (2014). Hot-electron-induced dissociation of H_2 on gold nanoparticles supported on SiO_2. *J. Am. Chem. Soc.* 136: 64.

17 Larsen, M.A.B., Thøgersen, J., Stephansen, A.B. et al. (2016). Transient IR spectroscopic observation of singlet and triplet states of 2-nitrofluorene: revisiting the photophysics of nitroaromatics. *J. Phys. Chem. A* 120: 28.

18 Zaki, A., Dave, N., and Liu, J. (2012). Amplifying the macromolecular crowding effect using nanoparticles. *J. Am. Chem. Soc.* 134: 35.

19 Bhardwaj, N., Pandey, A., Avasthi, D.K., and Mohapatra, S. (2016). Ion beam engineering of morphological, structural and optical properties of Au/SnO_2 hybrid nanostructured thin films. *J. Alloys Compd.* 680: 155.

20 Zhang, C., Chen, J., Wang, S. et al. (2020). Metal halide perovskite nanorods: shape matters. *Adv. Mater.* 32: 2002736.

21 Atwater, H.A. and Polman, A. (2010). Plasmonics for improved photovoltaic devices. *Nat. Mater.* 9: 205.

22 Park, N.-G. (2016). Crystal growth engineering for high efficiency perovskite solar cells. *CrstEngComm* 18: 5977.

23 Green, M.A., Ho-Baillie, A., and Snaith, H.J. (2014). The emergence of perovskite solar cells. *Nat. Photon.* 8: 506.

24 Zhiping, H., Zhengzheng, L., Zijun, Z. et al. (2021). Advances in metal halide perovskite lasers: synthetic strategies, morphology control, and lasing emission. *Adv. Photon.* 3: 034002.

25 Guo, J., Fu, Y., Lu, M. et al. (2020). Cd-rich alloyed $CsPb_{1-x}CdxBr_3$ perovskite nanorods with tunable blue emission and Fermi levels fabricated through crystal phase engineering. *Adv. Sci.* 7: 2000930.

26 Zhou, H., Yuan, S., Wang, X. et al. (2017). Vapor growth and tunable lasing of band gap engineered cesium lead halide perovskite micro/nanorods with triangular cross section. *ACS Nano* 11: 1189.

27 Li, Z.-J., Hofman, E., Davis, A.H. et al. (2018). General strategy for the growth of $CsPbX_3$ (X = Cl, Br, I) perovskite nanosheets from the assembly of nanorods. *Chem. Mater.* 30: 3854.

28 Li, F., Lu, J., Zhang, Q. et al. (2019). Controlled fabrication, lasing behavior and excitonic recombination dynamics in single crystal $CH_3NH_3PbBr_3$ perovskite cuboids. *Sci. Bull.* 64: 698.

29 Huang, X., Li, H., Zhang, C. et al. (2019). Efficient plasmon-hot electron conversion in Ag-$CsPbBr_3$ hybrid nanocrystals. *Nat. Commun.* 10: 1163.

30 Zhang, X., Xu, B., Wang, W. et al. (2017). Plasmonic perovskite light-emitting diodes based on the Ag-$CsPbBr_3$ system. *ACS Appl. Mater. Interfaces* 9: 4926.

31 Wu, X., Jiang, X.-F., Hu, X. et al. (2019). Highly stable enhanced near-infrared amplified spontaneous emission in solution-processed perovskite films by employing polymer and gold nanorods. *Nanoscale* 11: 1959.

32 Bai, Y., Hao, M., Ding, S. et al. (2022). Surface chemistry engineering of perovskite quantum dots: strategies, applications, and perspectives. *Adv. Mater.* 34: 2105958.

33 Li, G., Wang, H., Zhang, T. et al. (2016). Solvent-polarity-engineered controllable synthesis of highly fluorescent cesium lead halide perovskite quantum dots and their use in white light-emitting diodes. *Adv. Funct. Mater.* 26: 8478.

34 Ye, S., Yu, M., Yan, W. et al. (2017). Enhanced photoluminescence of $CsPbBr_3$@Ag hybrid perovskite quantum dots. *J. Mater. Chem. C* 5: 8187.

35 Juan, F., Xu, F., Wang, M. et al. (2020). Photoluminescence enhancement of perovskite $CsPbBr_3$ quantum dots by plasmonic Au NRs. *Chem. Phys.* 530: 110627.

36 Hsieh, Y.-H., Hsu, B.-W., Peng, K.-N. et al. (2020). Perovskite quantum dot lasing in a gap-plasmon nanocavity with ultralow threshold. *ACS Nano* 14: 11670.

37 Shi, Z., Li, Y., Li, S. et al. (2018). Localized surface plasmon enhanced all-inorganic perovskite quantum dot light-emitting diodes based on coaxial core/shell heterojunction architecture. *Adv. Funct. Mater.* 28: 1707031.

38 Li, X., Zhu, J., and Wei, B. (2016). Hybrid nanostructures of metal/two-dimensional nanomaterials for plasmon-enhanced applications. *Chem. Soc. Rev.* 45: 3145.

39 Jasuja, K. and Berry, V. (2009). Implantation and growth of dendritic gold nanostructures on graphene derivatives: electrical property tailoring and Raman enhancement. *ACS Nano* 3: 2358.

40 Huang, X., Zhou, X., Wu, S. et al. (2010). Reduced graphene oxide-templated photochemical synthesis and in-situ assembly of Au nanodots to orderly patterned Au nanodot chains. *Small* 6: 513.

41 Huang, X., Li, S., Wu, S. et al. (2012). Graphene oxide-templated synthesis of ultrathin or tadpole-shaped Au nanowires with alternating hcp and fcc domains. *Adv. Mater.* 24: 979.

42 Huang, X., Li, S., Huang, Y. et al. (2011). Synthesis of hexagonal close-packed gold nanostructures. *Nat. Commun.* 2: 292.

43 Govinda Rao, B., Matte, H.S.S.R., and Rao, C.N.R. (2012). Decoration of few-layer graphene-like MoS_2 and $MoSe_2$ by noble metal nanoparticles. *J. Cluster Sci.* 23: 929.

44 Yu, X., Cai, R., Jiao, J. et al. (2018). Enhanced thermal stability of boron nitride-coated Au nanoparticles for surface enhanced Raman spectroscopy. *J. Alloys Compd.* 730: 487.

45 Wei, T., Xu, J., Kan, C. et al. (2022). Au tailored on g-C_3N_4/TiO_2 heterostructure for enhanced photocatalytic performance. *J. Alloys Compd.* 894: 162338.

46 He, D., Wang, S., Lei, L. et al. (2015). Core-shell particles for controllable release of drug. *Chem. Eng. Sci.* 125: 108.

47 Deng, T.-S., van der Hoeven, J.E.S., Yalcin, A.O. et al. (2015). Oxidative etching and metal overgrowth of gold nanorods within mesoporous silica shells. *Chem. Mater.* 27: 7196.

48 Fang, W., Yang, J., Gong, J., and Zheng, N. (2012). Photo- and pH-triggered release of anticancer drugs from mesoporous silica-coated Pd@Ag nanoparticles. *Adv. Funct. Mater.* 22: 842.

49 Chen, Y., Chen, H., Zeng, D. et al. (2010). Core/shell structured hollow mesoporous nanocapsules: a potential platform for simultaneous cell imaging and anticancer drug delivery. *ACS Nano* 4: 6001.

50 Gao, W., Kan, C., Ke, S. et al. (2021). Au nanobipyramids with Pt decoration enveloped in TiO_2 nanoboxes for photocatalytic reactions. *Nanoscale Adv.* 3: 4226.

51 Pougin, A., Dodekatos, G., Dilla, M. et al. (2018). Au@TiO_2 core-shell composites for the photocatalytic reduction of CO_2. *Chem. Eur. J.* 24: 12416.

Part III

Applications of Plasmonic Nanostructures

9

Hot Electron Effect on Optoelectronic Device

In the Future – Plasmon-Induced Hot Electrons Do the Impossible

9.1 Introduction

Surface plasmons (SPs) are an electromagnetic oscillation mode formed by the strong interaction of light and free electrons on metal surfaces, which provides an excellent platform for designing and constructing new-type optoelectronic devices at nanometer scale. In appropriate metal nanostructures (typically Au, Ag, and Pt), if the incident light matches the resonant frequency of the free electrons on the metal surface, it will oscillate collectively, known as surface plasmon resonance (SPR), including local surface plasmon resonance (LSPR) and propagating surface plasmon polaritons (SPPs), giving rise to intense and highly localized electromagnetic fields. Recently, the research on plasmonic has formed a prosperous field with many applications, including nanolasers [1–3], metamaterials and metasurfaces [4, 5], sub-diffraction limited imaging [6], SERS [7–9], which benefits from its high spatial localization and strong near-field enhancement characteristics. Compared with high temperatures and a strong electric field, it has unique advantages in exciting hot carriers, including hot electrons and holes [10, 11]. Following the light absorption and SPR excitation in metal nanostructures, the collective oscillated electrons are easily damped through re-emitting photons in a radiative way or transferring energy to electrons in the conduction band through Landau damping [12] in a nonradiative way, resulting in intraband or interband (d bands) transitions in the femtosecond timescale [13], producing well-known "hot carriers," which is defined as the carriers in a nonequilibrium state with larger energy than that in a thermal equilibrium state. Subsequently, hot electrons will redistribute their energy in many low-energy electrons through electron–electron scattering and electron–phonon scattering processes with a relaxation time of around 100 fs to 1 ps [14], and finally convert into heat.

For a long time, non-radiative plasmon decay has been considered to be a parasitic process with negative effects, which limits the performance of plasmonic devices,

Plasmonic Metal Nanostructures: Preparation, Characterization, and Applications,
First Edition. Caixia Kan.

such as shortening the propagation distance of plasmonic waveguides [15] and leads to light absorption in metal. However, in recent research, the hot electrons generated by the plasmon nonradiative decay have emerged as attractive application prospects in many fields, such as photothermal [16, 17], photovoltaic [18], photocatalysis [19–21], photoemission [22, 23], and photodetection [24, 25]. This chapter will summarize the application of the hot electron effect on optoelectronic devices from the perspective of photoemission and photodetection and demonstrate their functional diversity. Also, we discussed the fundamental issues of these new optoelectronic devices based on the hot electrons effect that deserve more attention, such as the improvement of quantum yield, as well as related challenges and opportunities.

9.2 Light-Emitting Device and Modulation

9.2.1 Antenna-Coupled Light Emission Based on Hot Electron

Optical antenna driven by electrical generators is significant for constructing the subwavelength optical sources and realizing the on-chip integration and wireless signal communication at the nanometer scale. In 2015, J. Kern et al. [26] demonstrated a directly electrically driven in-planar optical antenna through the broadband quantum-shot noise of hot electrons tunneling across its feed gap for the first time. They combined a top-down focused ion beam to construct a “T” shaped Au antenna and an AFM to manipulate Au nanoparticles (NPs) for forming a crucial lateral tunnel junction in the feed gap, as shown in Figure 9.1a. Figure 9.1b shows the energy band diagram of the tunnel junction under bias voltage. Hot electrons tunnel from the high barrier side to the low barrier side, driven by the applied voltage, and release energy as photons. The white light scattering spectrum of a NP-loaded antenna is shown by the red solid line in Figure 9.1c. Two resonance peaks can be observed, one at 770 nm on the lower energy side and a second at 590 nm on the higher energy side. Due to its characteristics of stronger field confinement and enhancement, only the lower energy resonance can significantly enhance light emission. The spectra of the emitted photons under different applied voltages are also demonstrated by open colored symbols in Figure 9.1c. Each spectrum has the characteristics of a peak and a high energy cutoff, accompanied by the blue shift and growing intensity of the peak with the increasing applied voltage.

To reveal these novel physical phenomena, they multiplied the spectral density of the tunneling current fluctuation with the far-field scattering spectrum of the antenna obtained from the experiment to model the electroluminescence (EL) spectrum, as shown by the colored solid lines in Figure 9.1c. The calculated results based on this model agree with the experimental values of EL spectra at different bias voltages. A typical far-field emission spot overlaid with an SEM micrograph of the antenna area is depicted in Figure 9.1d. The Airy pattern with an FWHM of ~350 nm shows that this device can act as an electrically driven subwavelength photon source. Coincidentally, M. Buret et al. [28] exploited in-plane tunneling electron-fed optical antenna constructed from 100 nm wide × 4 μm long Au nanowires to achieve

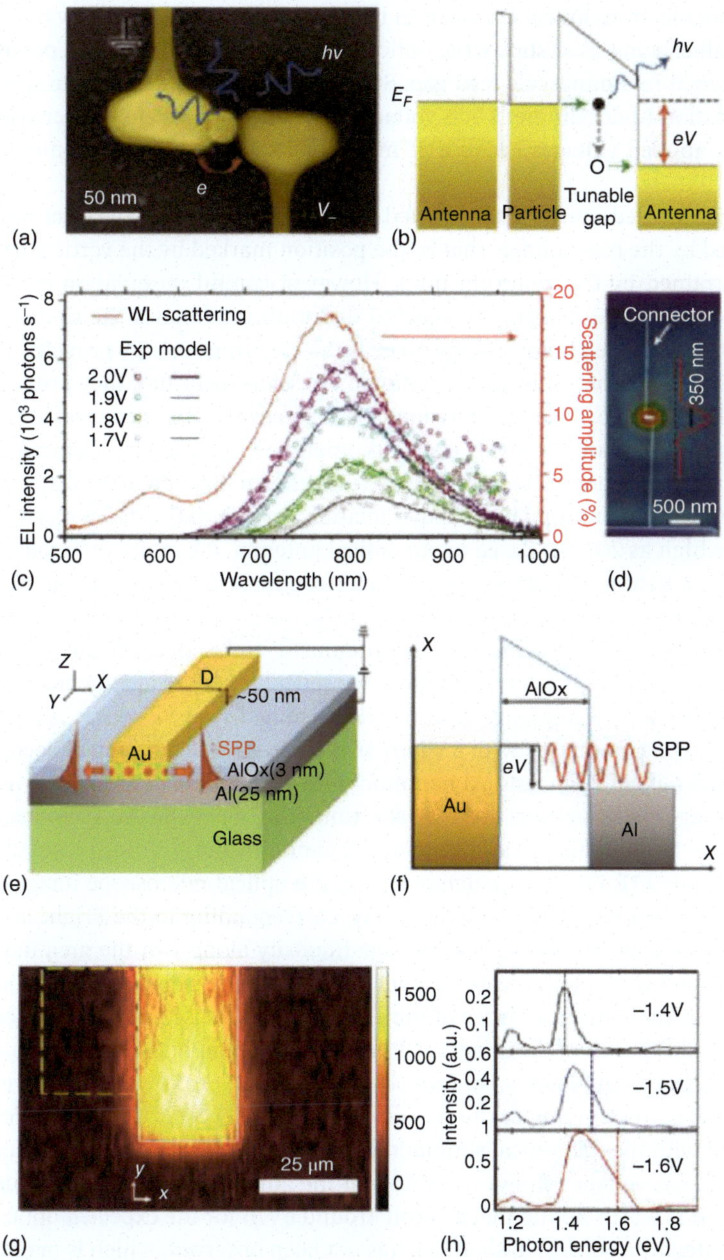

Figure 9.1 (a) SEM image of the device; (b) schematic diagram of tunneling junction energy level with applied voltage; (c) electroluminescence spectra at different bias voltages; (d) the subwavelength emission spot with an FWHM of 350 nm overlapped with SEM image. (e) Schematic diagram of the device consisting of Au antenna, Al film, and AlO_x tunneling barrier; (f) energy level diagram of tunneling junction under bias voltage; (g) optical image of the emitted SPP under a biased voltage of 1.6 V; (h) experimental SPP emission spectra under different applied voltages. Source: (a,d) Reproduced with permission from Kern et al. [26] @2015, Springer Nature. (g) Reproduced with permission from Zhang et al. [27] © 2019, Springer Nature/CC BY 4.0.

spontaneous hot electron optical emission in the same year. A close-up view of the junction area after being used shows the optical response and the electrical potential are self-aligned in a nanoscale feed gap. Subsequently, they collected the spectra of light emission under different bias voltages. The emission intensity increases with increased applied voltages because of more injected charges in the antenna feed gap.

Generally, the released energy of the emitted photons cannot exceed the electronic energy provided by the bias voltage, that is, the position marked by the vertical bar, which is constrained by the quantum limit. However, careful observation of the spectra shows that a small part of the spectral distributions violates the so-called quantum limit; that is, the photon energy exceeds the electronic energy provided by the bias voltage. To interpret this unconventional emission spectrum, they propose a model based on the spontaneous emission of hot electrons. Hot electrons in the antenna feed gap could not effectively exchange energy with phonons, resulting in their collision with the antenna borders to achieve the thermalization of the distribution and spontaneously emitting blackbody radiation corresponding to an electron temperature as high as 2000 K. Based on an optical antenna, it can not only realize electrically driven light emission through hot electron inelastic tunneling but also obtain SP emission.

In 2019, C. Zhang et al. [27] achieved nanoantenna SP emission by inelastic electron tunneling and proposed a corresponding theoretical model in agreement with the experiment. As shown in Figure 9.1e, the electrically driven SP source is composed of Au nano patch antenna with a width of 128 nm, AlO_x insulating dielectric layer with a thickness of 3 nm, and Al nanofilm with a thickness of 25 nm. Driven by the bias voltage, the electrons are inelastic tunneling from the Au nanopatch antenna through the AlO_x dielectric layer to the Al nanofilm and emitting plasmons, as shown in Figure 9.1f. Figure 9.1g demonstrates the optical microscope image of SP emission of the Au nanopatch antenna array, corresponding to the bright area marked by a white solid rectangle. The average intensity along y in the area indicated in the dashed rectangle decreases exponentially with the negative x direction, indicating the plasmon characteristics of the emitted light. Subsequently, they collected the emission spectra of the antenna SPs at three different bias voltages of 1.4, 1.5, and 1.6 V, as shown in Figure 9.1h. Two obvious emission peaks can be observed at 1.2 and 1.5 eV, corresponding to the two modes of the Fabry-Perot cavity, which were consistent with the theoretical simulation results. The blue shift phenomenon with increasing bias voltages shows the ability of the antenna to control the emission spectrum. It is worth noting that a cutoff around eV/h for the expected optical emission indicated by a vertical dashed line has not been observed, which is mainly attributed to the spectrum broadening and shifting by the finite temperature and antenna resonance.

To improve the efficiency of light emission, many efforts have been made, including improving the stability of the tunneling junction, changing the manipulated form of excitation, and replacing the electrode type. In 2015, M. Parzefall et al. [29] proposed to exploit stable and high-quality hexagonal boron nitride (h-BN) to replace the insulating dielectric layer prepared by the top-down method to build

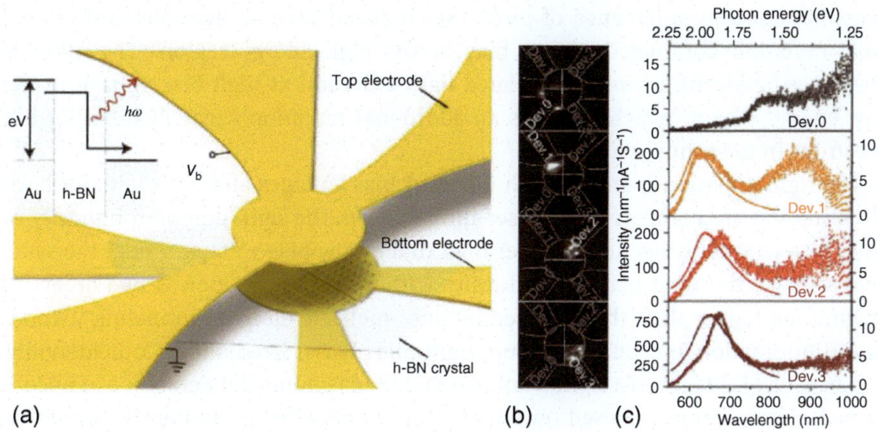

Figure 9.2 (a) Schematic diagram of four devices; (b, c) the light emitted images and corresponding spectra of the four devices at a bias voltage of 2.5 V. Source: Reproduced with permission from Parzefall et al. [29] © 2015, Springer Nature.

a tunneling barrier to avoid the device instability caused by grain formation and defects. Figure 9.2a presents the overall structure of the four devices, composed of a circular Au bottom electrode (1 nm Ti/50 nm Au), ~3 nm-thick multilayer h-BNs, and 15 nm-thick Au top electrode. Then the bottom electrode is divided into four quarter-circle electrodes and led separately. Finally, three of them are nanostructured by focusing ion beam milling and leaving one as the reference (Dev. 0). The nanostructured devices 1, 2, and 3 are formed by optical antenna arrays of rectangular slots with dimensions of 150×50, 250×50 and $350 \times 50\,\text{nm}^2$, respectively.

The inset of Figure 9.2a illustrates the physical process of tunneling electrons exciting SPs and then decaying into far-field radiation. Figure 9.2b demonstrates the real space image of light emission for Dev. 0, 1, 2, and 3 under the bias voltage of 2.5 V. The light emission of the unstructured Dev. 0 is mainly localized at the edge, and its corresponding emission spectrum in Figure 9.2c shows the characteristics of wideness and continuous reduction toward the high energy side. On the contrary, the nanostructured devices have obvious light emissions in the entire antenna area. It can be observed that their corresponding light emission spectra were enhanced obviously, especially in the spectral domain of the transmission peaks. In comparison, the external electron-to-photon conversion efficiency of the latter is increased by two orders of magnitude than that of the former from $QE_{D0} = 4 \times 10^{-7}$ to $QE_{D3} = 2.5 \times 10^{-5}$ at a bias voltage of 2.5 V. In addition to the above method, the light emission efficiency through the inelastic tunneling of electrons can also be effectively enhanced by manipulating the excitation.

In 2015, X. Wang et al. [30] proposed that the light emission caused by inelastic tunneling can be enhanced significantly through exploiting a 634 nm laser to irradiate the tunneling junction. The sketch of the corresponding enhancement mechanism is explained by phenomenological model. Under laser illumination, the PL spectrum of tunnel junction can be divided into two processes. One is the recombination luminescence of pure laser-induced excited electrons and d-band hole radiation dominated at low bias in the high energy region. The other is the inelastic electron tunneling-related light emission at high bias voltage in the low energy region, which provides an additional radiation decay channel for hot electrons in the tunneling junction.

The light emission intensity with different bias voltages and with/without laser illumination are compared. For laser illumination, the light emission intensity of high bias voltage is five times higher than that of low bias voltage. Under the same bias voltage, the light emission intensity with laser illumination is two orders of magnitude higher than that induced by pure inelastic electron tunneling without laser illumination, indicating the important role of laser irradiation. Coincidentally, L. Cui et al. [31] realized a significant increase in plasmonic light emission at upconverted photon energies based on electrical and optical co-excitation by exploiting plasmonic tunnel junctions, with the enhancement ratio exceeding three orders of magnitude than excited alone in recent. The schematic diagram of light-emitting devices under three types of external stimuli, namely EL, photoluminescence (PL), and electro-photoluminescence (EPL), is proposed. The up-converted light emission spectra of the devices under three types of excitation demonstrate that the photon energy is significantly greater than the excitation energy (1.58 eV). More interestingly, compared with the case where the electric field and the light field are excited separately, the light emission intensity of the device has been greatly improved under the electro-optic synergy. These beneficial efforts paved the way for the practical process of this type of device.

Based on the above milestone progress, the more urgent demand is how to make more effective and full use of the hot electron light emission devices based on tunneling junctions to achieve multi-functional applications. In 2018, P. Wang et al. [32] constructed the nanoscale tunneling junctions with a density of 10^{11} per square centimeter. They exploited the radiative decay of tunneling-induced SP to monitor the dynamics of chemical reactions *in situ*. Figure 9.3a shows the schematic diagram of the tunneling junction arrays, composed of Au NRs, Al_2O_3 insulating layer, and Ta_2O_5 coated glass substrate. The areal density of Au NRs is as high as 10^{10} to 10^{11} per square centimeter. A large number of hot electrons make tunneling junctions become an efficient platform for providing oxidation and reduction reactions, which can be used to monitor the reaction progress in real time through changes in the intensity of light emission. The illustration in Figure 9.3b shows the photograph of the device under the bias voltage of 2.5 V, accompanied by light solid emission of an red color. Subsequently, the device is placed in a gas environment control unit. When the gas environment gradually changes from 2% H_2 in N_2 to air, that is, the oxidation reaction occurs, and the emission intensity gradually increases until twice the initial value, as shown in Figure 9.3b. When 2% H_2 in N_2 is reintroduced for reduction reaction, the light emission intensity decreases to the original value again, indicating good recoverability and repeatability, as shown in Figure 9.3c. It not only provides an important method for *in-situ* monitoring of redox reactions but also provides a new strategy for dynamically regulating the light emission of tunneling junctions.

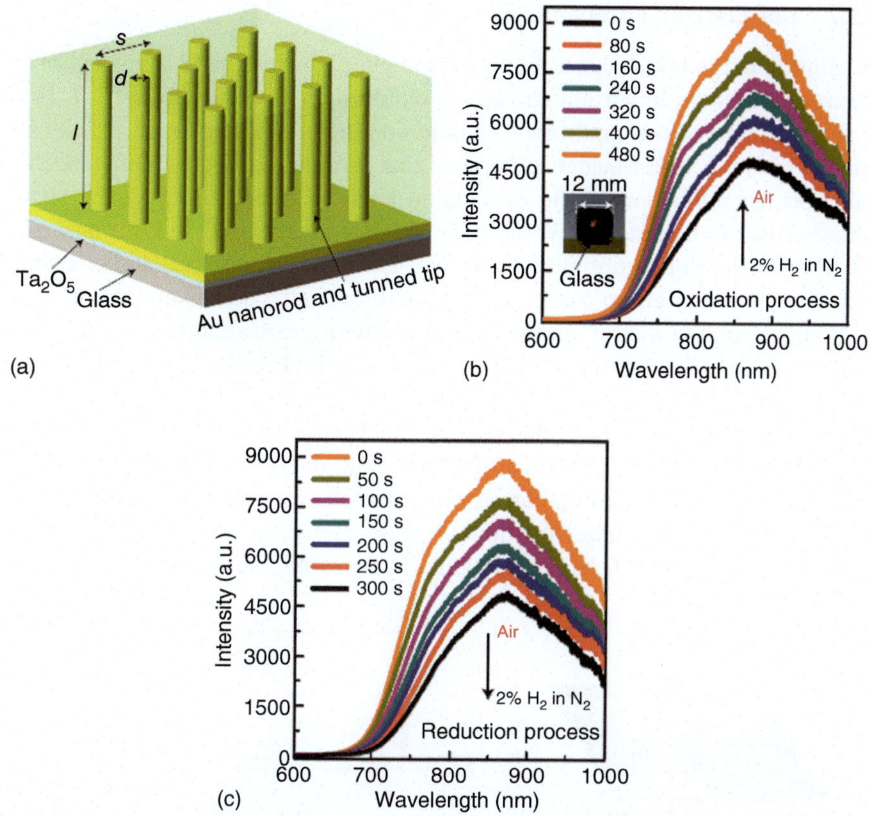

Figure 9.3 (a) Schematic diagram of nanorod arrays; the evolution of emission spectra of nanorod arrays in oxidation (b) and reduction (c) processes. Source: Reproduced with permission from Wang et al. [32] @2017, Springer Nature.

In addition to the effective utilization of light emission intensity, P. Grimm et al. [33] realized a nanoscale color-switchable organic light-emitting antenna through the device composed of an Au antenna and organic semiconductor zinc phthalocyanine (ZnPc) in 2022. As the bias voltage is applied to the device, the obvious light emission as the diffraction-limited spot localized in the gap of the Au nano-antenna is observed, which is a compatible electrically driven sub-wavelength point light source. The color of its light emission is closely related to the polarity of the applied voltage, which is marked with different false colors. The light emission spectra with positive and negative applied voltages, respectively, demonstrate that the emission spectrum peak of negative (positive) voltages is at 810 nm (910 nm). In order to explore the physical mechanism, the schematic energy diagram in each scenario is proposed, suggesting the formation of excitons near the cathode. While the scattering resonance peaks of electrode antennas with different aspect ratios are located at 800 and 900 nm, respectively. The corresponding light emission spectrum is the convolution of the scattering resonance of the cathode antenna and the ZnPc EL spectrum, which provides a feasible strategy for manipulating the color of the emitted light.

9.2.2 Hot-Electron Intraband Emission

In addition to the hot electrons inelastic tunneling, the intraband transition of hot electrons is also an important strategy to obtain light emission. In 2015, T. Haug et al. [34] observed the continuous emission on metal NPs aggregate surfaces and provide the most direct evidence to prove that it originated from the hot electron radiation recombination in the conduction band. Figure 9.4a shows the spatial emission distribution excited by 86 fs laser pulses at a wavelength of 770 nm through the objective lens of an upright microscope. The luminescent spectrum of a single emissive site covers an up-conversion region (blue) and a down-conversion region (red) superimposed a laser spectrum depicted with green solid line, as shown in Figure 9.4b. In order to compare more intuitively, pulsed and continuous-wave (cw) lasers are utilized for low and high irradiance excitation, respectively. Figure 9.4c plots the emission spectra under pulsed (black) and cw (red) excitation at the irradiance of 100 W cm^{-2} (solid) and 1 kW cm^{-2} (dashed). Under pulsed excitation, both down-conversion and up-conversion emission covering visible spectrum can be observed, whereas under continuous wave excitation only down-conversion emission can be detected. As the irradiance is increased, the former shows obvious

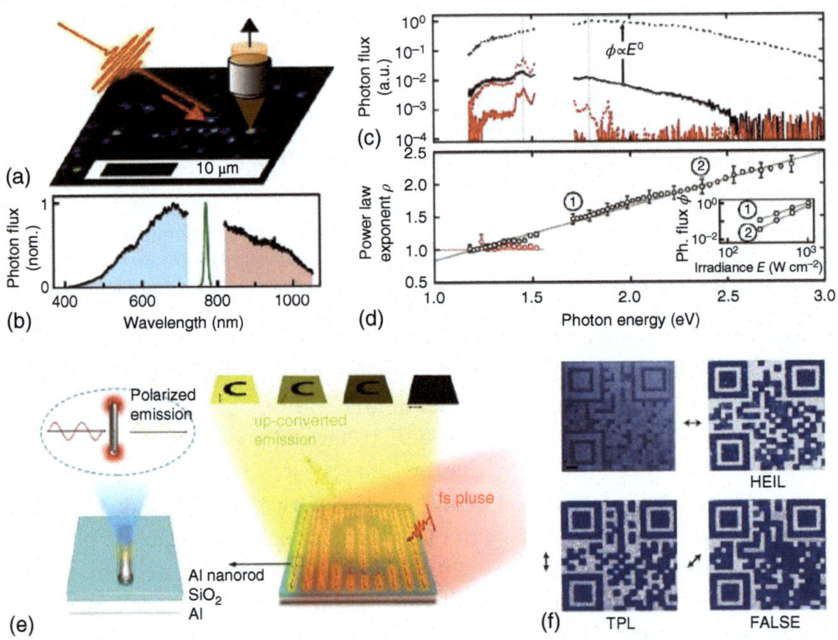

Figure 9.4 (a) The emission micrograph; (b) emission spectrum of intraband transitions above (blue region) and below (red region) excitation energy; (c) intraband emission spectra pumped by different excitation sources and power densities; (d) dependence of power-law exponent on the photon energy. (e) Schematic of the Al MIM nanostructure for extremely polarized up-converted broadband emission; (f) application demonstration on two-dimensional (2D) code anti-counterfeiting. Source: (d) Reproduced with permission from Haug et al. [34] @2015, American Physical Society. (f) Reproduced with permission from Zhang et al. [35] © 2020, John Wiley & Sons.

blue shift, while the emission peak of the latter remains unchanged. It indicates that the spectral variation of the power-law exponent $p(\nu)$ is a function of the emitted photons frequency ν, as plotted in Figure 9.4d. Under cw excitation, the extracted power exponent $p(\nu)$ is always about equal to 1, whereas under pulsed excitation, the power exponent $p(\nu)$ depends on the emitted photon frequency ν, and presents a linear feature. It is a straightforward property of intraband recombination for hot carriers.

In order to achieve tunable intraband emission wavelength of hot electrons, a new strategy of LSPR scattering assisted excitation has been proposed. In 2016, K.Q. Lin et al. [36] prepared silver nanorods with different aspect ratios and achieved wavelength-adjustable hot electron intraband emission. The LSPR scattering spectra and PL spectra of silver NPs with different aspect ratios were measured. The obvious PL peaks (solid lines) in the range of 560–700 nm were observed, which follow the envelope of the LSPR scattering peaks (dotted lines) in similar to the behavior of Au NRs reported in the literature [37]. Meanwhile, the PL emission spectrum also demonstrates obvious polarized characteristics. When the polarization of the incident laser is perpendicular to the axis of the nanorod, the PL peak disappears, indicating a close correlation with the longitudinal LSPR mode. This also directly proves that the PL emission controlled by LSPR originates from the radiative recombination of electron–hole pairs following the intraband transition.

Combined with the high polarized characteristics of the hot electron intraband emission, it can be exploited to realize new applications in information security, such as high-contrast optical image encryption and anti-counterfeiting. In 2021, Y. Zhang et al. [35] prepared the Al metal–insulator–metal (MIM) nanostructure arrays and realized the highly polarized up-converted hot electron intraband luminescence excited by femtosecond pulse laser, as shown in Figure 9.4e. On the basis of optical image encryption, the cross-structural Al nanorod arrays were utilized to demonstrate the anti-counterfeiting function of optical images, as shown in Figure 9.4f. Two encoded images can be simultaneously achieved with polarization along the longitudinal direction of the nanorod, respectively, excited by circularly polarized or linearly polarized femtosecond laser along the diagonal direction of the cross structure. When the direction of the analyzer is along the longitudinal direction of the nanorods, two true QR codes, "HEIL" and "TPL," are observed. When the analyzer's direction is along the diagonal direction of the cross structure, the "FALSE" QR code image will be observed with two true QR codes overlapping.

9.3 Hot-Electron Transfer Induced by Plasmon

Excepting for the capacity to generate light emission through direct inelastic tunneling and intraband transition, the interaction between hot electrons and excitons of semiconductor materials also has an important impact on the optical properties of their hybrid systems. In 2013, K. Wu et al. [38] synthesized colloidal CdS-Au composite structure as a model system to demonstrate the charge transfer

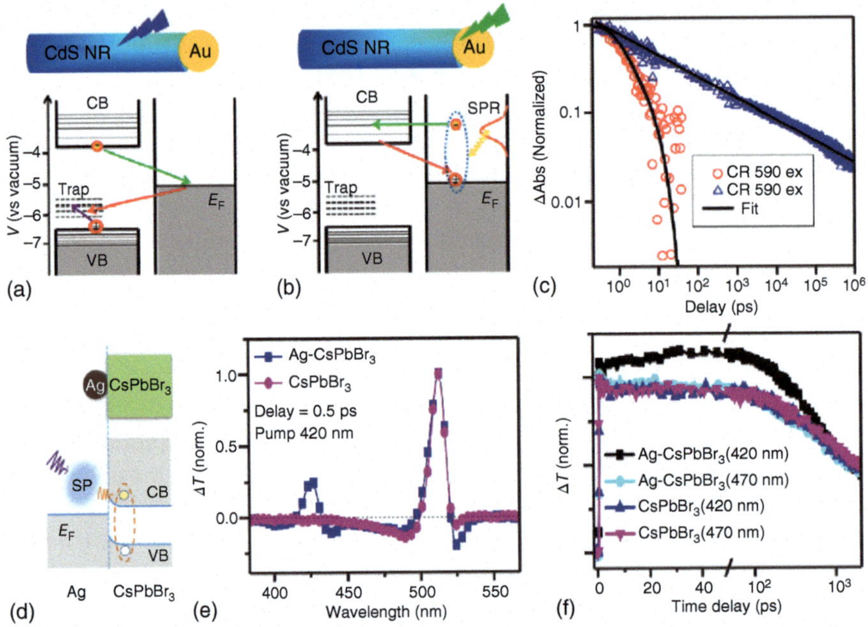

Figure 9.5 (a, b) Schematic diagram of charge separation under different mechanisms; (c) comparison of 1σ electron dynamics excited at 400 and 590 nm. (d) Schematic diagram of plasmon-induced resonant energy transfer (PIRET) process in Ag-$CsPbBr_3$ hybrid nanocrystals (NCs); transient absorption spectra (e) pumped at the wavelength of 420 nm and kinetic curves (f) probed at 514 nm from $CsPbBr_3$ and Ag-$CsPbBr_3$ (NCs). Source: (a–c) Reproduced with permission from Wu et al. [38] @2013, American Chemical Society. (h) Reproduced with permission from Huang et al. [39]/with permission of Springer Nature/Public Domain CC BY 4.0.

process at the metal/semiconductor interface. For excitation sources with different energies, the mechanism of interfacial charge separation and migration is essentially different, as shown in Figure 9.5a,b. Under the excitation of a high-energy light source (400 nm), the interband transition occurs on the side of CdS nanorod, and the photogenerated excitons are separated by electron transfer to Au tip. However, the excitation of low energy light source (590 nm) can only generate the plasmons in Au tips, and then decay into hot electron–hole pairs, followed by the injection of hot electrons to CdS nanorods and recombination with the holes in Au tips.

Since the two different excitation conditions involve the transfer of electrons from the CdS conduction band (CB) to the unfilled level of Au, it is of great significance to compare the injected hot electrons dynamics of the two. Figure 9.5c compares the dynamics of 1σ electron in CdS-Au NRs under the excitations of 400 nm (blue triangles) and 590 nm (red circles), respectively. In the former case, excitons are randomly localized along CdS nanorods due to hole trapping, resulting in a broad distribution of electron transfer rates with a half-life of 4.85 ps, whereas the dynamics of the latter can be well fitted by a two-exponential function with a half-life of 1.83 ps. Furthermore, the transient absorption signal of the 1σ electrons intraband transition is exploited to quantify the quantum yield of the plasmon-induced hot electron transfer (PHET) in CdS-Au NRs, which is 2.75 (±0.07)%. However, the extremely low quantum yield caused by ultrafast electron–electron scattering has seriously hindered the practical process of the devices.

In 2015, K. Wu et al. [40] further proposed a new pathway to solve the problem of quantum yield, which is called plasmon-induced interfacial charge-transfer transition (PICTT). This model enables the decay of a plasmon by directly generating an electron in semiconductor and a hole in metal, avoiding the loss of hot electrons via electron–electron scattering. They compare the theoretically predicted values of quantum yields in three cases of the traditional PHET mechanism (dashed lines) with the experimental values of the PICTT mechanism (red open circles). The quantum yield of the traditional PHET mechanism is closely dependent on the pump energy, whereas that of the PICTT mechanism is independent of the pump energy. As long as the excitation energy exceeds the absorption threshold (~0.85 eV), the same plasmon is excited with quantum yield of 24%, which is an order of magnitude higher than that of the traditional PHET mechanism.

In order to further improve the quantum efficiency of interfacial hot excitation transfer, X. Huang et al. [39] proposed a new pathway of PIRET based on the hybrid structure consisting of Ag NPs and perovskite quantum dots (QDs) in 2019. As shown in Figure 9.5d, the plasmonic energy transfers to the part of perovskite QDs and generates the electron–hole pairs (EHP) because of the interaction between the plasmons and excitons. Figure 9.5e compares the normalized transient absorption spectra of hybrid Ag-$CsPbBr_3$ and $CsPbBr_3$ under the excitation of 420 nm, where the enhanced excited-state absorption (ESA) signal at 526 nm should be attributed to the local-plasmon-mode-driven hot electron transfer (HET) from Ag to $CsPbBr_3$. In order to estimate the quantum efficiency of PIRET process, the bleaching signal at 514 nm normalized at 2 ns from $CsPbBr_3$ and hybrid Ag-$CsPbBr_3$ nanocrystals under resonant (420 nm) and nonresonant (470 nm) pumps was collected, as shown in Figure 9.5f. The difference of XB signal tracks implies that the EHP created by PIRET and the charge separation state produced by HET under resonant pump both contributed to the signal with the estimated quantum efficiency of $\sim 50 \pm 18\%$.

In addition, the design of different structures is also an important means to optimize and improve the quantum efficiency of HET. In 2013, M.W. Knight et al. [41] designed an embedded plasmonic nanostructure, and successfully realized the enhancement of hot electron emission. The schematic diagram of the fully embedded devices is formed by three Schottky interfaces with Si substrate to support electron transmission. The SEM image shows the different device with a width of 120 ± 10 nm embedded ~5 nm (blue frame), 15 nm (green frame) and 25 nm (red frame) in the substrate, respectively. For devices with the same width and different embeddedness, the photocurrent responsivity is significantly positively correlated with the embedding depth, which is ascribed to the larger optical cross section dominated by the increased transverse dimension. On the other hand, the measured dipole peak energy is in good agreement with the calculated value of FDTD, accompanied by an obvious redshift phenomenon with increasing of

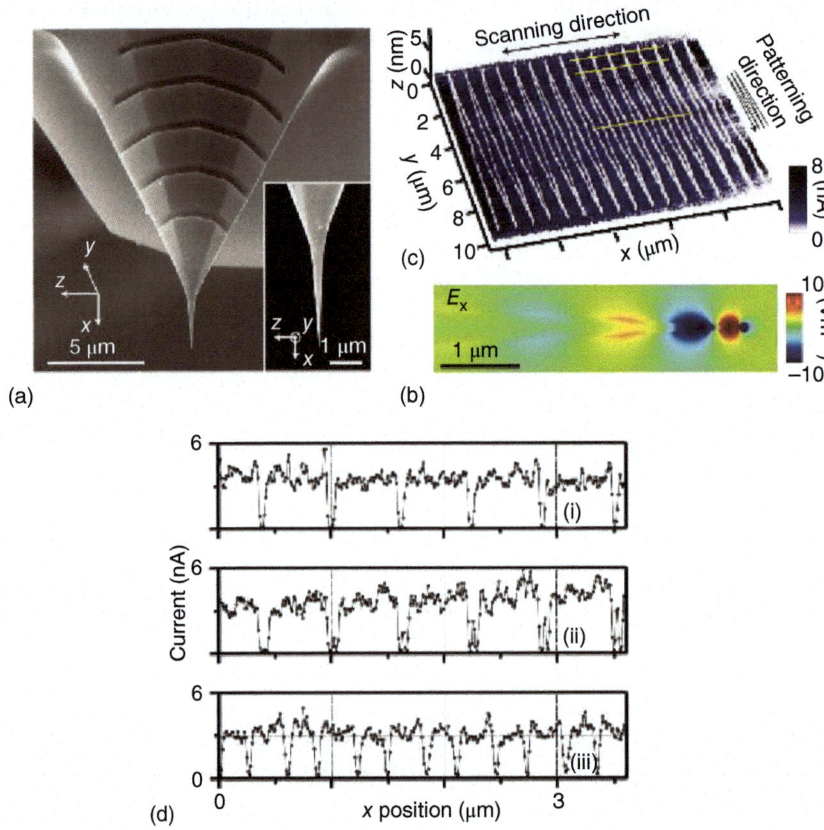

Figure 9.6 (a) SEM image of grating and cone fabricated on an octagonal pyramid; (b) electric field distribution at the tip; (c) photocurrent imaging overlaid on three-dimensional topography; (d) the corresponding photocurrent intensity profiles of three yellow lines on image (i). Source: Reproduced with permission from Giugni et al. [42] © 2013, Springer Nature.

nanowire width. The highest photocurrent response ratio of the embedded devices has been improved by about 25 times compared with the Au pad, indicating a more effective efficiency of HET.

In addition to improving the efficiency of hot electron transfer by increasing the contact interface, it is also an effective strategy for designing quantum efficiency devices by using structural design to extremely localize the distribution space of hot electrons. In 2013, A. Giugni et al. [42] fabricated the tapered tip based on the Schottky junction with nanoscale dimensions that can supply the focusing of SPs with a plasmon-hot electron conversion efficiency of ~30%, as shown in Figure 9.6a. The numerical simulation results in Figure 9.6b show that the x component of the electric field amplitude is along the grating and reaches its maximum at the tip. This highly localized electric field in space has greatly improved the efficiency of plasmon-hot electron conversion and is emerging as an extraordinary application prospect in the new type of hot electron nanoscope. Figure 9.6c shows that the photocurrent image with high-resolution topography is based on a GaAs deposited with a continuous folded oxide layer, and achieving the high-resolved photocurrent and topographic information simultaneously. The representative scan lines (marked with yellow solid lines) in Figure 9.6d contain the quantitative information of the current intensity and resolution, with an average spatial resolution of less than 50 nm.

9.3.1 Light Emission Enhanced by Hot-Electron

Luminescent intensity is an important parameter to judge the properties of materials. In this section, the influence of the introduced hot electrons on the radiation intensity of semiconductor luminescent materials is mainly presented. In 2016, X. Huang et al. [43] synthesized Cu doped ZnO nanorods modified by Au NPs, as shown in Figure 9.7a. After the modification of Au NPs with a diameter of 10 nm, the near-band edge (NBE) emission at 377 nm of Cu-doped ZnO was enhanced by two orders of magnitude with an enhancement factor of up to 280, accompanied by a significant decrease in green emission at 502 nm, as observed in Figure 9.7b. In order to clarify the mechanism of luminescent enhancement, time-resolved PL spectra of Cu-doped ZnO nanorods were measured, as shown in the inset of Figure 9.7b. The decay curve of Cu doped ZnO nanorods without Au NPs modification (black solid line) satisfies the biexponential function fitting with the time contents of $\tau_1 = 0.52 \pm 0.03$ ns and $\tau_2 = 2.45 \pm 0.09$, respectively. The former

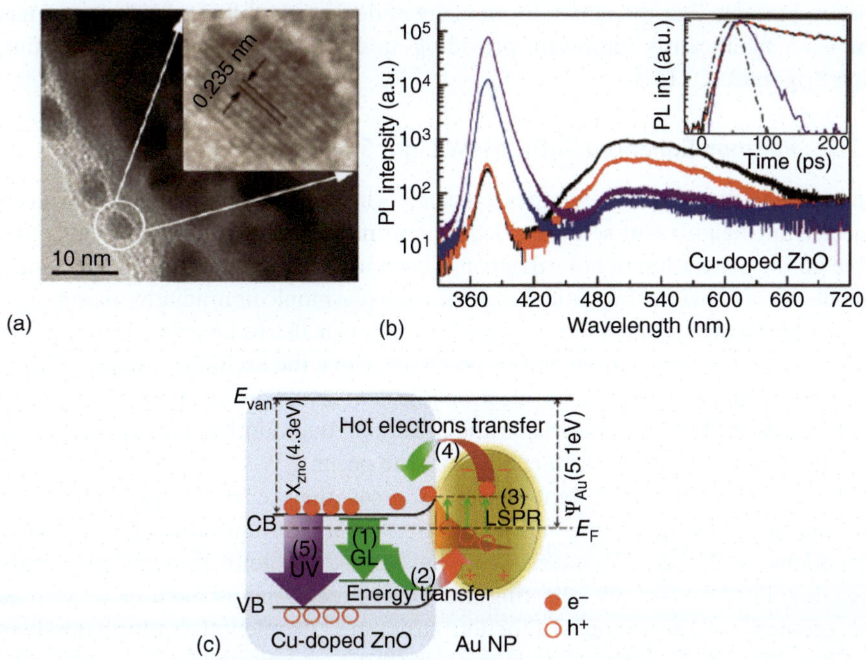

Figure 9.7 (a) TEM image of Au/Cu-doped ZnO nanowire inserted with HRTEM image of Au nanoparticle; (b) PL spectra of Cu-doped ZnO nanowires decorated with different sized Au nanoparticles; (c) schematic diagram of energy band for emission enhancement. Source: Reproduced with permission from Huang et al. [43] © 2016, John Wiley & Sons, Inc.

can be attributed to non-radiative recombination through the surface states of ZnO, whereas the latter is derived from the intrinsic free exciton radiation recombination of ZnO. The decay curve of Cu-doped ZnO nanorods modified with Au NPs exhibits a single exponential decay behavior with an estimated lifetime of 18 ps. The Purcell factor of up to 136 arisen from the sharp decay of the lifetime suggests that the enhancement of NBE emission originates from the HET process assisted by the LSPR. Figure 9.7c demonstrates the detailed energy band diagram of Au-modified Cu-doped ZnO.

In general, the LSPR of Au generated by the coupling of photons and electrons (energy transfer process) is then attenuated into the plasmonic-induced hot electrons and injected into the conduction band of Cu-doped ZnO (hot electrons transfer process), which increases the spontaneous emission rate of the excitons. Recently, H. Wang et al. [44] also exploited Au antenna to realize the enhancement of CdSe/CdS QDs, and designed a high-brightness deep-red QD light-emitting diode (QLED). SEM images of low-magnification and enlarged Au antenna on the surface of self-assembled CdSe/CdS core-shell QDs film are shown in Figure 9.7d,e. The fluorescence confocal measurement shows that the Au antenna locally significantly enhances the luminous intensity of the self-assembled QDs film. The profile of PL intensity corresponding to the dotted line in the fluorescence confocal image gives a sharp contrast between the area with and without the Au antenna. The mechanism of PL enhancement is attributed to the local Kirchhoff's law. Furthermore, the QLED modified by Au antenna is designed and fabricated. Similarly, the EL spectra presents that the emission intensity of devices modified with Au antennas has been significantly improved, providing an effective strategy for constructing high-brightness QLEDs.

9.3.2 Plasmon Enhanced ZnO:Ga MW Light Source

Nonradiative decay of SPs generates hot carriers in metal nanostructures that can be injected into neighboring semiconductor micro/nanostructures. Such working principle on the generation of hot carriers has been utilized to construct electronic and optoelectronic devices. In particular, during the plasmonic nonradiative damping, a nonthermal distribution of the electron–hole pair can also be created with the aid of thermalizing the lattice in several picoseconds. Once the excitation energy is large enough, the higher the rate of hot electron injection across the metal–semiconductor contact potential barrier, the greater the likelihood that nonthermal distribution of hot electrons in the nearby semiconductors can occur.

In the electrically driven light-emitting sources, the excited energy is harvested into plasmonic energy by the surface electrons profited from intraband/interband transitions, and then is converted to be emitted photons through plasmonic decaying. Furthermore, by adjusting the sizes of metal nanostructures, as well as the geometry of the tunneled junction structures, wavelength-tuning emissions can be achieved Ga-doped ZnO microwire (MW), as reported by Jiang group in recent years. [45, 46] Bare ZnO:Ga MW based device and LED were constructed through depositing different plasmonic nanostructures on the surface of individual ZnO:Ga MW.

Figures 9.8a,b are the incandescent-type light source composed of Au NRs decorated with single ZnO:Ga MW (AuNRs@ZnO:Ga MW). By introducing Au NRs deposition, the intrinsic lighting from the bare ZnO:Ga MW emitter can be completely suppressed. Benefiting from Au NRs with controlled aspect ratio, wavelength-tunable incandescent-type lighting was achieved, with the dominant emission peaks tuning from visible to near-infrared spectral regions [45]. ZnO:Ga MW-based filament emitter was lightened with light centered in the wire. By contrast, another ZnO:Ga MW was also selected, and then Au NRs partially deposited on the MW were introduced, as demonstrated in Figure 9.8b. When the applied bias reached a certain value, such as 44 V, bright and visible emission can be observed, with the red lighting regions located the section deposited Au NRs. The most interesting finding is that bright and green emission can also be observed, with the emission regions located at the bare segment. It is reasonable to deduce that the dominating emission wavelength of the emitted photons is relevant to the energies of the hot electrons, and then redshifts monotonically after the relaxation procedure. The generation and injection of hot electrons with energy up to the plasmon frequency can be utilized to dominate the characteristics of the emitted photons, such as the emission wavelength and spectral line width. Despite that, the Joule heating effect played a significant role in the incandescent-type emission characteristics, but it cannot be compatible with the thermal radiation mechanism (blackbody radiation theory).

To exploit modulation of Au-nanorods on the emission characteristics, it can be found that Au-nanorod plasmons induced generation and injection of nonequilibrium hot electrons can be responsible for the wavelength-tunable incandescent-type lighting features, as present by Figure 9.8c. Being electrically driven, the hottest spots can also be formed toward the center of the MW prepared with Au NR decoration due to the Joule heating effect. Thus, a sufficiently strong electric field ($\sim 10^6\,\mathrm{V\,m^{-1}}$) could be produced toward the hottest spots owing to the extremely poor heat dissipation, which is induced by high crystallization quality of the as-synthesized MWs. Then, the transported electrons in the MW channel can be accelerated and become energized electrons, which are spatially localized toward the center of the MW. Because of the inelastic collision, the interplay of the energized electrons transported in the MW channel and free electrons localized at the surface of Au-nanorod can occur. In consequence, collective oscillation of the free electrons in Au NRs can be excited. Finally, through nonradiative plasmonic decay, Au NR plasmons can transfer their energies into oscillating higher-energy electrons, leading to formation of hot electrons.

Usually, Au-nanorods are not stable in high temperatures. Therefore, alloyed plasmonic NRs of high stability were applied on the ZnO:Ga MW-based light source [46]. Figure 9.8d–g is 1 MW-based Si-LED without and with AuAg NRs decorating the surface of MW. Upon electrical lighting at a forward bias voltage, red light-emission from the single MW Si-LED was captured, with the emission regions located along the wire. EL images of the single MW-based Si LEDs were captured via a CCD. Figure 9.8e shows the optical microscope EL images of the single ZnO:Ga MW-based LED, with the emission regions distributed along the

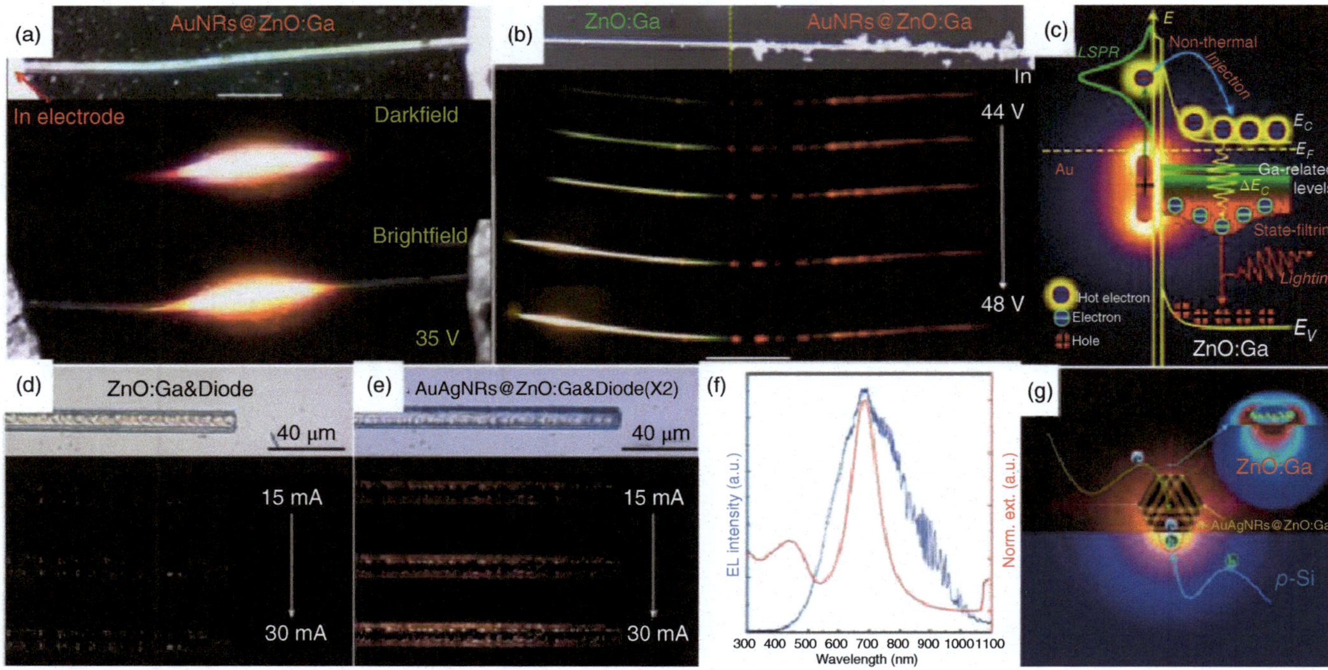

Figure 9.8 (a) Micrographs of bright visible light emitting from an electrically driven single AuNRs@ZnO:Ga MW-based incandescent-type light source (scale bar, 200 μm). (b) Optical microscopic images of light emitting from electrically driven single ZnO:Ga MW prepared with partial Au NRs decoration (scale bar, 300 μm). (c) Schematic diagram of the working principle of bright visible light emitting from electrically biased single AuNRs@ZnO:Ga MW-based incandescent-type light source. (d, e) Comparison of plasmon enhanced high-performance single MW-based Si-LED without and with AuAgNRs. (f) The comparison of normalized extinction spectrum of AuAg NRs (red line), and EL spectrum of a single AuAgNRs@ZnO:Ga MW-based heterojunction LED with the forward current of 30.0 mA (blue line). (g) Working mechanism of AuAg NRs plasmon-enhanced red radiation from single AuAgNRs@ZnO:Ga MW-based Si-LEDs. Source: (a–c) Reproduced with permission from Sun et al. [45] © 2020, Optica Publishing Group. Reproduced with permission from Liu et al. [46] © 2020, Royal Society of Chemistry.

wire. Whereas, bright red light emission could be noticed with the naked eye. By incorporating AuAgNRs, an obvious enhancement of lighting brightness and emission regions can be seen. The normalized optical extinction spectrum of AuAg-NRs and PL spectrum of a single ZnO:Ga MW are depicted in Figure 9.8f. From the figure, it can be seen that the plasmon-energy of the deposited AuAg NRs matches well with that of the main peak wavelengths of the single MW-based Si-LED, leading to the enhancement of red radiation output from the single MW-based Si-LED. Due to the well matching AuAg NRs plasmon and the Si-LEDs, the incorporating AuAg NRs can facilitate the radiative recombination rate of the electron–hole at the ZnO:Ga/Si interface, leading to plasmon enhanced red-radiation.

For understanding the working principle of AuAgNRs and their plasmonic effect on the EL emission behavior of single ZnO:Ga MW-based Si-LEDs, numerical calculations and simulation were carried out. By depositing AuAg NRs, the EL performances of the Si-LED could be modulated. Under electrical bump, the output radiation from the single AuAgNRs@ZnO:Ga MW-based Si-LED is absorbed by the deposited AuAg NRs, and localized plasmons of the AuAg NRs can be excited. Correspondingly, a strong coupling interaction between the electron–hole pair and AuAg NRs plasmons can occur toward the AuAgNRs-ZnO:Ga interface, leading to the enhancement of the light output. Figure 9.8g. Under electrical illumination, the radiative recombination of electron–hole can occur at the ZnO:Ga/Si interface.

More recently, an electrically driven single-mode microlaser was realized by M.M. Jiang group. [47] The laser device is composed of ZnO:Ga MW with platinum nanoparticles (Pt NPs, $d \sim 130$ nm) covering, a magnesium oxide (MgO) nanofilm, a Pt nanofilm, and a p-type GaN substrate. To further optimize the junction interface quality, particularly the electrical properties, Pt buffer layer was further deposited on the MgO/p-GaN substrate, which facilitate carrier transport in the as-designed LEDs and improve the electrical contact between the ZnO:Ga MW and p-type GaN film. The schematic of the device is shown in Figure 9.9a. In the device configuration, ITO and Ni/Au working as electrodes are responsible for the current injection. The SEM observation reveals that physically isolated PtNPs are uniformly distributed on the side surfaces of ZnO:Ga MW, and the average diameter is approximatively measured to about 130 nm (see Figure 9.9b).

In addition to improving the efficiency of hot electron transfer by increasing the contact interface, it is also an effective strategy for designing quantum efficiency devices by using structural design to extremely localize the distribution space of hot electrons. In 2013, A. Giugni et al. [42] fabricated the tapered tip based on Schottky junction with nanoscale dimensions that can supply the focusing of SPs with a plasmon-hot electron conversion efficiency of ~30%, as shown in Figure 9.6a. The numerical simulation results in Figure 9.6b show that the x component of the electric field amplitude is along the grating and reaches the maximum at the tip. This highly localized electric field in space has greatly improved the efficiency of plasmon-hot electron conversion and is emerging as an extraordinary application prospect in the new type of hot electron nanoscope. Figure 9.6c shows that the photocurrent image with high-resolution topography is based on a GaAs deposited with a continuous folded oxide layer, and achieving the high-resolved photocurrent and topographic information simultaneously. The representative scan lines (marked with yellow solid

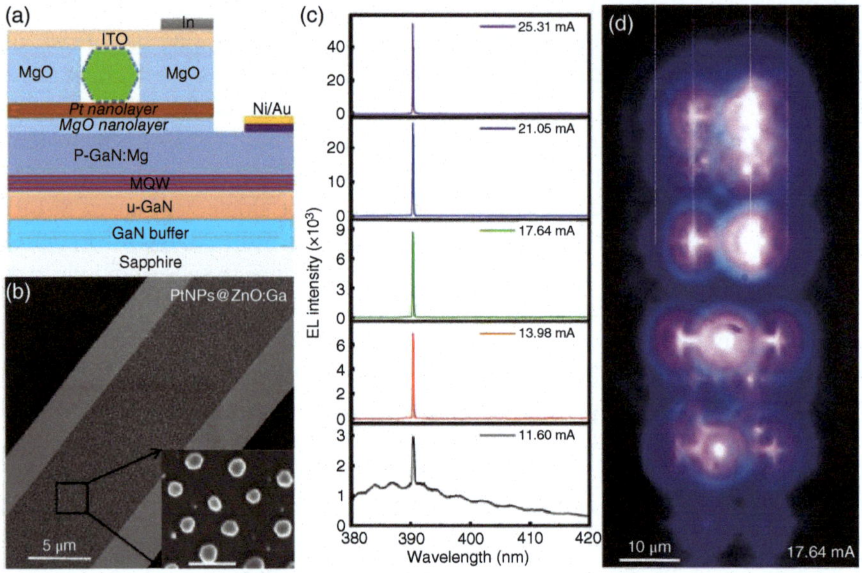

Figure 9.9 (a) Schematic illustrating the cross section of the laser device architecture based on the n-PtNPs@ZnO:Ga MW/Pt/MgO/p-GaN heterojunction. (b) SEM image of a single ZnO:Ga MW covered by Pt NPs. Inset: Enlarged SEM image of the Pt NPs, which are deposited on the side surface of ZnO:Ga MW. The diameter of the deposited Pt NPs is extracted to about 130 nm. (c) Current dependent EL spectra of the as-fabricated single MW-based LED, as the injection current is increased from 11.6 to 25.31 mA. (d) Optical microscopic CCD image of the emission from the fabricated device. Source: Reproduced from [47]/with permission of Springer Nature / Public Domain CC BY 4.0.

lines) in Figure 9.6d contain the quantitative information of the current intensity and resolution, with an average spatial resolution of less than 50 nm.

As the injection current reaches 11.6 mA, a sharp emission mode appearing at 390.5 nm can be seen, which is amplified due to optical oscillation in the MW microresonator. As the injection current increases above 11.6 mA, the corresponding EL spectra are plotted in Figure 9.9c. More noticeably, no other oscillation peaks are found even at higher injection levels. There is little variations of the line shape and peak position of the EL spectra by increasing the injection current. Figure 9.9d illustrates the CCD image of the recorded light emission pattern at an operating current above the threshold (~17.64 mA). From the figure, the strongest lighting spots are distributed along the straight and sharp edges of the hexagon-shaped ZnO:Ga wire. The EL emission is so bright that it can be visibly observed by naked eyes under normal lighting conditions. In the study of single-mode laser, they found that the incorporation of Pt/MgO buffer nanofilms can integrate the depletion layer, laser medium, and microresonator in the single ZnO:Ga MW, realizing multimode lasing actions in the UV region. By manipulating the cavity resonant modes, the incorporation of Pt NPs, potentially showing superabsorber behavior, can lock a desired lasing mode in the higher energy side of multimode lasing in the as constructed single-bare MW-based microlaser.

9.3.3 Other Application for Light Modulation

The introduction of hot electrons cannot only effectively improve the luminescent intensity of semiconductor materials, but also have an important impact on their crystal structure, optical bandgap, exciton binding energy and other structural and optical parameters, especially in 2D material systems. In 2014, Y. Kang et al. [48] decorated Au NPs on the surface of a MoS_2 Monolayer, and realized the transformation of MoS_2 lattice from 2H phase to 1T phase under CW laser illumination. Under the 488 nm incident light, the electrons on the surface of Au NPs oscillate collectively to produce plasmons, and then decay to form hot electrons that cross the 0.8 eV Schottky barrier and transfer into the MoS_2 monolayer to achieve an effective hot electron doping, as shown in Figure 9.10a. The three typical Raman peaks of 1T phase MoS_2 at 156, 226, 323 cm^{-1} in the Raman spectra of Figure 9.10b provide direct evidence for the 2H-to-1T transition. In order to further verify that the phase transition is derived from the hot electron doping induced by plasmon, the super continuum laser (400–900 nm) is exploited to investigate the dependence of PL shift on the excitation wavelength, as shown in Figure 9.10c. The PL shift curves of hexagonal MoS_2 modified with Au NPs and Au NRs and absorption spectra (dashed lines) of the corresponding metal structures in solutions are compared, where Au NPs are previously used, and the length and width of Au NRs are 35 and 10 nm respectively. It can be observed that the PL shift curve of Au NPs modified MoS_2 exhibits a clear peak at 520 nm, which is consistent with its dipole plasmon resonance at 520 nm. The peaks of PL shift curve for Au NRs modified MoS_2 correspond to a longitudinal plasmon mode at 650 nm and a transverse resonance at 520 nm, respectively, further suggesting that the mechanism causing phase transition is hot electron doping. Subsequently, they observed that the hot electrons injection induced by plasmons can significantly affect the exciton binding energy of MoS_2 monolayer [49]. The structure of the device is composed of a 30 nm Au film, 30 nm SiO_2 insulation layer and MoS_2 monolayer modified with Au NPs. The dependence of the energy for PL peak and absorption feature A on the incident laser power demonstrate that both of them appear redshift with increasing of the incident power. The exciton binding energy can be calculated by the formula of $E_b = E_g - E_{PL}$ [50], where E_g represents the band gap of MoS_2 monolayer.

Due to the influence of doping on exciton energy, the absorption peak energy cannot directly reflect the band gap. Hence, the energy difference between absorption and PL peak is exploited to indirectly reflect the change of exciton binding energy. As the incident laser power increases from 0.01 to 5 mW, the exciton binding energy obtains an increment of 15 meV due to the doping of hot electrons. The detailed transfer process and energy band structure evolution under hot electrons injection are proposed. The tunable exciton binding energy expressed by the hollow arrow is closely related to the relative position of the doping level (white dashed line) and the conduction band, both of which are controlled by the hot electron doping. Therefore, hot electron doping plays a decisive role in the active control of exciton-binding energy.

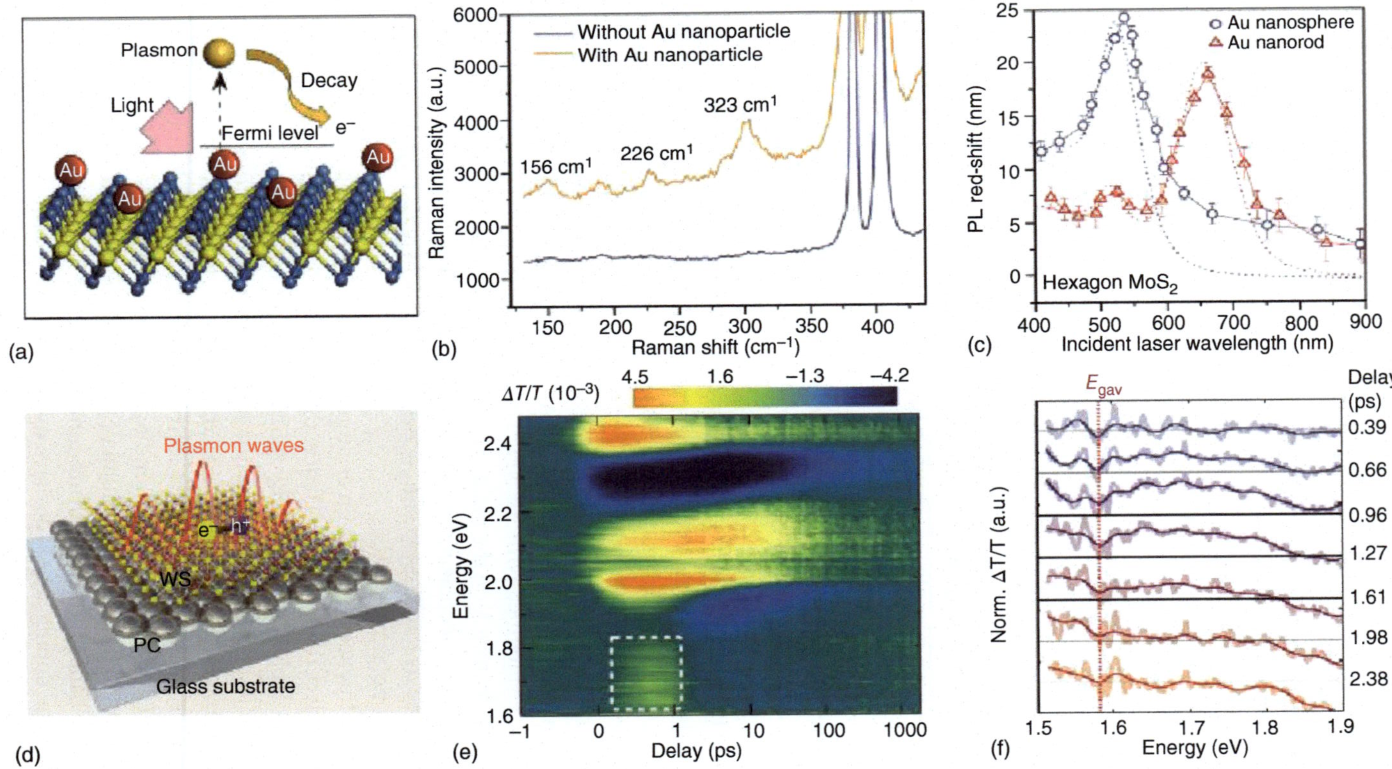

Figure 9.10 (a) Schematic diagram of hot electron generation and interaction with MoS_2; (b) Raman spectra of MoS_2 coupled with/without Au nanoparticles; (c) dependence of PL red-shift on the incident laser wavelength for hexagon MoS_2 with Au nanosphere and nanorod. (d) Schematic diagram of a plasmonic crystal (PC)/Al_2O_3/WS_2 hybrid structure; (e) intensity plot of $\Delta T/T$ spectra for the plasmonic crystal (PC)/Al_2O_3/WS_2 hybrid structure; (f) normalized $\Delta T/T$ spectra at different delay time. Source: Reproduced with permission from Kang et al. [48] © 2014, John Wiley & Sons, Inc.

Recently, Y.H. Chen et al. [51] realized the active control of the band gap in 2D semiconductor WS_2 through the coherent doping of plasmonic hot electrons. The WS_2 monolayers are integrated into the self-assembled plasmonic crystal on the glass substrate shown in Figure 9.10d, which enables the resonant coupling between the semiconductor excitons and SPs. In order to confirm the extraordinary discovery, the transient optical responses of the hybrid plasmonic crystal WS_2 at high-pump power was measured with much more significant broadening maxima, as shown in Figure 9.10e. The $\Delta T/T$ spectra at different delay time extracted from white dashed frame is shown in Figure 9.10f, indicating that the renormalized bandgap starts at $E_g \approx 1.6$ eV, which is ~550 meV lower than that of WS_2 monolayers. It is concluded that the coherent doping of plasmonic hot electrons is the dominantly responsible for the band gap narrowing at low pump power density.

9.4 Hot-Electron Photodetection

9.4.1 Traditional Schottky Barrier-Based Photodetection

The participation of hot electrons has not only made a lot of important research progress in the fields of new nanolight sources, information security, nanoscope, and light modulation but has also emerged with extraordinary application potential and prospects in the field of light detection. Based on the hot electron effect, optical detection devices with different structures have been designed, such as traditional Schottky, metamaterial and metasurface, graphene, planar distributed Bragg reflection, and optical antennas, to meet different functional requirements. In this section, we will discuss various types of photodetectors based on hot electrons.

The photodetectors based on the traditional metal–semiconductor Schottky barrier are first introduced. In 2014, A. Pescaglini et al. [52] realized near-infrared photodetection based on the hybrid structure of Au NR (NR)-ZnO nanowire (NW) through the effective hot electrons injection. The representative SEM image and an enlarged SEM illustration (red dashed frame) of the ZnO NW-based device modified with Au NRs show the disconnection between Au NRs on the surface of ZnO NW ensuring the normal operation of the device without short circuit. The photocurrent spectra of the hybrid device illuminated at the wavelengths of 650 and 850 nm, manifests a fast growth and decay time. The biexponential fitting results show that the fast process of photoresponse is dominant in hybrid devices, which is 40 times faster than the response time of bare ZnO NW-based devices. The distinct response characteristics for the hybrid and bare devices indicate that the carrier dynamics of the two devices are essentially different. The generation and injection of hot electrons modulated by plasmons at the interface of Au NR and ZnO NW is the main factor in the photoconductivity behavior of hybrid devices. To further investigate the physical mechanism of its enhancement and rapid response, the absorption cross section of the interface between Au NR and ZnO NW with a simulation illustration of the electric field enhancement factor is demonstrated. An obvious plasmonic resonance peak related to the longitudinal mode of Au NR at 650 nm is observed.

Under the excitation of the incident laser, the electric field is localized at the interface and generates a large number of high-energy hot electrons, which are injected into the semiconductor across the Schottky barrier to form the photoresponse. After that, T. Gong et al. [53] fabricated an angle-independent, vertical-structured hot electron photodetector based on transparent conductive oxides electrode. The incident light is absorbed by Au through transparent conductive oxides and Al_2O_3 layer, resulting in plasmons that then decay into high-energy hot carriers, which are collected by the electrode across the insulating barrier to form photocurrent. The photoresponse of the device and Au absorption spectrum illuminated at different wavelengths inserted with a photograph of the device manifest that the photocurrent increases significantly for short illumination wavelengths in the measured range of 400–700 nm, which is similar to the tendency of the Au absorption spectrum. It confirms that stronger light absorption in the short wavelength region generates more hot electrons, corresponding to higher photoresponse.

In addition, the absorption of the device is not dependent on different incident angles. In the range of measured angles, it still increases with the reduced wavelength, which is beneficial to the demand for illumination at different angles. In 2015, W. Wang et al. [54] designed a type of hot electron-based near-infrared photodetector by exploiting new 2D semiconductor material with a photogain of 10^5 and photoresponsivity of 5.2 A/W at 1070 nm. This device is composed of Au electrodes, interdigitated resonant (RWs) and non-resonant (NRWs) Au nanostructures, and bilayer MoS_2, as shown in the micrograph of Figure 9.11a. To gain insight into the photogain mechanism and hot electron injection efficiency, the photoresponse at visible and near-infrared frequencies is compared, as shown in Figure 9.11b. The growth region of both satisfies the bi-exponential function fitting, while the decay region satisfies the single exponential function fitting. Similar time parameters indicate the same trapping mechanism and photogain for both above and below-band gap illumination. Moreover, the effect of photothermoelectric effect on photocurrent is also clarified. Two individual devices sharing the same MoS_2 flake are fabricated, as shown in Figure 9.11d. A thick Al_2O_3 film is deposited on one device as a barrier for hot electron injection, while the other is not covered. The former exhibits no obvious photocurrent, while the latter exhibits a clear photoresponse. It indicates that the photoresponse generated by photothermoelectric effect is negligible compared with hot electron injection at the current illumination intensity, thus confirming the physical mechanism of device operation.

9.4.2 Photodetection Based on Metamaterials

Although the formation of hot electrons by nonradiative decay of SPs is widely used in optoelectronic devices, the extremely low quantum efficiency of traditional hot electron devices leads to poor electron injection in low light absorption cases, which hinders its further application. Improving optical absorption through the design of metal nanostructures is an effective strategy to change this situation. In 2014, W. Li et al. [55] integrated metal metamaterials on Si substrates to build the 1D and 2D perfect absorbers, and realized the broadband hot electron photodetectors with a

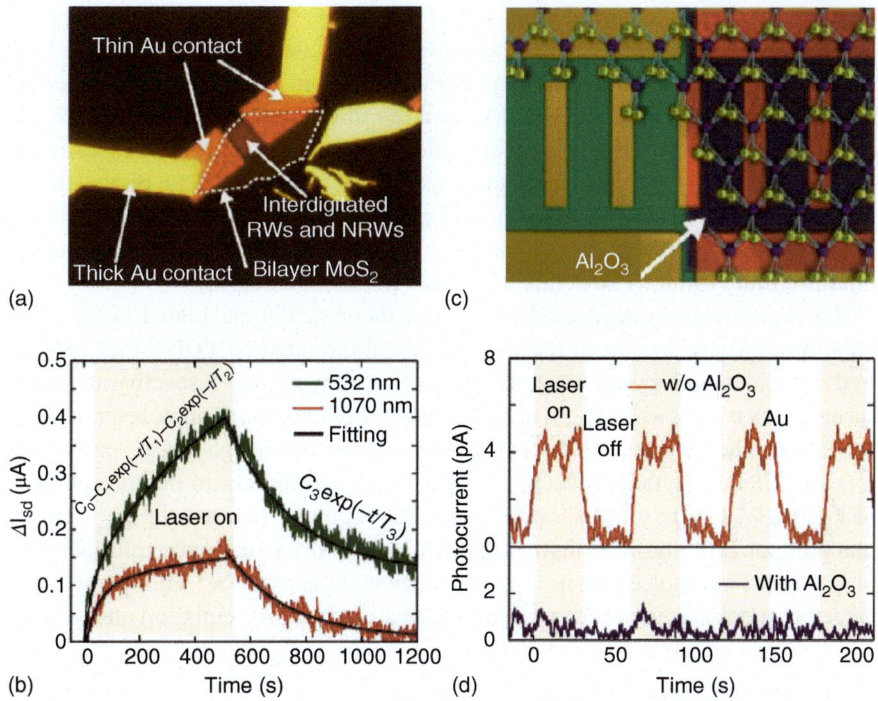

Figure 9.11 (a) Optical microscope image of the device; (b) photoresponse of the device illuminated at the wavelengths of 532 and 1070 nm; (c) photocurrent of the devices with/without Al_2O_3. Source: Reproduced with permission from Wang et al. [54] @2015, American Chemical Society.

high photoresponsivity. The structural diagram of 1D metamaterial perfect absorber (MPA) is demonstrated, where the etched Si grid is covered with thin Au film. Three devices (D1, D2, and D3) with dimensions of $L = 160$, 170, 170 nm, $P = 320$, 320, 340 nm, and $H = 120$ nm are designed, which demonstrate a broadband optical absorption with a maximum value of 95% and a FWHM of 400 nm in the near infrared region.

In order to characterize the HET efficiency of the devices, the photoresponse spectra of three 1D MPA photodetectors were measured across the wavelength range from 1200 to 1500 nm. The calculated and experimental values are in good agreement and present a wide photoresponsivity spectra with peak values of 3.37, 3.05, and 2.75 mA/W for these three devices, respectively, which is significantly higher than the previous reports [56]. The schematic diagram of HET process is proposed, including three steps. Firstly, the plasmons nonradiatively decay into hot electrons. Secondly, hot electrons across the interface of metal–semiconductor before thermalization. Finally, hot electrons inject into the conduction band of semiconductor to form photocurrent.

In order to seek special application requirements, metal metamaterials are further designed for efficient and highly selective circularly polarized light (CPL) detection. In 2015, their group constructed a ultra-compact and integratable CPL detector in succeed by exploiting chiral plasmonic metamaterials through hot electron injection [57]. Figure 9.12a shows the structural diagram of chiral metamaterials, which is composed of the periodic array of Ag chiral meta-molecules, the dielectric spacer (such as PMMA), and the thick Ag backplane. Figure 9.12b,c demonstrates the prepared left-handed (LH) metamaterial arrays and right-handed (RH) metamaterial arrays to verify the discrimination ability of CPL photodetectors to left-handed circularly polarized (LCP) light and right-handed circularly polarized (RCP) light. Typical SEM images and unit cells in enlarged illustrations reflect the unique chiral symmetry structure of LH and RH metamaterials.

The experimentally measured (dots) and theoretically calculated (solid curve) photoresponsivity spectra of the two devices illuminated by LCP (blue) and RCP (red) are shown in Figure 9.12e,f, which match well with their respective absorption spectra with a peak wavelength of 1340 nm and the maximum photoresponsivity of 2.2 mA/W. The polarization discrimination ratio of the designed CPL photodetectors for LCP and RCP light reaches up to 3.4 with a difference in photoresponsivity of 1.5 mA/W. Figure 9.12d shows the excellent contrast in scanning photocurrent mapping for LCP and RCP light, suggesting the ability to integrate multiple chiral plasmonic meta-molecules in a single ultra-compact device. This type of CPL detectors promotes great potentials in polarization imaging, emission and sensing.

9.4.3 Photodetection Based on Graphene

Since graphene is discovered and won the 2010 Nobel Prize in Physics, it has been widely used in all walks of life. Its unique optical and electrical behavior enables it to be used in some special scenes. In 2019, Y. Yu et al. [58] realized ultrafast photoelectric conversion in the near-infrared based on the PHET process by using $WO_{2.9}$/graphene hybrid structure. Figure 9.13a demonstrates the schematic diagram of device structure, which is composed of Si substrate coated with SiO_2 film, monolayer graphene, $WO_{2.9}$ NP arrays and Au electrodes (source and drain). SEM illustration exhibits that the plasmonic tungsten oxide nanostructures (TONs) are periodically and uniformly distributed on the surface of graphene by means of ultrathin anodic aluminum oxide (AAO) templates. Figure 9.13b presents the transient photoresponse of devices operated by PHET and common interband electronic transition (CET) based on PbS QDs-graphene hybrid structure. The typical response time of the latter is in the order of seconds (~3.7 seconds), while that of the former is accelerated by three orders of magnitude with a response time of milliseconds (1.8 ms) under 1550 nm excitation.

In order to investigate the origin of ultrafast response time, the dynamic process of PHET model is analyzed and drawn as shown in Figure 9.13c. The excited laser interacts with electrons trapped in the defect states to generate local surface plasmons (LSPs) and decay into hot electrons. And then, hot electrons are rapidly transferred from $WO_{2.9}$ to graphene (<150 fs), thus avoiding electron trapping and carrier interband recombination in $WO_{2.9}$ nanostructures, which is an important strategy for achieving efficient and fast photoelectric conversion in hybrid NIR structures. Apart from hybrid structures with semiconductors, the hybrid structures

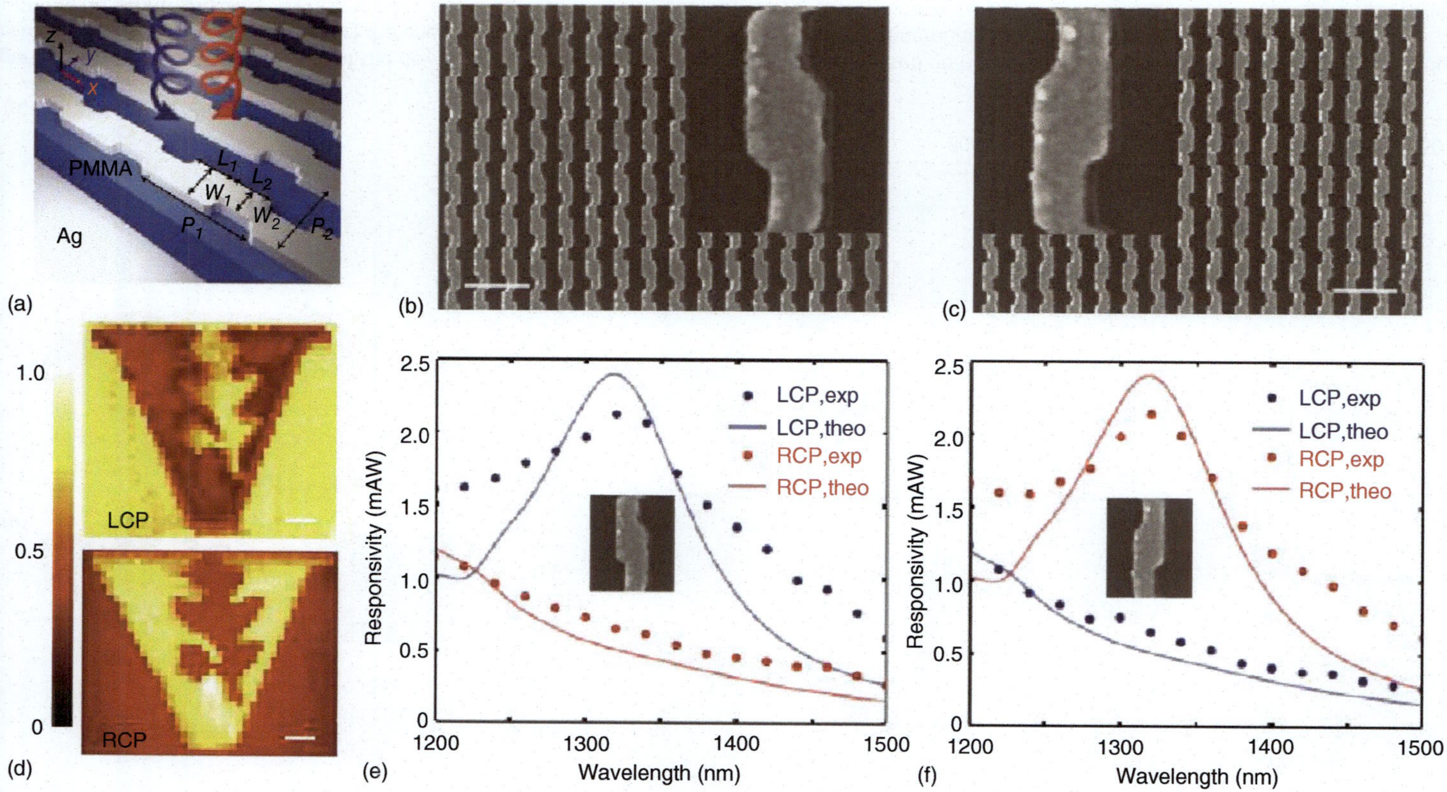

Figure 9.12 (a) Schematic of the chiral plasmonic metamaterial; SEM images of the left-handed (LH) (b) and right-handed (c) metamaterial, Scale bar, 500 nm; (d) Scanning photocurrent maps of the metamaterial under LCP (up) and RCP (below) illumination, Scale bar, 10 μm; the experimental and theoretical responsivity of LH (e) and RH (f) metamaterial under LCP and RCP illumination. Source: Reproduced with permission from Li et al. [57]

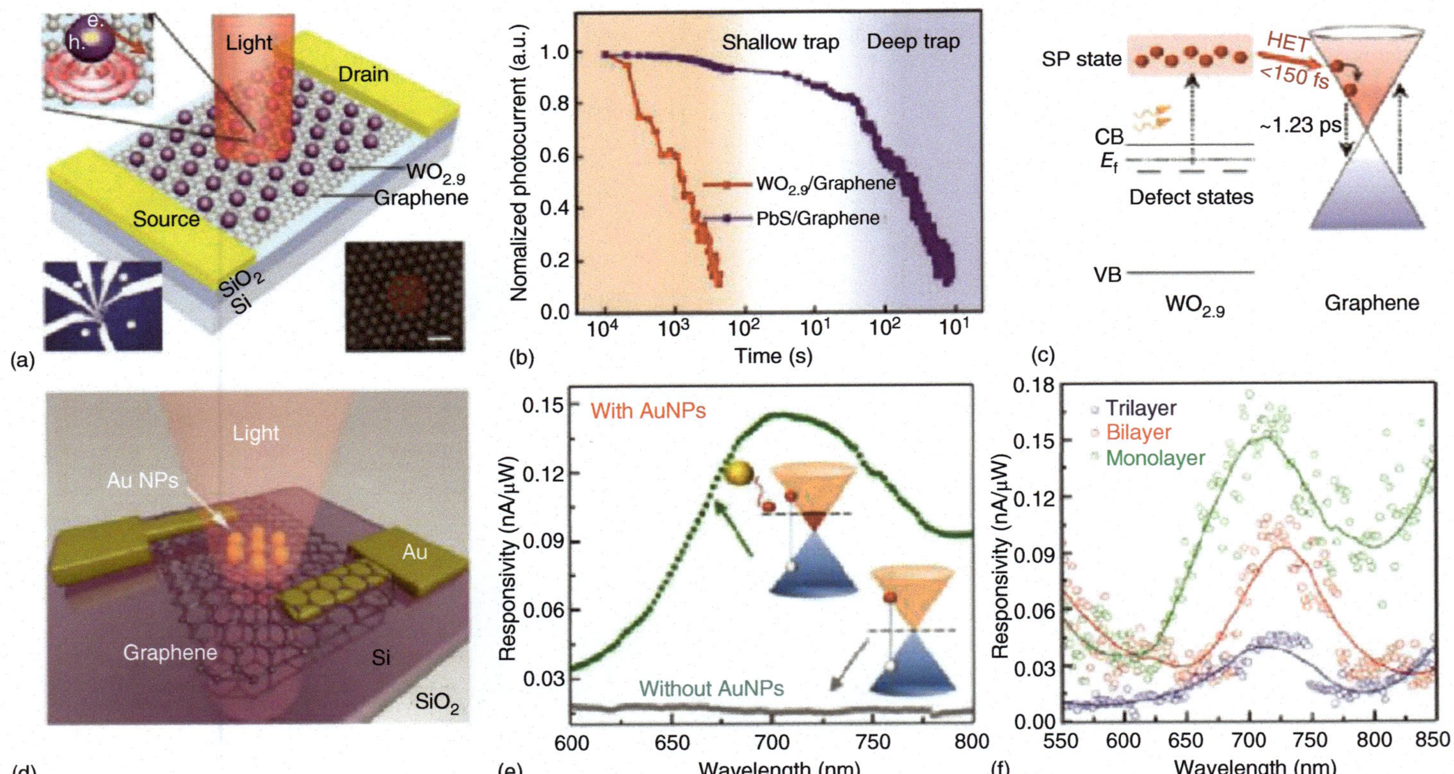

Figure 9.13 (a) Schematic diagram of the device architecture; (b) transient photoresponse based on hot-electron transfer (HET) and common band-edge electron transfer (CET) mechanisms; (c) the energy-levels schematic diagram of hot electron generation and transfer. (d) Schematic diagram of Au-graphene hybrid structure; (e) responsivity of the devices with/without Au nanoparticles; (f) responsivity of the devices with different graphene layers. Source: (a–c) Reproduced with permission from Yu et al. [58] @2019, Wiley-VCH Verlag GmbH & Co. KGaA, Weinheim. (e, f) Reproduced with permission from Du et al. [59] © 2017, John Wiley & Sons, Inc.

formed by graphene and metal NPs can also be utilized for photodetection. In 2017, Du et al. [59] designed the vertical Au-graphite hybrid nanostructures to achieve plasmonic hot electron tunneling photodetection. The structural diagram of the device is shown in Figure 9.13d. Monolayer graphene is loaded between the Au top electrode and the bottom electrode with a 5 nm-Ti adhesive layer. The Au NPs coated with tannic acid are deposited on the graphene in the middle of two electrodes, and excited by the incident laser. Figure 9.13e presents the dependence of responsivity on the wavelength of incident laser. It can be observed that the responsivity of Au-graphene hybrid structural device is much higher than that of bare graphene device, and reaches the maximum value of 0.16 nA μW^{-1} at 710 nm.

Similarly, the same phenomena have been observed in the hybrid structures consisting of bilayer and trilayer graphene, as shown in Figure 9.13f. The responsivity has a maximum value at ~710 nm, which is close to the value of the absorption peak for Au NPs. Therefore, the reason for the photoresponse of the hybrid structure can be attributed to the hot electrons generated by the decay of the Au-LSPRs transferring to graphene through inelastic tunneling to form photocurrent, which is closely related to the coupling efficiency between the incident laser and the free electrons of metal.

9.4.4 Planar Hot-Electron Photodetection

Effective absorption of the incident light is a feasible approach to high-efficiency photodetection. The plane structure of traditional devices determines that most of the incident light cannot be utilized. A distributed Bragg reflector (DBR) is a special optical structure that can greatly limit the light of a specific wavelength to its interior. DBR is a special optical structure, where the light with specific wavelength is strongly confined. Meanwhile, a type of SPs called Tamm plasmons (TPs) will be formed at the interface between metal and DBR [60]. Its unique in-plane wave-vector characteristic inside the light cone enables it to be excited by free space light without additional vector-matching processing. In 2017, Zhang et al. [61] introduced the distributed Bragg reflector into the plane hot electron (HE) photodetector (PD) based on TP resonance, and more than 87% of the incident light is absorbed in succeed. The structural diagram of TP-HE PD is proposed, consisting of SiO_2 substrate, metal–semiconductor–metal (MSM) heterostructure, and DBR. In this configuration, hot electrons decayed by TPs generated at the metal/DBR interface will be collected by MSM, forming photocurrent.

The dependence of the photoresponsivity of TP-HE PD and grating-based SP-HE PD on the applied voltage at the resonant wavelength is compared. The photoresponsivity of the former is as high as 13.7 nA/mW, more than twice that of the latter (6.5 nA/mW). Compared with the gating-based systems, the incident light can only interact with the top metal layer and be localized at the metal/DBR interface to effectively generate TPs in the former configuration. The extremely strong space confinement of TPs and the effective utilization of incident light promote the generation of hot electrons, thus improving the photoresponse.

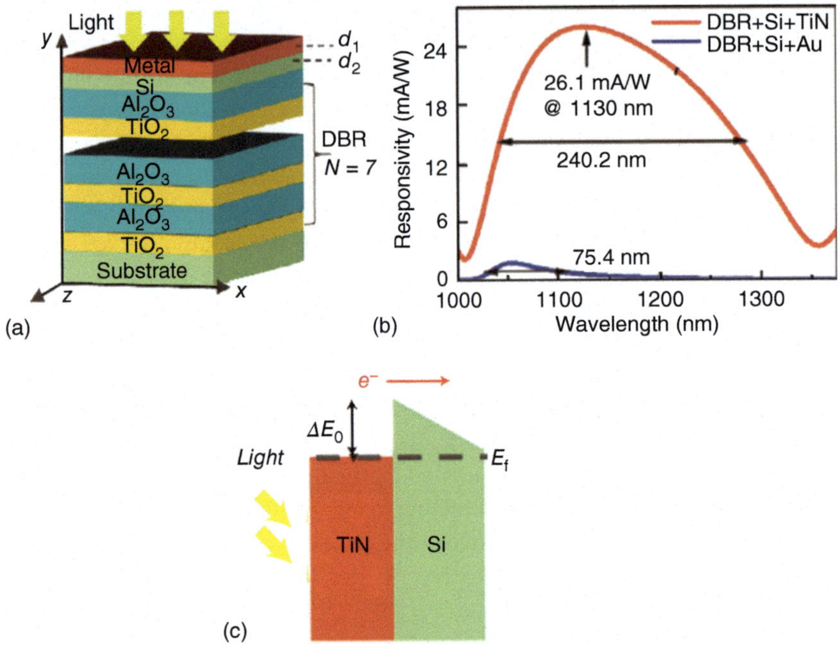

Figure 9.14 (a) Schematic diagram of the planar hot-electron photodetector; (b) responsivity of the devices based on DBR/Si/TiN and DBR/Si/Au structures; (c) energy band diagram of the TiN/Si structure. Source: Reproduced with permission from Wang et al. [62] @2020, The Royal Society of Chemistry.

Recently, J. Wang et al. [62] replaced the Au layer by using TiN layer with broadband TPs response to realize a planar TP-HE photodetector with wide response band. Figure 9.14a demonstrates the structural diagram of the device, which is composed of TiN/Si heterojunction, seven pairs of DBR, and Si substrate. In this configuration, incident light is from the top and strongly localized between the DBR and the top structure, exciting TP resonance at the TiN/Si interface. The introduction of TiN layer replacing Au is the key to achieving a broadband response. Figure 9.14b compares the dependence of optical responsivity for both the configuration of DBR/Si/TiN and DBR/Si/Au with the same thickness of metal layer on the wavelength of incident light, which exhibits similar characteristics with the absorption spectra of the two devices. Compared with the latter, the photoresponsivity of DBR/Si/TiN configuration reaches the maximum value of 26.1 mA/W at 1130 nm with the corresponding FWHM of 240.2 nm, making the response bandwidth to be increased by 3.2 times. It confirms that the high and broad photoresponse of the former originates from the more favorable absorption of TiN for the incident light because of its larger imaginary part in the near infrared region. Figure 9.14c shows the energy band diagram of TiN/Si heterostructure to describe the process of photocurrent generation. After absorbing photons, TiN generates TPs resonance at the interface with Si and then decays into hot electrons, which cross the potential barrier and inject into the conduction band of adjacent Si semiconductors to form photocurrent.

9.4.5 Photodetection Based on Optical Antennas

In Section 9.2.1, we introduced the application of optical antenna as a light emission unit based on the inelastic tunneling of hot electrons in the realization of nanoscale light sources. Meanwhile, it also shown important application value in the field of photodetection. In 2011, M.W. Knight et al. [56] exploited metallic nanoantenna to harvest the incident photons for the first time and excite resonant plasmons, which then decayed into hot electrons crossing the potential barrier at the interface of nanoantenna and semiconductor, resulting in the photocurrent. This active optical antenna photodetector is composed of ITO conductive layer, SiO_2 insulation layer, Si substrate and independent rectangular Au NR arrays. The incident light interacts with the free electrons on the Au surface passing through the ITO electrode to generate SPs, and then decays into hot electrons across the Schottky barrier and injects into the semiconductor to form a detectable photocurrent.

In order to demonstrate the detection performance of the device, the photoresponsivity with different Au NR lengths is measured. The optimum wavelength and relative intensity of the photoresponsivity for the different devices are highly dependent on the geometry of the Au nanostructures. As the length of Au NRs increases, the intensity of photoresponsivity decreases accompanied by the redshift of the optimum response wavelength, which is consistent with the tendency of the longitudinal dipole absorption resonance of the plasmon mode. This also determines that the polarization dependence of the photocurrent follows the polarization characteristics of the longitudinal dipole resonance, resulting in 90% attenuation of the photocurrent when the light polarization is perpendicular to the nanorod axis. The linear relationship between incident light power and photocurrent suggests that the conversion between single photon and single hot electron is dominant.

Apart from metal/semiconductor structures mentioned before, planar MIM structures can also serve for photodetection. In 2014, H. Chalabi et al. designed a plasmonic nanostripe antenna, which successfully realized photodetection and improved the performance of the devices [63]. Figure 9.15a presents the structural diagram, SEM image, and photocurrent image of the device, which is an MIM junction consisting of a wide Au bottom electrode, Al_2O_3 insulation layer, and a series of nanoscale grid top electrodes on the whole. Its working mechanism can be detailed into five steps, as shown in Figure 9.15b. These five steps can be summarized as follows: the metal electrode absorbs the incident photons to produce hot electrons, and the part of hot electrons with energy higher than the potential barrier can cross to the other electrode to form photocurrent. In this case, both the grid top electrode and the bottom electrode can absorb photons to generate hot electrons and form photocurrent. The former is the forward current ($I_{forward}$), while the latter is the backward current ($I_{backward}$). These two opposite currents produce a net current $I_{net} = I_{forward} - I_{backward}$, which is closely related to metal material, barrier height and applied bias voltage. Figure 9.15c demonstrates the dependence of the square root of the response on the incident photon energy under different applied bias voltages. It can be observed that for each applied bias the responsivity has an obvious onset and subsequent linear increase with photon energy, which can be explained by the modified Fowler equation [64].

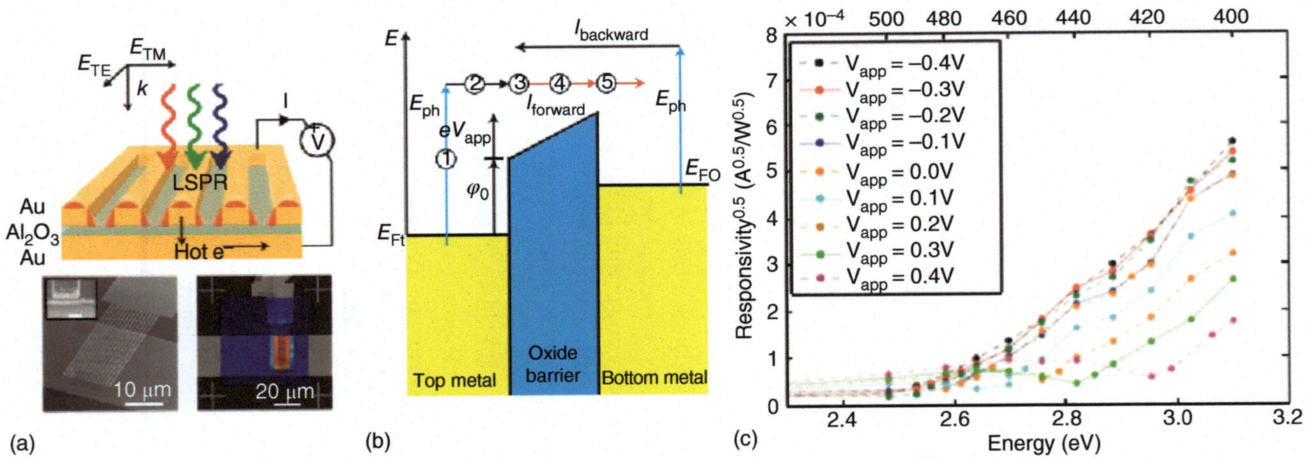

Figure 9.15 (a) Schematic illustration, SEM image and photocurrent image of the constructed device; (b) energy band diagram of the planar metal-oxide-metal structure; (c) responsivity of the device under different applied voltages. Source: (a) Reproduced with permission from Chalabi et al. [63] © 2014 American Chemical Society.

9.4.6 Other Structures for Hot Electron Photodetection

In addition to several common structures reviewed previously, some special nanostructures also be used to construct hot electron photodetectors, such as nanocones, NPs, plasmonic-crystal, nanogratings. In 2017, C.V. Hoang et al. [65] fabricated a hot electron photodetector based on localized and propagating dual-plasmons, which is composed of 100 μm water layer, 25 nm TiO_2 active layer embedded with 4 nm-diameter Au NPs and 200 nm Au film. Figure 9.16a shows the cross-sectional TEM image, the structure diagram and the working principle of the device under the excitation of the LSPs. The Rayleigh scattering is generated when the incident light is irradiated on Au NPs, and leading to the excited SPPs of Au film. If the incident light is at on-resonance wavelength, the LSPs of Au NPs will be excited, where the net photocurrent is dominated by hot electrons degraded by the LSPs. For off-resonance wavelengths, the net photocurrent is only provided by hot electrons degraded by the SPPs of the Au film, which is the working principle for manipulating the photocurrent direction. Meanwhile, the current responsivity and polarity-switching wavelength can also be tailored in different methods, such as NP size, gap, potential, and barrier height. The NP size and gap tuning are demonstrated in Figure 9.16b,c. The shift of LSPR induced by the variation of NP size and gap is the key to the manipulated methods. In addition, hot electron photodetector based on single Si nanocone was built by B. Desiatov et al. [66] in 2015. Figure 9.16d demonstrates the structural diagram of the device. Light is incident and propagates from the back of the Si substrate to the active pixel area, where it is concentrated by the silicon pyramid toward the apex, resulting in high spatial localization and field enhancement for the light field at the apex because of the extremely small tip size of the nanocone at ~50 nm, as shown in Figure 9.16e.

In order to demonstrate the performance of the device, the IV characteristics under different wavelength illumination are measured, as shown in Figure 9.16f. Without light illumination, the IV curve demonstrates a nonlinear, asymmetric rectification behavior. On the contrary, the photogenerated current gradually increases, and the minimum current value drifts with increasing incident photon energy, which can be attributed to the rapid growth of the probability of internal light emission. Also, the dependence of photocurrent on incident light power at the same illumination wavelength is shown in the inset of Figure 9.16f, presenting a linear relationship. The pyramid-shaped structure provides an excellent platform for the accumulation of light field, the generation of hot electrons, and enhancing the photodetection performance of such devices.

Similarly, the grating structures also provide a suitable configuration for the strong coupling between the incident light and the plasmons propagation. In 2013, A. Sobhani et al. [67] manufactured periodic spacing slits in a metal film, forming the nanograting structures with strong resonance absorption, as illustrated by Figure 9.17a. The periodic parameters of the structure, such as the interslit distance (D), the slit width (W) and metal thickness (T), jointly determine the frequency of the resonant absorption. Figure 9.17b shows the SEM image of the Au gating structure, with the structural parameters of $T = 200$ nm, $D = 950$ nm,

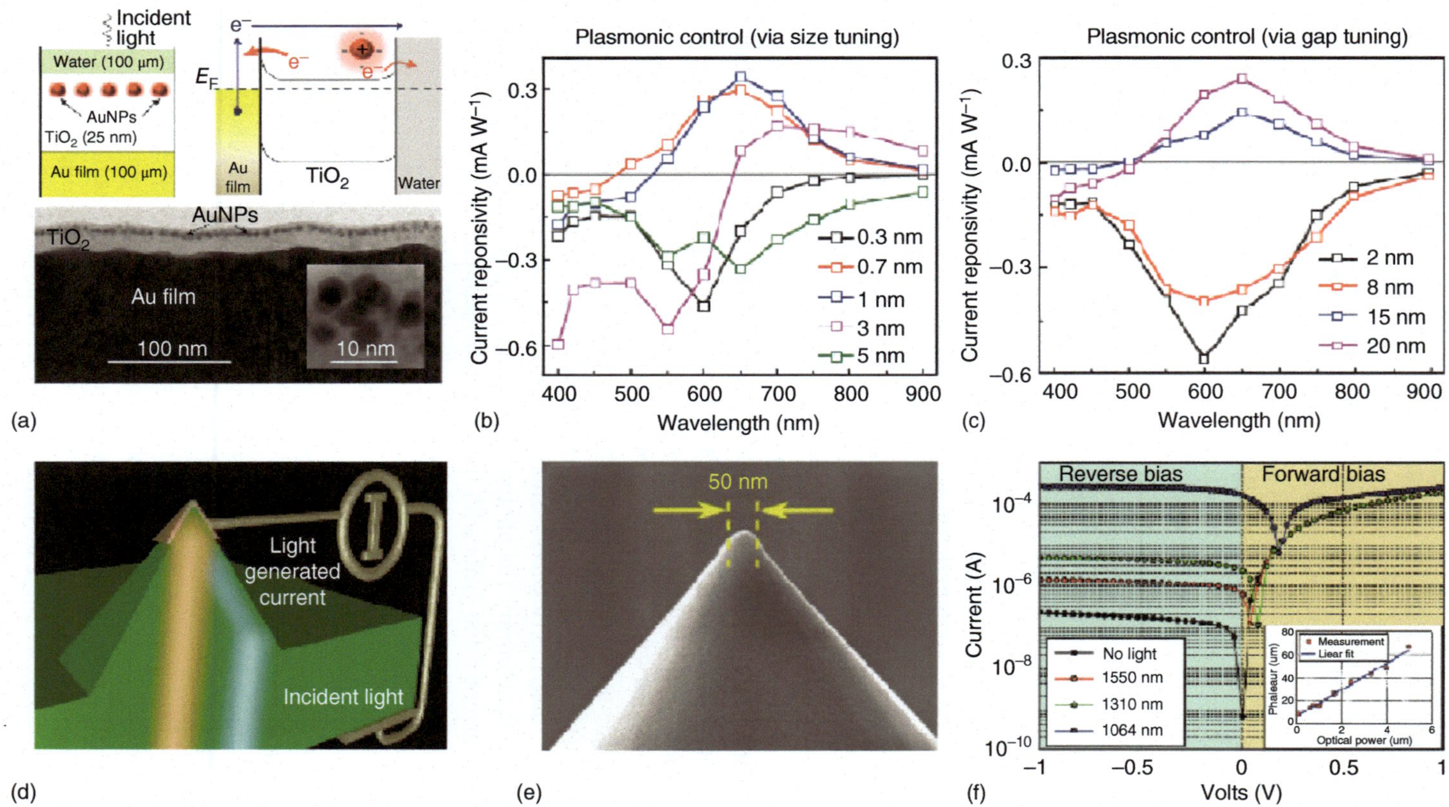

Figure 9.16 (a) Schematic illustration, energy band diagram and cross-sectional TEM image of the device; current responsivity of the devices with different sizes (b) and gaps (c). (d) Schematic diagram of the device; (e) SEM image of the enlarged nanotip; (f) IV characteristics of the pyramid Schottky device at different illuminated wavelengths. Source: (a–c) Reproduced with permission from Hoang et al. [65] © 2017, Springer Nature/CC BY 4.0. (d–f) Reproduced with permission from Desiatov et al. [66] © 2015 Optical Publishing Group.

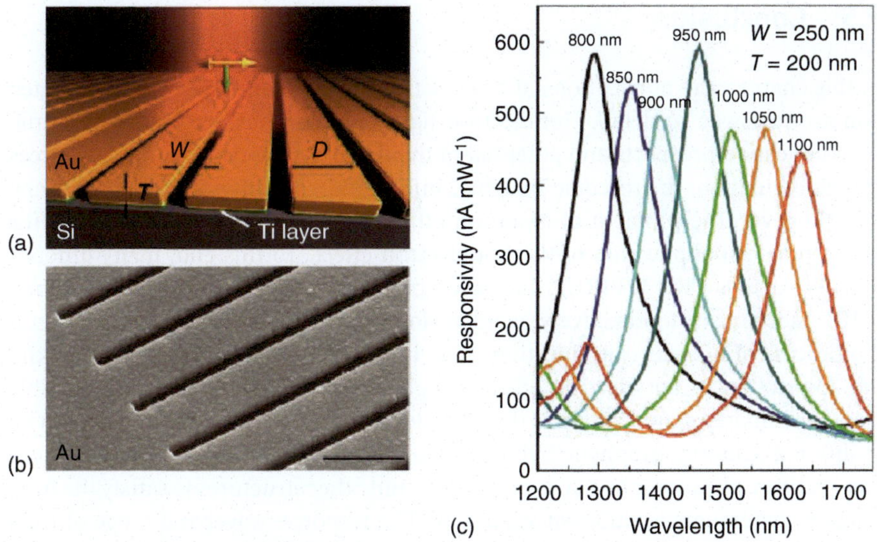

Figure 9.17 Schematic diagram (a) and SEM image (b) of the gold grating; (c) responsivity of grating-based photodetectors with different interslit distances. Source: Reprinted with permission from Sobhani et al. [67]; © 2013 Springer Nature.

$W = 250$ nm. The ultranarrow FWHM of 54 ± 4 meV indicates the strong interaction between light and matter, which is independent of the gating thickness. Meanwhile, the resonant wavelength of the gating can be effectively tuned for a given gating thickness by controlling the interslit distance, as indicated by Figure 9.17c. As D increases from 800 to 1100 nm, the resonant wavelength is adjusted from 1295 to 1635 nm with a stable peak FWHM of 53 ± 6 meV. The relationship of the interslit distance and resonant wavelength presents a linear dependence, which is consistent with the FDTD simulation results, suggesting an outstanding light detection capability in the near-infrared region. In 2015, G. Konstantatos et al. [68] demonstrated a new type of large-scale hot electron photodetector based on plasmonic-crystal structure. The intense color contrast observed by the naked eyes is determined by the different structural period parameters of the device. A cross-sectional SEM image shows that the device structure is composed of the conformal Au, TiO_2 and ITO layers deposited on the imprinted photoresist.

In order to investigate the structure dependence of such photodetectors, devices with different period parameters were constructed, and their photoresponsivity was also measured. The responsivity of devices with different structure periodic parameters shows obvious differences, which are closely related to the resonance modes at different frequencies caused by the corresponding crystal configuration. The predicted responsivity (red solid line) calculated from Fowler's equation is in good agreement with the experimental value in the low energy band, but has obvious divergence in the high energy band, which is related to the increased contribution of interband transition.

9.5 Conclusion

In this chapter, the applications of hot electrons in light emission and light detection are reviewed in detail. Hot electron light emission has emerged with attractive application prospects and potential in the fields of new-type nanolight sources, optical regulation, microscopic imaging, information security, and chemical sensing. However, the problem of quantum efficiency is still an important challenge in the practical application of the hot-electron effect. To this end, many different models, such as PICTT, PIRET, have been proposed on the basis of the traditional PHET mechanism to effectively avoid various competitive energy dissipations in the process of hot electron injection, thus improving quantum yield and making the application of hot electron effect in light emission and modulation possible. Another important application is thermal electron-based photodetectors, involving traditional Schottky, graphene hybrids, planar DBR, metamaterials, optical antennas, gratings, plasmonic crystals, nanocones, and other structures to satisfy the functional requirements in different situations. The response waveband range of these photodetectors is not restricted by the intrinsic physical structure properties of the material, but only related to the interface properties, so that it can effectively expand their functional application and carry out flexible control. The special physical properties of hot electron effect provide a new strategy for building new optoelectronic devices and developing new technologies.

References

1 Noginov, M.A., Zhu, G., Belgrave, A.M. et al. (2009). Demonstration of a spaser-based nanolaser. *Nature* 460 (7259): 1110–1112.
2 Oulton, R.F., Sorger, V.J., Zentgraf, T. et al. (2009). Plasmon lasers at deep subwavelength scale. *Nature* 461: 629–632.
3 Nezhad, M.P., Simic, A., Bondarenko, O. et al. (2010). Room-temperature subwavelength metallo-dielectric lasers. *Nat. Photonics* 4 (6): 395–399.
4 Valentine, J., Zhang, S., Zentgraf, T. et al. (2008). Three-dimensional optical metamaterial with a negative refractive index. *Nature* 455 (7211): 376–379.
5 Yu, N., Genevet, P., Kats, M.A. et al. (2011). Light propagation with phase discontinuities: generalized laws of reflection and refraction. *Science* 334 (6054): 333–337.
6 Fang, N., Lee, H., Sun, C., and Zhang, X. (2005). Sub-diffraction-limited optical imaging with a silver superlens. *Science* 308 (5721): 534–537.
7 Nie, S. and Emory, S.R. (1997). Probing single molecules and single nanoparticles by surface-enhanced Raman scattering. *Science* 275 (5303): 1102–1106.
8 Kneipp, K., Wang, Y., Kneipp, H. et al. (1997). Single molecule detection using surface-enhanced Raman scattering (SERS). *Phys. Rev. Lett.* 78 (9): 1667–1670.
9 Lu, J., Xu, C., Nan, H. et al. (2016). SERS-active ZnO/Ag hybrid WGM microcavity for ultrasensitive dopamine detection. *Appl. Phys. Lett.* 109 (7): 073701.

10 Tagliabue, G., DuChene, J.S., Abdellah, M. et al. (2020). Ultrafast hot-hole injection modifies hot-electron dynamics in Au/p-GaN heterostructures. *Nat. Mater.* 19 (12): 1312–1318.

11 Tagliabue, G., DuChene, J.S., Habib, A. et al. (2020). Hot-hole versus hot-electron transport at Cu/GaN heterojunction interfaces. *ACS Nano* 14 (5): 5788–5797.

12 Li, X., Xiao, D., and Zhang, Z. (2013). Landau damping of quantum plasmons in metal nanostructures. *New J. Phys.* 15 (2): 023011.

13 Brongersma, M.L., Halas, N.J., and Nordlander, P. (2015). Plasmon-induced hot carrier science and technology. *Nat. Nanotechnol.* 10 (1): 25–34.

14 Inouye, H., Tanaka, K., Tanahashi, I., and Hirao, K. (1998). Ultrafast dynamics of nonequilibrium electrons in a gold nanoparticle system. *Phys. Rev. B* 57 (18): 11334–11340.

15 Khurgin, J.B. (2015). How to deal with the loss in plasmonics and metamaterials. *Nat. Nanotechnol.* 10 (1): 2–6.

16 Zhou, L., Tan, Y., Wang, J. et al. (2016). 3D self-assembly of aluminium nanoparticles for plasmon-enhanced solar desalination. *Nat. Photonics* 10 (6): 393–398.

17 Coppens, Z.J., Li, W., Walker, D.G., and Valentine, J.G. (2013). Probing and controlling photothermal heat generation in plasmonic nanostructures. *Nano Lett.* 13 (3): 1023–1028.

18 Clavero, C. (2014). Plasmon-induced hot-electron generation at nanoparticle/metal-oxide interfaces for photovoltaic and photocatalytic devices. *Nat. Photonics* 8 (2): 95–103.

19 Linic, S., Christopher, P., and Ingram, D.B. (2011). Plasmonic-metal nanostructures for efficient conversion of solar to chemical energy. *Nat. Mater.* 10 (12): 911–921.

20 Mukherjee, S., Libisch, F., Large, N. et al. (2013). Hot electrons do the impossible: plasmon-induced dissociation of H_2 on Au. *Nano Lett.* 13 (1): 240–247.

21 Mubeen, S., Lee, J., Singh, N. et al. (2013). An autonomous photosynthetic device in which all charge carriers derive from surface plasmons. *Nat. Nanotechnol.* 8 (4): 247–251.

22 Novotny, L. and van Hulst, N. (2011). Antennas for light. *Nat. Photonics* 5 (2): 83–90.

23 Cui, L., Zhu, Y., Abbasi, M. et al. (2020). Electrically driven hot-carrier generation and above-threshold light emission in plasmonic tunnel junctions. *Nano Lett.* 20 (8): 6067–6075.

24 Berini, P. (2014). Surface plasmon photodetectors and their applications. *Laser Photonics Rev.* 8 (2): 197–220.

25 Lee, Y.K., Jung, C.H., Park, J. et al. (2011). Surface plasmon-driven hot electron flow probed with metal-semiconductor nanodiodes. *Nano Lett.* 11 (10): 4251–4255.

26 Kern, J., Kullock, R., Prangsma, J. et al. (2015). Electrically driven optical antennas. *Nat. Photonics* 9 (9): 582–586.

27 Zhang, C., Hugonin, J.-P., Coutrot, A.-L. et al. (2019). Antenna surface plasmon emission by inelastic tunneling. *Nat. Commun.* 10 (1): 4949.
28 Buret, M., Uskov, A.V., Dellinger, J. et al. (2015). Spontaneous hot-electron light emission from electron-fed optical antennas. *Nano Lett.* 15 (9): 5811–5818.
29 Parzefall, M., Bharadwaj, P., Jain, A. et al. (2015). Antenna-coupled photon emission from hexagonal boron nitride tunnel junctions. *Nat. Nanotechnol.* 10 (12): 1058–1063.
30 Wang, X., Braun, K., Zhang, D. et al. (2015). Enhancement of radiative plasmon decay by hot electron tunneling. *ACS Nano* 9 (8): 8176–8183.
31 Cui, L., Zhu, Y., Nordlander, P. et al. (2021). Thousand-fold increase in plasmonic light emission via combined electronic and optical excitations. *Nano Lett.* 21 (6): 2658–2665.
32 Wang, P., Krasavin, A.V., Nasir, M.E. et al. (2018). Reactive tunnel junctions in electrically driven plasmonic nanorod metamaterials. *Nat. Nanotechnol.* 13 (2): 159–164.
33 Grimm, P., Zeißner, S., Rödel, M. et al. (2022). Color switchable subwavelength organic light-emitting antennas. *Nano Lett.* 22 (3): 1032–1038.
34 Haug, T., Klemm, P., Bange, S., and Lupton, J.M. (2015). Hot-electron intraband luminescence from single hot spots in noble-metal nanoparticle films. *Phys. Rev. Lett.* 115 (6): 067403.
35 Zhang, Y., Han, J., Shi, L. et al. (2021). Extremely polarized and efficient hot electron intraband luminescence from aluminum nanostructures for nonlinear optical encoding. *Laser Photonics Rev.* 15 (1): 2000339.
36 Lin, K.-Q., Yi, J., Hu, S. et al. (2016). Intraband hot-electron photoluminescence from single silver nanorods. *ACS Photonics* 3 (7): 1248–1255.
37 Yorulmaz, M., Khatua, S., Zijlstra, P. et al. (2012). Luminescence quantum yield of single gold nanorods. *Nano Lett.* 12 (8): 4385–4391.
38 Wu, K., Rodríguez-Córdoba, W.E., Yang, Y., and Lian, T. (2013). Plasmon-induced hot electron transfer from the Au tip to CdS rod in CdS-Au nanoheterostructures. *Nano Lett.* 13 (11): 5255–5263.
39 Huang, X., Li, H., Zhang, C. et al. (2019). Efficient plasmon-hot electron conversion in Ag-$CsPbBr_3$ hybrid nanocrystals. *Nat. Commun.* 10 (1): 1163.
40 Wu, K., Chen, J., McBride, J.R., and Lian, T. (2015). Efficient hot-electron transfer by a plasmon-induced interfacial charge-transfer transition. *Science* 349 (6248): 632–635.
41 Knight, M.W., Wang, Y., Urban, A.S. et al. (2013). Embedding plasmonic nanostructure diodes enhances hot electron emission. *Nano Lett.* 13 (4): 1687–1692.
42 Giugni, A., Torre, B., Toma, A. et al. (2013). Hot-electron nanoscopy using adiabatic compression of surface plasmons. *Nat. Nanotechnol.* 8 (11): 845–852.
43 Huang, X., Chen, R., Zhang, C. et al. (2016). Ultrafast and robust UV luminescence from Cu-doped ZnO nanowires mediated by plasmonic hot electrons. *Adv. Opt. Mater.* 4 (6): 960–966.

44 Wang, H., Guo, Y., Hao, H. et al. (2021). Bright CdSe/CdS quantum dot light-emitting diodes with modulated carrier dynamics via the local Kirchhoff law. *ACS Appl. Mater. Interfaces* 13 (47): 56476–56484.

45 Sun, Z.P., Jiang, M.M., Mao, W.Q. et al. (2020, 2020). Nonequilibrium hot-electron-induced wavelength tunable incandescent-type light sources. *Photonics Res.* 8: 91–102.

46 Liu, Y., Jiang, M.M., Tang, K. et al. (2020). Plasmon-enhanced high-performance Si-based light sources by incorporating alloyed Au and Ag nanorods. *CrstEngComm* 22: 6106–6115.

47 Zhou, X.B., Jiang, M.M., Xu, K. et al. (2022). Electrically driven single microwire-based single-mode microlaser. *Light Sci. Appl.* 11: 198.

48 Kang, Y., Najmaei, S., Liu, Z. et al. (2014). Plasmonic hot electron induced structural phase transition in a MoS_2 monolayer. *Adv. Mater.* 26 (37): 6467–6471.

49 Li, Z., Xiao, Y., Gong, Y. et al. (2015). Active light control of the MoS_2 monolayer exciton binding energy. *ACS Nano* 9 (10): 10158–10164.

50 Lin, Y., Ling, X., Yu, L. et al. (2014). Dielectric screening of excitons and trions in single-layer MoS_2. *Nano Lett.* 14 (10): 5569–5576.

51 Chen, Y.-H., Tamming, R.R., Chen, K. et al. (2021). Bandgap control in two-dimensional semiconductors via coherent doping of plasmonic hot electrons. *Nat. Commun.* 12 (1): 4332.

52 Pescaglini, A., Martín, A., Cammi, D. et al. (2014). Hot-electron injection in Au NR-ZnO nanowire hybrid device for near-infrared photodetection. *Nano Lett.* 14 (11): 6202–6209.

53 Gong, T. and Munday, J.N. (2015). Angle-independent hot carrier generation and collection using rransparent conducting oxides. *Nano Lett.* 15 (1): 147–152.

54 Wang, W., Klots, A., Prasai, D. et al. (2015). Hot electron-based near-infrared photodetection using bilayer MoS_2. *Nano Lett.* 15 (11): 7440–7444.

55 Li, W. and Valentine, J. (2014). Metamaterial perfect absorber based hot electron photodetection. *Nano Lett.* 14 (6): 3510–3514.

56 Knight, M.W., Sobhani, H., Nordlander, P., and Halas, N.J. (2011). Photodetection with active optical antennas. *Science* 332 (6030): 702–704.

57 Li, W., Coppens, Z.J., Besteiro, L.V. et al. (2015). Circularly polarized light detection with hot electrons in chiral plasmonic metamaterials. *Nat. Commun.* 6 (1): 8379.

58 Yu, Y., Sun, Y., Hu, Z. et al. (2019). Fast photoelectric conversion in the near-infrared enabled by plasmon-induced hot-electron transfer. *Adv. Mater.* 31 (43): 1903829.

59 Du, B., Lin, L., Liu, W. et al. (2017). Plasmonic hot electron tunneling photodetection in vertical Au-graphene hybrid nanostructures. *Laser Photonics Rev.* 11 (1): 1600148.

60 Sasin, M.E., Seisyan, R.P., Kalitteevski, M.A. et al. (2008). Tamm plasmon polaritons: slow and spatially compact light. *Appl. Phys. Lett.* 92 (25): 251112.

61 Zhang, C., Wu, K., Giannini, V., and Li, X. (2017). Planar hot-electron photodetection with Tamm plasmons. *ACS Nano* 11 (2): 1719–1727.

62 Wang, J., Zhu, Y., Wang, W. et al. (2020). Broadband Tamm plasmon-enhanced planar hot-electron photodetector. *Nanoscale* 12 (47): 23945–23952.

63 Chalabi, H., Schoen, D., and Brongersma, M.L. (2014). Hot-electron photodetection with a plasmonic nanostripe antenna. *Nano Lett.* 14 (3): 1374–1380.

64 Fowler, R.H. (1931). The analysis of photoelectric sensitivity curves for clean metals at various temperatures. *Physiol. Rev.* 38 (1): 45–56.

65 Hoang, C.V., Hayashi, K., Ito, Y. et al. (2017). Interplay of hot electrons from localized and propagating plasmons. *Nat. Commun.* 8 (1): 771.

66 Desiatov, B., Goykhman, I., Mazurski, N. et al. (2015). Plasmonic enhanced silicon pyramids for internal photoemission Schottky detectors in the near-infrared regime. *Optica* 2 (4): 335–338.

67 Sobhani, A., Knight, M.W., Wang, Y. et al. (2013). Narrowband photodetection in the near-infrared with a plasmon-induced hot electron device. *Nat. Commun.* 4 (1): 1643.

68 García de Arquer, F.P., Mihi, A., and Konstantatos, G. (2015). Large-area plasmonic-crystal hot-electron-based photodetectors. *ACS Photonics* 2 (7): 950–957.

10

Applications in Catalysis and Energy

Decay of plasmonic excitations creates highly nonequilibrium hot carriers that can alter the activation barrier or catalyze reactions

10.1 Introduction

The continuous consumption of unsustainable fossil fuels will inevitably lead to energy depletion and serious environmental pollution. While solar energy is a clean and abundant energy source that can be converted to electricity in photovoltaic devices, chemical energy using solar concentrators. The development of technologies and high-efficiency catalyst is the utmost significant scientific challenge to resolve solar energy conversion. The redox processes induced by catalysts as the conversion platform recently showed many promising results as an alternative method that can produce value-added chemicals (e.g. hydrogen, hydrocarbons, and oxygenates) from simple, abundant feedstocks (e.g. water and carbon dioxide) [1, 2]. Despite the extensive effort to create electrocatalytic materials with a high turnover rate, the current outcome still lacks the demonstrative performance for viable industrial applications.

With the rapid development of nanotechnology, nanocatalysis is one of the important applications in harvesting energy from sunlight and creating new clean energy. Today, scientists can prepare nanocatalysts with superior characteristics regarding activity, selectivity, durability, and recoverability. Among the nanocatalysis, plasmonic nanostructures are promising candidates for harvesting solar energy originating from the localized SPR absorption over a wide UV–vis–NIR region [3]. Therefore, they were defined as plasmon-enhanced catalysts.

Since the Middle Ages, spherical Au and Ag nanoparticles have been embedded in the stained glass windows of churches and cathedrals with brilliant colors for their Surface plasmon resonance (SPR) effect in the visible light region [3]. The SPR effect is the collective oscillation of free electrons in metal nanoparticles driven by the electromagnetic field of incident light. Some typical plasmonic metal nanoparticles, such as Au, Ag, and Cu, absorb visible and infrared light wavelength regions. This unique

Plasmonic Metal Nanostructures: Preparation, Characterization, and Applications,
First Edition. Caixia Kan.

characteristic has been of great interest for the capacity to efficiently harvest solar energy. Resonant optical excitation of surface plasmons also produces energetic hot electrons that can be collected to facilitate chemical reactions. Thus, incorporating plasmonic metal nanostructures into photocatalytic systems holds great promise for dramatically improving the efficiency of sunlight absorption and solar energy conversion.

Major progress was recently achieved with the development of plasmon-mediated energy conversion technologies such as plasmon-mediated photochemical reactions, photovoltaics, and chemical synthesis [4, 5]. Various anisotropic Au and Ag nanostructures have been extensively investigated because of their strong and tunable SPR effect in the UV–vis–NIR region. This characteristic allows these nanomaterials to function as the nanoantenna in the semiconductor photocatalytic systems, where they can harvest and convert light energy into electrons that can participate in the catalytic reactions.

The first paper about metal nanoparticles was published in 1943, in which the poly(vinyl) alcohol-protected Pd nanoparticles as hydrogenation catalysts were prepared by reduction of the metal salts [6]. This reminiscent research was promoted by Paul Sabatier [7], who discovered the catalyzed hydrogenation using finely dispersed Ni particles prepared upon the reduction of nickel oxide or hydroxide by H_2. Another great breakthrough in the twentieth century is the oxidation of CO to CO_2 by nanocatalysts. In 1987, Haruta discovered that the catalytic activity of Au nanoparticles for oxidation of CO to CO_2 by O_2 was boosted when the Au nanoparticles were smaller than 5 nm [8]. Haruta's dramatic finding reduced the direction of nanocatalysis research to the smallest nanoparticles. There are exceptions, however, for instance, the plasmonic excitation by visible light irradiation of large Au nanoparticles generates hot electrons that activate semiconductors such as TiO_2 for substrate transformation. In addition, decreasing the nanocatalyst's size down to single atom has attracted much recent attention because, in this case, no atoms located at the interior of the nanoparticle are lost [9]. Moreover, supported single-atom catalysts are better defined than sub-nanometer-sized catalysts that are more or less polydispersed and are, therefore less selective.

Nanocatalysis requires concepts from heterogeneous and homogeneous catalysis, materials science, surface science, bioinorganic chemistry, and organometallic chemistry. Hence, a more fundamental understanding of the catalytic reaction, novel materials, and strategies that can outperform the current state-of-the-art catalysts are required to introduce renewable energy technology at a global scale. Here, some applications of plasmonic nanostructures in electrocatalysis, photocatalysis, and energy about of will be introduced.

10.2 Electrocatalysis

Metallic nanoparticles are key building blocks in nanoscience and nanotechnology due to their unique optical, electronic, and catalytic properties that differ from their bulk counterparts. As far as electrocatalysis is concerned, although bulk single-crystalline gold surfaces are known to exhibit high electrocatalytic performance (e.g. Au(111) toward CO oxidation in alkaline solution), nanoparticulated gold can offer advantages such as much higher surface-to-volume ratios, facile control over crystalline surface exposed using mild low-cost wet chemistry synthesis, and lower material usage by attaching nanoparticles onto carbon or glass electrode surfaces. For instance, Au nanoparticles could exhibit a current density of about three times that of bulk Au in the catalysis toward methanol oxidation. This high mass activity originated from 3D nanoparticle surfaces versus 2D planar bulk metal surfaces. Moreover, nanocatalysts could be readily assembled on rigid, flexible, and nonconductive substrates, which facilitated the custom design of electrocatalytic electrodes. Hence, metal nanoparticles offer a low-cost strategy to electrocatalysts in comparison with the corresponding bulk metal. In addition, control over nanoparticle morphologies indicates the potential of programming tailor-made designer electrocatalysts in the future. Excitingly, recent progress in wet chemistry synthesis has allowed for precise adjustment of the particle size and shape forging the so-called "nanoparticle periodic table," composed of 0D, 1D, 2D, polygonal, branched, and hollow particles. In addition, nanoparticles with low-index and high-index crystalline facets exposed could be controlled, enabling tailoring of the catalytic selectivities. Bimetallic/multimetallic nanocatalysts with core shells or heterojunctions or alloyed structures could be generated at a large-scale, offering novel strategies to design novel multifunctional electrocatalysts.

10.2.1 Key Parameters of Structure Influenced Electrocatalysis

10.2.1.1 Size Effects

It is commonly accepted that size is a primary factor influencing electrocatalytic activity. Note that the smaller the particle size is, the larger the percentage of surface atoms (Figure 10.1). The surface-bonded atoms are not fully coordinated, which can influence molecular absorption. For nanoparticles with diameters of less than 3 nm, there will be more corners and edges, and a relatively higher density of low coordination sites might enhance the interaction between catalysts and the intermediate molecules in electrolyte solutions, which increases the efficiency of the electrocatalytic reaction. Also, the smaller the particle size is, the higher the surface-to-volume ratios and the higher amount of atoms with uncoordinated dangling bonds. Thus, better catalytic activities are expected for smaller particle sizes. However, there is a critical particle size below which electrocatalytic activity can decrease and even disappear due to the transition of molecule-like unusual physical properties such as the metal-to-insulator transition and Coulomb blockade effect [12]. Three to five-nanometer Pt and Pd nanoparticles have been reported to be the most active catalysts.

It was found that larger particle size led to the reduction of onset potential. Moreover, there existed a critical particle size of ~4 nm exhibiting the maximum mass-specific catalytic current density (Figure 10.1) [10]. In 2010, it was reported that Pt nanospheres with a diameter of 3 nm showed the maximum catalytic activity toward MOR [11]. The mechanism was explored by X-ray absorption near-edge

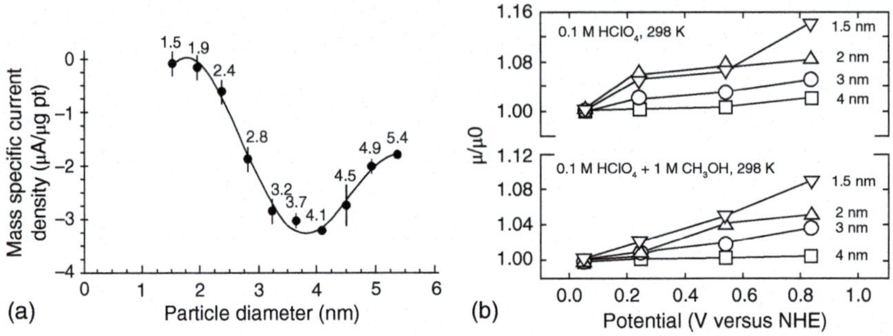

Figure 10.1 Size effects on nanoparticle-based electrocatalysts. (a) Mass-specific current densities for the reduction of oxygen at 0.85 V versus reversible hydrogen electrode (RHE) as a function of particle size in 0.5 M $HClO_4$ (aq.) for Pt particles supported on carbons. Source: Reproduced with permission from Anastasopoulos et al. [10]; © 2013 Wiley-VCH Verlag GmbH & Co. KGaA, Weinheim. (b) Comparison of the change in the absorption peak obtained from the X-ray absorption near-edge structure spectra as a function of potential for the 1.5–4 nm Pt nanocatalysts. Source: Reproduced with permission from Yoo et al. [11]; © 2010 Royal Society of Chemistry.

structure spectra, as shown in Figure 10.1b. On the surface of 3 nm catalysts, the binding of –OH was neither too weak nor too strong, which indicated that the interplay of the surface electronic properties at the metal–solution interface was critical to the catalytic activity toward MOR.

10.2.1.2 Crystalline Facets

The surface crystalline structure influences the activity and selectivity of electrocatalytic reactions because different crystalline facets have different surface energies. For the *fcc* metallic nanoparticles, the surface energy increases in the order of $\gamma_{(111)} < \gamma_{(200)} < \gamma_{(220)} < \gamma_{(hkl)}$ (Figure 10.2a), where (*hkl*) represents high-

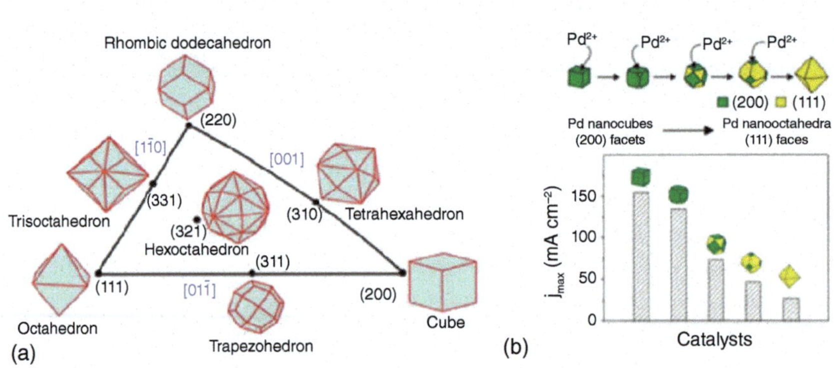

Figure 10.2 Crystalline structure effects on nanoparticle-based electrocatalysts. (a) Unit stereographic triangle of polyhedral nanoparticles with different crystal planes. Source: Reproduced with permission from Zhou et al. [13]; © 2011 Royal Society of Chemistry. (b) Maximum current densities of FAOR over Pd nanoparticles enclosed by (111) and (200) facets in different proportions. Source: Reproduced with permission from Jin et al. [14]; © 2012 Royal Society of Chemistry.

index facets [13]. The high-index (*hkl*) facets possess a high density of atomic steps, edges, and kinks, which can be active sites for breaking chemical bonds. It was reported that high-index Pt tetrahexahedra exhibited an enhanced activity of up to 400% of the low-index ones. Moreover, molecular absorption is highly sensitive to crystalline facets. Hence, catalytic selectivity could be controlled by adjusting the fractions of crystalline facets exposed. It was found that Pt(111)-dominant ultrafine nanowires could substantially boost the oxidation of formic acid and methanol, showing greater specific catalytic activities than the corresponding Pt/C catalyst. In addition, crystalline structures could also affect the catalytic pathways of electron transfer in oxidation reduction reaction (ORR). In a comparison of nanodecahedra enclosed with (111) facets to nanocubes enclosed with (200) facets, it can be seen that the direct four-electron process of ORR occurred in the former one, whereas, the two-step ORR occurred in the latter one.

Encouraging progress in synthetic chemistry renders it possible to control the proportions of crystalline facets exposed to electrolytes, hence, building up a correlation between the surface structure and electrocatalytic activity. For example, Pd polyhedra with different facet portions could be synthesized through a seed-mediated growth method, and the enclosed facets could be continuously transformed from (200) (Pd cubes) to (111) (Pd octahedra). The detailed crystalline structural evolution and their influences on specific current densities are shown in Figure 10.2b, demonstrating that the (200) facets led to maximum catalytic activities [14]. The results were consistent with the ones derived from single-crystal electrodes; the highest oxidation rate was observed on Pt(200) facets with open steps and lower step densities. However, the catalytic peak potential increased as the fraction of the exposed (200) facets increased (Figure 10.2b, inset). For an ideal formic acid oxidation reaction (FAOR) electrocatalyst, the highest current density should be generated at the lowest possible potential. As a compromise, Pd nanocubes with slight truncations at the corners were regarded as the best choice after balancing the two requirements.

It has been demonstrated that nanoparticles with high index facets could exhibit significantly enhanced activities. Compared with low-index facets, high-index facets possess a higher density of low-coordinated atoms on steps, edges, and kinks, which are recognized as catalytically more active sites. Since the first synthesis of tetrahexahedral Pt nanoparticles enclosed by (730), (210), and (520) facets through electrochemical deposition, the preparation approaches have been significantly extended toward the pursuit of shape-controllable nanocrystals at a large scale. For example, Xia and coworkers reported the generation of Pt concave nanocubes bound with (510), (720), and (830) facets through a seed-mediated growth method. These high-index catalysts showed significantly enhanced activity toward ORR than their low-index counterparts [15]. It was claimed that the specific activities of Pt concave nanocubes were 3.1–4.1, 1.9–2.8, and 3.6 times higher than those of Pt cubes, Pt cuboctahedra, and 3.2 nm Pt nanoparticles on Pt/C catalysts, respectively.

10.2.1.3 Composition Effects

Multimetallic nanomaterials allow the researchers to (i) increase the catalytic activity by manipulating the surface facets, (ii) reduce the metal loading by replacing parts of the precious metals with non-precious metals, and (iii) improve the stability by incorporating more inert metals. For bimetallic/multimetallic nanocatalysts, a critical parameter is the combination, controlling the species of the incorporated metal and the ratio of each component. To date, Au, Ni, Ru, Cu, Pd, Co, Ti, and Fe have been successfully incorporated into Pt-based multimetallic structures, which greatly enhanced the durability of the resulting catalysts.

The species of the incorporating metal determines the overall electrocatalytic performance of multimetallic nanoparticles, including the activity, stability, and even cost. For instance, although Pt is considered the best choice of ORR, the aggregation of Pt nanoparticles in an electrochemical environment has been an issue limiting their durability and preventing their commercialization [16]. It has been found that the addition of gold and iron elements to form trimetallic Au–Fe–Pt nanostructures by an organic solvent method exhibited enhanced durability and activity. The core–shell structures were 2.5 times more active than monometallic Pt in the catalyzed reduction of oxygen [17]. The activity enhancement was attributed to Fe alloying, which altered the electronic structure of Pt and further improved the reaction kinetics by controlling the binding energy between the catalyst surface and the key reaction intermediates. The durability of the core–shell catalysts was also superior, and no obvious activity loss was observed after 60 000 potential cycles. Au contributed mainly to the electrochemical stability, which made the formation of subsurface oxides less energetically favorable and thus suppressed the dissolution of Pt. Later, a series of investigations were undertaken to understand the electrocatalysis of multimetallic nanowires. For example, core–shell FePtM/FePt (M = Pd, Au) nanowires were synthesized through a seed-mediated method, and their electrocatalytic activity and stability toward ORR were compared with those of previously prepared bimetallic FePt, FePtPd, and Fe nanowires [18]. The FePtPd/FePt nanowires showed the highest mass activity, which was 12.5 times higher than that for the commercial Pt catalysts.

Moreover, no activity decrease was observed after 5000 potential sweeps. Compared to Au, Pd is less expensive. Moreover, by downshifting the d-band center of Pt, Pd doping also facilitated the absorption, activation, and desorption of O_2.

The ability to fine-tune the elemental ratio of bimetallic/multimetallic nanoparticles contributed greatly to the understanding of composition-dependent electrocatalysis, leading to the rational design of multifunctional designer electrocatalysts. For instance, PtNi bimetallic electrocatalysts with adjustable ratios were synthesized by an organic solvothermal synthesis method, and their compositional profile was investigated by atomic-level microscopy analysis. The electronic properties of the nanosegregated Pt were modified by Ni doping, where less blocking species were absorbed, and thus the electrocatalytic activity was enhanced. As shown in Figure 10.3, PtNi nanoparticles showed the highest mass activity toward ORR compared with Pt_3Ni, $PtNi_2$, and $PtNi_3$, although their Pt loading in PtNi particles was not the highest [19]. The elemental analysis showed that the Pt-skeleton structures in PtNi nanoparticles were responsible for the high activities because they provided the largest specific surface area.

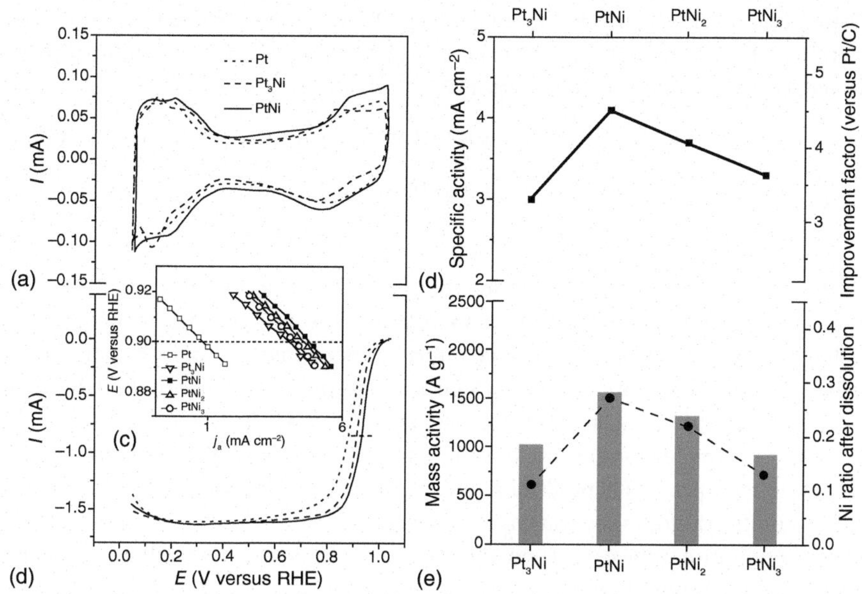

Figure 10.3 Electrochemical characterizations by RDE in 0.1 M $HClO_4$ of Pt_xNi_{1-x} nanoparticle catalysts compared to commercial Pt/carbon: (a) Cyclic voltammetric in argon purged electrolyte, (b) polarization curves for the ORR and (c) corresponding Tafel plots. The half-wave potentials in (b) are marked with a dashed line. (d) Summary of the specific activities for the ORR at 0.9 V and the improvement factors versus the Pt/carbon. (e) The mass activity of PtNi alloys with various ratios. Source: Reproduced with permission from Wang et al. [19]; © 2011 Wiley-VCH.

Moreover, compositional ratio control was also realized on trimetallic catalysts. The alloyed face-centered tetragonal FePtAu nanoparticles with tunable ratios could be synthesized by a solution-phase process. The electrocatalytic activities of $Fe_{42}Pt_{44}Au_{14}$, $Fe_{43}Pt_{37}Au_{20}$, and $Fe_{34}Pt_{34}Au_{32}$ toward FAOR were compared [20]. It was found that $Fe_{34}Pt_{34}Au_{32}$ NPs with the most Au loading showed the highest activity and stability. Au segregation promoted the dehydrogenation of formic acid, which enhanced the reaction efficiency. At the same time, the formation of CO was inhibited by Au. Therefore, the possibility of CO-poisoning was reduced.

10.2.2 Types of Electrocatalysis

10.2.2.1 Oxygen Reduction Reaction

The electrocatalytic ORR is a challenging reaction in metal–air battery and fuel cell technology (for example, polymer electrolyte membrane [PEM] fuel cells) because of its slow rate and high overpotentials. The volume of research, in the context of electrochemical energy technologies, conducted on the ORR over the past several decades has been tremendous. In PEM fuel cells, the catalysts at the anode and cathode typically consist of Pt nanoparticles supported on conducting carbon (for example, carbon black).

Depending on the electrode material, the pathway of the ORR can be either the direct four-electron reduction to H_2O (Eq. (10.1)) or the competitive, indirect pathway via H_2O_2 as an intermediate (Eqs. (10.2) and (10.3)).

$$O_2 + 4H^+ + 4e^- = 2H_2O \quad E_0 = +1.23\,\text{V versus RHE} \tag{10.1}$$

$$O_2 + 2H^+ + 2e^- = H_2O_2 \quad E_0 = +0.695\,\text{V versus RHE} \tag{10.2}$$

$$H_2O_2 + 2H^+ + 2e^- = 2H_2O \quad E_0 = +1.78\,\text{V versus RHE} \tag{10.3}$$

Early isotope measurements revealed that H_2O_2 formation occurs without O—O bond cleavage, whereas H_2O formation requires the O—O bond to break [21]. Much experimental and computational effort has been invested in investigating the detailed ORR mechanism on various electrode materials, revealing that the binding energy of oxygenate intermediates on the catalyst surface may serve as a suitable controlling parameter for ORR selectivity and activity. According to these studies, the two-electron process is predominantly catalyzed on Hg, Au (except Au(200) in alkaline solution), graphite, and most metal oxides. By contrast, the four-electron pathway predominantly occurs on pure Pt as well as Pt-based alloys, resulting in H_2O as the main product.

PEM fuel cell electrodes typically contain nanocatalysts of Pt nanoparticles supported on carbon. In terms of sustainability and cost efficiency, studies on fuel cell catalyst development have focused on the synthesis of new nanostructured electrocatalysts to reduce the Pt content. For example, it is well established that both specific and mass activity can be enhanced by alloying Pt with other transition metals such as Ni, Fe, or Co. Furthermore, it has been shown that an improvement in activity can be achieved by maximizing the exposure of certain facets using shape-controlled nanoparticles.

In recent years, several innovative surface and subsurface alloy structural concepts have been proposed for active bimetallic ORR catalysts. The most popular are Pt-alloy skin and skeleton catalysts, Pt-ML catalysts, and dealloyed Pt core–shell catalysts [17, 22–24]. Dealloyed Pt core–shell structures show higher activity and improved stability in the ORR than other Pt-rich alloys and standard Pt/C catalysts. The most active and stable performance of a Pt alloy cathode catalyst during a realistic PEM fuel cell test was recently reported using a size-controlled dealloyed $PtNi_3$ core–shell nanoparticle catalyst. Furthermore, Wang et al. reported similar, well-defined core–shell nanocatalysts obtained by stepwise Pt-ML depositions on Pd and Pd_3Co nanoparticles with controlled particle sizes (4–5 nm) and enhanced activity for O_2 reduction. Their results suggested that Pt(111) facets are the most conducive on small particles, even more so if they are moderately compressed.

The structure sensitivity of PtNi alloys was investigated by Stamenkovic et al. using a particular Pt-skin structure, which greatly improved ORR activity on the surface of $Pt_3Ni(111)$ facets: ORR activity was ten times higher than that of a Pt(111) surface and 90 times higher than that of commercial Pt/C nanoparticle catalysts a so-called dream electrocatalyst [25]. The key characteristic of such a skin surface is an oscillatory near-surface compositional segregation of Pt and Ni across the outermost three layers of the (111) surface, that is, Pt is segregated at the outermost surface, and Ni

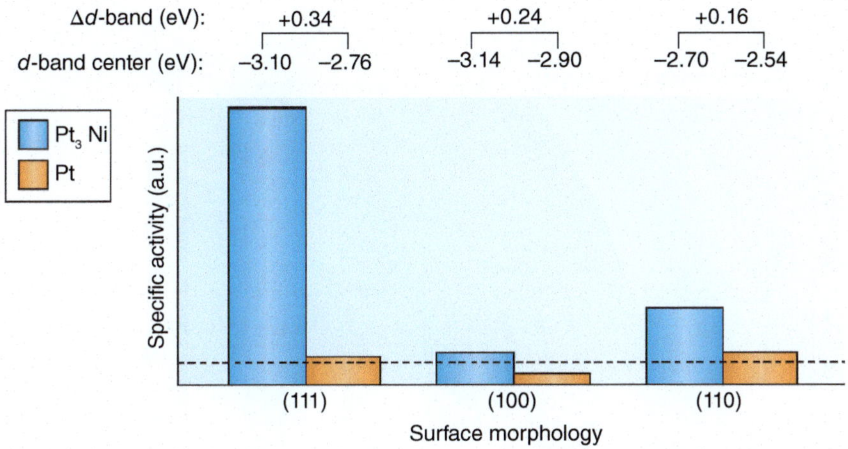

Figure 10.4 Influence of surface morphology and electronic surface properties on the kinetics of the ORR. Rotating ring disk electrode measurements for the ORR on Pt_3Ni (*hkl*) surfaces compared with the corresponding Pt(*hkl*) surfaces are shown. Source: Adapted from Stamenkovic et al. [25].

is segregated at the second layer. The exceptional catalytic activity is related to the combined effects of the atomic structure of the surface (that is, the compact (111) surface) and the composition near the surface (there is high Ni content in the second subsurface layer), which induces a significant downward shift of the d-band center of the outermost Pt surface, thus leading to weakened adsorption of oxygenated intermediate species and high ORR activities. Furthermore, Stamenkovic et al. investigated the synergy between the surface geometry and the surface electronic structure of Pt_3Ni (*hkl*) for the ORR. The order of catalytic activity in the ORR for the different low-index surfaces can be found in Figure 10.4.

Inspired by the exceptional ORR activity reported on an extended Pt-skin (111) surface, different studies have investigated approaches for shaped nanoparticles with well-defined facets for the ORR. The impressive performance evolution of shaped PtNi nanoparticles is summarized in Figure 10.5 [26]. For example, the groups of Zou and Yang synthesized octahedral Pt_3Ni nanoparticles based on surfactant-directed reduction in organic solution using tungsten carbonyl or CO as a shape-directing agent [27, 28]. These particles showed a fourfold increase in specific activity compared with pure Pt/C electrocatalysts. Furthermore, a record activity (17 times higher mass activity and 51 times higher specific activity) was reported by Choi et al., who used a slightly modified method to obtain particles with less residual surfactant on the surface [29]. A different surfactant-free approach for obtaining shaped Pt-alloy nanoparticles was presented by Carpenter et al., who used DMF as both the solvent and reductant under solvothermal conditions. The obtained octahedral PtNi nanoparticles with diameters in the range of 12–15 nm exhibited a mass activity six times higher and a specific activity ten times higher than that of Pt/C. Through modification of the synthesis approach by milder heating, octahedral PtNi nanocatalysts with a size of 9 nm were achieved, showing an ORR mass activity

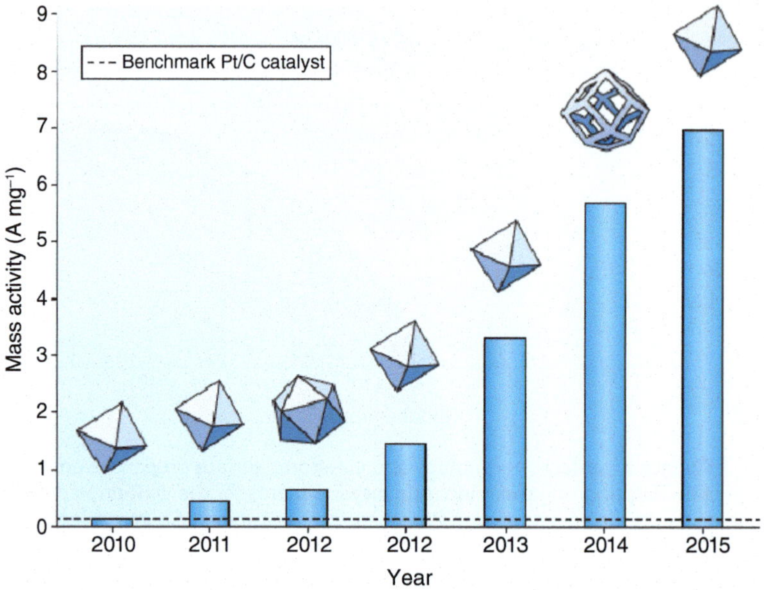

Figure 10.5 Performance evolution of shaped PtNi nanoparticles. Source: Adapted from Strasser [26].

ten times higher than that of Pt/C by tuning the Pt:Ni composition via the reaction time [30, 31]. An isotropic growth mechanism for both metals was proposed, in which the formation of a Pt-rich phase in hexapod-like concave nanocrystals was followed by the deposition of Ni into concave surfaces, resulting in PtNi octahedra with Pt-rich frames and Ni-rich (111) facets.

Chen et al. presented nickel-depleted PtNi nanoframes that are even more active than the previously reported octahedral electrocatalysts, with a mass activity 36 times higher and a specific activity 22 times higher than that of Pt/C [32]. They developed a Pt-skin structure from $PtNi_3$ polyhedra, which were converted to Pt_3Ni nanoframes by etching in nonpolar solvents, followed by dispersion of the particles on carbon and subsequent thermal treatment (370–400 °C in argon). Recently, Huang et al. reported the record-breaking activity of octahedral PtNi nanoparticles doped with transition metals (such as V, Cr, Mn, Fe, Co, Mo, W, and Re) [33]. From this selection of the transition metals, Mo-doped PtNi nano-octahedra (with an edge length of 8 nm) showed great enhancements in specific and mass activity in the ORR, respectively, compared with the commercial Pt/C catalysts. Based on computational data, the origin of this exceptional activity was suggested to be Mo atoms on surface sites near the vertices of the nano-octahedra, which modify the binding strength of reactive intermediates while forming stable Mo—Pt and Mo—Ni bonds.

In summary, the most promising emerging materials concepts for active and selective PEM fuel cell catalysts are dealloyed Pt–Ni core–shell nanocatalysts as well as Pt-based nanoparticles with well-defined geometric shapes. Current studies give testament to the high level of control and craftsmanship that is needed for the chemical preparation of bimetallic nanoparticles with well-defined shapes to achieve a high activity in the ORR. However, concrete evidence of sufficient long-term morphological and performance stability under realistic fuel cell conditions is still lacking, making dealloyed cuboctahedral $PtNi_3$ core–shell nanocatalysts the initial size of which is carefully controlled the most stable and viable alloy catalysts for PEM fuel cells to date.

10.2.2.2 Carbon Dioxide Reduction

The electrocatalytic conversion of CO_2 is a promising means to convert CO_2 into valuable products such as CO, formate, CH_4, C_2H_4, and alcohol. Compared with other conversion methods, the electrocatalytic reduction of CO_2 is favorable because the reaction occurs under ambient temperature and pressure and in neutral pH. Moreover, because the electrocatalytic splitting of H_2O is intrinsically coupled to the electrocatalytic reduction of CO_2, the latter does not require H_2 as a feedstock. However, CO_2 electroreduction is not yet industrially viable owing to its high overpotential (about 1.0 V) and low selectivity for desirable products compared with the evolution of H_2 (namely, the hydrogen evolution reaction [HER]).

Among all pure-metal catalysts, Cu is unique in that it can selectively produce hydrocarbons. Recently, it has been shown that various metals beyond Cu, as well as solid nonmetallic carbon-based catalysts, can also produce hydrocarbons, although at a very low rate. The differing selectivity of these metals appears to be due to the binding energies of intermediates from the CO_2 electroreduction reaction on different metal surfaces, with Cu having the optimal electronic structure for the rate-limiting CO protonation step. In addition, Cu has optimal binding energy for –COOH and CO intermediates, such that the Cu surface can stabilize –COOH without being poisoned by CO. By contrast, metals such as Pt, Ni, Fe, and Co are suitable HER catalysts and may be poisoned at the surface by adsorbed CO*, which explains their low hydrocarbon selectivity.

Studies on single crystals have clearly shown that both the reactivity and selectivity of Cu during CO_2 electrochemical reduction are dependent on the crystal facets. The ratio between CH_4 and C_2H_4 depends strongly on the crystal orientation, with Cu(111) favoring the formation of CH_4 and Cu(200) favoring the formation of C_2H_4. Steps in the crystal surface may improve activity even further by increasing the binding strength of reactants. Indeed, Hori et al. showed that steps in the (200) surface increased C_2H_4 selectivity, with an optimal selectivity of 58.9% on Cu(711) surfaces [34]. Both experimental and theoretical studies have been carried out to unravel the mechanism behind the selectivity change on (111) and (200) surfaces. It was found that Cu(200) surfaces facilitate the formation of C_2H_4 through CO dimerization, whereas Cu(111) surfaces favor CO protonation, which is a key step toward CH_4 formation, although C_2H_4 can also be formed via this pathway.

In a recent study of size-controlled nanoparticles synthesized through reverse micelle encapsulation, it was determined that the activity toward CO_2 electroreduction over Cu nanoparticles greatly increased for nanoparticles below 5 nm in size owing to the low-coordinated sites on the surface [35]. This activity increase originated from increased H_2 evolution and CO formation over smaller nanoparticles. In addition, it was found that the selectivity for hydrocarbons was suppressed, compared with bulk Cu, with decreasing nanoparticle size. By comparison, larger Cu nanoparticles have shown much higher selectivity for hydrocarbons. Some examples include glassy carbon-supported Cu nanoparticles (7 nm) and Cu nanoparticles (12 nm, 40% by weight) supported on Vulcan carbon [36, 37]. To understand size-dependent trends, it is also essential to consider changes in the size and shape of nanoparticles during the reaction. For example, the 7-nm nanoparticles grow to 23 nm during the reaction, which may explain their different selectivity from the smaller nanoparticles.

The nanoparticle shape may also have an important role in the CO_2 reduction reaction, although this parameter has not been much explored. Roberts et al. recently investigated Cu nanocubes grown by successive oxidation–reduction cycles on polycrystalline Cu in the presence of KCl, and found an improvement in C_2H_4 formation. Chen et al. synthesized Cu nanocubes using a similar method and found enhanced C_2H_4 selectivity and high stability for CO_2 electroreduction [38]. The mechanisms behind the improved activity of these structures may involve the (200) facets on the cubes, which are known to favor C_2H_4 selectivity, the nanostructured Cu cube surface, or even Cl^- or Cu^{2+} species that may remain on the catalyst during the reaction.

Other metal nanoparticles have also shown size dependence in CO_2 electroreduction activity and selectivity. For size-controlled Au nanoparticles, it was found that the activity increased significantly below 5 nm in size, primarily because of the increase in low-coordinated sites on the nanoparticle surface (Figure 10.6a). Smaller nanoparticles were found to be selective for H_2 owing to weaker COOH* binding on low-coordinated sites under high H* coverage conditions. Similarly, Zhu et al. studied 4–10-nm Au nanoparticles and nanowires with lengths of 15–500 nm [39]. They reported that larger nanoparticles and longer nanowires were more selective for CO,

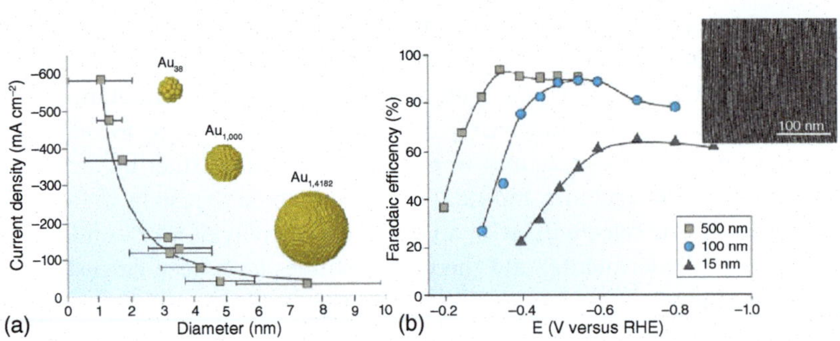

Figure 10.6 Nanostructure effects on activity and selectivity of Au catalysts for CO_2 electroreduction. (a) Nanoparticle size-dependent current density during CO_2 electroreduction over micelle-synthesized Au nanoparticles supported on glassy carbon at −1.2 V versus RHE. (b) Faradaic efficiency for CO_2 electroreduction to CO for Au nanowires 500, 100, and 15 nm in length supported on carbon. TEM image of the 500 nm-long Au nanowires is also shown inset of (b). Source: Reproduced with permission from Zhu et al. [39]; © 2014 American Chemical Society.

which they explained in terms of the greater number of edge sites that favor CO_2 reduction compared with corner sites. The Faradaic efficiency of CO production for nanowires of different lengths is shown in Figure 10.6b.

A common strategy to improve the catalytic activity of polycrystalline materials is to fabricate rough surfaces with high geometric current densities. These rough surfaces can be achieved by dealloying a less noble material from an alloy precursor. Lu et al. have used this method to produce a nanoporous Ag catalyst with a large surface area and highly active sites for the reduction of CO_2, which exhibited a 92% Faradaic efficiency of CO production [40]. Other nanoporous materials have also shown high CO_2 reduction efficiency. For example, Cu nanofoams showed improved formate production and suppression of CO, CH_4, and C_2H_4 compared with a smooth polycrystalline Cu foil, probably because of the suppression of the pathway toward the CO* intermediate.

Interestingly, these nanofoams showed small Faradaic efficiencies for C_2H_6 and C_3H_6 production. In addition, tuning the composition of porous nanostructured AuCu alloy catalysts resulted in 28% Faradaic efficiency toward methanol and ethanol at a low overpotential. The unique C_2, C_3, and alcohol selectivity of these catalysts could result from the increased residence time of CO_2 reduction intermediates in the porous structure.

10.2.2.3 Ethanol Oxidation

Ethanol electrooxidation is a promising reaction for energy production in fuel cells, but similar to CO_2 reduction, and it is currently hindered by the poor selectivity exhibited in common electrocatalyst materials. Instead of complete oxidation to CO_2, which requires the breaking of the C—C bond either in ethanol or the acetaldehyde intermediate, many catalysts produce only acetaldehyde and acetic acid, the latter of which cannot be further oxidized to CO_2. In addition, ethanol oxidation catalysts commonly have a high Pt content, making them expensive and unfeasible to scale up. Furthermore, the poisoning of active sites by strongly adsorbed CO and CH_x intermediate species can hinder the activity of Pt catalysts for this reaction. As we detail in this section, the results from recent studies of structure-dependent selectivity in ethanol oxidation suggest that nanostructured and bimetallic materials may be the key to optimizing the selectivity for CO_2.

Alloying Pt with other metals to create binary or ternary catalysts is a well-established strategy to enhance the catalytic performance for ethanol electrooxidation. Although a range of binary and ternary nanocatalysts are being explored for this reaction, we focus on studies of model catalysts that have involved various measurement techniques to elucidate the mechanism behind the activity and selectivity of these materials. One of the main challenges with Pt ethanol oxidation catalysts is the poisoning of the surface by the accumulation of adsorbed CO. Alloying with Sn can improve the activity of Pt via a dual mechanism whereby Sn stabilizes OH, which can further oxidize CO. The structure and oxidation state of these catalysts is also important because SnO_2 enhances C—C cleavage, unlike alloyed Sn. Alternatively, the activity of Pt-alloy catalysts can be considerably enhanced by replacing the Sn component with Ni, which has a superior O-donating effect.

The Pt–Rh–Sn system is one of the most promising ternary catalysts studied for ethanol electrooxidation, because it can catalyze total ethanol oxidation with high efficiency at potentials below 0.5 V versus RHE [41]. It was recently shown that $PtRhSnO_2/C$ catalysts have high activity and selectivity for CO_2 owing to the electronic interactions between the metals. Because Pt withdraws electrons from Rh, the interaction between Rh and ethanol becomes strong enough to break the C—C bond. Conversely, SnO_2 contributes to the catalytic performance by strongly adsorbing H_2O, thereby minimizing the formation of activity-blocking PtOH and RhOH. The structure of the Pt–Rh–Sn system may be crucial for its activity because specific active-site ensembles may be necessary to facilitate the reaction. Adding Ir to Pt–SnO_2 catalysts can also cleave C—C bonds and lower the overpotential of ethanol oxidation.

Another mechanism that may affect bimetallic ethanol oxidation catalysts involves geometric effects due to lattice-mismatch-induced strain. This has a major impact on the activity of epitaxial Pt MLs on various single crystals. Surfaces such as Au(111), which exert tensile strain on Pt, are more active for ethanol and methanol oxidation. However, these surfaces are only selective for the partial oxidation products acetaldehyde and acetic acid [42]. By contrast, surfaces that induce compressive strain on the Pt ML, such as Pd(111), Ir(111), Rh(111), and Ru(0001), suppress ethanol oxidation activity. Pt MLs deposited on Au nanoparticles, tested for ethanol electrooxidation by Loukrakpam et al., were found to have higher activity and higher efficiency for CO_2 reduction than Pt sub-ML/Au nanoparticles. This result was rationalized regarding these catalysts having the optimal balance between tensile strain induced by lattice mismatch with the Au support and compressive strain induced by finite-size effects. In summary, ethanol electrooxidation is highly sensitive to the structure and composition of nanocatalysts.

Kan and coworkers have fabricated Au nanoframe@Pd array hollow nanostructures to promote electrocatalytic performance (Figure 10.7) [43]. The Au nanoframe@Pd arrays exhibit an enhanced electrocatalytic performance toward ethanol electrooxidation due to an abundance of catalytic active sites. The ethanol electrooxidation in an alkaline environment was carried out to test the electrocatalytic performance of Au NF@Pd arrays. In comparison with Au NBP@Pd arrays and Au NF@Pd nanostructures, the combination of Au NF and Pd array makes our Au NF@Pd arrays 13.7 times higher specific activity and 4.1 times higher mass activity than the commercial Pd/C electrocatalyst. Meanwhile, the Au NF@Pd arrays also showed better stability than the other three electrocatalysts. Our work provides a versatile strategy for constructing Au–Pd hollow nanostructures with super-large particular surface area and its superiority in EOR, which can be extended to the other metal electrocatalysts and promote the development of electrocatalysts.

10.2.3 Key Performance Indicator in Electrocatalysis

Electrochemistry has become an important topic for nanomaterials because of applications in energy storage and conversion. In the electrolytic water and fuel

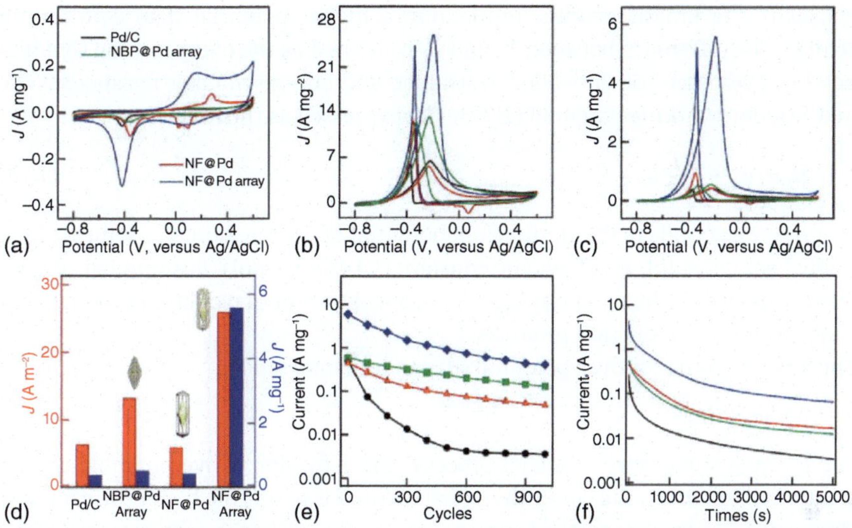

Figure 10.7 CV curves of commercial Pd/C, Au NBP@Pd array, Au NF@Pd and Au NF@Pd array in a N_2-saturated (a) aqueous KOH and (b, c) mixture of ethanol and KOH, respectively. The currents shown in (a, c) and (b) are normalized by the Pd mass loaded and the ECSA values, respectively. (d) Specific and mass activities towards EOR. (e) Cycling measurements and (f) CA curves using commercial Pd/C, Au NBP@Pd array, Au NF@Pd and Au NF@Pd array as electrocatalysts. The scan rate was fixed at 50 mV s^{-1}. Source: Reproduced with permission from Yun et al. [43]; © 2022 Royal Society of Chemistry.

cells, important reactions, such as OER, ORR, HER, and HOR, require highly active catalysts. The establishment of reasonable activity measurement and correct evaluation of catalytic properties of catalysts are of great significance for finding the best catalyst. During the electrocatalysis process, the adsorption/desorption of reactants and products essentially takes place on the surface or in the vicinity. Some measurement parameters, such as over-potential, specific activity, mass activity, and Tafel-slope, are important parameters for catalytic activity [44–47].

10.2.3.1 Over-Potential

In electrochemistry, over-potential is the potential difference (voltage) between a half-reaction's thermodynamically determined reduction potential and the potential at which the redox event is experimentally observed. In an electrolytic cell, the existence of over-potential implies the cell requires more energy than thermodynamically expected to drive a reaction. In a galvanic cell, the existence of over-potential means less energy is recovered than thermodynamics predicts. In each case, the extra/missing energy is lost as heat. The quantity of over-potential is specific to each cell design and varies across cells and operational conditions. Over-potential is experimentally determined by measuring the potential at which a given current density (typically small) is achieved.

Potential in the equilibrium state is the theoretical operating potential (E_t) required for electrocatalytic or photocatalytic reactions under ideal conditions. In actual reactions, however, the working potential (E_i) needs to overcome the obstacles of the kinetic process and exhibit a higher potential than equilibrium potential. The difference between E_t and E_i is defined as the overpotential that was needed to overcome the activation resistance and charge transfer resistance. The actual reaction working potential (E_i) can be expressed as the following:

$$E_i = E_0 + \frac{RT}{nF} \ln \frac{C_o}{EC_R} \tag{10.4}$$

where E_0 corresponds to the equilibrium potential of the reaction, T is temperature, R is ideal gas constant, F is Faradaic constant (96 485 C/mol), n is number of electrons transferred in the reaction, C_O is the concentration of oxidation products, C_R is concentration of reduction products.

The basic formula for over-potential (η) can be expressed as:

$$\eta = E_i - E_t \tag{10.5}$$

For a simple one-electron transfer process, this definition of overpotential is clear and rational. In theory, the closer the over-potential is to 0 V, the better the catalyst's performance. Under the specified current density, the lower the over-potential of the catalyst, the stronger its catalytic ability for the target reaction. The catalyst's performance is usually evaluated under the condition of a current density of 10 mA cm^{-2}. Therefore, it is important to specify the specific current density when comparing the over-potential of different catalysts.

10.2.3.2 Electrochemical Surface Area

Electrochemical surface area (ECSA) is the actual surface area of a unit mass catalyst participating in electrochemical catalytic reactions. The accurate estimation of the ECSA is exceptionally critical. Known and mature ECSA characterization methods for Pt/C catalysts have been established and have been widely available for many decades. However, there are issues remaining for ECSA characterization of Pt-alloy nanocatalysts due to the significant difference in properties, such as composition, particle size, and shape, compared with Pt nanocatalysts. In literature, the definition of the ECSA may vary depending on the conditions. Therefore, the ECSA characterization approaches for Pt/C catalysts cannot be transplanted to Pt-alloy nanocatalysts since the characterization accuracy will no longer meet the requirement.

ECSA cannot only be used as an indicator to measure the number of active sites of a catalyst but also as an important parameter (such as a specific activity) in other indicators to evaluate catalyst activity. On the other hand, ECSA is also an important parameter in the basic research of catalysts, such as the classic Butler–Volmer equation describing the kinetic electron transfer in the redox process for electrodes [48].

$$j_0 = j_0^0 \alpha_c L_c \left(\frac{P_r}{P_r^0} \right)^{\gamma} \exp\left[-\frac{E_c}{RT} \left(1 - \frac{T}{T_0} \right) \right] \tag{10.6}$$

where j_0 is exchange current density, j_0^0 is the reference exchange current density, α_c presents the ECSA, L_c is the loaded catalyst, P_r is the cathode side gas pressure, P_r^0 is reference pressure, γ is ORR dynamic coefficient, E_c is ORR activation energy barrier, R is ideal gas constant, T and T_0 are working and reference temperatures.

Usually, characterization methods of ECSA for Pt-alloy catalysts involve physical and electrochemical methods. The physical method covers XRD and TEM, by which the particle sizes can be calculated or observed directly. Physical methods are conducted in non *in situ* working environments, assuming all particle surfaces have electrocatalytic activity. This ECSA is geometric-specific surface area with significant differences compared to real catalysts. While electrochemical methods involve hydrogen underpotential deposition (H-upd), carbon monoxide stripping (CO-stripping), and underpotential deposition of metals. Among the electrochemical methods, H-upd and CO-stripping were mature in accurately measuring the ECSA of Pt/C catalyst. However, the actual ECSA for an alloyed catalyst cannot be calculated by adding the respective metal surface area since the contribution of different metal components to the catalytic activity is different.

At present, accurately measuring the ECSA of the electrode materials is very challenging because the reaction process is related to many factors, such as electrode, catalyst shape, electrolyte reaction, surface adsorption, and non-Faraday reaction. In the H-upd CO-stripping electrochemical methods, the active area of alloyed catalysts is usually underestimated, leading to overestimating the specific activity of alloyed catalysts compared with Pt/C catalysts. Therefore, it is essential to develop an innovative electrochemical method with high accuracy.

10.2.3.3 Tafel-Slope

It is preferable to define a current range relative to the limiting steady-state value instead of determining a suitable potential range for meaningful Tafel analysis giving either a transfer coefficient or, equivalently, a "Tafel-slope." In a Tafel plot, the log of the current, $\log_{10}|i|$, is plotted against the applied potential, E. If we assume the reaction is fully irreversible and well described by the Tafel equation.

$$i = FAK^{0}_{\frac{A}{B}} \exp\left(\frac{\alpha F\eta}{RT}\right) C_{\mathrm{A}} \tag{10.7}$$

It can be seen for a one-electron transfer process that such a semilog plot will have a slope that equals to $-\alpha F/(RT \times \ln 10)$ and an intercept (at zero overpotential, f) equal to $\log_{10}(\mathrm{FAk_0C_A})$. Originally electrocatalysis experiments were performed using a form of current-interrupt technique known as the commutator method. A desired current density was driven to occur on a working electrode. The cell was subsequently disconnected, and the rapidly changing cell potential was measured against a reference cell. Measurement of the potential of the disconnected cell as a function of the disconnection time allowed the original cell potential to be inferred by extrapolation. Hence in reporting data in terms of current/potential graphs on the y-axis, the potential (the measured variable) was plotted against the current (the controlled variable).

A linear Tafel plot of log10$|i|$ versus E requires that, on the basis of Eq. (10.7), the process is fully irreversible and that the surface concentrations of the electroactive species remain constant throughout the potential range of analysis. In practice, the replenishment of the active species via diffusion is slow, and distortions arise from mass-transport limitations. Consequently, it is necessary to "correct" for

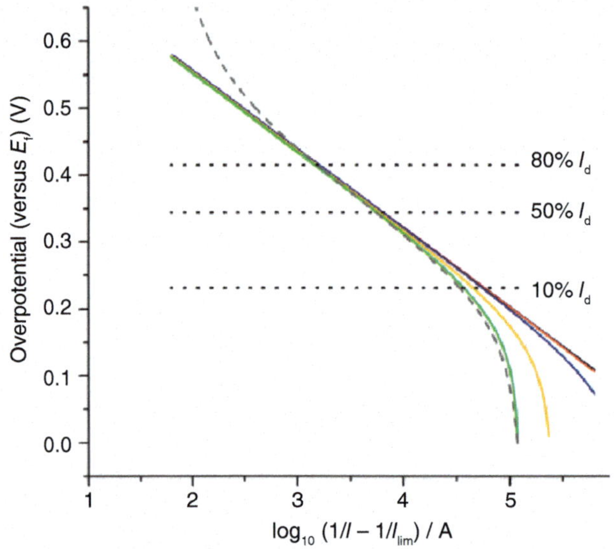

Figure 10.8 Simulated mass-transport corrected Tafel plots for a rotating disk electrode. Solid lines represent the electrode with constant capacitances with variable roughness factors R_f. $R_f = 0$ (black), $R_f = 1$ (red), $R_f = 10$ (blue), $R_f = 50$ (yellow), and $R_f = 100$ (green). The dashed line stands for the electrode with a variable capacitative current (gray dashed). Source: Reproduced with permission from Mayrhofer et al. [49]; © 2008 Elsevier B.V. or its licensors or contributors.

these changes in the redox active species at the electrode surface. Here, Figure 10.8 presents an example mass-transport corrected Tafel plot for a simulated RDE experiment. It is assumed that the total current can be expressed simply as the sum of the Faradaic and non-Faradaic contributions ($i_{tot} = i_{farad} + i_{cap}$). For the solid lines in Figure 10.8, the contribution of the capacitance is assumed to be a constant. Alternatively, the dotted line assumes that the capacitance of the electrode varies linearly as a function of the applied potential. It also gives a clear indication as to how sensitive the data will be to the quality of this correction. It is on this basis that Mayrhofer et al. previously proposed that the kinetically useful part of a rotating disk electrode cyclovoltammogram is the mass-transport corrected current between 10% and 80% of the steady-state value, and only the 10–50% current regime is useable [49].

10.3 Photocatalysis

Solar energy is an abundant energy source that can be converted to electricity in photovoltaic devices and chemical energy by photocatalysts. The efficiency of solar conversion in photocatalytic devices depends on four processes: light absorption, charge separation, charge migration, and charge recombination. Semiconductors (or combined with Pt family metal) used in photocatalytic devices have low solar-energy-conversion efficiency due to limitations in one or more of these four processes. The rapid development of plasmonic nanostructures responding to a broad vis–NIR light has offered a new opportunity to overcome the limited efficiency of photocatalysts. It is found that incorporating plasmonic metal nanostructures into semiconductors can enhance photocatalytic activity through extending light absorption to longer wavelengths, increasing light scattering, and exciting electron-hole pairs in the semiconductor by transferring the plasmonic energy from the metal to the semiconductor. In this part, the plasmon-induced hot-electron and the latest photocatalytic applications of these morphologies are presented. Some of the challenges for the development of metal nanoparticle photocatalysts are also provided.

10.3.1 Plasmon-Induced Hot-Electron Transfer

During the excitation and decay process of SPR, the re-emissive resonant photons, charge carriers, and heat can be used as driving forces for the excitation of adsorbates and some chemical reactions through different mechanisms [50]. The radiative decay process of SPR enables the intramolecular excitation of molecules whose energy gap between the highest occupied molecular orbital (HOMO) and lowest unoccupied molecular orbital (LUMO) matches the energy of the re-emissive photons from plasmonic materials (Figure 10.9a). In fact, this excitation mechanism is quite similar to the electromagnetic enhancement mechanism in SERS, another important application of LSPR. When molecules are adsorbed on a plasmonic nanostructure surface, the energy gap between HOMO and LUMO (or the CB and VB positions for semiconductors) is smaller than those in the gas and liquid phase, thanks to the chemical interaction between adsorbates and plasmonic nanostructures. Then, the energy of the intense re-radiated photons is obtained by the adsorbed molecules via the vibronic energy exchange resulting from photon absorption.

Another critical excitation mechanism is the charge transfer mechanism, including direct and indirect charge transfer from the plasmonic metal surface to the adsorbates. In the indirect charge transfer excitation process, hot electron–hole

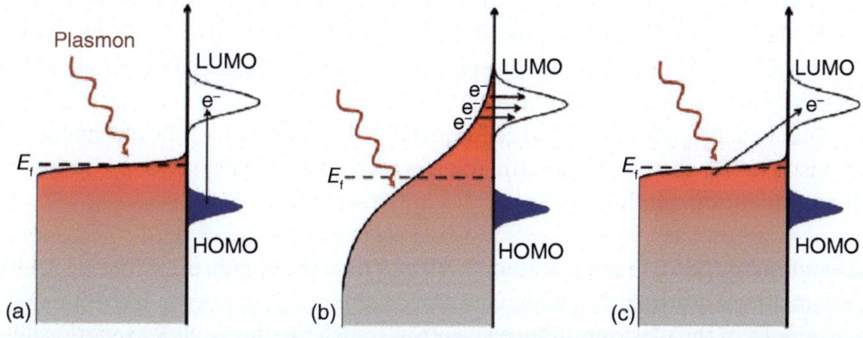

Figure 10.9 Excitation and decay mechanisms of LSPR on the metal surface: (a) Plasmon-induced direct intramolecular excitation mechanism, (b) indirect charge transfer mechanism, and (c) direct charge transfer mechanism. Source: Reproduced with permission from Li et al. [50]; © 2020 Wiley-VCH.

pairs are first generated in the metal nanostructures with nonthermal distribution. After rapid thermalization to Fermi-Dirac distribution, the electrons with higher energy can be transferred into the LUMO of the adsorbates (Figure 10.9b). Classical plasmon-driven photoreactions, e.g. the dissociation of H_2 and O_2, are driven by this mechanism. Notably, the relaxation of the initially nonthermal hot carriers will dramatically reduce the number of qualified hot electrons. Comparatively, in the direct charge transfer process, also known as chemical interface damping, hot electrons are directly injected into the LUMO of the adsorbates, leaving the hot holes at the metal surface (Figure 10.9c). The hot electrons relax to thermal distribution but deposit energy in the adsorbed molecules. It must be pointed out that the direct charge transfer mechanism requires strong interaction between plasmonic nanostructure and adsorbates, which is not easy to achieve over coinage metals. In realistic SPR-driven/assisted chemical reactions, these two charge transfer processes sometimes exist and contribute to the reactivity simultaneously.

Semiconductors are known to have distinct catalytic characteristics from the metal surfaces and provide additional active sites to facilitate the transfer of accumulated hot electrons to adsorbed molecules to form active intermediate species. However, different from metals, semiconductors have fewer electron-accepting states for the hot electrons to transfer back from the activated intermediates, leading to their significantly extended lifetimes on the semiconductor surfaces. But for the plasmonic metal/semiconductor complex system, the heterogeneities enable the charge-transferring process under SPR excitation. There are several complicated strategies for transferring the hot electrons to the semiconductor. The conventional strategy is a plasmon-induced hot-electron-transfer mechanism. In this process, a photo-excited plasmon (blue ellipsoid) in the metal decays into a hot electron–hole pair (solid and open red circles in the dotted ellipsoids) through Landau damping, followed by injection of the hot electron into the CB of the semiconductor. The second one is the direct interfacial charge transfer transition pathway, in which optical excitation of an electron in the metal directly transfers into the CB of the semiconductor. The plasmon-induced interfacial charge-transfer transition enables the decay of a plasmon by directly exciting an electron from the metal to a strongly coupled acceptor. The plasmon-induced interfacial charge-transfer transition enables the decay of a plasmon by directly exciting an electron from the metal to a strongly coupled acceptor. And it was proposed and experimentally demonstrated for the high quantum efficiency of Au-to-CdSe charge separation in colloidal quantum-confined CdSe–Au NR heterostructures [51].

The charge transfer from metal surface to semiconductor will be allowed only if the band energy of CB in semiconductor is lower than the energy of excited electrons in plasmonic nanostructures, which is 1–4 eV higher than the Fermi level of metal. For example, in the plasmon-induced electron-transfer pathway, SPR excitation significantly increases the yield of hot electrons and holes. The hot electrons with high potential energy on the plasmonic metals induce fast and efficient transfer of hot electrons to the semiconductors as shown in Figure 10.10a,b. The Schottky barrier at the metal/semiconductor interface helps trap the transferred hot electrons in the

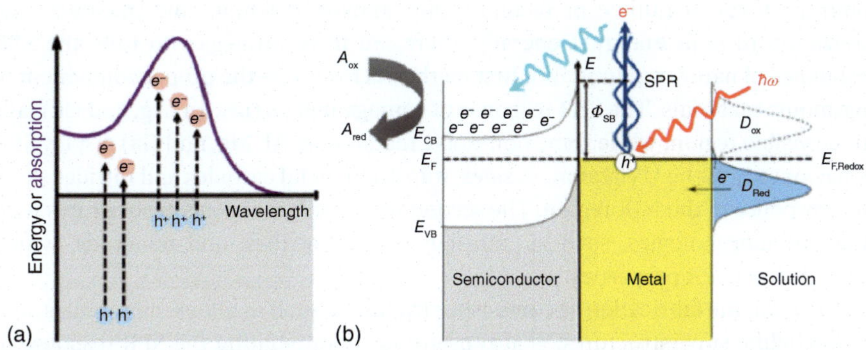

Figure 10.10 (a–c) Metal-to-semiconductor charge-separation pathways. Source: Reproduced with permission from Wu et al. [51]; © 2015 American Association for the Advancement of Science. (d) Hot carrier generation and the corresponding absorption spectrum of plasmonic metals. e^- and h^+ represent hot electrons and holes, respectively, which are generated after SPR dephasing. (e) Strategy in a plasmonic metal/semiconductor heterostructure for efficiently capturing and converting visible light to facilitate the surface reduction reactions. SPR-generated hot electrons with sufficient energy can transfer to the CB of the semiconductor. Source: Reproduced with permission from Zhang et al. [52]; © 2018 American Chemical Society.

CB of semiconductors by delaying them from traveling back to the plasmonic metals. This strategy effectively prolongs the lifetimes of hot electrons transferred to the CB of semiconductors and therefore makes them capable of fostering various surface chemical reactions. The φ_{SB} at the interface helps trap the transferred hot electrons and make them available for facilitating surface chemical reactions. D_{ox} and D_{red} represent the oxidation and reduction states, respectively, of an electron donor in solution. A_{ox} and A_{red} represent the oxidation and reduction states, respectively, of an electron acceptor in solution. E_{CB} stands for the bottom of CB, E_{VB} stands for the top of the valence band (VB), and E_F stands for Fermi level [52].

In the nonradiative decay process, the energy of hot electrons not involved in charge transfer will be finally dissipated to heat and then used for driving the chemical reactions (i.e. plasmonic heating strategy). In fact, utilization of the photothermal property of plasmonic nanostructures is also frequently adopted in the biomedical field (i.e. photothermal therapy). In this mechanism, plasmon is only utilized as a localized heating source and thus enhances the activity of reactions obeying the rules in conventional thermally driven heterogeneous catalysis. The plasmonic heating efficiency depends on the optical property of the catalyst and the choice of light source and reactor. Notably, when the reaction is evaluated with a batch reactor and intensive light source, some systems can be heated up to >300 °C over the plasmonic nanostructures.

Plasmon-induced catalytic reactions can be divided into two categories: (i) direct catalysis, where plasmonic nanostructures act as both light absorbers and catalytically active sites, and (ii) indirect catalysis, where plasmonic nanostructures harvest light while the adjacent semiconductors or metal species catalyze the reaction. In both situations, it is important to tune the optical property of plasmonic nanostructures to initiate or facilitate the catalytic reactions and maximize the utilization of light energy. Generally, there are three strategies to tune the SPR resonance of nanostructures. The first method is to choose the proper component of plasmonic materials. The SPR response of coinage metals (i.e. Au, Ag, and Cu) falls in the visible region, while some non-noble metals (e.g. Al, Mg, and Ga) display SPR behavior only in the UV region. Besides, transition metal carbides and nitrides show SPR response in the NIR region. The second way is to fabricate nanostructures with different morphologies, such as nanospheres, nanocubes, and nanorods, which show distinct SPR properties.

Moreover, the fabrication of composite structures, such as alloys, nanorattles, and core–satellite superstructures, is also useful for manipulating the SPR response of plasmonic nanostructures. The third strategy is to tune the particle size of plasmonic nanostructures. It is evidenced that increasing the size of plasmonic NPs leads to the redshift of the SPR absorption peak. For a specific chemical reaction, researchers may select a proper strategy to construct plasmonic nanostructures with targeted optical response.

10.3.2 Types of Photocatalysis

10.3.2.1 Photocatalytic Decomposition of Organic Compounds

A vast majority of the publications discussing plasmon-enhanced photocatalysis have focused on reactions involving the photocatalytic decomposition of organic compounds. These are typically dye molecules (e.g. methyl orange) whose decomposition rates can be easily observed by optical means. In photocatalytic degradation, photogenerated electrons reduce adsorbed oxygen (O_2) to form superoxide ($HO_2\cdot$) radicals, and photogenerated holes react with H_2O to form OH· radicals. Hydroxyl radicals subsequently oxidize the organic pollutants resulting in mineralization and complete degradation. These pollutants are broken down into small molecules, such as CO_2, H_2O, and NH_3, which are not particularly toxic. While most scientific works study the decomposition of dye molecules, this basic process applies to industrial pollutants, such as perchloroethane, pesticides, and pharmaceutical drugs. Enhanced photodegradation of organic compounds has been observed after adding Au or Ag nanoparticles to several semiconductor photocatalysts, including TiO_2, ZrO_2, ZnO, Cu_2O, SnO_2, CeO_2, Fe_2O_3, SiO_2, $KNbO_3$ under both UV and visible illumination.

Kan and coworkers designed and fabricated a new Au–Pt@TiO_2 nanocomposite, in which Au nanobipyramids (Au NBPs) decorated with platinum (Pt) clusters were enveloped in mesoporous TiO_2 nanoboxes with nanocavities (Figure 10.11a) [53]. Au NBPs provide strong SPR absorption and localized field enhancement restricted to the cavities of TiO_2 nanoboxes. The Pt nanoclusters decorated on the surface of Au NBPs can effectively modulate the charge movement and energy transfer in the photocatalytic process. The enhanced electric field provides a local thermal effect for the photocatalytic reaction and promotes the injection process of hot electrons, which facilitates carrier separation. The nanoboxes with nanocavities can effectively manage the usage of localized energy and provide space for reaction. Under the

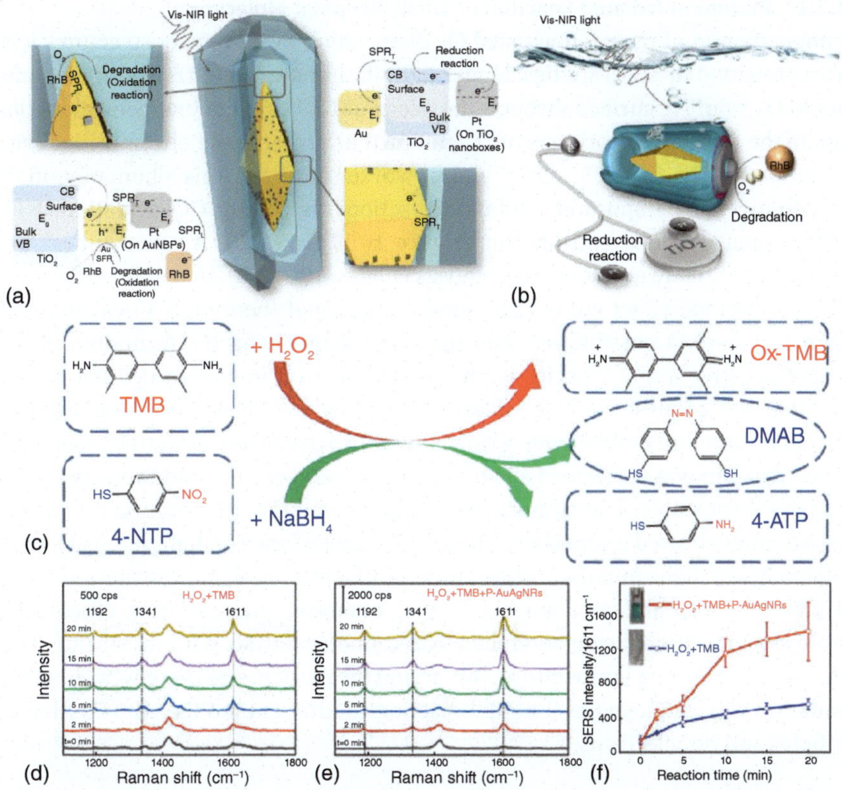

Figure 10.11 (a) Photocatalysis schematic diagram and energy band diagram; (b) Model analogy diagram of the platinum-modified nanoboxes. Source: Reproduced from Gao et al. [53]/with permission of Royal Society of Chemistry/Public Domain CC BY 4.0. (c) Chematic illustrating of the oxidation of TMB by $H_2O_2$and the reduction of 4-NTP by $NaBH_4$ using P-AuAg NRs nanocatalysts. (d, e) The SERS spectra of TMB were recorded at different time intervals during the peroxidase-like catalytic reaction in the presence of H_2O_2 and (f) the corresponding SERS intensities at 1611 cm^{-1} versus reaction time. Source: Reproduced with permission from Ke et al. [54]; © 2021 Elsevier B.V.

cooperative effects, the photocatalytic performance was remarkably improved along with durability and stability. For the Au NBP–Pt@TiO_2 nanoboxes, the rhodamine-B degradation efficiency was ~6.5 times that of Au NBP@TiO_2 nanoboxes.

In addition, they also fabricate highly porous Au–Ag alloy nanorods (P-AuAg NRs) by critically dealloying Ag atoms from homogeneous solid Au–Ag alloy nanorods (AuAg NRs) (Figure 10.11b–e) [54]. Combining the merits of the tunable plasmonic properties of noble metal nanorods, excellent stabilities of alloys, and superior catalytic activities of porous structures, we use the P-AuAg NRs as a Raman probe for the *in-situ* monitoring of the catalytic oxidation of TMB and reduction of 4-NTP. We also compare their composition-dependent catalytic activities. The results show that P-AuAg NRs possess superior chemical stability and higher catalytic activity than core–shell structures due to synergistic structural and chemical mechanisms.

10.3.2.2 Plasmon-Mediated Reaction of Small Nonpolar Molecules

The phenomenon of plasmon-assisted O_2 dissociation over Ag nanostructures was also first reported in SERS studies. Under light illumination, electrons can be transferred to O_2 from Ag surface through the decay of LSPR. These energetic electrons populate the unoccupied orbitals of O_2 and further promote the electron-mediated dissociation of O_2. In 2011, Linic and coworkers utilized this phenomenon to drive commercially important oxidation reactions, including CO oxidation, NH_3 oxidation, and ethylene epoxidation (Figure 10.12) [55, 56]. Ag nanocubes supported on Al_2O_3 were employed as catalysts (Figure 10.12a,b). In combination with the kinetic isotope effect study and density functional theory (DFT) calculations (Figure 10.12c–e), a possible reaction mechanism involving the formation of the transient negative ion O_2-species is proposed, as shown in Figure 10.12f,g. It was found that the catalyst could couple low-intensity visible photons and thermal energy and therefore realize long-term stability of activity and selectivity since the catalytic reaction was performed at a relatively low temperature and the structure of the catalyst kept stable. The authors also demonstrated that the reaction rate on Ag nanostructures showed a superlinear power law dependence on light intensity, even at relatively low light intensity. More importantly, plasmonic nanostructures, as a new kind of photocatalyst, possessed higher photocatalytic quantum efficiencies than those of traditional semiconductor-based photocatalysts with increasing light intensity and operating temperature. By varying the particle size of Ag NPs, ethylene epoxidation was employed as a model reaction to differentiate the contribution of near fields and hot electrons.

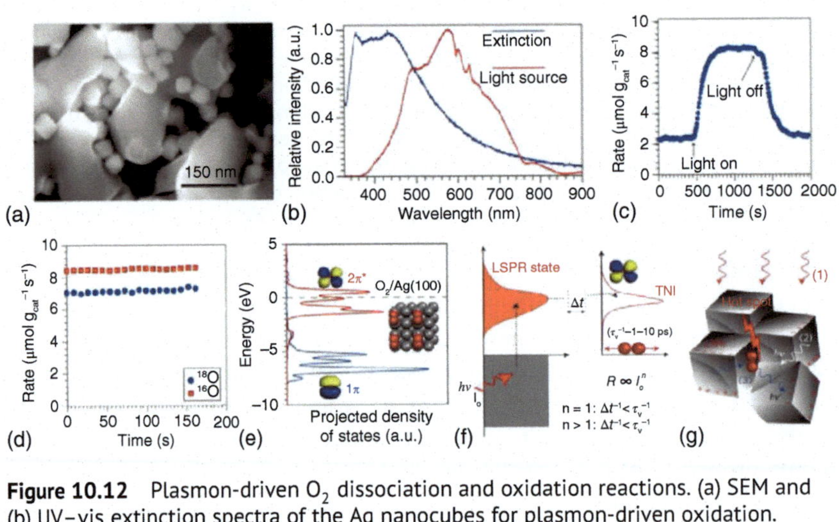

Figure 10.12 Plasmon-driven O_2 dissociation and oxidation reactions. (a) SEM and (b) UV–vis extinction spectra of the Ag nanocubes for plasmon-driven oxidation. (c) The rate of C_2H_4 epoxidation (μmol g^{-1} s^{-1}) at 177 °C in the dark and with visible light irradiation. (d) The reaction rate for plasmon-driven oxidation at 177 °C for $^{16}O_2$ and $^{18}O_2$. (e) Molecular density of states on O_2 adsorbed Ag(200) surface, which was calculated by DFT. (f, g) The schematic of hot-electron-driven O_2 dissociation and oxidation reaction. Source: Reproduced with permission from Christopher et al. [55]; © 2012 Springer Nature.

In the field of heterogeneous catalysis, H_2 cannot be dissociated by Au or Ag NPs until plasmon-driven H_2 dissociation over plasmonic metal nanostructures was first discovered by Halas's group [57]. When Au/TiO_2 nanocomposites were submitted to visible light illumination, dissociation of H_2 was observed under atmospheric pressure and room temperature, evidenced by the formation of HD molecules from the dissociation of H_2 and D_2. The dependence of the dissociation rate on metal loading, particle size, and laser wavelength were studied systematically. The authors further demonstrated the plasmon-induced dissociation of H_2 on Au/SiO_2 and even aluminum nanocrystals at room temperature.

Chemoselective reduction of organic compounds by molecular H_2 is an important method for the industrial production of agrochemicals, pharmaceuticals, and various fine products [57]. For traditional thermally driven hydrogenation processes, supported metal NPs such as Pt, Pd, and Ni are benchmark catalysts [58]. As discussed above, plasmonic nanostructures can dissociate H_2 molecules under visible light illumination, which makes it possible to achieve LSPR-induced chemoselective hydrogenation over plasmonic photocatalysts. In modern agriculture and chemical industry, the transformation of N_2 to NH_3 is a crucial process. Given the difficulty of activating and dissociating the nonpolar N—N bond, the industrial Haber–Bosch process requires harsh reaction conditions and high energy input. In the past few years, solar-induced ammonia synthesis has drawn intensive research interest as the reaction can be performed under mild conditions. As the chemical bond between coinage metal and nitrogen weakens, it is hard to achieve direct plasmon-driven photocatalytic N_2 transformation. However, plasmonic metal NPs have been incorporated as cocatalysts in some plasmonic metal/semiconductor systems to enhance light absorption and provide hot electrons for N_2 activation.

Naturally, the biological room-temperature N_2 fixation is achieved on nitrogenase enzyme on which N_2 is adsorbed and activated over the MoFe protein and then converted into ammonia over another Fe protein. This "working in tandem" mechanism inspires the design of photocatalysts. Wang and coworkers reported an ultrathin Au/TiO_2 nanosheet hybrid that catalyzes the N_2 fixation in a "working in tandem" mechanism, which exhibits remarkable catalytic activity under visible light illumination with methanol as sacrificing agent [59]. The hot electrons generated on Au NPs inject into the CB of TiO_2 and then enter the antibond orbital of N_2 to form NH* and further NH_3.

In addition, it is reasonable to fabricate an antenna reactor with plasmonic metal and catalytically active metal (e.g. Ru, Co, Fe) for traditional thermally driven NH_3 synthesis. Very recently, Xiong and his coworkers reported plasmon-assisted ammonia synthesis in pure water under mild conditions (0.2 MPa and room temperature) through a dissociative mechanism [60]. Using polyol reduction method, the prepared $AuRu_{0.31}$ catalyst contains a gold-rich core as light absorber and some branched-out Ru whiskers as sites for N_2 adsorption and activation, as shown in Figure 10.13a,b. When subjected to white light illumination ($400\,mW\,cm^{-2}$), $AuRu_{0.31}$ achieves the optimal ammonia production rate of $101.4\,\mu mol\,g^{-1}\,h^{-1}$ without the use of any sacrificial agent (Figure 10.13c). The light intensity and wavelength dependency of catalytic performance are investigated, and the nitrogen

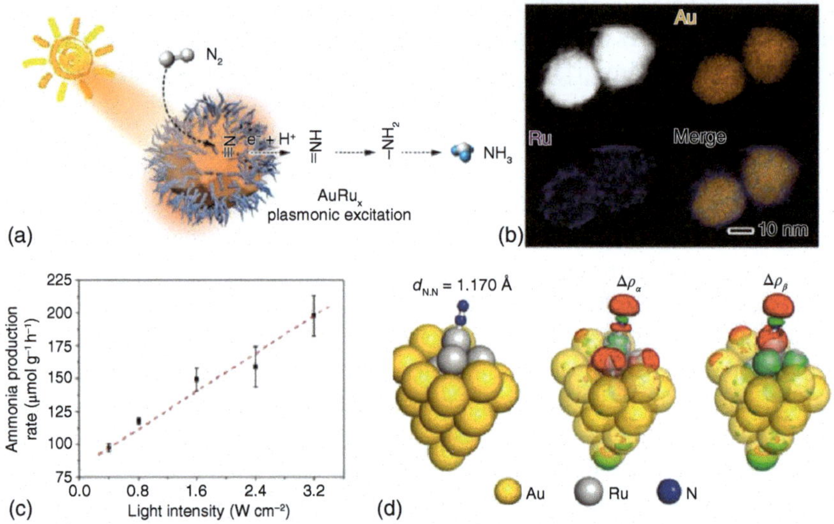

Figure 10.13 (a) Plasmon-driven N_2 fixation over AuRu catalyst. (b) STEM and corresponding EDS mapping of AuRu photocatalyst. (c) Photocatalytic reaction rates over AuRu catalyst under distinct illumination intensity. (d) Optimized structures of N_2 adsorbed on AuRu cluster. Source: Reproduced with permission from Hu et al. [60]; © 2019 American Chemical Society.

resource is verified to be N_2 with an isotope experiment (Figure 10.13d). Based on a series of *in situ* characterizations, the authors claimed that N_2 was reduced in a dissociative mechanism rather than an associative mechanism, with enhanced selectivity toward NH_3 by limiting the production of N_2H_4.

10.3.3 Thermal and Nonthermal Effects in Plasmon-Enhanced Catalysis

Light absorbed by metal nanostructures through intraband or interband transitions excites hot carriers or electron–hole pairs, respectively. Particularly strong light–matter interactions are observed when LSPs of metal nanostructures are excited. Photoexcitation generates mobile free carriers, and electron–phonon scattering generates heat, both of which contribute to the catalytic activity of these nanostructures. Recently, the plasmonic community has generated excitement by demonstrating that transferring hot carriers into antibonding orbitals of surface adsorbates at the rate-determining step may select products and accelerate chemical reactions by weakening certain chemical bonds. By catalyzing reactions under mild conditions, plasmon-enhanced catalysis offers a compelling remedy to the exorbitant energy requirements of traditional thermal catalysis while simultaneously extending the lifetime of catalysts. However, absorbed light also heats these metal nanoparticles, so claims about the beneficial effects of hot carriers must first convincingly distinguish the effects of "nonthermal" hot carriers from those of "thermal" hot surfaces.

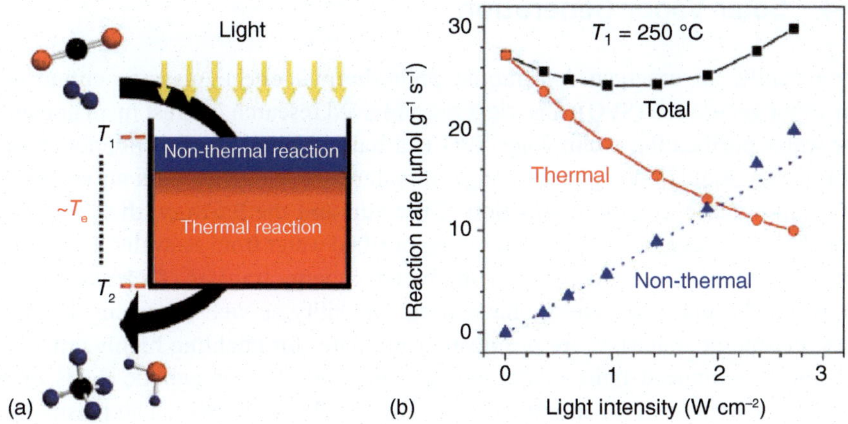

Figure 10.14 (a) Temperature gradient model of the catalyst bed and the equation to calculate equivalent temperature T_e from T_1 and T_2. The nonthermal reaction only occurs in the top illuminated part of the catalyst bed, while thermal reaction occurs throughout. (b) Total (black squares), thermal (red circles), and nonthermal (blue triangles) reaction rates at $T_1 = 250\,°C$ as a function of Iuv. The nonthermal reaction rate is the difference between the total and effective thermal reaction rates. The reactions are carried out with 50 sccm CO_2, 150 sccm H_2, and 50 sccm Ar. Source: Reproduced with permission from Zhang et al. [61]; © 2018 American Chemical Society.

Unfortunately, it is quite difficult to measure the temperature of individual metal nanoparticles, so this discrimination has primarily depended on calculations that estimate the degree of photothermal heating or on simple measurements of the catalyst's thermal activity prior to illumination.

Liu and coworkers demonstrate a more comprehensive experimental exploration and discrimination of thermal and nonthermal catalytic activities under illumination (Figure 10.14) [61]. They used Rh nanoparticles, already well-known as an effective thermal catalyst and recently demonstrated as an effective plasmon-enhanced catalyst through the discovery of its UV plasmonic behavior. Specifically, the CO_2 hydrogenation reaction is investigated, and by using much smaller Rh nanoparticles on a more active TiO_2 support, we dramatically enhance the activity of the catalyst to create conditions under which thermal and nonthermal catalytic behaviors are at work simultaneously and cooperatively. Measurements and a model describing the temperature profile of the catalyst bed are proposed to obtain the effective thermal and nonthermal reaction rates under illumination with and without external heating. By simultaneously measuring the total reaction rate and the temperature gradient of the catalyst bed, the effective thermal reaction rate may be extracted. The residual nonthermal rate of the plasmon-enhanced reaction is found to grow with a superlinear dependence on illumination intensity, and its apparent quantum efficiency reaches ~46% on an Rh/TiO_2 catalyst at a surface temperature of 350 °C. Heat and light work synergistically in these reactions: the higher the temperature, the higher the overall nonthermal efficiency in plasmon-enhanced catalysis.

10.4 Solar Vapor Generation

Very recently, an emergent plasmonic photothermal effect, plasmon-enhanced solar vapor generation (SVG), has triggered a revived research interest in an ancient solar water purification technology. SVG is a natural photothermal phenomenon that induces liquid vapor phase change of water by solar energy. As an essential part of the water cycle, the interaction of the Sun and the Earth (with >70% area covered by the ocean) is endlessly providing purified water from seawater. However, such a natural SVG process is rather inefficient (energy transfer efficiency <20%) because of the low solar energy harvesting capability as well as severe thermal losses. Plasmonic materials are promising candidates for enabling highly efficient SVG due to enhanced light–matter interactions and electromagnetic fields in a tiny volume, beneficial for minimizing optical and thermal losses. The plasmonic materials-enabled solar evaporation process can be divided into three sub-processes: solar energy harvesting by plasmonic resonant absorption, light-to-heat conversion via electron relaxation, and heat transfer to water as latent heat via phonon scattering. In order to realize highly efficient solar evaporation, light absorption, and light-to-heat conversion are two major processes most deeply investigated thus far.

10.4.1 Plasmon-Enhanced Absorption for Solar Evaporation

In order to improve solar evaporation efficiency, the input solar energy should be effectively harvested by plasmonic nanostructures. Therefore, broadband plasmonic solar absorbers are highly desired. As overviewed above, SPRs are strongly affected by material, size, shape, surrounding environment, and assembling profiles, resulting in efficient absorption in a particular wavelength range. Noble metals such as Au and Ag have been extensively investigated in solar energy conversion due to their strong plasmonic responses in the vis–NIR region, which are well-matched with the solar spectrum. In the past few years, many other metals have attracted much attention in this field as well, such as Al, Cu, Mo, Co, Ni, Pt, Pb, Ti, W, and T_e. There have been significant efforts to achieve broadband and efficient absorption through plasmonic hybridization between different building blocks.

In 2013, the Halas group reported the first plasmon-assisted solar evaporation, which inspired exciting development in the following five years [62]. They dispersed SiO_2/Au NPs in water to enhance the evaporation process. These NPs possess a SiO_2 core of 120 nm in diameter and an Au nanoshell of 40 nm in thickness. The plasmons associated with the inner core and outer shell mix and hybridize. They found that the thinner shell layers, the stronger the hybridization interaction. This leads to an obvious red-shift resonance for the "bright" plasmon in a large spectral wavelength region determined by the shell thickness and the overall particle radius [63]. Because of plasmon hybridization, the plasmon resonance-induced absorption cross-section can be tuned to overlap within the solar spectrum.

In the past few years, there have been explosive developments of plasmonic solar evaporators with various broadband light-harvesting structures. M.S. Zielinski et al. demonstrated a particular technique to fabricate Ag/Au bimetallic hollow mesoporous plasmonic nanoshells as a promising material for solar evaporation to produce vapor with high efficiency [64]. It was shown that for the 5 nm hollow Ag/Au alloy shells, the reported absorption peak was ~1280 nm. Although it is far from the irradiant peak of the solar spectrum (~500 nm for 5800 K blackbody irradiance), the utilized broadband region of the solar spectrum in the NIR regime was still enough to enhance the water vaporization efficiency, provided that the plasmonic nanostructures are carefully designed. The vapor generation kinetics is significantly improved due to the NP architecture, surface stabilization with a short PEG ligand, improved buoyancy, and material composition.

Apart from the core/shell structure, there are a couple more idealized structures for polarization-independent broadband light absorption. Kyuyoung Bae et al. developed one large-area flexible thin-film black gold membrane consisting of self-aggregated metallic nanowire bundle arrays, which incorporated adiabatic nanofocusing to reach an average absorption of 91% at 400–2500 nm [65]. This membrane was fabricated by self-aggregation of collapsed nanowires after appropriate pore widening of the anodic aluminum oxide (AAO), then sputtered with a gold film of 40 nm thickness. The unit of the nanowire's structure was shaped like the David star, with a triangular cross-section of each nanowire. Then these hexagonal units merge and align to a funnel-shaped bundle, as shown in Figure 10.15a. Both the small taper angles and gradually changed nanogaps between the aggregated nanowires are proven to contribute to broadband absorption in the vis–NIR regions. Figure 10.15b shows their experimental measurement of total reflection, transmission, and absorption spectra for the black Au membrane on an Al tape.

The incorporation of other 3D porous matrices with plasmonic NPs also performs an appreciably enhanced solar absorption. Natural wood has attracted much attention from researchers due to its naturally mesoporous and hierarchical structure. Uniform deposition of plasmonic metal NPs into the natural wood pores led to the preparation of plasmonic wood by the Hu group, which turned black. Ultimately, due to the plasmonic effect of the metal NPs and the waveguide effect of the wood microchannels, the plasmonic wood was reported to have a highly enhanced light absorption ability (~99%) over the 200–2500 nm wavelength range. In addition, because of the multiple scattering and the absorption of the particular microstructures in the wood, the efficient absorption was angle-independent. As a result, the naturally hydrophilic plasmonic wood with numerous microchannels for water transport performed impressive high-efficiency solar evaporation.

Among various articles about plasmonic evaporators (utilizing plasmonic materials) reported thus far, the Zhu group reported a series of three-dimensional (3D) plasmonic NPs for solar evaporation in 2016, as illustrated in Figure 10.16 [66, 67]. Unlike the conventional designs for plasmonic absorbers, which commonly employ flat optical substrates, they employed a 3D nanoporous substrate as the scaffold of the solar evaporator. Such porous substrate is ideal for solar evaporation with three advantages. Firstly, the porous substrate greatly changes the distribution profile of the metallic NPs, from a continuous metal film to a well-defined close-packed 3D NP assembly, greatly increasing the optical absorption path. Secondly, the high porosity greatly decreases the filling ratio of metal (compared with film or other

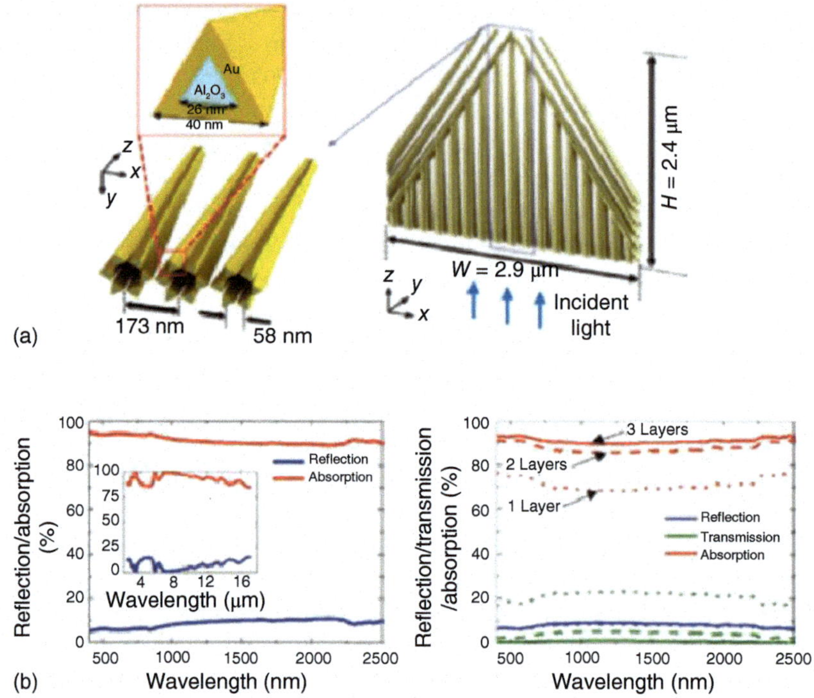

Figure 10.15 (a) Schematic of flexible thin-film black gold membrane consisting of self-aggregated metallic nanowire bundle arrays and (b) the optical properties of a single layer (left) and multilayer membrane (right). Source: Reproduced with permission from Bae et al. [65]; © 2015 Springer Nature Limited.

larger filling ratio structures), which decreases the effective refractive index as well as surface reflectance. Thirdly, the highly porous structure combined with proper surface wettability makes the entire structure self-floatable on the water surface, further reducing the thermal loss to underlying bulk water. Therefore, it provides point-of-use design rules for highly efficient solar evaporators, which include efficient anti-reflectivity, high density of optical modes, and strong light coupling and scattering channels for efficient absorption with a well-defined porous structure. In their work, the porous substrate provided an impedance match for efficient antireflection and strong light coupling to the optical modes. It acted as an efficient light trapper by strongly scattering the incident light into the nanopores. The randomly sized and distributed metal NPs in these pores enabled a high density of hybridized localized SPR to absorb a wide wavelength range of light efficiently.

Based on the above design principles, the Zhu group has also explored optimal cost-effective 3D Al NPs (Figure 10.16b–d) [66, 67]. By combining the plasmonic hybridization effect, porous substrate effect, the intrinsic self-limiting oxidation effect, and intrinsic Al absorption in the NIR region (~800 nm), the Al/AAO absorber enabled an efficient solar absorption up to 96.5% (Figure 10.16e),

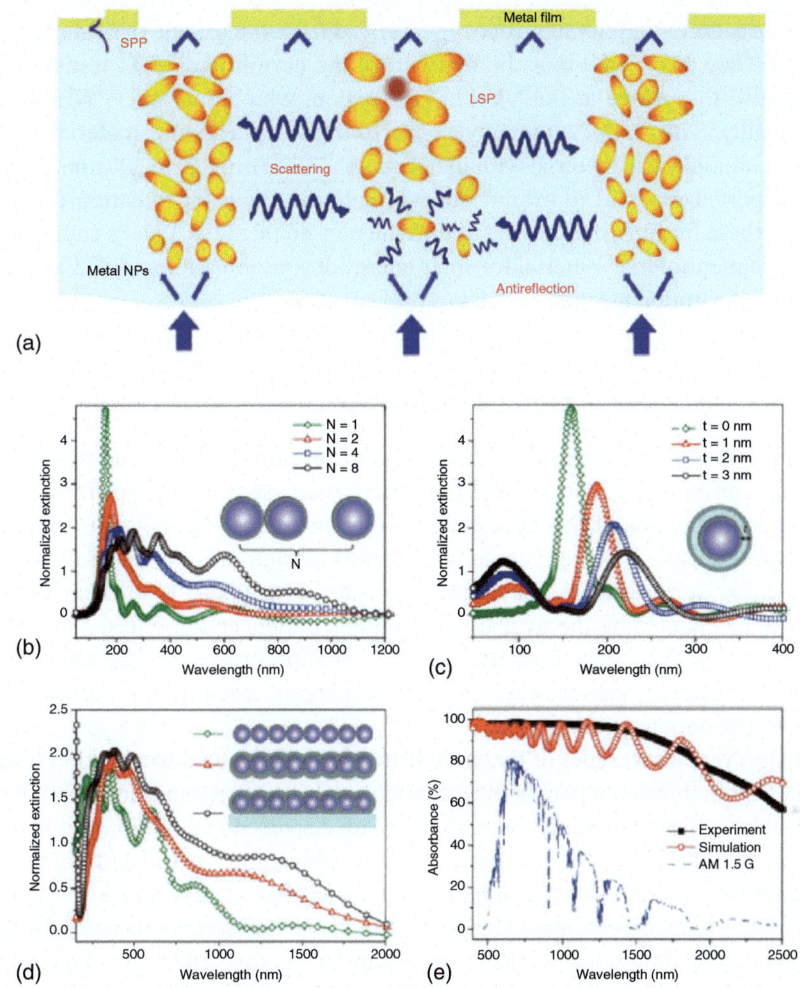

Figure 10.16 (a) Schematic of hybridized plasmonic structure to enhance solar absorption; dependence of calculated normalized extinction cross-section of Al nanoparticles on (b) the particle number (*N*), (c) the thickness (*t*) of the alumina oxidation layer, and (d) the substrate. (e) Experimental and simulated absorption of Al-based plasmonic absorbers. The normalized spectral solar irradiance density of air mass 1.5 global (AM 1.5 G) tilt solar spectrum is shown by the blue dashed line. Source: Reproduced with permission from Zhou et al. [66]; © 2016 American Association for the Advancement of Science. Source: Reproduced with permission from Zhou et al. [67]; © 2016 Springer Nature Limited.

beneficial for highly efficient solar evaporation when combined with other advantages mentioned above.

Apart from the complex hybridization of different plasmonic resonant modes to increase the light absorption mentioned above, a novel broadband absorber made of self-assembled tellurium (T_e) NPs with a wide size distribution, was designed by the Yang group, combining both plasmonic-like and all-dielectric Mie-type resonances [68]. Perfect absorption (more than 85%) was achieved in the entire spectrum of solar radiation (300–2000 nm), derived from the unique optical duality of T_e NPs. They discovered that the real part of the permittivity of T_e transforms from negative to positive at the UV–vis–NIR region, which endows T_e NPs with optical duality. The T_e NP gradually converts from a plasmonic material to a high-index all-dielectric material with increasing the sizes from 10 to 300 nm. It was the first reported material to exhibit optical duality in the solar radiation region. Therefore, these findings demonstrated that the non-noble metal T_e NPs could be a promising nanophotonic material for solar energy absorption and provided insight into perfect absorption assisted by various optical modes.

10.4.2 Heat Localization for Solar Evaporation

For the entire solar evaporation process, light absorption is merely the first step. Light–heat conversion is the second most important issue. Heat localization is widely regarded as one of the most crucial components of high-efficiency SVGs, which can be enabled by both material/structure design and system design for plasmon-based solar evaporators. From the material/structure level, heat localization can be enabled by the near-field plasmonic enhancement effect. Once the energy of the incident light converts to the oscillation energy of free electrons, there will be different passageways: electron-to-photon, electron-to-electron, and electron-to-phonon, respectively.

In the past years, two types of material/structure-induced heat localization have emerged. The first one is a plasmonic nanofluid under certain conditions [69]. For the typical bulk heating system where the NPs are dispersed in a liquid, heat localization can be achieved by fine-tuning the particle concentration. In addition, there are three distinct regimes for plasmonic nanofluid-based solar evaporation. At low concentrations or small volumes (Regime I), NPs serve as absorbers instead of scatterers, and Beer–Lambert law works in this regime. Therefore, all scattered photons are lost from the beam, and light transmits through the solution. Because of the linear relationship between particle concentration and absorption efficiency, the linear dependence on particle concentration in the log scale profile is presented. For higher concentrations (or multiple scattering regimes, Regime II) where scattering length is comparable to characteristic dimensions of the nanofluid, NPs serve as scatterers and absorbers. In this case, multiple scattering plays a crucial role and increases the average path length of the photons, which increases the absorption coefficient and concentrates light into a smaller penetration depth.

Furthermore, at extremely high NP concentration (Regime III), photon absorption is saturated, and part of the incident photon will be backscattered off the fluid surface. Note that plasmonic nanofluid in both Regimes I and III will encounter tremendous heat loss due to large optical and thermal dissipation. Regime II can be regarded as an interfacial-like solar evaporation by heating an infinite volume of bulk water, which has inspired the second type of plasmonic structure-induced local heating systems.

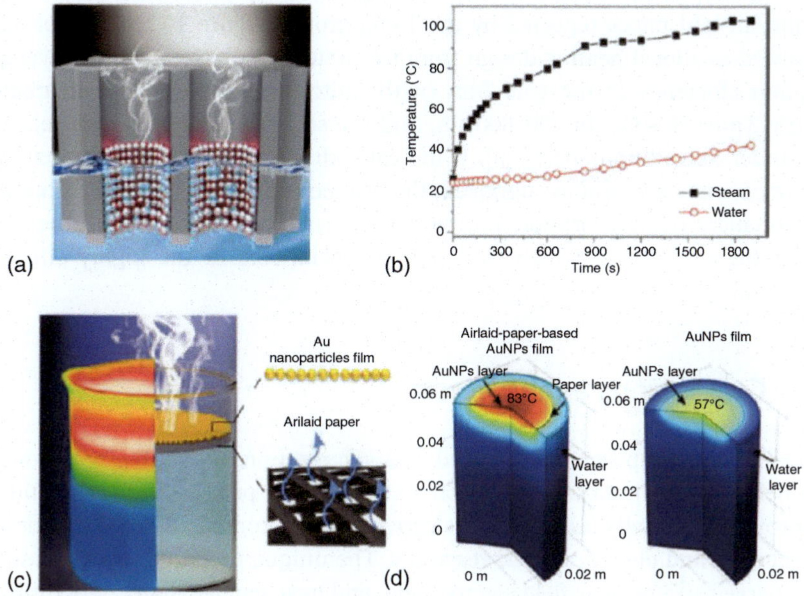

Figure 10.17 (a) Schematics of the self-floated solar evaporator based on the assembly Al nanoparticles on porous AAO membrane and (b) temperature changes of the steam and water over time (24 °C ambient temperature and 48% humidity). Source: Reproduced with permission from Zhou et al. [67]; © 2016 Springer Nature Limited. (c) Schematic illustration of the structure of air-laid paper-based Au NPs film evaporation system and (d) the simulated temperature distribution comparison of air-laid paper-based Au NPs film evaporation system with a maximum temperature of 83 °C and free-floating Au NPs film with a maximum temperature of 57 °C. Source: Reproduced with permission from Liu et al. [70]; © 2020, John Wiley & Sons, Inc.

Unlike the plasmonic nanofluid system, the second type features integrated structures hosted by thin porous substrates. Such solar evaporators are self-floating on air/water interfaces due to their hydrophobic surface and/or low density. Suppressed thermal dissipation is commonly suggested due to the reduced effective thermal conductivity. Figure 10.17a,b illustrates one of the representative integrated plasmonic evaporators with material/structure-induced heat localization [67]. By assembling the close-packed metal NPs in a highly porous substrate, the evaporator can integrate into one structure the interfacial effects of self-floating substrates, as well as the NP-induced nearfield enhancement effects as in nanofluids. This leads to a suppressed heat dissipation to bulk water and, thus, a distinct temperature difference between vapor and underneath water. Note that the thermal conductance loss for the material/structure designed system is still distinct due to direct contact between plasmonic materials and bulk water, especially for low solar irradiance. Researchers have been attempting to further decrease thermal conductance loss by introducing thicker thermal insulating layers into the plasmonic absorbers, such as porous membranes, air-laid paper, wood, and biofoam.

One of the representatives is the double-layer solar evaporators constructed with gold film/air-laid paper reported by the Deng group (Figure 10.17c), which can be regarded as a local heating design from the system level. Note that the average absorption efficiency of the gold film (with uniform gold particles of diameter ~17.4 ± 0.3 nm) is ~84% in 400–800 nm, and the solar thermal efficiency reaches 77.8% under Sun illumination [70]. It indicates that apart from the explanations from the optical viewpoint as suggested, the unique textile-like porous hydrophilic paper enhances light absorption and reduces thermal conductance with the bulk water at the same time, which may play a crucial role in the highly efficient evaporation.

10.5 Conclusions and Outlook

Plasmon-driven and plasmon-enhanced catalytic reactions are at the sharp corner in the fields of photocatalysis and electrocatalysis, as it provides new possibilities to improve the activity as well as selectivity when compared with traditional photocatalytic and electrocatalytic reactions. The unique properties of LSPR, such as hot charge carriers, near-field electric field, and heat, can facilitate the activation of reactants, especially the inert nonpolar molecules, and change the selectivity by tuning the reaction pathway. Under light illumination, varieties of crucial reactions in energy conversion and storage, such as hydrogenation and oxidation of organic compounds, CO_2 reduction, N_2 fixation, and some decomposition reactions, have been realized with the assistance of plasmonic nanomaterials under facile conditions. Plenty of electrocatalytic reactions, including HER, OER, ORR, and AOR, can also be enhanced by LSPR. Thus, remarkable research efforts have been devoted to the rational design and synthesis of plasmonic materials and related hybrid materials.

Coinage metal nanostructures are conventional plasmonic materials. In contrast, the LSPR property of other metals (e.g. Mg, Al, Ga) and heavily doped semiconductors has also been investigated in the past few years. Despite the great success achieved in the plasmon-driven and -enhanced catalysis, researchers are still exploring new catalysis for the cost of performance and sources. ZnO nanostructures are prominent photocatalyst candidates to be used in photodegradation owing to the fact that they are low-cost, nontoxic, and more efficient in absorption across a large fraction of the solar spectrum.

In the fundamental study, despite the current exciting results, the overall reaction efficiencies remain low, even though very large plasmon-to-electron conversion efficiencies have been proved. The mismatch between the short lifetimes of hot electrons and the slow kinetics of surface reactions is the major limitation to the efficiencies of plasmonic-enhanced or driven reactions. Thus, the key issue that needs to be addressed in the area is how to effectively capture, stabilize, and accumulate the plasmon-generated hot electrons and make them available for participating in subsequent chemical reactions. Acquiring insights into the precise physical locations of active sites on the plasmon-enhanced catalysts would be greatly important in the design and optimization of high-performance catalysts. Additionally, despite numerous examples of hot-electron-driven processes, little is known about these plasmon-generated hot holes. All the fundamental studies will be illustrated to guide the further development of advanced plasmonic photocatalysts.

For the characterizations, current state-of-the-art characterization techniques have been able to probe catalytic systems at the nanometer or even sub-nanometer scale, with time intervals of nanoseconds or even femtoseconds and high energy resolution. However, it remains a major challenge to apply these techniques for characterizing catalysts under the temperature and pressure conditions at which catalysis happens. The dynamic picture of *in-situ* characterization of catalysts and catalytic processes should be constructed under realistic reaction conditions.

For the practical application, many important issues need to be solved in the commercialization of nanocatalysts, such as their scale-up synthesis and stability under industrial reaction conditions and the regeneration and recycling of deactivated catalysts. Anyway, it is expected that nanotechnology will play an increasingly important role in catalysis, making the design of highly efficient catalysts feasible in applications.

References

1 Seh, Z.W., Kibsgaard, J., Dickens, C.F. et al. (2017). Combining theory and experiment in electrocatalysis: insights into materials design. *Science* 355 (6321): eaad4998.

2 Jiao, Y., Zheng, Y., Jaroniec, M., and Qiao, S.Z. (2015). Design of electrocatalysts for oxygen- and hydrogen-involving energy conversion reactions. *Chem. Soc. Rev.* 44 (8): 2060–2086.

3 Kreibig, U. and Vollmer, M. (1995). *Optical Properties of Metal Cluster*. Berlin: Springer.

4 Hou, W. and Cronin, S.B. (2013). A review of surface plasmon resonance-enhanced photocatalysis. *Adv. Funct. Mater.* 23 (13): 1612–1619.

5 Yu, S., Wilson, A.J., Kumari, G. et al. (2017). Opportunities and challenges of solar-energy-driven carbon dioxide to fuel conversion with plasmonic catalysts. *ACS Energy Lett.* 2 (9): 2058–2070.

6 Rampino, L.D., Kavanagh, K.E., and Nord, F.F. (1943). Relationship between particle size and efficiency of palladium-polyvinyl alcohol (Pd-PVA) catalysts. *PNAS* 29 (8): 246–256.

7 Sabatier, M.P. (1927). The inversion of reactions in catalysis. *J. Chem. Technol. Biotechnol.* 46 (31): 702–708.

8 Haruta, M., Kobayashi, T., Sano, H., and Yamada, N. (1987). Novel gold catalysts for the oxidation of carbon monoxide at a temperature far below 0 °C. *Chem. Lett.* 16 (2): 405–408.

9 Wang, A., Li, J., and Zhao, T. (2018). Heterogeneous single atom catalysis. *Nat. Rev. Chem.* 2: 65–81.

10 Anastasopoulos, A., Davies, J.C., Hannah, L. et al. (2013). The particle size dependence of the oxygen reduction reaction for carbon-supported platinum and palladium. *ChemSusChem* 6 (10): 1973–1982.

11 Yoo, S.J., Jeon, T.-Y., Lee, K.-S. et al. (2010). Effects of particle size on surface electronic and electrocatalytic properties of Pt/TiO_2 nanocatalysts. *Chem. Commun.* 46 (5): 794–796.

12 Chen, S., Ingram, R.S., Hostetler, M.J. et al. (1998). Gold nanoelectrodes of varied size: transition to molecule-like charging. *Science* 280 (5372): 2098–2101.

13 Zhou, Z.-Y., Tian, N., Li, J.-T. et al. (2011). Nanomaterials of high surface energy with exceptional properties in catalysis and energy storage. *Chem. Soc. Rev.* 40 (7): 4167–4185.

14 Jin, M., Zhang, H., Xie, Z., and Xia, Y. (2012). Palladium nanocrystals enclosed by (100) and (111) facets in controlled proportions and their catalytic activities for formic acid oxidation. *Energy Environ. Sci.* 5 (4): 6352–6357.

15 Wang, J., Gong, J., Xiong, Y. et al. (2011). Shape-dependent electrocatalytic activity of monodispersed gold nanocrystals toward glucose oxidation. *Chem. Commun.* 47: 6894–6896.

16 Porter, N.S., Wu, H., Quan, Z., and Fang, J. (2013). Shape-control and electrocatalytic activity-enhancement of Pt-based bimetallic nanocrystals. *Acc. Chem. Res.* 46 (24): 1867–1877.

17 Wang, C., van der Vliet, D., More, K.L. et al. (2010). Multimetallic $Au/FePt_3$ nanoparticles as highly durable electrocatalyst. *Nano Lett.* 11 (3): 919–926.

18 Guo, S., Zhang, S., Su, D., and Sun, S. (2013). Seed-mediated synthesis of core/shell FePtM/FePt (M = Pd, Au) nanowires and their electrocatalysis for oxygen reduction reaction. *JACS* 135 (37): 13879–13884.

19 Wang, C., Chi, M., Wang, G. et al. (2011). Correlation between surface chemistry and electrocatalytic properties of monodisperse Pt_xNi_{1-x} nanoparticles. *Adv. Funct. Mater.* 21 (1): 147–152.

20 Zhang, S., Guo, S., Zhu, H. et al. (2012). Structure-induced enhancement in electrooxidation of trimetallic FePtAu nanoparticles. *JACS* 134 (11): 5060–5063.

21 Markovic, N., Schmidt, T., Stamenkovic, V., and Ross, P. (2001). Oxygen reduction reaction on Pt and Pt bimetallic surfaces: a selective review. *Fuel Cells* 1 (2): 105–116.

22 Sasaki, K., Naohara, H., Choi, Y. et al. (2012). Highly stable Pt monolayer on PdAu nanoparticle electrocatalysts for the oxygen reduction reaction. *Nat. Commun.* 3: 1115.

23 Cui, C., Gan, L., Heggen, M. et al. (2013). Compositional segregation in shaped Pt alloy nanoparticles and their structural behaviour during electrocatalysis. *Nat. Mater.* 12: 765–771.

24 Oezaslan, M., Hasché, F.d.r., and Strasser, P. (2013). Pt-based core-shell catalyst architectures for oxygen fuel cell electrodes. *J. Phys. Chem. Lett.* 4 (19): 3273–3291.

25 Stamenkovic, V.R., Fowler, B., Mun, B.S. et al. (2007). Improved oxygen reduction activity on $Pt_3Ni(111)$ via increased surface site availability. *Science* 315 (5811): 493–497.

26 Strasser, P. (2015). Catalysts by platonic design. *Science* 349 (6246): 379–380.

27 Wu, J.B., Zhang, J.L., Peng, Z.M. et al. (2010). Truncated octahedral Pt_3Ni oxygen reduction reaction electrocatalysts. *JACS* 132 (14): 4984–4985.

28 Zhang, J., Yang, H., Fang, J., and Zou, S. (2010). Synthesis and oxygen reduction activity of shape-controlled Pt_3Ni nanopolyhedra. *Nano Lett.* 10 (2): 638–644.

29 Choi, S.-I., Xie, S.F., Shao, M. et al. (2013). Synthesis and characterization of 9 nm Pt-Ni octahedra with a record high activity of 3.3 A/mgPt for the oxygen reduction reaction. *Nano Lett.* 13 (7): 3420–3425.

30 Cui, C., Gan, L., Li, H.-H. et al. (2012). Octahedral PtNi nanoparticle catalysts: exceptional oxygen reduction activity by tuning the alloy particle surface composition. *Nano Lett.* 12 (11): 5885–5889.

31 Gan, L., Cui, C.H., Heggen, M. et al. (2014). Element-specific anisotropic growth of shaped platinum alloy nanocrystals. *Science* 346 (6216): 1502–1506.

32 Chen, C., Kang, Y., Huo, Z. et al. (2014). Highly crystalline multimetallic nanoframes with three-dimensional electrocatalytic surfaces. *Science* 343 (6177): 1339–1343.

33 Huang, X., Zhao, Z.P., Cao, L. et al. (2015). High-performance transition metal-doped Pt_3Ni octahedra for oxygen reduction reaction. *Science* 348 (6240): 1230–1234.

34 Hori, Y., Takahashi, I., Koga, O., and Hoshi, N. (2002). Selective formation of C_2 compounds from electrochemical reduction of CO_2 at a series of copper single crystal electrodes. *J. Phys. Chem. B* 106 (1): 15–17.

35 Reske, R., Mistry, H., Behafarid, F. et al. (2014). Particle size effects in the catalytic electroreduction of CO_2 on Cu nanoparticles. *JACS* 136 (19): 6978–6986.

36 Manthiram, K., Beberwyck, B.J., and Alivisatos, A.P. (2014). Enhanced electrochemical methanation of carbon dioxide with a dispersible nanoscale copper catalyst. *JACS* 136 (38): 13319–13325.

37 Baturina, O.A., Lu, Q., Padilla, M.A. et al. (2014). CO_2 electroreduction to hydrocarbons on carbon-supported Cu nanoparticles. *ACS Catal.* 4 (10): 3682–3695.

38 Chen, C.S., Handoko, A.D., Wan, J.H. et al. (2015). Stable and selective electrochemical reduction of carbon dioxide to ethylene on copper mesocrystals. *Catal. Sci. Technol.* 5 (1): 161–168.

39 Zhu, W., Zhang, Y.-J., Zhang, H.Y. et al. (2014). Active and selective conversion of CO_2 to CO on ultrathin Au nanowires. *JACS* 136 (46): 16132–16135.

40 Lu, Q., Rosen, J., Zhou, Y. et al. (2014). A selective and efficient electrocatalyst for carbon dioxide reduction. *Nat. Commun.* 5 (1): 3242.

41 Erini, N., Loukrakpam, R., Petkov, V. et al. (2014). Ethanol electro-oxidation on ternary platinum-rhodium-tin nanocatalysts: insights in the atomic 3D structure of the active catalytic phase. *ACS Catal.* 4 (6): 1859–1867.

42 Li, M., Liu, P., and Adzic, R.R. (2012). Platinum monolayer electrocatalysts for anodic oxidation of alcohols. *J. Phys. Chem. Lett.* 3 (23): 3480–3485.

43 Yun, Q.R., Xu, J., Wei, T.C. et al. (2022). Synthesis of Pd nanorod arrays on Au nanoframes for excellent ethanol electrooxidation. *Nanoscale* 14 (3): 736–743.

44 Bligaard, T., Bullock, R.M., Campbell, C.T. et al. (2016). Toward benchmarking in catalysis science: best practices, challenges, and opportunities. *ACS Catal.* 6 (4): 2590–2602.

45 Voiry, D., Chhowalla, M., Gogotsi, Y. et al. (2018). Best practices for reporting electrocatalytic performance of nanomaterials. *ACS Nano* 12 (10): 9635–9638.

46 Wei, C., Rao, R.R., Peng, J. et al. (2019). Recommended practices and benchmark activity for hydrogen and oxygen electrocatalysis in water splitting and fuel cells. *Adv. Mater.* 31 (31): 1806296.

47 Li, D., Batchelor-McAuley, C., and Compton, R.G. (2020). Some thoughts about reporting the electrocatalytic performance of nanomaterials. *Appl. Mater. Today* 18 (10): 100404.

48 Zhang, H., Zhou, F., and Pan, M. (2023). Review on electrochemical active surface area characterization methods of Pt alloy catalysts. *Chin. Sci. Bull.* 68 (5): 448–456.

49 Mayrhofer, K.J.J., Strmcnik, D., Blizanac, B.B. et al. (2008). Measurement of oxygen reduction activities via the rotating disc electrode method: from Pt model surfaces to carbon-supported high surface area catalysts. *Electrochim. Acta* 53 (7): 3181–3188.

50 Li, S.W., Miao, P., Zhang, Y.Y. et al. (2020). Recent advances in plasmonic nanostructures for enhanced photocatalysis and electrocatalysis. *Adv. Mater.* 33 (6): 2000086.

51 Wu, K., Chen, J., MaBride, J.R., and Lian, T. (2015). Efficient hot-electron transfer by a plasmon-induced interfacial charge-transfer transition. *Science* 349 (6248): 632–635.

52 Zhang, Y.C., He, S., Guo, W.X. et al. (2018). Surface-plasmon-driven hot electron photochemistry. *Chem. Rev.* 118 (6): 2927–2954.

53 Gao, W.J., Kan, C.X., Ke, S.L. et al. (2021). Au nanobipyramids with Pt decoration enveloped in TiO_2 nanoboxes for photocatalytic reactions. *Nanoscale Adv.* 3 (14): 4226–4234.

54 Ke, S.L., Kan, C.X., Zhu, X.Z. et al. (2021). Effective fabrication of porous Au-Ag alloy nanorods for in-situ Raman monitoring catalytic oxidation and reduction reactions. *J. Mater. Sci. Technol.* 91: 262–269.

55 Christopher, P., Xin, H., Marimuthu, A., and Linic, S. (2012). Singular characteristics and unique chemical bond activation mechanisms of photocatalytic reactions on plasmonic nanostructures. *Nat. Mater.* 11 (12): 1044–1050.

56 Christopher, P., Xin, H., and Linic, S. (2011). Visible-light-enhanced catalytic oxidation reactions on plasmonic silver nanostructures. *Nat. Chem.* 3 (6): 467–472.

57 Mukherjee, S., Libisch, F., Large, N. et al. (2013). Hot electrons do the impossible: plasmon-induced dissociation of H_2 on Au. *Nano Lett.* 13 (1): 240–247.

58 Lin, L., Yao, S., Gao, R. et al. (2019). A highly CO-tolerant atomically dispersed Pt catalyst for chemoselective hydrogenation. *Nat. Nanotechnol.* 14 (4): 354–361.

59 Yang, J., Guo, Y., Jiang, R. et al. (2018). High-efficiency "working-in-tandem" nitrogen photofixation achieved by assembling plasmonic gold nanocrystals on ultrathin titania nanosheets. *JACS* 140 (27): 8497–8508.

60 Hu, C., Chen, X., Jin, J. et al. (2019). Surface plasmon enabling nitrogen fixation in pure water through a dissociative mechanism under mild conditions. *JACS* 141 (19): 7807–7814.

61 Zhang, X., Li, M.E.R., Zhang, D. et al. (2018). Plasmon-enhanced catalysis: distinguishing thermal and nonthermal effects. *Nano Lett.* 18 (3): 1714–1723.

62 Neumann, O., Urban, A.S., Day, J. et al. (2013). Solar vapor generation enabled by nanoparticles. *ACS Nano* 7 (1): 42–49.

63 Brinson, B.E., Lassiter, J.B., Levin, C.S. et al. (2008). Nanoshells made easy: improving Au layer growth on nanoparticle surfaces. *Langmuir* 24 (24): 14166–14171.

64 Zielinski, M.S., Choi, J.W., La Grange, T. et al. (2016). Hollow mesoporous plasmonic nanoshells for enhanced solar vapor generation. *Nano Lett.* 16 (4): 2159–2167.

65 Bae, K., Kang, G., Cho, S.K. et al. (2015). Flexible thin-film black gold membranes with ultrabroadband plasmonic nanofocusing for efficient solar vapour generation. *Nat. Commun.* 6: 10103.

66 Zhou, L., Tan, Y., Ji, D. et al. (2016). Self-assembly of highly efficient, broadband plasmonic absorbers for solar steam generation. *Sci. Adv.* 2 (4): e1501227.

67 Zhou, L., Tan, Y., Wang, J. et al. (2016). 3D self-assembly of aluminium nanoparticles for plasmon-enhanced solar desalination. *Nat. Photonics* 10: 393–398.

68 Ma, C., Yan, J., Huang, Y. et al. (2018). The optical duality of tellurium nanoparticles for broadband solar energy harvesting and efficient photothermal conversion. *Sci. Adv.* 4 (8): eaas9894.

69 Hogan, N.J., Urban, A.S., Ayala-Orozco, C. et al. (2014). Nanoparticles heat through light localization. *Nano Lett.* 14 (8): 4640–4645.

70 Liu, Y., Yu, S., Feng, R. et al. (2015). A bioinspired, reusable, paper-based system for high-performance large-scale evaporation. *Adv. Mater.* 27 (17): 2768–2774.

11

Applications in SERS and Sensor

A molecular vibration "fingerprint" can be detected by SERS and plasmonic sensor

11.1 Introduction

Research on metal nanoparticles has been boosted by a wide variety of applications. As one of the most important and promising applications, they are widely used in the SERS substrates for high performance sensor and quantitative detection of chemical species with molecular specificity and single-molecule sensitivity [1]. SERS is an ultrasensitive analytical technique for describing the fingerprint information through amplifying the inelastic Raman scattering of molecules on or near the SERS-active substrates (most often plasmonic metallic nanomaterials). In 1974, Fleischmann et al. accidentally found the enhanced Raman scattering signal of pyridine adsorbed on a rough Ag electrode, which was attributed to an increase in the number of excitable molecules on the large surface [2]. This was the first SERS measurement, although it was not recognized as such at that time. In 1977, Van Duyne and Creighton soon carefully devised a procedure for measuring the surface enhancement factor and discovered this phenomenon independently [3]. They recognized that this marked enhancement could not be explained simply by the increased surface area. In 1978, Moskovits first explained the effect of surface plasmons on roughened Ag electrodes on SERS enhancement and also predicted that the same effect may occur with Ag colloids covered with an adsorbate. This prediction was successfully verified by Creighton et al. using Ag and Au colloids in 1979 [4]. Subsequently in 1985, Moskovits was aware of the relation of the intensity of SERS with the electromagnetic field enhancement due to SPR effect of metal nanoparticles [5, 6] was a strong stimulus for the field. For the rapid development of SERS during the last two decades, in particular for "real-life" applications, advances in nanofabrication and Raman instrumentation were equally important [7, 8].

It is now well known that SERS is a phenomenon in which the Raman signals of molecules are enormously enhanced when they are very close to certain SERS-active

Plasmonic Metal Nanostructures: Preparation, Characterization, and Applications,
First Edition. Caixia Kan.

nanostructures. Due to many complex effects on the SERS spectra, understanding of the SERS mechanism has remained divided until now. Fundamentally, the SERS signal is generated due to two different mechanisms: chemical mechanism and electromagnetic mechanism [9, 10]. The chemical mechanism refers to the electronic interaction between substrates and adsorbed molecules, which depends on the binding affinity between molecules and substrates. Electromagnetic mechanism provides the contribution through enhancing the electromagnetic field around plasmonic structures generated upon the excitation of SPR under the incident light irradiation. Raman scattering signals of molecules on the surface of plasmonic nanomaterials are thus amplified mainly due to the SPR effect.

Currently, a wide range of state-of-the-art plasmonic nanomaterials have been conceived to further progress the SERS enhancement efficiency [11–13]. Owing to the electromagnetic field enhancement associated with "hotspots" of plasmonic nanostructures, extremely low intrinsic Raman cross section of molecules is dramatically increased, hence enabling their detection down to single-molecule level.

In this chapter, we first introduce several typical SERS substrates. These structures present strong electromagnetic fields and chemical enhancement due to the structural advantages attributed to the sharp edge and tips, multiple hot spots, or strong coupling. Moreover, the exploration of SERS applications in the molecular detection, the chemical reaction monitoring, and sensing are carried out.

11.2 Typical SERS Substrates

The SERS technique has been employed in various applications including biochemical sensing and ultrasensitive detection by using plasmonic nanoparticles as SERS substrates [14]. Plasmonic nanospheres, such as Au and Ag, are the most commonly investigated nanomaterials due to their simple synthesis and easy functionalization. However, they are not suitable for the SERS substrate due to the SPR in a narrow visible region and high light attenuation. Usually, single nanosphere can only generate low enhancement on the order of 10^2–10^3. In contrast to the single nanosphere, nanosphere dimers or oligomers can generate higher enhancement efficiency than single nanosphere because of the intensified electromagnetic enhancement due to strong interparticle SPR coupling effects in the nanogap and new SPR absorbance in the NIR region [15, 16]. Other coupling nanostructure dimer or trimer, such as nanocube dimer, end-to-end nanorods and nanocube-nanosphere dimer, as described in Chapter 2. An alternative approach is to use anisotropic nanoparticles where the sharp edges, tips and protrusions can produce strong localized electromagnetic fields for SERS enhancement. Now, the anisotropic structures, including nanoplates, nanocubes, nanobipyramids, nanoframes, nanotips, and nanostars, are attractive for biological and biomedical applications due to their capability of providing high SERS activity from single nanoparticle [17–19]. These anisotropic nanostructures and coupled nanostructures are ideal SERS substrates. Generally, strong electromagnetic fields (or defined as "hot spots") could be created in the nanogaps between nanostructures. Among the plasmonic nanostructures, Au nanostructures are of particular interest because of their stability, nontoxicity, and easy bioconjugation ability. Here, a special emphasis is given some SERS-active plasmonic Au nanostructures.

11.2.1 Edge and Tip Hot Spot

The most common plasmonic nanomaterials used for preparing SERS substrates are Au, Ag, and Cu with different shapes, such as spheres, rods, wires, cubes. Among the plasmonic metals, Ag nanocubes have attracted a great attention owing to their highest quality factor of SPR across vis-NIR wavelength range and outstanding features, such as atomically flat facets, sharp corners, and large contact areas between nanoparticles [20–22]. Here, one typical work about the size and dimension dependent SERS properties of Ag nanocubes from J. F. Li group was introduced [23]. Ag nanocubes with the size range from nanoscale to microscale and the dimension range from "0D" to "3D" were prepared and used as SERS substrates (Figure 11.1a). Figure 11.1b presents a schematic illustration and SEM images of the seed-mediated method for preparing Ag nanocubes with different sizes. The extinction spectra (Figure 11.1c) of the Ag nanocubes' aqueous solutions were measured. Figure 11.1c shows the extinction spectra of Ag nanocubes as a function of size. It is clear that the quadrupolar and dipolar LSPR modes gradually redshift with the continuous increase in size of the nanocubes.

They first prepared Ag nanocubes monolayer as SERS substrate for probing MBA molecule. The SERS collecting spots consisting of a single-monolayer of ordered nanocubes were presented in Figure 11.1d, e. The SERS intensities of MBA disclose that the SERS performance is a function both of the size of the nanocube and wavelength of the laser. It was found that the optimal combinations of the substrate with the excitation line were that 80 nm nanocubes for excitation with the 532 nm laser, 110 nm nanocubes for the 638 nm laser and 130 nm nanocubes for the 785 nm laser. To further elucidate the size-dependent SERS activity, they used FDTD to simulate the EM field enhancement of nanocubes for specific experimental excitation, and the simulation results coincided well with the experimental results. Then Ag nanocube hierarchical nanostructures were fabricated from zero-dimension (0D) to three-dimension (3D), and the influence of the dimension on the SERS properties were investigated. Figure 11.1f shows for the SERS spectra using 0D, 1D, and 2D Ag nanocubes hierarchical nanostructures as SERS substrates. It is obvious that the Normalized Raman intensity of the 1D nanostructures is about eight times higher than the 0D one, which is due to the strong coupling effect in the nanogap of the dimer. The normalized intensity of the 2D nanostructure is further increased by 4.5 times compared to the 1D one, as a result of the increased number of hotspots.

Furthermore, 3D Ag nanocube hierarchical nanostructures were also fabricated by the layer-by-layer assembly of Ag nanocube monolayers. When the "3D" nanostructures consist of more than two monolayers of Ag nanocubes, the normalized intensity from the calculated spectra shows the Raman enhancement declines significantly due to the limited penetration depth of the incident light. Such results

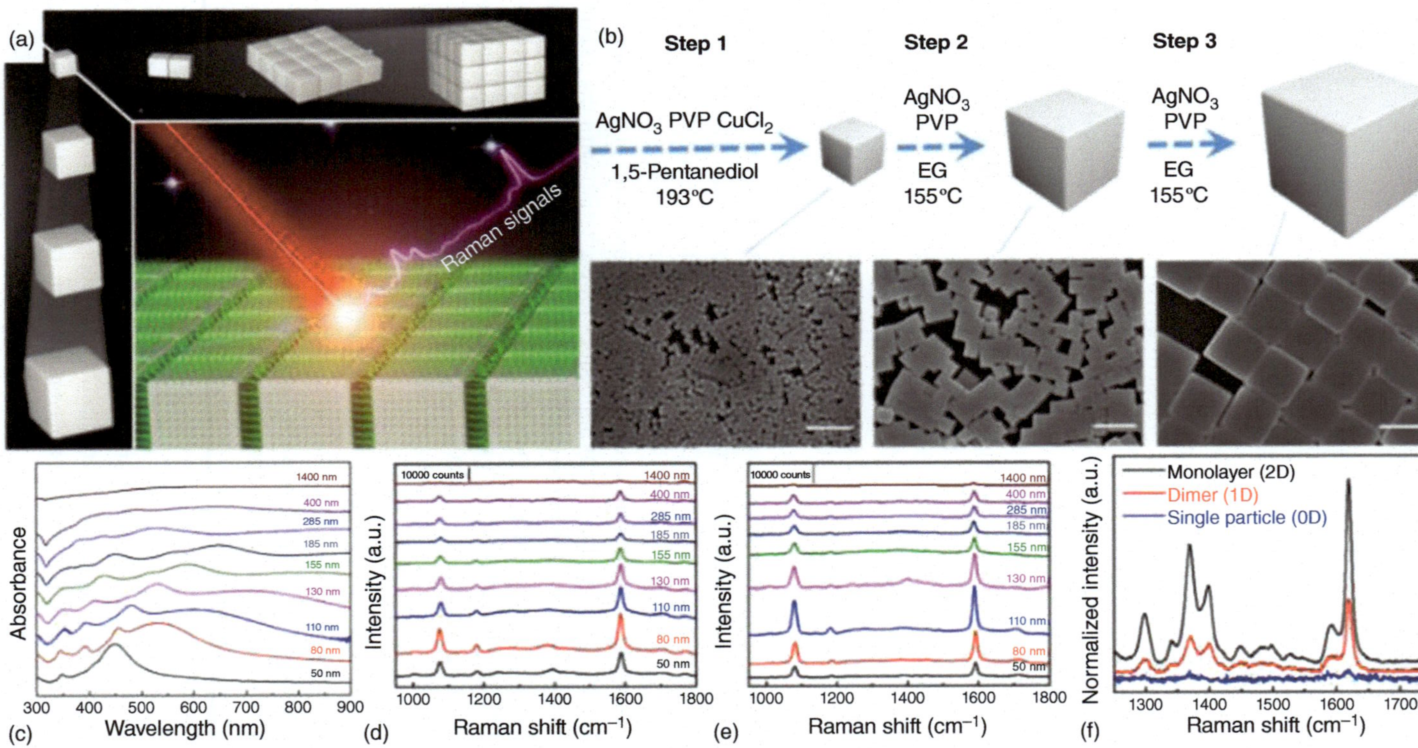

Figure 11.1 (a) Schematic of the SERS study using Ag nanocubes hierarchical nanostructures with different sizes and different dimensions as the substrate. (b) Schematic of the three-step synthesis of Ag nanocubes with different sizes. All the scale bars are 1 μm. (c) Extinction spectra of Ag nanocubes with different sizes. Raman spectra of SERS substrate with different Ag nanocube sizes using lasers of (d) 532 nm, (e) 638 nm. (f) Normalized Raman spectra of Ag nanocube hierarchical nanostructures with different dimensions. Source: Reproduced with permission from Lin et al. [23]; @2019 Elsevier.

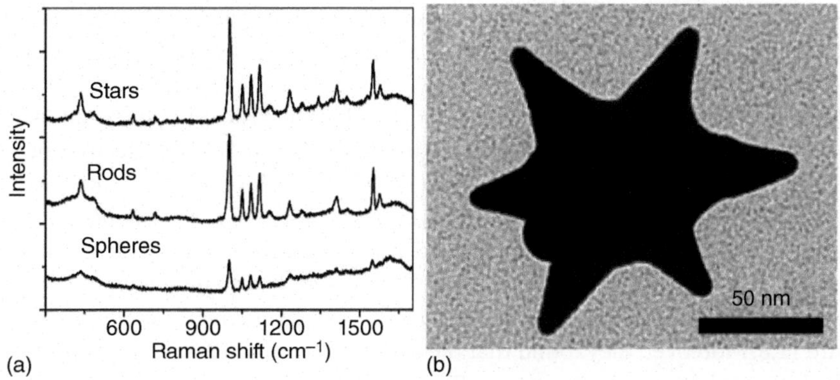

Figure 11.2 (a) Raman spectra of 2-MPy on Au nanostars, nanorods and Au nanospheres, respectively. (b) TEM image of Au nanostar. Source: Reproduced with permission from Esenturk and Walker [25]; © 2009, John Wiley & Sons, Inc.

clearly indicate that increasing of the thickness of the SERS substrate is not able to infinitely improve its Raman enhancement.

Au nanostars are a type of anisotropic plasmonic nanostructure with a star-shape morphology, in which several nanoscale branches or protrusions are protruded from solid spherical nanoparticles [24]. Au nanostars are especially attractive due to their particular structural and optical properties. The strong electromagnetic field generated at the sharp tips of Au nanostars can lead to an extraordinary field enhancement and also the plasmons focused at the core couples with the tips in Au nanostar structures, making it an outstanding candidate for numerous spectroscopic applications. Walker and coworker [25] compared the Raman enhanced effect of 2-mercaptopyridine adsorbed on Au nanosphere, nanorods and nanostars. They found that the SERS activity of Au nanostars was remarkably stronger than that of nanospheres of similar size (Figure 11.2). The Liz-Marzán group extensively studied the SERS activities of Au nanostars for numerous applications [26, 27].

For Au nanostars, there are two distinct SPR features centered around 650–1000 and 500–600 nm. The former SPR band is strong and assigned to the dipolar resonances of tips. While the second one band is weaker and originated from the central core particle. Although Au nanostars are expected to display broad SPR band due to the unavoidable shape poly-dispersity, very high electromagnetic fields is localized at each tip and strongly depended on the tip aperture angle. It is also found that the detection ability depends on not only the roughness, but also on the light source [28]. Meng et al. measured the SERS spectra directly from the Au nanostars colloids by using the 4-MBA ligands as probe molecules. They found that distinct 4-MBA SERS signals can be clearly observed from sharp-branched nanostars and rough Au nanostars under a 1064 nm excitation. Moreover, the sharp branches on Au nanostars provide extra SERS activities as compared to the ones with a rough surface. The enhancement for 4-MBA can be attributed to the electromagnetic enhancement at the vicinity of sharp Au nanostars, as well as the possible chemical enhancement related to the charge transfer from Au to 4-MBA molecules. Considering the excitation wavelength, it is very likely that the additional SERS enhancement of sharp Au nanostars originates from the strong electromagnetic enhancement around the sharp branches, as vividly expressed by the right cartoon expression.

Compared to isolated nanostars, it is reported that the SERS enhancement factor can be strongly increased by decreasing the interparticle gap. Usually, electron-beam lithography (EBL) is a popular approach to fabricate coupled nanostructures with a controllable interparticle spacing, in which case strong electric field can be induced. Chirumamilla and coworkers [29] applied nanostar dimers with an interparticle spacing below 10 nm as enhancement substrate of Raman signal for adenine at very low concentrate (~1 pM) and obtained obvious Raman intensity, as illustrated in Figure 11.3. Moreover, they found that the enhancement in SERS intensity increase as the interparticle spacing decreases (Figure 11.3d).

In addition to EBL approach, DNA origami offers a novel nanostructure with features of precise addressability and defined size for the creation of various

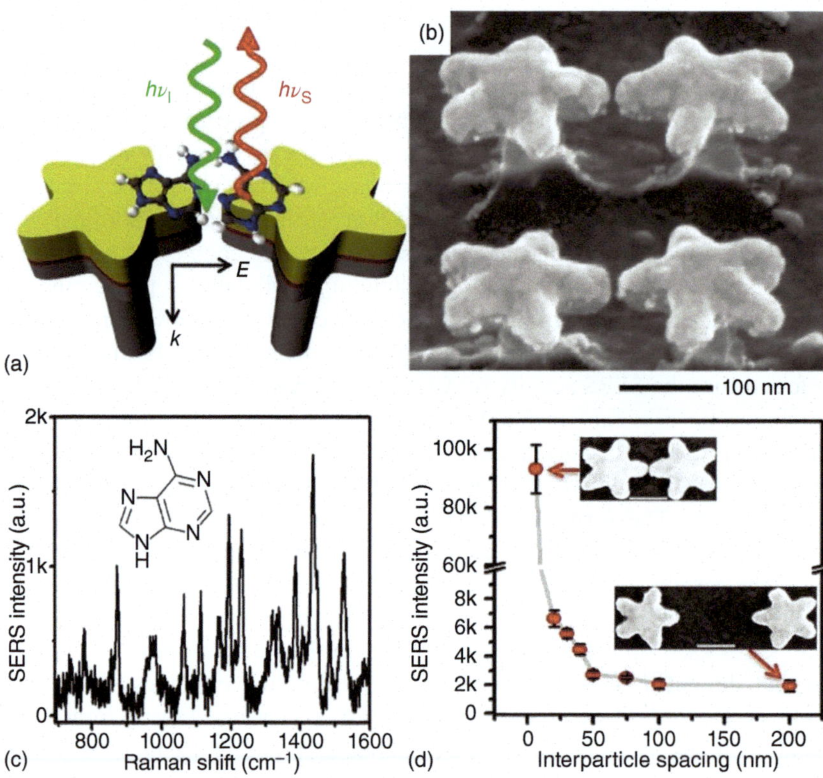

Figure 11.3 (a) Schematic representation of the SERS working principle for a free-standing 3D nanostar dimer. E and k are the electric field and wavevector, respectively. $H\nu_i$ and $h\nu_s$ represent the incoming and scattered photon energy, respectively. (b) SEM images of nanostar dimer arrays. (c) Raman spectrum recorded from adenine that was chemisorbed from a solution at 1-pM concentration. (d) Dependence of the SERS intensity on the IPS while keeping the pillar height constant at 150 nm. Source: Reproduced with permission from Chiruumamilla et al. [29]; © 2014, John Wiley & Sons, Inc.

functional structures. Tanwar et al. reported the investigation on DNA origami directed Au nanostar dimers for single molecule SRES detection [30]. They created Au nanostar dimers with tunable interparticle gap and controlled stoichiometry assembled on DNA origami with Texas red dye attached in the junction of the dimerized origami to act as a Raman reporter. Using 532 nm laser excitation on dimer structures having interparticle gaps of 7 and 13 nm, the mean enhancement factors of Texas red dye were found to be 9×10^9 and 6×10^9. While using 633 nm laser excitation for the same Au nanostar dimers, the SERS enhancement factors of single Texas red dye molecule located in the conjunction region in are 2×10^{10} and 8×10^9, respectively. Higher value of enhancement factor for 633 nm laser can be attributed due to close resonance between wavelength of laser source and SPR of Au nanostars. The distance dependence of enhancement factors was also found to be in good agreement with that of 4-ATP in Au nanostar dimer, as reported by Ma et al. [31]. The highly enhanced electromagnetic field generated by the plasmon coupling between sharp tips and cores of two Au nanostars in the wide conjunction region allows the accommodation and specific detection of large biomolecules. Such DNA-directed assembled nanoantennas with controlled interparticles separation distance and stoichiometry, and well-defined geometry, can be used as excellent substrates in single-molecule SERS spectroscopy.

Similar with Au nanostars, other metals such as Ag, Pt, Pd, Rh and some alloys have also been grown into metallic nanostars, not only due to their different and interesting optical response, but also for their applications in different fields [32, 33]. Regarding optical properties and enhanced spectroscopies, theoretical predictions and proof-of-concept measurements have been carried out, which promise a bright future for noble metal nanostars in a variety of nanoplasmonic applications in connection with the highly focalized electromagnetic fields at tip apexes when certain plasmon modes are excited.

11.2.2 Nanocavity and Porous Nanostructure

As reported, SERS enhancement is dependent on the electromagnetic field which is determined by the nature and morphology of metal nanomaterial. Nanocavities provide an additional degree of freedom to allow the regulation of SPR modes by changing the thickness of shell. The resonance wavelength can be tuned from the visible region into infrared. Moreover, The interaction between inner-core and outer-shell induces nanostructures with additional plasmon resonance modes due to the plasmon hybridization [34]. The compound morphology endows nanostructure with more excellent properties. For example, multilayered metal nanoshells exhibit more exciting optical performances than convenient single-layered nanoshells. And the theory results indicated that multilayered nanoshells can focus the light on the inner-core surface, resulting in a significant near-field enhancement [35, 36], which is promising to serve as a SERS substrate.

Halas and coworkers [37] illustrated that Au and Ag nanoshells as SERS substrates can yield much larger SERS enhancement. They also studied the effect of nanoparticle densities and tunable plasmon absorption of gold nanoshells as SERS substrates.

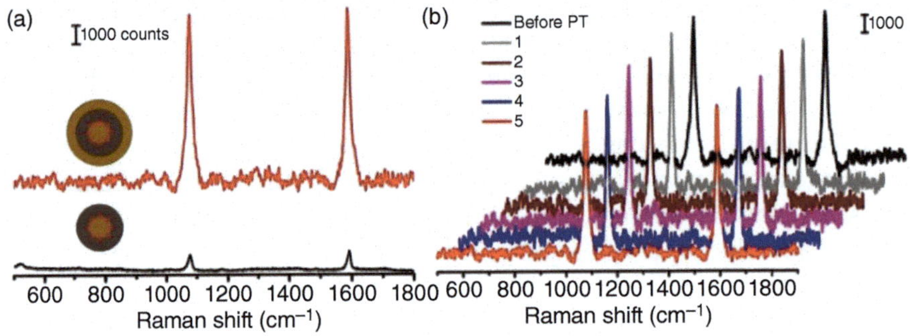

Figure 11.4 (a) Raman spectra of MBA-labeled multilayered Au nanoshells and Au@organosilicas with the same amount of inner Au nanoparticles. (b) Raman spectra of MBA-labeled multilayered Au nanoshells before laser heating and after five repeated cycles of laser heating. Source: Adapted from Gao et al. [38].

The particle density can be adjusted by the deposition time. Large Raman scattering enhancement factors of 10^9–10^{10} were observed for *p*-mercaptoaniline. Although the SERS intensity is a linear function as nanoparticle density increase, the SERS response is independent of the particle density, which is attributed to the entire particle surface area. Each individual nanoshell can be used as standalone SERS nanosensors. Moreover, the coupling between the inner-core and outer-shell can tune the SPR peak to near-infrared region (biological window). Gao et al. [38] synthesized uniform-sized, monodispersed and stable multilayered core–shell nanostructures, which express remarkable SERS signal compared with Au@organosilicas, as shown in Figure 11.4a. And stable SERS signal under repeated cycles of laser irradiation can be detected (Figure 11.4b). Therefore, multilayered nanoshell can be applied into SERS detection in vivo and photothermal therapy.

Another typical nanostructure with many hot spots is porous nanostructure. Unlike solid NPs, porous or hollow metal nanostructures exhibit excellent catalytic properties due to their high surface-to-volume ratios and abundant catalytic active sites. Porous Au–Ag alloy nanorods (P-AuAg NRs) represent a group of catalysts whose aspect ratios and compositions can be tuned to achieve the optimal catalytic performance toward specific reactions. For example, longitudinal surface plasmon resonance peaks of the P-AuAg NRs can be regulated across visible to near-IR regions. Ke et al. reported the precise control of nanoporous in rod-like alloy nanostructures and used the P-AuAg NRs as a Raman probe for *in-situ* monitoring of the catalytic oxidation of tetramethylbenzidine (TMB) and reduction of 4-nitrothiophenol (4-NTP) [39, 40]. Figure 11.5a, b show the schematic illustration and corresponding TEM images for the synthesis process of P-AuAg NRs by a silica protected annealing method.

Enhanced SERS detection of 4-NTP adsorbed on P-AuAgNR$_{741}$ (P-AuAgNR exhibiting SPR$_L$ at 741 nm) was compared to the detection of the non-resonant probe molecule upon adsorption on Au@Ag NRs, S-AuAg NRs, and Au NRs (Figure 11.5c). Au@Ag NRs generated the strongest SERS signal of 4-NTP among tested nanorods, which could be attributed to the stronger electric field interaction

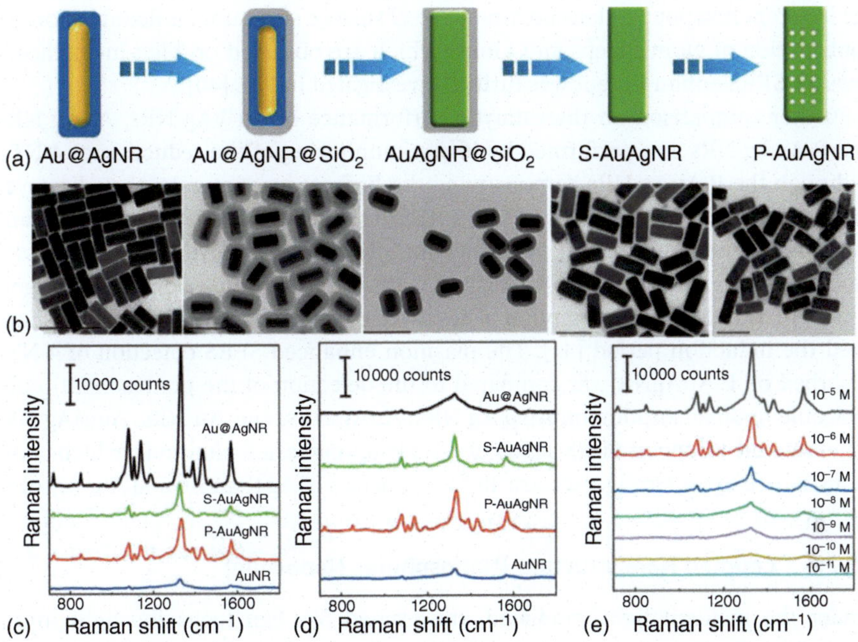

Figure 11.5 Schematic illustration on the synthesis process of P-AuAg NRs (including core–shell Au@Ag NRs, mesoporous SiO_2 coated Au@Ag NRs, alloyed AuAg NRs within mesoporous SiO_2, solid alloyed AuAg NRs and porous alloyed AuAg NRs) and (b) corresponding TEM images. SERS spectra of 10^{-6} M 4-NTP adsorbed on Au@Ag NRs, P-AuAgNR$_{741}$, S-AuAg NRs and Au NRs before (c) and after (d) etching in H_2O_2 (1.5%, v/v) and $NH_3{\cdot}H_2O$ (1%, v/v). (e) SERS spectra of 4-NTP adsorbed on P-AuAgNR$_{741}$ with decreasing 4-NTP concentration from 10^{-5} to 10^{-11} M. Source: Reproduced with permission from Ke et al. [39]; © 2021 Royal Society of Chemistry.

of Ag than Au and Au–Ag alloy. Nevertheless, the distinct bands and SERS intensity at the highest-intensity band (1334 cm^{-1}) of the 4-NTP adsorbed on Au@Ag NRs significantly diminished upon exposure to an aqueous mixture of H_2O_2 and $NH_3{\cdot}H_2O$ (Figure 11.5d), which could be addressed to oxidative etching of Ag in basic solution catalyzed by H_2O_2. In contrast, Au NRs and P-AuAgNR$_{741}$ produced similar bands (Figure 11.5c,d) and stable SERS signals before and after etching. While the SERS signals generated in the presence of Au NRs were weak in both cases, P-AuAgNR$_{741}$ demonstrated enhanced SERS detection of 4-NTP upon the etching. The significant SERS enhancement observed for P-AuAgNR$_{741}$ was attributed to abundant SERS hotspots generated by nanopores and distinct corners, and residual silver. The enhanced SERS detection of 4-NTP in the presence of P-AuAgNR$_{741}$ was further evaluated by reducing the concentration of 4-NTP from 10^{-5} to 10^{-11} M. As shown in Figure 11.5e, SERS signals gradually dissipated with decreasing 4-NTP concentration.

Increasing porosity of P-AuAg NRs is hypothesized to augment electromagnetic enhancement by generating a large number of hotspots, and chemical enhancement by generating a high surface area for adsorption of Raman reporter molecules on the inherent hotspots. Details on atomic-level surface structure for deducing on the contribution of atomic steps and kinks (which are observed on high-index facets) towards SERS enhancement was further investigated in Ref. [40].

Further comparison on the catalytic performance of Au@Ag NRs, AuAg NRs, and P-AuAg NRs was performed by *in-situ* monitoring of the reduction of 4-NTP. Although the P-AuAg NRs demonstrated the highest reaction rate, they displayed the longest induction time, while AuAg NRs displayed the lowest reaction rate and the smallest induction period. This was likely because the P-AuAg NRs possessed more complex nanopores that were filled with a large amount of dissolved oxygen. As for AuAg NRs, the consumption of dissolved oxygen during N_2H_4 addition shortened the induction period [41]. The plasmon enhanced SERS detection of 4-NTP adsorbed on P-AuAgNR was compared to the detection of the nonresonant probe molecule upon adsorption on Au@Ag NRs, AuAg NRs, and Au NRs. Au@Ag NRs generated the strongest SERS signal of 4-NTP among tested nanorods, which could be attributed to the stronger electric field interaction of Ag than Au and Au–Ag alloy.

11.2.3 Coupled Nanostructure (Nanospheres Nanocube)

Under the incident light irradiated, the excited SPs can focus the light into a nanoscale edge, tips, or gaps, leading to the enhancement of local EM field by 2–5 orders. The plasmonic nanomaterials serve as optical antennae to transmit the local Raman signals from near field to the far field. It has been reported the SERS intensity of molecules located at the nanogaps of interparticle can be extremely enhanced. For example, the gap of Au nanosphere dimer is reduced from 10 to 2 nm, the SERS enhancement factor arise from 10^5 to 10^9 [42, 43]. However, the local EM field and SERS intensity cannot be infinitely increased as the gap size further decreases due to the quantum tunneling effects. In this case, the nonlocal effect plays a dominant role, which can induce the generation of hotspots on the nanoparticle surface. The generated hotspots can trigger the catalysis and change the physical and chemical properties of adsorbed molecules. Considering the different nature of EM coupling in plasmonic nanostructures, Ding et al. classified three generations of SERS hotspots, involved in single plasmonic structures, coupled plasmonic nanostructures, and hybrid structures [43].

Lee and Yoon assembled heterodimers where Au nanocubes with edge length of 84 nm are linked to Au nanospheres with diameter 55 nm through their vertices or edges via 1,8-octanedithiol. Regioselective disintegration of CTAB bilayers on the vertices and edges of Au NCs using acetonitrile permits the formation of Au NC-AuNS dimers in high yield. Strong plasmonic interactions between the Au NCs and AuNSs produce the longitudinal plasmon coupling band at 790 nm, significantly red-shifted from the surface plasmon resonance band of the isolated Au NCs ($\lambda = 581$ nm) or AuNSs ($\lambda = 534$ nm). FDTD simulations indicated that localized electric fields confined to the nanogaps between the Au NCs and AuNSs also generate a strong SERS signal. In Figure 11.6, the strong Raman peak of the C–S stretching mode from the transform of 1,8-octanedithiol (720 cm^{-1}), marked with a star, and the absence of a peak from the gauche form (620–660 cm^{-1}) indicate

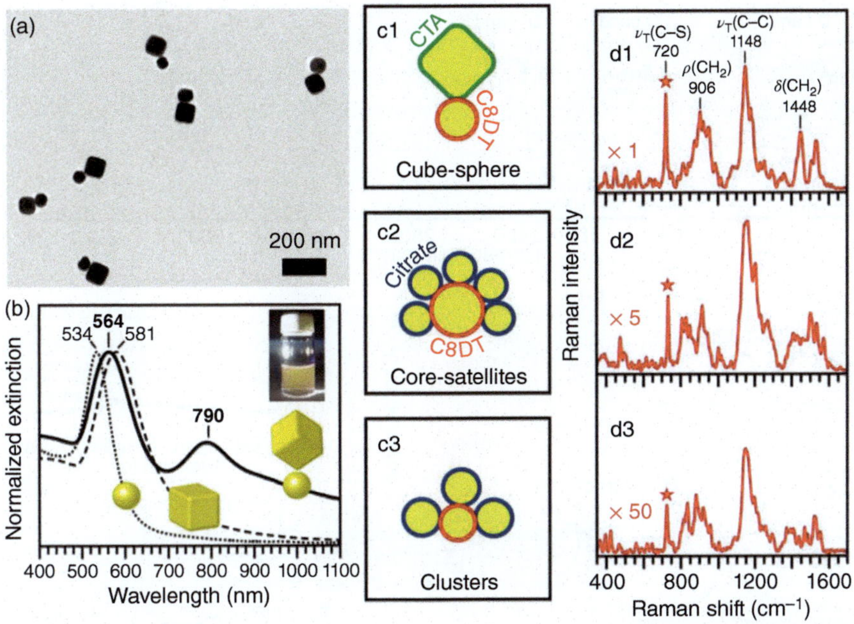

Figure 11.6 (a) TEM images of the prepared Au NC-AuNS dimers, (b) UV–Vis extinction spectrum for the Au NC-Au NS dimers dispersed in solution (solid line). For comparison, the UV–Vis extinction spectra for the Au NSs (dotted line) and Au NCs (dashed line) used to assemble these dimers are included after being normalized to the 564 nm band intensity. The inset is a photograph of the sample containing the dispersed Au NC-Au NS dimers. (c1–c3) Assembly schemes of three nanoassemblies and corresponding Raman spectra, for the nanoassemblies supported on glass slides: (a) Au NC (84 nm)-Au NS (55 nm) dimers, (b) Au NS core (92 nm)-AuNS satellite (55 nm) assemblies, and (c) AuNS clusters that consist of one 55 nm Au NS and three 58 nm Au NSs. In the Raman spectra, ν_T, ρ, and δ denote the stretching (of transform), rocking, and bending vibrational modes, respectively. Source: Reproduced with permission from Lee and Yoon [44]; © 2015 American Chemical Society.

that 1,8-octanedithiol constructs highly ordered SAM structures between Au NCs and AuNSs [44]. The intense Raman signal observed for the Au NC-AuNS dimers, despite the small number of 1,8-octanedithiol molecules confined to the nanogap region between the Au NC vertex and AuNS surface, confirms that strong SERS occurs in the dimers. For comparison, they prepared two nanoassemblies consisting of only spherical nanoparticles: core-satellites and clusters. They observed the Raman spectrum for 1,8-octanedithiol from the Au NC-Au NS dimers with a 1×10^{10} enhancement factor (EF), which is much larger than that for assemblies consisting of only Au nanosphere, such as core-satellites (EF = ~10^8) and clusters (EF = ~10^7).

11.2.4 Other SERS Substrates

Advanced 2D materials are continuously developed to act as new SERS substrates [45]. Graphene with a perfect 2D crystalline structure made of sp^2 bonded

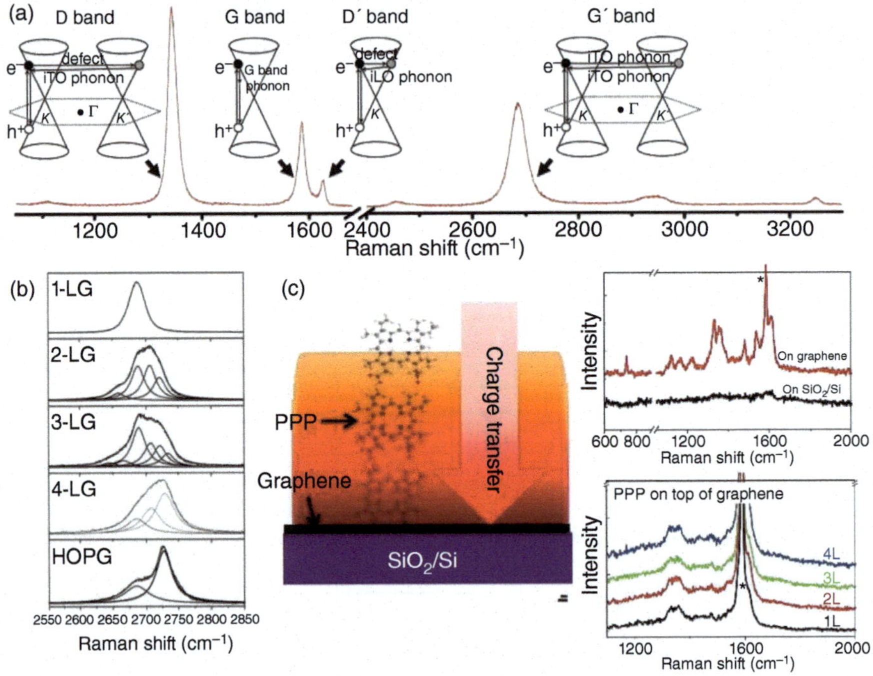

Figure 11.7 (a) Typical Raman spectrum of defect-containing graphene measured with a 514.5 nm laser. Illustrations of the first-order Raman scattering process of G band and second-order Raman scattering processes of D, D′, G′ band in graphene are shown together. (b) Layer-dependent G′ band of 1, 2, 3, 4-layer graphene and HOPG at 514.5 nm excitation. (c) Left: Schematic illustration of different layers of PPP on graphene. Top right: Comparisons of Raman effect of *PPP* deposition on graphene (red line) and on the SiO_2/Si substrate (black line). Bottom right: Raman spectra of PPP with different layers on the top of graphene. Source*:* (a, b) Reproduced with permission from Malard et al. [46]; @2009, Elsevier. (c) Reproduced with permission from Ling and Zhang [47]; @2010, Wiley-VCH.

carbon atoms expresses some unique advantages for SERS. The defect-free graphene shows two characteristic peaks of G band at ~1585 cm^{-1} and G′ band at ~2685 cm^{-1}, as shown Figure 11.7a. Moreover, several other peaks in disordered graphene or at the edges would appear at ~1345 cm^{-1} (D band) and 1625 cm^{-1} (D′ band). In the consideration of Raman features of graphene, the number of graphene layers can be distinguished, and the type of defects and the degree of doping can be monitored [46, 48]. Ferrari et al. found that the G′ band of graphene with different layers shown a different shape, position and intensity (Figure 11.7b). For monolayer structure, the G′ peak is a perfect single Lorentzian feature, while it shows a broad band as the layer numbers increases. Graphene possesses a high structure-sensitive performance.

Moreover, graphene with large surface area represents superior molecule adsorption ability, it can effectively quench the fluorescence of dyes molecules. Raman signal of molecular can be enhance by graphene labeling, where low concentration of dye molecules can be monitored. Feng and coworkers [49] illustrated that the Raman signal of RhB on CVD N-doped graphene can be observed even at the low concentration to 5×10^{-11} M. The phenomenon is attributed to the closer level between the lowest unoccupied molecular orbital of RhB molecules and the Fermi level of N-doped graphene. Graphene substrates are usually applied to explore the interaction between graphene and molecules.

Due to smooth surface, high optical transmission in the visible range and surface plasmon in the scope of terahertz, graphene substrates serve as a charge transfer SERS substrate without electromagnetic enhancement. Ling et al. [47] discovered some emerging bands of Raman signals for protoporphyrin *IX* (*PPP*) on graphene substrate, while no clear Raman signal was found on a SiO_2/Si substrate (Figure 11.7c). The intriguing and inspiring discovery was originated from the graphene-molecule interaction, called by graphene-enhanced Raman scattering. Moreover, they found that the GERS intensities of *PPP* with one-four layers on the graphene have no obvious change, which illustrated that the first layer on the graphene plays an important role and contributes the dominant part of Raman enhancement. Yoon et al. [50] systematically investigated the Raman scattering of graphene transferred onto another substrate via different methods. It was found that poly methyl methacrylate transferred graphene induced six-times Raman signal enhancement in comparation of that obtained by thermal release tape. Although graphene can induce an excellent charge transfer, it is not ideal Raman candidate due to low Raman cross section and the richness of Raman peak signatures.

It has been known that the enhancement mechanism of Raman signal is influenced by the electromagnetic mechanism and chemical mechanism. Noble metal nanostructures normally display high electromagnetic fields. The combination of graphene and noble metallic nanoparticles can synergistically bring improve SERS and serve as a high-performance SERS substrate. Different types of composite substrates consisted of graphene and metal nanostructures have been designed, mainly including graphene covered metal nanostructures substrates, graphene separated metal nanostructures substrates, graphene-metal nanostructures-graphene sandwich substrates. In these cases, graphene provides a flat platform for the uniform deposition of nanoparticles to protect nanoparticles from oxidizing or corrosion. Noble metallic nanoparticles-graphene hybrid substrates exhibited superior enhancement of Raman signal of graphene, due to that the electromagnetic enhancement of metal nanostructures could penetrate single-atom graphene. Among these three types, graphene-metal nanostructures-graphene sandwich substrates present good long-term stability because graphene layers protect metal nanostructures from oxidation in top and bottom directions [51].

Representative investigation on graphene covered metal nanostructures mainly involved in graphene-covered Au/Ag nanoarrays [52, 53]. Wang and coworkers demonstrated single molecule detection of dopamine and serotonin using a graphene-Au nanopyramid platform. By comparing SERS of single layer graphene onto the Au tipped surface, flat Au surface, and flat SiO_2 surface, they obtained graphene on Au tips presented an obvious enhancement, which was attributed to the much more hotspots. It has been known that the more intrinsic hotspots metal nanostructure arrays, the higher enhancement factor could achieve.

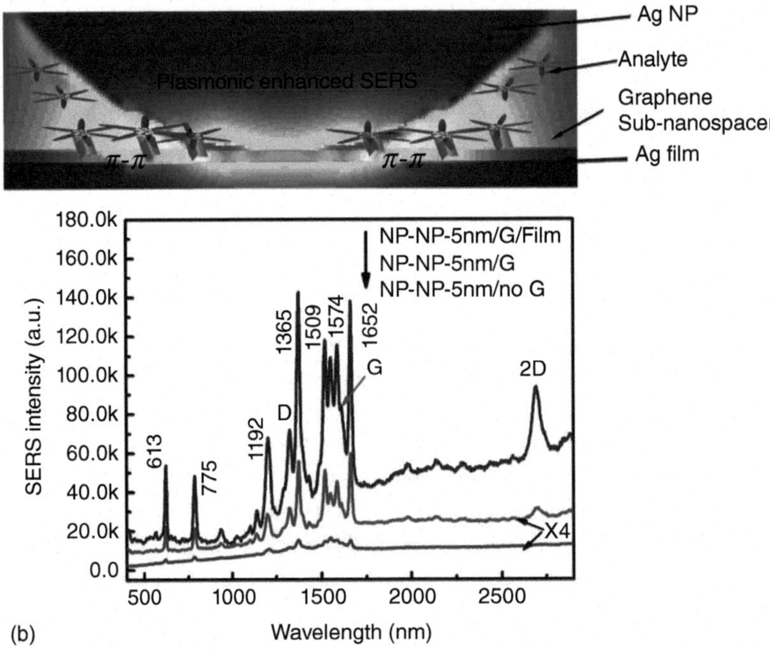

Figure 11.8 (a) Schematic illustration of detection of organic molecule by using G-NFG system. (b) SERS performance of G-NFG (NP-NP-5 nm/G/Film) system compared with that of NP-NP-5 nm/no G and NP-NP-5 nm/G structures in detecting R6G with 1×10^{-9} M in ethanol. Source: Reproduced with permission from Li et al. [54]; © 2014, John Wiley & Sons, Inc.

For graphene separated metal nanostructures, graphene placed in the middle of the two metallic layers provide a charge transfer-sensitive gap. The plasmon coupling of metallic layers generated strong electromagnetic field, and graphene induced additional chemical enhancement. Li and coworkers [54] found a remarkable SERS by introduced monolayer graphene as nanospacer into metal film-metal nanoparticle coupling system (G-NFG system). Compared the Ag nanoparticles with interparticle gap (5 nm) system on silica glass (NP-NP-5 nm), the Raman intensity of R6G on the G-NFG substrate is approximate to 10 times higher (Figure 11.8). Zhao and coworkers [53] fabricated an Au nanoparticles-graphene-Ag nanoparticles structure on quart exhibited high enhancement factor over 10^9 for the detection of RhB, R6G and TNT. The synergistic effects between plasmonic metal nanostructures and graphene have particular contribution in high-performance SERS substrates.

11.3 SERS for Detection and Sensor

As it is known that peak position of surface plasmon resonance (SPR) spectroscopy of metallic nanoparticles is very sensitive not only to the shape and size of nanoparticles, but also to the local environment and configure, including the adsorbate molecular on the nanoparticles and coupling state of nanoparticles. However, the SPR (usually broad band) red or blue shifting does not present any signals for molecular. SERS technique led by the electromagnetic-field enhancement of SPR effect is a powerful technique for detecting chemical and biological molecules [55, 56]. In this section, we mainly introduce the molecular detection and real-time monitor of catalytic reaction by means of SERS.

11.3.1 SERS for a Molecular Vibration "Fingerprint"

Remarkably, signal from the species adsorbed on the surface of metal nanostructure can be enhanced greatly, while that from the free species in the solution still remains unchanged [57]. Therefore, SERS can provide abundant information about target molecules on the surface or interface of nanoparticles. So far, the molecular detection by means of SERS based on nanoparticles has achieved to the single-molecule level [6, 57].

In Chapters 4 and 7, we introduced the fabrication of various of new composite nanostructures [39]. Some of them possess excellent SERS effect and can be used to detect molecules. For example, Au–Ag alloy nanostructures with high stability are a promising Raman enhanced substrate in a harsh environment. Figure 11.9a, b display the Raman spectra of 4-NTP attached on Au@Ag core-shell NR, solid Au-Ag alloy NRs (S-AuAg NRs), porous Au–Ag alloy NRs (P-AuAg NRs) and Au NRs before and after etching in H_2O_2 and $NH_3{\cdot}H_2O$. Obviously, SERS of Au NRs is relatively weak and the Raman intensity on Au–Ag alloy NRs is more stable than Au@Ag core–shell NRs, since Ag shell is apt to be etched by oxidation. Moreover, P-AuAg NRs with more hot spots present more elaborate SERS spectra of 4-NTP. P-AuAgNR as SERS probe can detect 4-NTP at low concentration to 10^{-10} M (Figure 11.9c).

It should be pointed out that this excitation wavelength at 785 nm inducing strongest SERS is not the SRP wavelength at 741 nm of P-AuAg NRs. This phenomenon is attributed to the strong local field effect of Ag in alloy NRs. It was reported that the maximum near-field enhancement effect generated in the lower energy region (long-wavelength region), compared to far field LSPR. Therefore, P-AuAg NRs with SRP wavelength at 741 nm presents the strongest SERS intensity. Similarly, SERS spectra of R6G on P-AuAg NRs with different resonance wavelength also show the same tendency.

Detailed exploration was also carried out on the SERS activities of different plasmonic nanostructures including AuBPs, AuBP-embedded composite nanostructures [58]. The enhanced intensity is dependent on the excitation light wavelength. Due to the excellent plasmon performance of Au, these composite nanostructures attached by 4-NTP under a certain wavelength (785 nm) light exciting, shown remarkable SERS activities shown in Figure 11.9d, while the feature peaks of 4-NTP cannot be detected at 633 or 514 nm excitation (Figure 11.9e). In consideration of the intensity of SERS, enhancement factors of 4-NTP modified different nanostructures can be obtained by the formula:

$$\mathrm{EF} = \frac{I_{\mathrm{SERS}}}{I_{\mathrm{NRS}}} \times \frac{N_{\mathrm{vol}}}{N_{\mathrm{surf}}}.$$

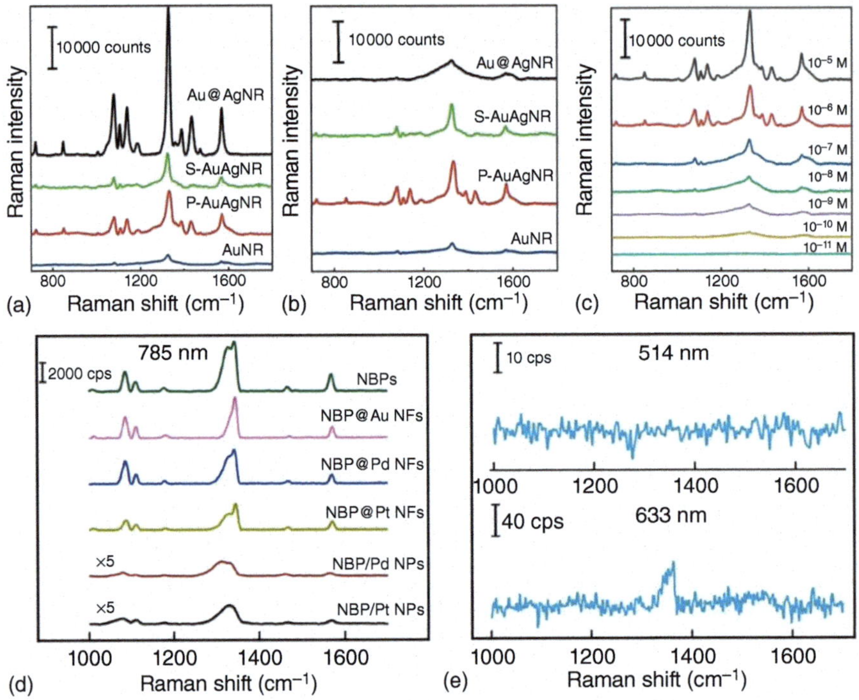

Figure 11.9 Raman spectra of 4-NTP attached on Au@Ag core-shell NR, solid Au–Ag alloy NRs (S-AuAg NRs), porous Au–Ag alloy NRs (P-AuAg NRs) and Au NRs (a) before and (b) after etching in H_2O_2 and $NH_3{\cdot}H_2O$. (c) Raman spectra of 4-NTP adsorbed on P-AuAg NR at decreasing 4-NTP concentration from 10^{-5} to 10^{-11} M. (d) Raman spectra of 4-NTP modified different nanostructures at 785 nm excitation. (e) Raman spectra corresponding Au NBP-embedded Au nanoframes at 514 and 633 nm excitations. Source: (a–c) Reproduced with permission from Ke et al. [39]; @2021 Royal Society of Chemistry. (d, e) Reproduced with permission from Zhu et al. [58]/with permission of Royal Society of Chemistry/Public Domain CC BY 4.0.

where I_{SERS} and I_{NRS} are the intensity areas of the characteristic band at 1572 cm^{-1} for pure 4-NTP and plasmonic enhanced 4-NTP samples, respectively. N_{surf} is the molecule number of 4-NTP adsorbed on the surface of nanostructures, and N_{vol} is the molecule number of 4-NTP dispersed in solution. The corresponding values are displayed in Table 11.1. When the wavelength is mismatched with SPR wavelength of nanostructures, the intensity is weak. The result demonstrated the advantage of SPR enhancement.

11.3.2 Real-Time Detection of Catalytic Reaction

The SERS as a sensitive surface spectroscopic techniques can apply into the real-time detection on the reaction in the interface of solid–liquid, solid–gas, or solid–solid, to insight into the nature of transient reaction intermediates and mechanism. Taking the reaction of 4-NTP by $NaBH_4$ as example, the variation of SERS signal can reflect reduction reaction 4-NTP to 4-ATP.

Table 11.1 Values of N_{surf}, I_{SERS}, N_{vol}, and I_{NRS} for Au NBP, Au NBP-embedded ultrathin metal (Au, Pd, Pt) nanoframe, and (AuNBP)/(Pd, Pt) nanostructure samples.

Sample	N_{surf}	I_{SERS}	N_{vol} (10^{22})	I_{NRS}	EF (10^7)
Au NBPs	2.27	27 054	6	588	12
Au nanoframes	5.16	16 392			3.3
Pd nanoframes	5.14	12 928			2.6
Pt nanoframes	5.34	12 968			2.4
NBP@Pd nanoframes	3.01	753			0.26
NBP@Pt nanoframes	3.51	2149			0.63

Source: Reproduced from Zhu et al. [58]/with permission of Royal Society of Chemistry/Public Domain CC BY 4.0.

The catalytic reaction from 4-NTP to 4-ATP has been a model reaction, which can be detected via the UV–Vis-NIR spectra. Due to the Raman shift difference between the nitro and amine functional groups, the reductant process from the nitro group to amine group can be detected *in-situ* via the adsorption of single molecular layer of 4-NTP on the surface of catalysis.

The whole reaction process contains three steps (Figure 11.10a) [40]. Firstly, the N—O bond in 4-NTP fractures to form nitro and hydroxylamine. And then, these two

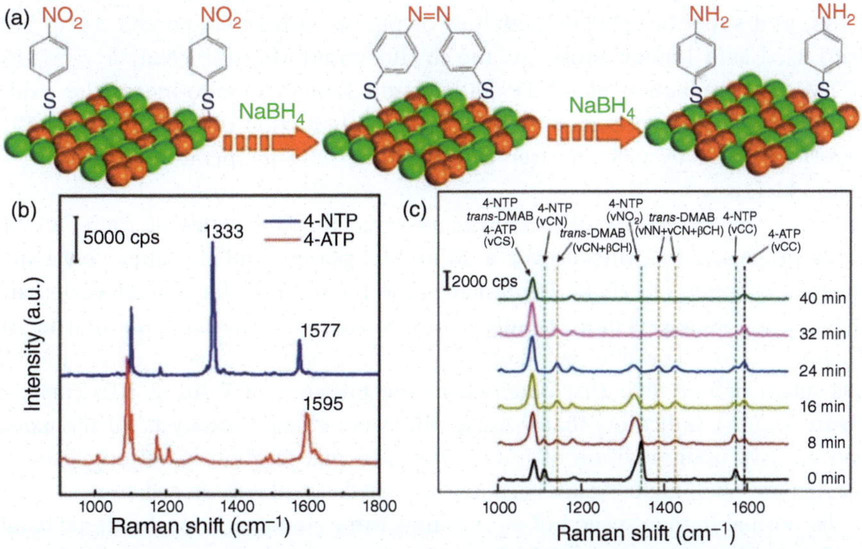

Figure 11.10 (a) Schematic diagram for the reduction of 4-NTP to 4-ATP. (b) Raman spectra of 4-NTR and 4-ATP. (c) Time-dependent SERS spectra of the reduction of 4-NTP attached on Au NBP embedded Au nanoframes by NaBH4 under excitation at 785 nm. Source: (a,b) Reproduced with permission from Ke et al. [40]; @2021 Elsevier. (c) Reproduced with permission from Zhu et al. [58]/with permission of Royal Society of Chemistry/Public Domain CC BY 4.0.

intermediates condense into DMAB. Finally, the hydrogenation leads to the breakfast of the azo bond and generates 4-ATP. The transformation involved in the bond vibration, fracture, and binding will present typical SERS signals. By combining the catalytic properties of metal atoms (Au, Pt, or Pd), AuBP-embedded nanoframes with multiple active sites can serve as a unique plasmon enhanced SERS probe to observe *in-situ* the reduction of 4-NTP to 4-ATP by $NaBH_4$ [58]. The Raman spectra of powdery 4-NTR and 4-ATP samples illustrated in Figure 11.10b are used to distinguish the products.

Figure 11.10c illustrate the time-dependent SERS spectra of the reduction of 4-NTP attached on nanostructures by $NaBH_4$ under excitation at 785 nm. At $t = 0$ min, there are three vibrational bands, which corresponds to C–N stretching (ν_{CN}, at 1108 cm^{-1}), O–N–O stretching (νNO_2, at 1336 cm^{-1}), and C–C stretching phenyl ring (ν_{CC}, at 1572 cm^{-1}). When the reaction progresses to 8 minutes, four new bands appear in the SERS spectrum, companying with the weaken characteristic bands at 1336 and 1572 cm^{-1}. Typically, the new band at 1595 cm^{-1} is ascribed to the νCC of 4-ATP. It has been demonstrated that the surface catalytic reaction of DMAB produced from 4-NTP via the SERS formation presented strong dependence on plasmonic substrates, irradiation wavelength and time in the case of surface plasmon enhanced catalytic reaction [59]. The hot electrons from the decay of SP provided the required electrons involved in the chemical reaction and can drive the surface catalytic reaction [60].

Similar results were achieved using porous nanostructures. Combining the merits of the plasmonic properties of noble metal NRs, excellent stabilities of alloys, and superior catalytic activities of porous structures, porous AuAg NRs were used as a Raman probe for the in-situ monitoring the catalytic oxidation of TMB and reduction of 4-NTP [40]. *In-situ* Raman monitoring of the oxidation of TMB in the presence of H_2O_2 and the reduction of 4-NTP with $NaBH_4$ were performed by 638 nm laser excitation at room temperature, as shown in Figure 11.11a.

During the peroxidase-like catalytic process, the intensities of these Raman peaks progressively climbed over a 20-minute period, and no other new bands emerged (Figures 11.11b, c). Remarkably, the intensity of the characteristic band at 1611 cm^{-1} increased more rapidly with P-AuAg NRs than that without catalysts (Figure 11.11d). The blue color of the nanocatalyst system without P-AuAgNR was much lighter than that obtained in the presence of P-AuAg NRs (inset of Figure 11.11d), indicating that P-AuAg NRs were efficient catalysts for the oxidation of TMB upon addition of H_2O_2. However, this reaction is not applicable to Au-Ag core–shell nanocrystals because their shells are slowly etched away.

The normal Raman spectra of 4-NTP show three characteristic vibrational bands at 1080, 1333, and 1577 cm^{-1}, which were attributed to the C–S stretching, O–N–O stretching, and phenol-ring mode [61], respectively. As the hydrogenation of 4-NTP proceeded, time-resolved SERS signals clearly demonstrated that the intensities of the bands at 1340 and 1572 cm^{-1} decreased gradually, accompanied by the emergence of a new band at 1590 cm^{-1} corresponding to the phenol-ring mode of 4-ATP (Figure 11.11e). Comparatively, Raman spectra have no obvious change

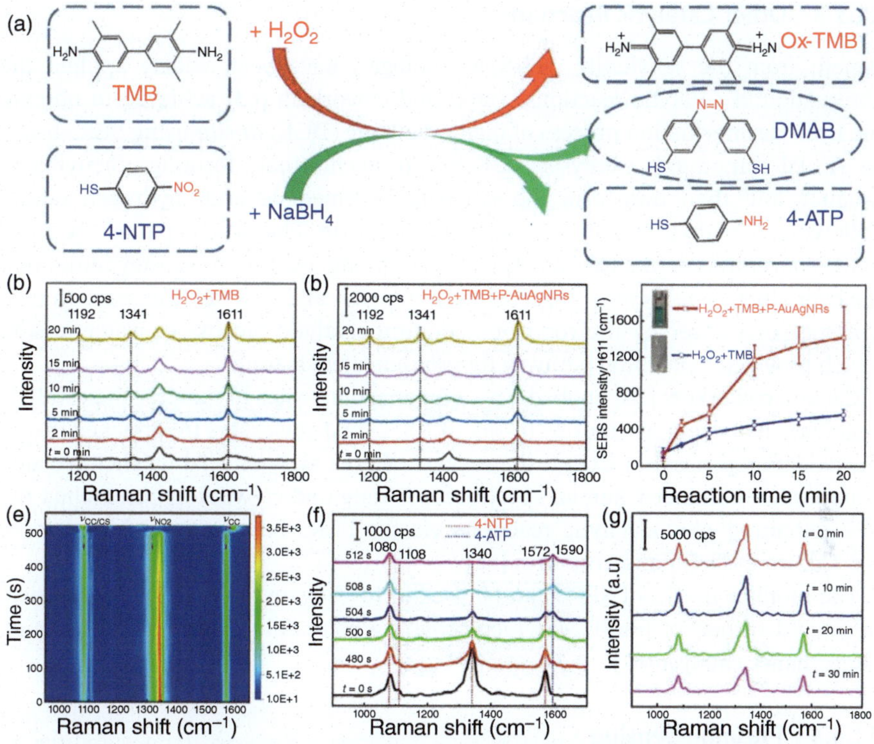

Figure 11.11 (a) schematic illustrating of the oxidation of TMB by H_2O_2 and the reduction of 4-NTP by $NaBH_4$ using P-AuAg NRs nanocatalysts. (b, c) The SERS spectra of TMB recorded at different time intervals during the peroxidase-like catalytic reaction in the presence of H_2O_2 and (d) the corresponding SERS intensities at 1611 cm^{-1} versus reaction time. (e) Color-coded intensity map of time-dependent SERS signals obtained from 4-NTP adsorbed on the surfaces of P-AuAg NRs (etching time = 30 minutes) at different reaction times. (f) Representative SERS signals acquired from 4-NTP functionalized on P-AuAg NRs at different reaction times (f) in the presence of $NaBH_4$ and (g) in the absent of $NaBH_4$. Source: Reproduced with permission from Ke et al. [40]; © 2021 ELSEVIER.

without the addition of $NaBH_4$ (Figure 11.11f), which excludes the possibility of photo-transformation for 4-NTP.

Apart from the reduction reaction, LSPR can also induce the oxidation reaction. For example, 4-ATP adsorbed on the Au (or Ag) substrate under the irradiation can be oxidized, leading to the generation of DMAB. Based on the SERS characteristic, it was demonstrated that the generation of DMAB was dependent on the reaction parameters, such as laser power, light wavelength, substrate material, surrounding atmosphere and pH value. The increased photoelectron energy is benefit for the generation of positive holes [62]. Further researches on the effect of surrounding atmosphere have been carried out by Huang and coworkers, which illustrated the oxide layer on the surface of Au–Ag nanostructures in O_2 atmosphere played a role of oxidizing agent. Therefore, SiO_2 coated on the surface of Au nanostructures will impede the oxidation reaction. As a result, the obtained SERS signal only reflect the nature of 4-ATP.

11.3.3 Other Catalytic Reaction

Benefit from the sensitivity, SERS technologies have been widely applied the observation of catalytic reactions. Halas and coworkers [63] achieved to observe the hydrodechlorination process of dichloroethene (DCE) *in-situ* using SERS based on Au-Pd compounds catalysis. DCE first to dechlorinate forming ethenylidene product, and then transform the ethylene intermediate after hydrogeneration. Following, the ethylene hydrogenates to form ethane. They also explored the effect of the concentration of DCE and H_2 on reaction rate. Furthermore, they found that the byproducts would cover the catalytic active sites and decrease the reaction rate. This research observed the room temperature catalytic process in water in-situ, which provided a potential pathway for groundwater cleanup.

Cai and coworkers [64] applied Au-CuS nanocomposites to monitor the catalytic degradation of dye molecules by *in-situ* SERS. Li and coworkers [65] tracked *in-situ* the oxidation process of CO on the surface of PtFe alloy and Pd by shell-isolated nanoparticle-enhanced Raman spectroscopy. Yang and co-workers researched the SPR enhanced photocatalytic reactions via the SERS based on the bifunctional Au–CdS core–shell nanostructures [66].

The monitor *in-situ* of SERS can be used to investigate the reaction mechanism of catalyst surface or interface, which provide a potential guidance to design high performance catalyst [67].

11.3.4 Plasmon Sensing

Since the first application of the SPR phenomenon for gas detection and biologic sensor in 1982, SPR sensing technology has been widely used for the detection of biological and chemical analytes, environmental monitoring, and medical diagnostics in the past two decades, including drug screening, food safety, and security [55], and pH sensing. Furthermore, the SPR based sensing approaches have grown in power, accuracy, selectivity, and sensitivity. They are also becoming more widely used. Increasing the electric field intensity around an SPR sensor causes it to become more sensitive to changes in the surrounding medium [68].

Two types of SPR sensors have been developed, each with a distinct design: the metallic grating SPR sensor and the racetrack resonator SPR sensor. For the design of the grating-based SPR hydrogen sensor, the sensitivity, as well as the width of the SPR curves and reflective amplitude, are all taken into account simultaneously, and a set of optimized structural parameters is presented [69]. The racetrack resonator SPR sensor has a broad linear detection range of analyte refractive index and a high extinction ratio. Furthermore, the plasmonic resonator architecture offers a promising realization for extremely effective modulators and wide band transmitters and receivers [70]. Au NRs have a number of characteristics that make them an excellent sensing platform, including their simplicity of synthesis, chemical stability, photostability, and biocompatibility.

Because of their great confinement and increase in the electromagnetic field near the surface, both the SPP and LSPR based structures are attractive platforms for optical sensors. In contrast to the SPP structure, which has a typical decay length of

100–600 nm, the LSPR structure has a substantially lower decay length of 5–60 nm. In this way, it is believed that the SPP structure is superior for the detection of bigger biomolecules, whilst the LSPR structure is superior for the detection of smaller biomolecules. Optical plasmonic sensors have previously demonstrated the capacity to detect changes in real time without the need for labels, as well as remarkable sensitivity. They have been used successfully in a variety of fields, including chemistry, biology, medicine, and food safety, among others. In general, the single mode of SPP and LSPR sensors, as well as their hybrid modes, serve as the primary driving forces behind these sensors. SPP and LSPR have the benefit of being adjustable in terms of wavelength, intensity, and phase [71].

Using the refractive index change-induced plasmon shift as a starting point, scientists have been able to fabricate ultrasensitive, light-weight, and remotely accessible chemical and biological sensors [72]. Due to their larger electric field enhancements and polarizabilities, Au NRs and other nanostructures with anisotropic morphology are more advantageous than their spherical counterparts for plasmonic sensing [73, 74] illustrated the refractive index sensitivity of AuNBP-embedded AuAg alloy nanostructures on the geometry and the composition. It has been known that plasmonic properties, including SPR modes and wavelength, are dependent on the dielectric function of metallic material, which varies with frequency. In this case, the frequency-related dielectric properties of alloy were calculated via the Drude-critical points model. Combining the Drude-critical points model and FDTD simulation, the sensitivity properties can be forecasted, as shown in Figure 11.12. The simulated results are matched with the experimented ones. With increasing the refractive index, SPR gradually red shifted. The wavelength shift was linear as a function of the refractive index. Moreover, the RI sensitivity was enhanced with increasing the Ag proportion, demonstrated the dependence of RI sensitivity on composition. Note that, the relationship was non-linear, which may be ascribed to overall effect of the complex dielectric function.

Among the many potential uses for metal nanoparticle assemblies, the one that has received the greatest attention and development has been their use in chemical and biochemical sensing applications [75]. As a result of their unique absorption and scattering properties, Au NRs have attracted considerable attention as instruments for biological sensing, imaging, drug administration, and therapies. The link between Au NR size, shape, and environment as a result of variations in dielectric and interparticle spacing makes these materials useful as sensors [76].

In recent two years, colloidal Au nanostructures have been used with high sensitivity to some typical molecular. One mature applied technology about Au colloid is pregnancy test that was used in family plan, as shown in Figure 11.13a. With the epidemic outbreak of global Novel Coronavirus 2019 (abbreviated as 2019-ncov), the State Food and Drug Administration approved the novel coronavirus antibody detection reagent. The main detection methods are colloidal gold method and magnetic particle chem-luminescence method. The product was presented in Figure 11.13b (http://www.snwimedical.com/info.php?class_id=102104&page=1; https://www.biochemtron.com/news/company/2020-05-16/1402.html). The 2019-ncov antigen test card (Colloidal Gold immunochromatography) contains colloidal

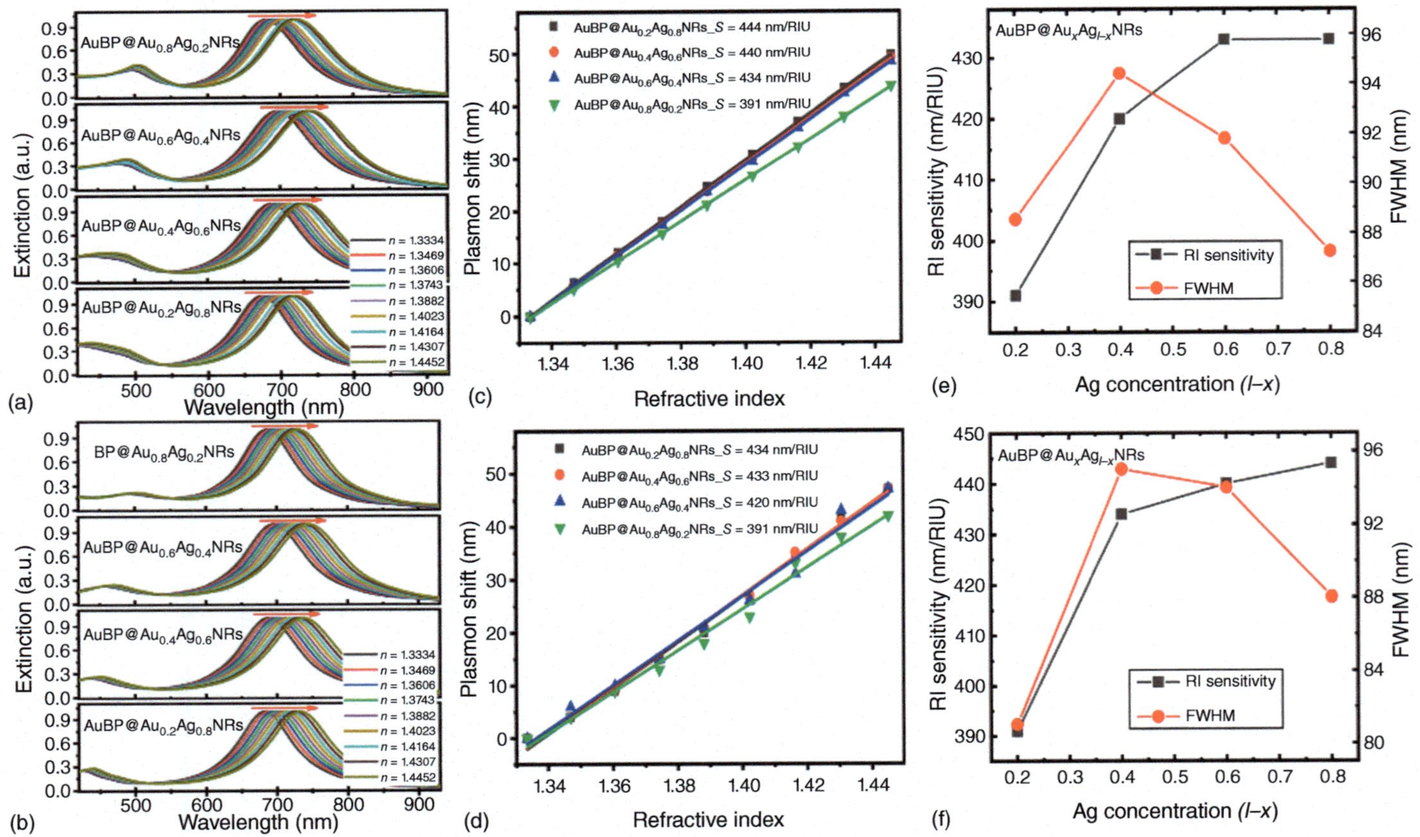

Figure 11.12 (a) The normalized extinction spectra of AuBP@Au_xAg_{1x}NRs (x: 0.2, 0.4, 0.6 and 0.8) synthesized experimentally with varying Ag concentrations dispersed in water-glycerol mixtures. (b) The FDTD calculated extinction spectra of the AuBP@Au_xAg_{1x}NRs (x: 0.2, 0.4, 0.6 and 0.8) with refractive index increasing from 1.3334 to 1.4452. The metal concentration dependence of the resonance wavelength shifts (c) in experiment and (d) in theory with the refractive index of the surrounding medium changing from 1.3334 to 1.4452. The slope of the linear plot increasing with Ag concentration represents the RI sensitivity. The RI sensitivity and the FWHM of the LSPR peaks (e) in experiment and (f) in theory as a function of the Ag concentration in the Au–Ag alloy nanostructures. Source: Reproduced with permission from Wang et al. [74]; @ Royal Society of Chemistry.

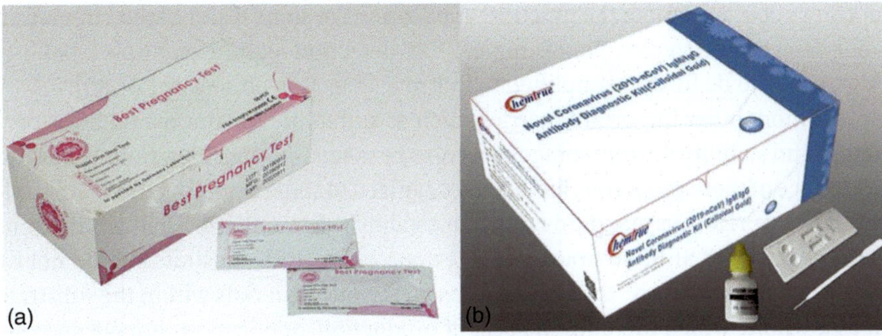

Figure 11.13 Two Au colloidal methods for home pregnancy test and 2019-ncov detections (http://www.snwimedical.com/info.php?class_id=102104&page=1; https://www.biochemtron.com/news/company/2020-05-16/1402.html).

Au labeled with novel coronavirus recombinant antigen and Au labeled rabbit IgG antibody. And it was coated with the mixture of anti-human IgM and anti-human IgG in the nitrocellulose-membrane. If there is a novel coronavirus antibody in the sample, the "novel coronavirus antigen colloidal gold, coronavirus antibody, anti-human IgM/IgG" complex is formed in the nitro cellulose membrane detection line to coagulate and display color, indicating a positive result.

11.4 Conclusion and Outlook

SERS technology has found a wide range of potential applications in both real-life chemical analysis and fundamental mechanistic studies. In this chapter, several typical SERS substrates based on plasmonic nanostructures were presented, which possessed strong electromagnetic fields and chemical enhancement due to the structural advantages attributed to the sharp edge and tips, multiple hot spots, or strong coupling.

In the SERS technology, the sensitivity is high enough to detect a single chemical bond, which provides a powerful mean for studying various chemical components and chemical reactions on the surface interface [77]. However, challenging subject are the high-performance, chemically stable probes with high efficiency.

Despite the fact that sensing techniques based on surface-enhanced spectroscopy and SPR effect have been extensively studied, the exploration of new methods that overcome existing drawbacks and improve the sensing performance is consistently pursued. It is worth pointing out that, these results have showed fundamental feasibility of new sensing methods, but there is still a long way to go before these can be translated into practical sensing devices. Taking the SERS for example, the laser power on a substrate of low SERS-activity has to be high, causing potential heat damage to the sample. In the SERS, the intrinsic spectral information is hidden or lost and the spectral analysis becomes complicated [78]. Moreover, the molecules are usually irreversibly adsorbed on the metal surface so that the SERS-substrates are not reusable, increasing the cost of SERS sensing devices. Also, refinement of assay workflows could promote the practical applications of some SERS-based circulating tumor detection strategies. For example, SERS detection requires multiple labelling, which needs to be further simplified to improve detection efficiency [79, 80].

In addition, other important metrics, such as suitability for automated mass production and stability during storage must also be taken into consideration. While the importance of these properties has been recognized to some degree by the SERS community, more rigorous standards must be established in order to facilitate the routine application of SERS in practice. For example, an SERS substrate should not be defined "cheap" simply by the apparent cost of the raw materials within the substrate but as a whole, which includes the cost of production, application, storage and disposal. To reach practical standards, the stability of the substrates should be assessed at least by months rather than days. These challenges can likely be addressed by smart substrate engineering combined with the discovery of novel enhancing materials, such as semiconductors and plasmonic polymers [81–83], which offer similar magnitudes of signal enhancement to Ag/Au but at a reduced cost.

Finally, standardized procedures and methodologies must be developed to determine substrate performance, as well as for the application of SERS in practice [84], since these are crucial preconditions for promoting SERS to non-specialist users.

References

1 Li, C., Huang, Y., Li, X. et al. (2021). Towards practical and sustainable SERS: a review of recent developments in the construction of multifunctional enhancing substrates. *J. Mater. Chem. C* 9 (35): 11517–11517.

2 Fleischmann, M., Hendra, P.J., and McQuillan, A.J.R. (1974). Raman spectra of pyridine adsorbed at a silver electrode. *Chem. Phys. Lett.* 26: 163–166.

3 Jeanmaire, D.L. and Van Duyne, R.P. (1977). Surface Raman spectroelectrochemistry (Part I): heterocyclic, aromatic, and aliphatic amines adsorbed on the anodized silver electrode. *J. Electroanal. Chem. Interfacial Electrochem.* 84: 1.

4 Creighton, J.A., Blatchford, C.G., and Albrecht, M.G. (1979). Plasma resonance enhancement of Raman scattering by pyridine adsorbed on silver or gold sol particles of size comparable to the excitation wavelength. *J. Chem. Soc., Faraday Trans.* 275: 790–798.

5 Moskovits, M. (1985). Surface-enhanced spectroscopy. *Rev. Mod. Phys.* 57: 783–826.

6 Kneipp, K., Wang, Y., Kneipp, H. et al. (1997). Single molecule detection using surface enhanced raman scattering (SERS). *Phys. Rev. Lett.* 78: 1667–1670.

7 Hamon, C. and Liz-Marzán, L.M. (2018). Colloidal design of plasmonic sensors based on surface enhanced Raman scattering. *J. Colloid Interface Sci.* 512: 834–843.

8 Wang, Z., Zong, S., Wu, L. et al. (2017). SERS-activated platforms for Immunoassay: probes, encoding methods, and applications. *Chem. Rev.* 117 (12): 7910–7963.

9 Yuan, Y.F., Panwar, N., Yap, S.H.K. et al. (2017). SERS-based ultrasensitive sensing platform: an insight into design and practical applications. *Coord. Chem. Rev.* 337: 1–33.

10 Schlücker, S. (2014). Surface-enhanced Raman spectroscopy: concepts and chemical applications. *Angew. Chem. Int. Ed.* 53: 4756–4795.

11 Li, D., Yao, D., Li, C. et al. (2020). Nanosol SERS quantitative analytical method: a review. *TrAC, Trends Anal. Chem.* 127: 115885.

12 Wang, A.X. and Kong, X. (2015). Review of recent progress of plasmonic materials and nanostructures for surface-enhanced Raman scattering. *Materials* 8 (6): 3024–3052.

13 Tripp, R.A., Dluhy, R.A., and Zhao, Y. (2008). Novel nanostructures for SERS biosensing. *Nano Today* 3 (3): 31–37.

14 Lin, X., Cui, Y., Xu, Y. et al. (2009). Surface-enhanced Raman spectroscopy: substrate-related issues. *Anal. Bioanal.Chem.* 394 (7): 1729–1745.

15 Lassiter, J.B., Sobhani, H., Fan, J.A. et al. (2010). Fano resonances in plasmonic nanoclusters: geometrical and chemical tunability. *Nano Lett.* 10: 3184–3189.

16 Ye, J., Wen, F.F., Sobhani, H. et al. (2012). Plasmonic nanoclusters: near field properties of the Fano resonance interrogated with SERS. *Nano Lett.* 12 (3): 1660–1667.

17 Schwenk, N., Mizaikoff, B., Cardenas, S., and Lopez-Lorente, A.I. (2018). Gold-nanostar-based SERS substrates for studying protein aggregation processes. *Analyst* 143: 5103–5111.

18 Wang, J., Koo, K.M., Wang, Y.L., and Trau, M. (2019). Engineering state-of-the-art plasmonic nanomaterials for SERS-based clinical liquid biopsy applications. *Adv. Sci.* 6: 1900730.

19 Keahey, P., Si, P., Razavi, M. et al. (2021). Spectral- and polarization-dependent scattering of gold nanobipyramids for exogenous contrast in optical coherence tomography. *Nano Lett.* 21 (20): 8595–8601.

20 Szuromi, P. (2007). SERS from sharp silver. *Science* 316: 21–21.

21 Moran, C.H., Rycenga, M., Xia, X. et al. (2014). Using well-defined Ag nanocubes as substrates to quantify the spatial resolution and penetration depth of surface-enhanced Raman scattering imaging. *Nanotechnology* 25 (1): 14007–14007.

22 Liu, L., Wu, Y., Yin, N. et al. (2019). Silver nanocubes with high SERS performance. *J. Quant. Spectrosc. Radiat.* 240.

23 Lin, Y., Zhang, Y.J., Yang, W.M. et al. (2019). Size and dimension dependent surface-enhanced Raman scattering properties of well-defined Ag nanocubes. *Appl. Mater. Today* 14: 224–232.

24 Chirico, G., Borzenkov, M., and Pallavicini, P. (2015). *Gold Nanostars: Synthesis, Properties and Biomedical Application*. Switzerland: Springer International Publishing.

25 Esenturk, E.N. and Walker, A.R.H. (2009). Surface-enhanced Raman scattering spectroscopy via gold nanostars. *J. Raman Spectrosc.* 40: 86–91.

26 Wang, Y., Serrano, A.B., Sentosun, K. et al. (2015). Stabilization and encapsulation of gold nanostars mediated by dithiols. *Small* 11: 4314–4320.

27 Andrés Guerrero-Martínez, A., Barbosa, S., Pastoriza-Santos, I., and Liz-Marzán, L.M. (2011). Nanostars shine bright for you: Ccolloidal synthesis, properties and applications of branched metallic nanoparticles. *Curr. Opin. Colloid Interface Sci.* 16: 118–127.

28 Meng, X.H., Dyer, J., Huo, Y.F., and Jiang, C.Y. (2020). Greater SERS activity of ligand-stabilized gold nanostars with sharp branches. *Langmuir* 36: 3558–3564.

29 Chiruumamilla, M., Gopalakrishnan, A., Toma, A. et al. (2014). 3D nanostar dimers with a sub-10-nm gap for single-/few-molecule surface-enhanced Raman scattering. *Adv. Mater.* 26: 2353–2358.

30 Tanwar, S., Haldar, K.K., and Sen, T. (2017). DNA origami directed Au nanostars dimers for single-molecule surface-enhanced Raman scattering. *J. Am. Chem. Soc.* 139: 17639–17648.

31 Ma, W., Sun, M.Z., Xu, L.G. et al. (2013). A SERS active gold nanostar dimer for mercury ion detection. *Chem. Commun.* 49: 4989–4991.

32 Jena, B.K., Mishra, B.K., and Bohidar, S. (2009). Synthesis of branched Ag nanoflowers based on a bioinspired technique: their surface enhanced Raman scattering and antibacterial activity. *J. Phys. Chem. C* 113: 14753–14761.

33 Watt, J., Young, N., Haigh, S. et al. (2009). Synthesis and structural characterization of branched palladium nanostructures. *Adv. Mater.* 21: 2288–2293.

34 Prodan, E., Radloff, C., Halas, N.J., and Nordlander, P. (2003). A hybridization model for the plasmon response of complex nanostructures. *Science* 302: 419–422.

35 Xu, H.X. (2005). Multilayered metal core-shell nanostructures for inducing a large and tunable local optical field. *Phys. Rev. B* 72: 073405.

36 Zhang, T.Y., Lu, G.W., Li, W.Q. et al. (2012). Optimally designed nanoshell and Matryoshka-nanoshell as a plasmonic-enhanced fluorescence probe. *J. Phys. Chem. C* 116: 8804.

37 Jackson, J.B. and Halas, N.J. (2004). Surface-enhanced Raman scattering on tunable plasmonic nanoparticle substrates. *Proc. Natl. Acad. Sci. U. S. A.* 52: 17930.

38 Gao, Y., Li, Y., Wang, Y. et al. (2015). Controlled synthesis of multilayered gold nanoshells for enhanced photothermal therapy and SERS detection. *Small* 11: 144–144.

39 Ke, S.L., Kan, C.X., Zhu, X.Z. et al. (2021). Synthesis of porous Au-Ag alloy nanorods with tunable plasmonic properties and intrinsic hotspots for surface-enhanced Raman scattering. *CrystEngComm* 23: 3467–3476.

40 Ke, S.L., Kan, C.X., Zhu, X.Z. et al. (2021). Effective fabrication of porous Au-Ag alloy nanorods for in-situ Raman monitoring catalytic oxidation and reduction reactions. *J. Mater. Sci. Technol.* 91: 262–269.

41 Neal, R.D., Hughes, R.A., Sapkota, P. et al. (2020). Effect of nanoparticle ligands on 4-Nitrophenol reduction: reaction rate, induction time, and ligand desorption. *ACS Catal.* 10: 10040–10050.

42 McMahon, J.M., Li, S., Ausman, L.K., and Schatz, G.C. (2012). Modeling the effect of small gaps in surface-enhanced Raman spectroscopy. *J. Phys. Chem. C* 116: 1627–1637.

43 Ding, S.Y., Yi, J., Li, J.F. et al. (2016). Nanostructure-based plasmon-enhanced Raman spectroscopy for surface analysis of materials. *Nat. Rev. Mater.* 1: 16021.

44 Lee, D. and Yoon, S. (2015). Gold nanocube-nanosphere dimers: preparation, plasmon coupling, and surface-enhanced Raman scattering. *J. Phys. Chem. C* 119: 7873–7882.

45 Venkata Subbaiah, Y.P., Saji, K.J., and Tiwari, A. (2016). Atomically thin MoS_2: a versatile nongraphene 2D material. *Adv. Funct. Mater.* 26: 2046–2069.

46 Malard, L.M., Pimenta, M.A., Dresselhaus, G., and Dresselhaus, M.S. (2009). Raman spectroscopy in graphene. *Phys. Rep.* 473: 51–87.

47 Ling, X. and Zhang, J. (2010). First-layer dffect in graphene-enhanced Raman scattering. *Small* 6: 2020–2025.

48 Das, A., Pisana, S., Chakraborty, B. et al. (2008). Monitoring dopants by Raman scattering in an electrochemically top-gated graphene transistor. *Nat. Nanotechnol.* 3: 210–215.

49 Feng, S.M., dos Santos, M.C., Carvalho, B.R. et al. (2016). Ultrasensitive molecular sensor using N-doped graphene through enhanced Raman scattering. *Sci. Adv.* 2: e1600322.

50 Yoon, J.C., Thiyagarajan, P., Ahn, H.J., and Jang, J.H. (2015). A case study: effect of defects in CVD-grown graphene on graphene enhanced Raman spectroscopy. *RSC Adv.* 5: 62772–62777.

51 Kim, Y.K., Han, S.W., and Min, D.H. (2012). Graphene oxide sheath on Ag nanoparticle/graphene hybrid films as an antioxidative coating and enhancer of surface-enhanced Raman scattering. *ACS Appl. Mater. Interfaces* 4: 6545–6551.

52 Liu, X.J., Wang, J.J., Wu, Y.C. et al. (2015). Compact shielding of graphene monolayer leads to extraordinary SERS-active substrate with large-area uniformity and long-term stability. *Sci. Rep.* 5: 17167–17178.

53 Wang, P., Xia, M., Liang, O. et al. (2015). Label free SERS selective detection of dopamine and serotonin using graphene-Au nanopyramid heterostructure. *Anal. Chem.* 87 (20): 10255–10261.

54 Li, X.H., Choy, W.C.H., Ren, X. et al. (2014). Highly intensified surface enhanced Raman scattering by using monolayer graphene as the nanospacer of metal film-metal nanoparticle coupling system. *Adv. Funct. Mater.* 24: 3114–3122.

55 Willets, K.A. and Duyne, R.P.V. (2007). Localized surface plasmon resonance spectroscopy and sensing. *Annu. Rev. Phys. Chem.* 58: 267–297.

56 Kneipp, J., Kneipp, H., Wittig, B., and Kneipp, K. (2010). Novel optical nanosensors for probing and imaging live cells. *Nanomed. Nanotechnol. Biol. Med.* 6: 214–226.

57 Nie, S. and Emory, S.R. (1997). Probing single molecules and single nanoparticles by surface-enhanced Raman scattering. *Science* 275: 1102–1106.

58 Zhu, X.Z., Xu, J., Zhang, H. et al. (2020). Gold nanobipyramid-embedded ultrathin metal nanoframes for in-situ monitoring catalytic reactions. *Chem. Sci.* 11: 3198–3207.

59 Sun, M., Hou, Y., and Xu, H. (2011). Can information of chemical reaction propagate with plasmonic waveguide and be detected at remote terminal of nanowire? *Nanoscale* 3: 4114–4116.

60 Dong, B., Fang, Y.R., Xia, L.X. et al. (2011). Is 4-nitrobenzenethiol converted to p, p′-dimercaptoazobenzene or 4-aminothiophenol by surface photochemistry reaction? *J. Raman Spectrosc.* 42: 1205–1206.

61 Van Schrojenstein, L.E.M., Deckert-Gaudig, T., Mank, A.J.G. et al. (2012). Catalytic processes monitored at the nanoscale with tip-enhanced Raman spectroscopy. *Nat. Nanotechnol.* 7: 583–586.

62 Xu, P., Kang, L.L., Mack, N.H. et al. (2013). Mechanistic understanding of surface plasmon assisted catalysis on a single particle: cyclic redox of 4-aminothiophenol. *Sci. Rep.* 3: 2997.

63 Heck, K.N., Janesko, B.G., Scuseria, G.E. et al. (2008). Observing metal-catalyzed chemical reactions in-situ using surface-enhanced Raman spectroscopy on Pd-Au nanoshells. *J. Am. Chem. Soc.* 130 (49): 16592–16600.

64 Cai, Q., Lu, S.K., Liao, F. et al. (2014). Catalytic degradation of dye molecules and in-situ SERS monitoring by peroxidase-like Au/CuS composite. *Nanoscale* 6 (14): 8117–8123.

65 Zhang, H., Wang, C., Sun, H.L. et al. (2017). In-situ dynamic tracking of heterogeneous nanocatalytic processes by shell-isolated nanoparticle-enhanced Raman spectroscopy. *Nat. Commun.* 8: 15447.

66 Yang, J.L., Xu, J., Ren, H. et al. (2017). In-situ SERS study of surface plasmon resonance enhanced photocatalytic reactions using bifunctional Au@CdS core-shell nanocomposites. *Nanoscale* 9 (19): 6254–6258.

67 Sukmanee, T., Wongravee, K., Kitahama, Y. et al. (2020). Distinguishing enantiomers by tip-enhanced Raman scattering: chemically modified silver tip with an asymmetric atomic arrangement. *Angew. Chem. Int. Ed.* 59 (34): 14564–14569.

68 Jana, J., Ganguly, M., and Pal, T. (2016). Enlightening surface plasmon resonance effect of metal nanoparticles for practical spectroscopic application. *RSC Adv.* 6: 86174–86211.

69 Lin, K., Lu, Y., Chen, J. et al. (2008). Surface plasmon resonance hydrogen sensor based on metallic grating with high sensitivity. *Opt. Express* 16: 18599–18604.

70 Wang, X., Wang, P., Chen, C. et al. (2010). Plasmonic racetrack resonator with high extinction ratio under critical coupling condition. *J. Appl. Phys.* 107: 124517.

71 Duan, Q., Liu, Y., Chang, S. et al. (2021). Surface plasmonic sensors: sensing mechanism and recent applications. *Sensors* 21: 5262.

72 Pérez-Juste, J., Pastoriza-Santos, I., Liz-Marzán, L.M., and Mulvaney, P. (2005). Gold nanorods: synthesis, characterization and applications. *Coord. Chem. Rev.* 249: 1870–1901.

73 El-Sayed, M.A. (2001). Some interesting properties of metals confined in yime and nanometer space of different shapes. *Acc. Chem. Res.* 34: 257–264.

74 Wang, X., Kan, C.X., Xu, J. et al. (2020). Dielectric function modelling and sensitivity forecast for Au-Ag alloy nanostructures. *Phys. Chem. Chem. Phys.* 22: 14932–14940.

75 Klinkova, A., Choueiri, R.M., and Kumacheva, E. (2014). Self-assembled plasmonic nanostructures. *Chem. Soc. Rev.* 43: 3976–3991.

76 Stone, J., Jackson, S., and Wright, D.W. (2011). Biological applications of gold nanorods. *WIREs Nanomed. Nanobiotechnol.* 3: 100–109.

77 Zhang, W., Cui, X., Yeo, B.-S. et al. (2007). Nanoscale roughness on metal surfaces can increase tip-enhanced Raman scattering by an order of magnitude. *Nano Lett.* 7: 1401–1405.

78 Lu, Y., Zhong, J., Yao, G., and Huang, Q. (2018). A label-free SERS approach to quantitative and selective detection of mercury(II) based on DNA aptamer-modified SiO_2@Au core/shell nanoparticles. *Sens. Actuator, B* 258: 365–372.

79 Hahm, E., Cha, M.G., Kang, E.J. et al. (2018). Multilayer Ag-embedded silica nanostructure as a surface-enhanced Raman scattering-based chemical sensor with dualfunction internal standards. *ACS Appl. Mater. Interfaces* 10: 40748–40755.

80 Pang, Y., Wan, N., Shi, L. et al. (2019). Dual-recognition surface-enhanced Raman scattering(SERS)biosensor for pathogenic bacteria detection by using vancomycin-SERS tags and aptamer-Fe_3O_4@Au. *Anal. Chim. Acta* 1077: 288–296.

81 Li, Z., Huang, X., and Lu, G. (2020). Recent developments of flexible and transparent SERS substrates. *J. Mater. Chem. C* 8: 3956–3969.

82 Yang, B., Jin, S., Guo, S. et al. (2019). Recent development of SERS technology: semiconductor-based study. *ACS Omega* 4: 20101–20108.

83 Yilmaz, M., Babur, E., Ozdemir, M. et al. (2017). Nanostructured organic semiconductor films for molecular detection with surface-enhanced Raman spectroscopy. *Nat. Mater.* 16: 918–924.

84 Bell, S.E., Charron, G., Cortes, E. et al. (2020). Towards reliable and quantitative surface-enhanced Raman scattering (SERS): from key parameters to good analytical practice. *Angew. Chem. Int. Ed.* 59: 5454–5462.

Index

Plasmonic Metal Nanostructures: Preparation, Characterization, and Applications, First Edition. Caixia Kan.